Geology of the USSR: A Plate-Tectonic Synthesis

Geodynamics Series

Geodynamics Series

1. Dynamics of Plate Interiors
   A. W. Bally, P. L. Bender, T. R. McGetchin, and R. I. Walcott (Editors)
2. Paleoreconstruction of the Continents
   M. W. McElhinny and D. A. Valencio (Editors)
3. Zagros, Hindu Kush, Himalaya, Geodynamic Evolution
   H. K. Gupta and F. M. Delany (Editors)
4. Anelasticity in the Earth
   F. D. Stacey, M. S. Patterson, and A. Nicholas (Editors)
5. Evolution of the Earth
   R. J. O'Connell and W. S. Fyfe (Editors)
6. Dynamics of Passive Margins
   R. A. Scrutton (Editor)
7. Alpine-Mediterranean Geodynamics
   H. Berckhemer and K. Hsü (Editors)
8. Continental and Oceanic Rifts
   G. Pálmason, P. Mohr, K. Burke, R. W. Girdler, R. J. Bridwell, and G. E. Sigvaldason (Editors)
9. Geodynamics of the Eastern Pacific Region, Caribbean and Scotia Arcs
   Rámon Cabré, S. J. (Editor)
10. Profiles of Orogenic Belts
    N. Rast and F. M. Delany (Editors)
11. Geodynamics of the Western Pacific-Indonesian Region
    Thomas W. C. Hilde and Seiya Uyeda (Editors)
12. Plate Reconstruction From Paleozoic Paleomagnetism
    R. Van der Voo, C. R. Scotese, and N. Bonhommet (Editors)
13. Reflection Seismology: A Global Perspective
    Muawia Barazangi and Larry Brown (Editors)
14. Reflection Seismology: The Continental Crust
    Muawia Barazangi and Larry Brown (Editors)
15. Mesozoic and Cenozoic Oceans
    Kenneth J. Hsü (Editor)
16. Composition, Structure and Dynamics of the Lithosphere-Asthenosphere System
    K. Fuchs and C. Froidevaux (Editors)
17. Proterozoic Lithospheric Evolution
    A. Kröner (Editor)
18. Circum-Pacific Orogenic Belts and Evolution of the Pacific Ocean Basin
    J. W. H. Monger and J. Francheteau (Editors)
19. Terrane Accretion and Orogenic Belts
    Evan C. Leitch and Erwin Scheibner (Editors)
20. Recent Plate Movements and Deformation
    K. Kasahara (Editor)

Geology of the USSR: A Plate-Tectonic Synthesis

Lev P. Zonenshain

Michael I. Kuzmin

Lev M. Natapov

Edited by Benjamin M. Page

Geodynamics Series
Volume 21

American Geophysical Union
Washington, D.C.

Published under the aegis of AGU Geophysical Monograph Board.

Library of Congress Cataloging-in-Publication Data

Zonenshaĭn, L. P. (Lev Pavlovich)
Geology of the USSR : a plate-tectonic synthesis / Lev P. Zonenshain, Michael I Kuzmin, Lev M. Natapov ; edited by Benjamin M. Page.
p. cm. — (Geodynamics series : v. 21)
Includes bibliographical references.
ISBN 0-87590-521-8
1. Plate tectonics—Soviet Union. 2. Geology—Soviet Union.
I. Kuz 'min, M. I. (Mikhael Ivanovich) II. Natapov, Lev M.
III. Page, Benjamin M. (Benjamin Markham), 1911– . IV. Title.
V. Series.
QE511.4.Z66 1990
551.1′36′0947—dc20 90-15478
CIP

Printed in the United States of America

CONTENTS

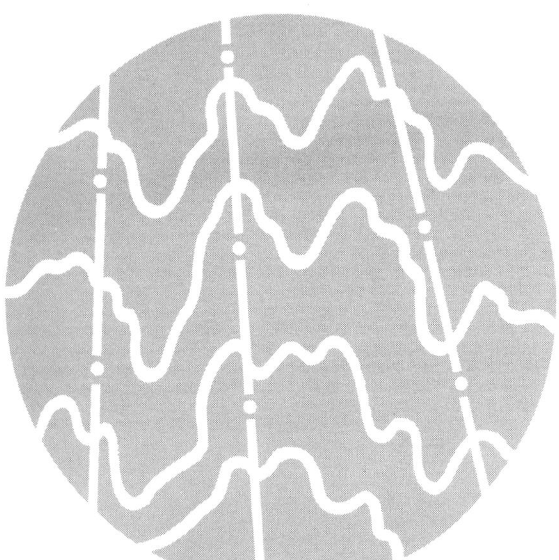

FOREWORD

The Soviet Union includes about one-seventh of the land area on Earth, and Soviet geologists most likely number more than one-seventh of the world's geologists. Moreover, Soviet geologists are probably considerably more familiar with the geological literature of North America and western Europe than the geologists from those regions are with the literature of the Soviet Union.

General summaries of Soviet geology in English have been available, but they are fairly old by now and do not reflect the more recent theoretical developments in the science, in particular the plate-tectonic paradigm. Zonenshain, Kuzmin, and Natapov have mastered that paradigm and are thoroughly familiar with its ramifications; between them they are also familiar with most parts of their vast country.

Because of the scope, uniqueness, and high quality of this work, we consider it to be a landmark volume. We know of no comparable, up-to-date summary of the plate-tectonic history of such a large region; the publication of this one should stimulate similar efforts in the rest of the world. The book, albeit lengthy is a condensed version of a larger treatise (in Russian) prepared during the course of many years and completed only recently. The formidable task of translating into English was accomplished by T. V. Bunayeva and L. P. Zonenshain.

Both of the undersigned have enjoyed friendships and valuable experiences in the U.S.S.R. One of us (Rodgers) has been privileged, by the friendly and helpful courtesy of hundreds of Soviet geologists, to visit the Soviet Union six times and to make field excursions into about a dozen different parts of the country (obviously only a smattering sample of the whole). We hope that by helping with the editing of this book we are repaying in a small way some of the kindnesses we have received. We wish to express appreciation and admiration for the authors' achievement in producing this volume.

John Rodgers
Yale University
New Haven, Connecticut

Benjamin M. Page
Stanford University
Stanford, California

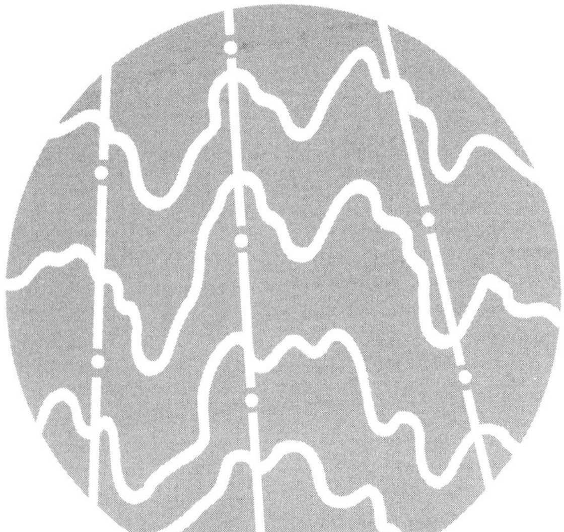

GEOLOGY OF THE USSR: A PLATE-TECTONIC SYNTHESIS

Lev P. Zonenshain

Institute of Oceanology, The USSR Academy of Sciences, Moscow, USSR

Michael I. Kuzmin

Institute of Geochemistry, Siberian Branch of The USSR Academy of Sciences, Irkutsk, USSR

Lev M. Natapov

Aerogeologia, The USSR Ministry of Geology, Moscow, USSR

Abstract. The territory of the USSR mainly comprises a large portion of continental northern Eurasia, and is bordered by the Arctic and Pacific Oceans. In this summary, available materials on geology, tectonics, petrology, paleomagnetics and plate kinematics are used to describe the geological evolution of the territory of the USSR from the plate-tectonic point of view. The territory of the USSR has a mosaic-like aspect, as it consists of two major continental masses--the East European and Siberian platforms--and numerous small microcontinental massifs incorporated in the Phanerozoic foldbelts. The mosaic implies that the USSR originated by amalgamation of many continental blocks during the Phanerozoic as a result of continental convergence, closure of intervening ocean basins, and eventual collision.

The Precambrian basement of the two ancient platforms (East European and Siberian) was also created by accretion when smaller continental blocks were successively joined in the Proterozoic along intricate suture zones. The final collisions within each of the two basements occurred in the Late Precambrian. Greenstone belts, which are characteristic features of the basement of both platforms, are interpreted as synforms in which greenstone successions tectonically lie over grey gneissic para-autochthon protruded to the surface as gneiss-granite domes. In the Late Precambrian and Phanerozoic, the ancient cratonic basement was repeatedly broken by aulacogens, which are considered to be failed branches of three-arm rifts. Evidently some pieces of the continental crust split off from previous larger continents along active spreading branches of these three-arm rifts.

Some foldbelts within the USSR occupy inner-continental positions, such as the Uralian, Central Asian, Mongol-Okhotsk, and Alpine-Himalayan belts. They have a strong imprint of long collisional events leading to their distinctive nappe structure. Some other foldbelts, like the Verkhoyansk-Kolymian, Koryak-Kamchatka, and Sikhote-Alin-Sakhalin belts have marginal continental positions. They are mostly belts formed by accretion of tectono-stratigraphic terranes to the Eurasian margin. The structure and development of each belt are considered in detail. The main emphasis is on description of key rock assemblages indicative of previous plate boundaries, especially of ophiolites, subduction-related volcanic and plutonic rocks, metamorphism, and nappe structures. Using paleomagnetic, paleoclimatic, kinematic, and geological data, plate-tectonic reconstructions are presented illustrating the evolution of each belt.

The Uralian belt evolved on the site of the Silurian-Devonian Uralian Paleo-Ocean and formed in the Late Paleozoic when East Europe, Kazakhstan, and Siberia collided. The Central Asian belt includes the very complicated structures of Central Kazakhstan, the Altai-Sayan region, and Tien-Shan. Central Kazakhastan and Altai-Sayan were forming during the whole Paleozoic, mostly by accretion. Tien Shan is a typical collisional foldbelt that formed in the Late Paleozoic when the Tarim block collided with Kazakhstan. The Mongol-Okhotsk belt originated in the Latest Paleozoic and Early Mesozoic as a result of collision of the small Amuria continent (which has been a part of the North China continent since the Permian) with Siberia. The Alpine-Himalayan belt was, as is well known, born from the closing of the Tethys Ocean. The Verkhoyansk-Kolymian and Sikhote-Alin-Sakhalin belts were formed mostly in the early-middle-Cretaceous when numerous terranes were amalgamated and attached to the Eurasian margin as a result of a rapid increase in convergence rate of the Izanagi/Kula and Eurasia plates. The structure of the Koryak-Kamchatka belt is still forming at present. Main events in its Late Mesozoic and Cenozoic history were related to successive accretions (to the continental margin) of numerous exotic terranes that were carried by the Kula plate from various sites in the Pacific Ocean.

Large sedimentary basins--the Pechora-Barents Sea, West Siberia, and Turan basins--were superimposed on the older Precambrian and Paleozoic basement in the Late Paleozoic and

Early Mesozoic. They originated by inner-continental rifting and are filled with sedimentary piles up to 10-15 km thick.

Present-day tectonics is mainly restricted to active plate boundaries, which are mostly on the periphery of the USSR. They chiefly belong to the convergent type, being related either to the active subduction within the Kuril-Kamchatka trench or to the India/Eurasia and Arabia/Eurasia collisional belts of diffuse seismicity in Central Asia, Pamirs, Kopet-Dag, and Caucasus.

The last chapter contains a synthesis of the plate-tectonic evolution of the territory of the USSR and provides 18 palinspastic maps showing possible plate interactions during the Phanerozoic.

Introduction

Numerous fundamental studies have been devoted to the geology of the USSR, e.g., Arkhangelsky [1947], Shatsky [1956], Nalivkin [1958], Khain [1977, 1979, 1984, 1985], Peive [1980], Bogdanov [1976], Koronovsky [1984], Milanovsky [1987]. Geologic and tectonic maps of the entire USSR have been published. However, even in the latest summary publications on geology and tectonics of the USSR, the material was considered from the viewpoint of the geosynclinal theory. Khain [1987] was the first to apply plate tectonics on a broad scale to the interpretation of geology of the USSR.

This study of the geological development of the USSR territory is not a summation of factual material; rather, the authors attempt to reinterpret existing published data on the geology of the USSR from the plate-tectonic point of view. In carrying out this study, we used ordinary plate-tectonic approaches, which were particularly important for reconstructing the geological history from the integration of several different kinds of data, e.g., kinematic, paleomagnetic, paleoclimatic, chronological, and geological data.

The kinematic data available for the Soviet Union territory are relevant only for the last 160 Ma; they allow us to reconstruct the interaction of the Eurasian plate with surrounding North American, Pacific basin, Indian, Arabian, and African plates. In some cases, it was possible to calculate motions of minor plates and subplates within Eurasia. Some efforts have been made to estimate absolute motions of continents for Early Mesozoic and even Paleozoic times by combining the trajectories of motions of the continents above hot spots with apparent polar wandering paths as small circles corresponding to Euler poles [Zonenshain et al., 1985].

Reliable paleomagnetic data on the USSR territory are on hand only for the two major continenal masses - East Europe and Siberia [Khramov, 1982]. Unfortunately, paleomagnetic data on minor blocks and foldbelts are not of a high quality. Only those related to Central Kazakhstan, Caucasus, Urals, and the Omolon massif seem to be reliable enough for use in compilation of palinspastic reconstructions.

Paleoclimatic data are used to show the positions of sedimentary sequences and appropriate minor blocks within certain latitudinal climatic zones.

To establish former plate boundaries and the mode of their interaction, key magmatic assemblages are of great importance. Suture zones are recognized from the distribution of ophiolites and from sharp changes of lithologies. These zones mark the amalgamation of different blocks. Ophiolites indicate the presence of ancient oceanic crust. Belts of calc-alkaline volcanism are good evidence for former subduction zones. Data on the polarity of volcanism allow restoration of the configuration of ancient Benioff zones. The study of chaotic complexes, in particular subduction melanges, and exotic terranes allows us to understand the geological development of certain regions. Former continental rifts, together with aulacogens, are traced by graben structures and distribution of bimodal volcanic associations. Interpreting an aulacogen as the failed arm of a three-arm junction of continental rifts, two arms of which have become the loci of opening of new oceanic areas, makes it possible to outline how the ancient continents broke up and how sedimentary complexes formed successively on passive continental margins.

From what is said above, it is clear that the authors have followed the principle of actualism. This principle cannot be applied to the entire history of the Earth, but the latest researches, along with data on the geochemical evolution of the mantle relative to Sr, Pb and Nd isotopes, show that the actualistic approach probably may be employed for the last 1.5-2 Ga. This is the period that is described in the present study. The authors have also discussed, albeit briefly, the structure of the Precambrian continental (ancient-platform) basements; the East European and Siberian. These continents were the nuclei to which numerous blocks of different nature were accreted during the Phanerozoic. Convergence and final collision of major continents produced the present-day geological structure of Eurasia, which is thus a composite continent.

The text is presented as follows: First, we consider the plate-tectonic evolution of individual regions, such as the East-European and Siberian platforms and the foldbelts of different ages: Uralian, Central Asiatic, Mongol-Okhotsk, Sikhote-Alin-Sakhalin, Verkhoyansk-Kolymian, Koryak-Kamchatka, Alpine-Himalayan. Next, palinspastic reconstructions are given for each region. Finally, we present a synthesis in the form of 18 reconstruction maps showing how the vast continental area of the USSR was amalgamated from numerous terranes throughout Phanerozoic time.

Acknowledgments. The authors wish to express their gratitude to V. E. Khain for helpful remarks and constant scientific support of this study. We also gratefully acknowledge fruitful cooperation with A. S. Monin and L. V. Tauson. Valuable criticism was obtained from Sh. A. Adamia, R. E. Aisberg, A. I. Al'mukhamedov, Z. Balla, B. V. Baranov, V. G. Belichenko, A. A. Belov, N. A. Bogdanov, Yu. A. Bogdanov, Ch. B. Borukaev, V. S. Burtman, R. G. Garetsky, N. L. Dobretsov, V. G. Kazmin, G. I. Kalyaev, L. A. Kiryukhin, A. L. Knipper, M. V. Kononov, A. P. Lisitsin, M. B. Lordkipanidze, M. S. Markov, N. V. Mezhelvovsky, G. L. Mitrophanov, V. M. Moralev, A. A. Mossakovsky, A. S. Perfiliev, D. M. Pechersky, Yu. M. Pushcharovsky, A. P. Stavsky, O. S. Stupka, S. M. Tilman, A. K. Uflyand, I. B. Filippova, V. D. Chekhovich, and S. I. Sherman.

Our special thanks go to B. M. Page who supported and edited this work from the very beginning and without whom the English version would never have been published,. Many valuable remarks and corrections were contributerd by J. Rodgers; his thorough reviewing of the entire manuscript revealed many errors and inconsistencies. We thank also the other reviewers, each of whom examined and improved parts of the manuscript: S. E. Box, B. C. Burchfiel, M. Churkin, R. G. Coleman, D. Doser, K. Fujita, W. B. Hamilton, and W. P. Harbert. The American Geophysical Union provided money for secretarial work and drafting. Mrs. B. Hayashida devoted many hours to the accurate computerizing of the text. We are indebted to T. V. Bunayeva, who translated the original Russian text into English.

Chapter I

TECTONIC FRAMEWORK

The main part of the territory of the USSR consists of ancient blocks of different dimensions cemented by foldbelts into a single continent. Figure 1 shows all exposures of the ancient basement older than 1000 Ma within the platforms and the foldbelts. The supposed continuation of the ancient basement under flat-lying or slightly deformed sedimentary cover is also shown, especially beneath the West Siberian and Turan basins and on the Arctic shelf. It is evident from Figure 1 that Eurasia is a composite continent and had an essentially different history in much of the Phanerozoic compared with the Mesozoic dispersal of Gondwana continents. The continental blocks that played the major role in the accretion of the USSR are the East European and Siberian platforms, which occupy the greater part of the country. These great platforms are bordered on the south by the African-Arabian, Indian, Tarim, and North-Chinese continental blocks. In addition, there are numerous microcontinents or massifs, such as the Barents massif or Barentsia, the Hyperborean platform or Arctida (perhaps including the independent blocks Kara and East Chukotka), Khanty-Mansiisk (hidden under the West Siberian plain), Kazakastan-Northern Tien Shan, Atasu-Mointy (Balkhashian), Ustyurt, Karakum, Tuva-Mongolian, Central-Mongolian, Khingan-Bureya, Omolon, etc. The dimensions of these blocks vary greatly, from about 500 to 3000 km, and their interaction with each other and with the great continents is responsible for a variety of foldbelts.

The northwestern part of the East European platform is over-thrust by the Caledonian Scandinavian foldbelt, which originated at the site of the Iapetus Ocean. It marks the zone of East Europe-North America collision.

In the northeast, the East European platform is bordered by the Late Precambrian Timan foldbelt which runs from the Scandinavian Caledonides to the Urals. It separates the proposed Barents massif from East Europe and may be considered as a suture between East Europe and the minor Barentsia continent, resulting from a collision at the end of Precambrian.

Vast areas of Inner Asia situated between the Siberian and East European platforms and between Siberia and the buried massifs of Central Asia and the Chinese platforms, belong to the Uralian and Central Asian foldbelts. The Uralian belt stretches north-south between the East-European platform on the one side and the ancient Khanty-Mansiisk and Kazakhstan-Northern Tien Shan massifs and the Siberian platform on the other. The Uralian paleo-ocean apparently existed in this place throughout the whole of Paleozoic time [Zonenshain et al., 1984].

The Central Asian belt includes the Tien Shan, Central Kazakhstan, the Altay-Sayan area, parts of Mongolia bordering Siberia, the Western Trans-Baikal area, and the Vitim plateau. The belt continuously evolved during Late Precambrian and Paleozoic times. It contains ophiolites ranging in age from 800 to 300 Ma. Folding and thrusting were diachronous in different parts of the belt. It is possible to distinguish Late Precambrian (Pre-Vendian), Cambrian (Salairian), Ordovician-Silurian (Caledonian time equivalent), Devonian-Early Carboniferous (Hercynian time equivalent), and Permian-Early Triassic (Late Hercynian) foldbelts. Available data on the history of the Central Asian belt allow the reconstruction of two paleo-oceans: a Paleo-Asian ocean, existing from Late Precambrian to Middle Paleozoic, and a Paleo-Tethys Ocean, from Silurian (or Late Ordovician) to Late Paleozoic [Zonenshain et al., 1976]. The latter ocean was connected with the Paleo-Tethyan basin of the Alpine-Himalayan belt, where the Tethys sensu stricto did not close until the early Mesozoic.

Toward the east, the Central Asian belt passes into the Mongol-Okhotsk belt. In contrast to the Central Asian belt, this belt is extremely narrow--from 10-20 to 100-200 km. It was formed as a result of collision of the Khingan-Bureya massif (Amuria) with Siberia in the end of the Paleozoic and beginning of the Mesozoic [Parfenov, 1984].

The eastern margin of the Soviet Union is divided into two parts by the long Okhotsk-Chukotka and Sikhote-Alin Cretaceous-Paleogene volcano-plutonic belts. To the west of those belts, folding terminated in the Middle Cretaceous. Two foldbelts--Verhoyansk-Kolymian and Sikhote-Alin--are distinguished there. The former resulted from collision of the Chukotka-Alaska block with Siberia and the Omolon and Okhotsk massifs, and the other by accretion of terranes arriving from the Pacific Ocean. The volcanic-plutonic belts mark the boundary of the Asiatic continent in Late Cretaceous time when they occupied the continental margin as belts of Andean-type. To the east, between the volcanic belts and the Pacific Ocean, is a collage of suspect terranes forming the Koryak-Kamchatka foldbelt adjacent to the northwestern Pacific marginal seas.

The Mediterranean or Alpine-Himalayan foldbelt extends along the southern boundaries of the USSR. As is well documented, this belt resulted from closure of the Tethys Ocean, when Africa, Arabia, and India converged and collided with Eurasia.

Present plate boundaries occur mainly on the margins of the territory of the USSR. The Gakkel Ridge (Arctic Mid-Ocean Ridge) in the Arctic Ocean is a spreading axis. A subduction zone consuming Pacific Ocean crust runs along the Kamchatka-Kuril Island arc. A wide belt of diffused seismicity and recent mountain building, extending through Central Asia and joining the Alpine and Himalayan belts, marks an incipient intracontinental plate boundary. This belt displays fracturing of the continental crust, probably as a result of the continuing collision of Africa, Arabia, and India with Eurasia.

Chapter II

EAST EUROPEAN PLATFORM

The East European platform is an acute-angled continental block, about 3000 km across, with a basement that formed 1600 Ma ago (Figure 2). Its boundaries mostly coincide with the thrust fronts of foldbelts; however, in the southeast, in the Peri-Caspian depression, an area underlain by Devonian oceanic crust is preserved like a bay within the continent.

To review the structure of the platform, we have used a number of summary studies [e.g., Khain, 1977; Gaal and Gorbatchev, 1987] as well as publications concerning the structure and isotopic age determinations for the Precambrian of the Baltic and Ukranian shields [Novikova, 1975; Voitovich, 1980; Martynova, 1980; Shcherbak and Bibikova, 1984; Kalayev, et al., 1984; Stupka, 1980, 1984; Sivoronov, et al., 1984; Lazarev, 1973; Bibikova, 1989].

PRECAMBRIAN BASEMENT

The Precambrian basement of the East European platform is characterized by a mosaic of angular blocks separated by suture zones. The blocks are from 100 to 300 km wide. The Baltic shield consists of the Murmansk, Kola, Belomorian, Karelian, Svecofennian, and Sveconorwegian blocks. The Ukranian shield is also made up of several blocks: Volyno-Podolian, Odessa-Belotserkovsky, Kirovograd, Pridneprovsky, and Pri-Azov. Blocks of the same type are believed to exist within the Voronezh and Volga-Uralian massifs.

The internal parts of the blocks are mainly composed of Archean domains of different compositions. These are chiefly of two types: (i) granite-greenstone (Kola and Karelian blocks of the Baltic shield, and the main parts of the Ukranian shield), and (ii) granite-gneiss (Murmansk and Belomorian blocks of the Baltic shield). These blocks suffered deformation and metamorphism 2500-2600 Ma ago [Khain, 1977; Bibikova, 1989]. An exception is the Svecofennian schist-gneissic block of the Baltic shield, which was mainly deformed at about 1700 Ma.

The protoplatform cover of Jatulian sedimentary rocks, 2200-1700 Ma, overlies in places the Archean metamorphic basement. Other Early Proterozoic rocks include volcanic and sedimentary assemblages which occur within suture zones separating the Archean blocks, for example, the Krivoy Rog zone of the Ukranian shield, the Imandra-Varzuga zone of the Baltic shield, the Laplandian granulite zone, and many others. They were deformed 1800-1700 Ma ago. By 1600 Ma the Archean blocks were sutured, thus producing the continental basement of the East-European platform, or craton.

Archean Granite-Greenstone and Gneissic Granite Domains

Granite-greenstone domains are more or less similar everywhere. A typical example is the Pridneprovsky block in the Ukranian shield (Figures 3 and 4) and the Karelian block in the Baltic shield (Figure 5). Cupolas or domes of highly metamorphosed granite and gneiss are separated by greenstone belts. The domes are either circular or slightly elongated, and are 40-60 km (in places 100 km) across. Their central parts expose migmatised rocks, generally grey gneisses. Available ages of the grey gneisses range from 3700 to 3000 Ma [Shcherbak and Bibikova, 1984); the mean value is 3650 Ma, and the oldest rocks are ultramafics of the Aul series, 3900 ± 200 Ma in age [Bibikova, 1989]..

The intracupola spaces are occupied by greenstone belts of various, often intricate shapes which are entirely determined by their intercupola position (Figures 3 and 4). They are usually 30-100 km long and 10-15 km wide. Such greenstone belts are recorded in practically all Precambrian massifs of the East European platform (Figure 2). Their isotopic ages are generally 3250-3000 Ma [Shcherbak and Bibikmova, 1984; Bibikova, 1989].

The overall composition of greenstone belts in all the blocks is rather similar, yet the rocks included are quite diverse. Lower portions are normally composed of mafic extrusive rocks of spilite-diabase composition. Ultramafic lavas--komatiites--are recognized in the Lopian of Karelia. These lavas alternate with iron formation, sometimes of commercial value, for instance the Kostomuksha deposit in the Gimolsky Formation in Karelia. The upper parts of greenstone belt sequences are often composed of silicic extrusive rocks: keratophyres, plagioporphyries, and felsic rocks with interbeds of quartzite and grit. The total thickness of sediments reaches 6-7 km. Volcanic piles include interlayers of serpentinites, peridotites, and gabbro-norites.

The rocks of granite-greenstone areas in both the Baltic shield and the Ukrainian shield suffered later metamorphism and deformation simultaneously and were intruded by granites and tonalites. In the Pridneprovsky block of the Ukranian shield, the age of most of the ancient tonalite-granodiorite bodies of the Dneprovsky complex is about 2970 Ma, while enderbites of the Dneprovsky-Bug formation in the Volyno-

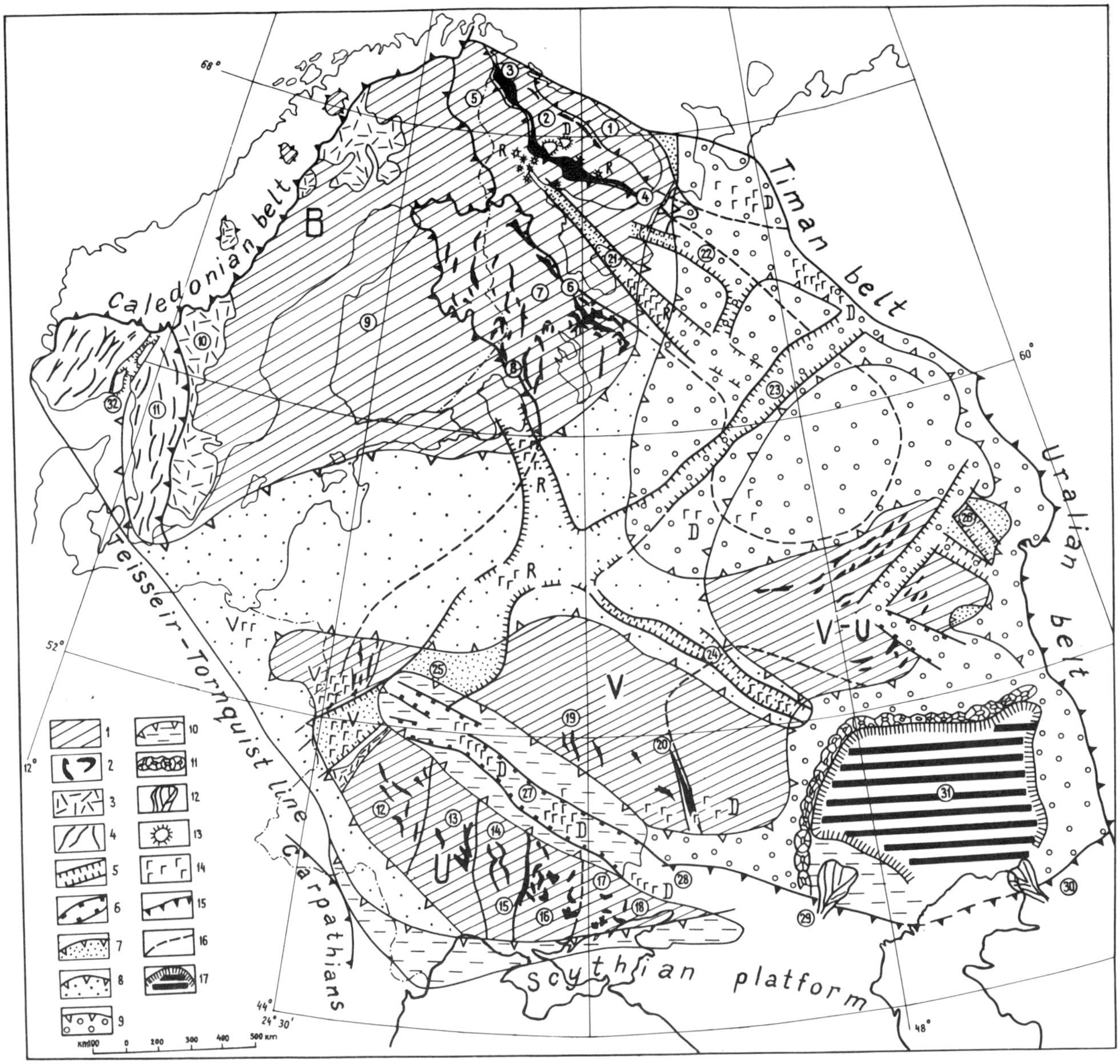

Fig. 2. Tectonics of the East European platform.
1, Shield; 2, Early Precambrian greenstone belt; 3, continental volcanic belt, 1600-1700 Ma; 4, Dalslandian orogenic belt, 800-1000 Ma (Sveco-Norwegian zone); 5, Riphean aulacogen; 6, Devonian aulacogen; 7, Riphean sedimentary cover; 8, Vendian Paleozoic sedimentary cover; 9, Upper Paleozoic sedimentary cover; 10, Mesozoic and Cenozoic sedimentary cover; 11, Carboniferous and Lower Permian barrier reef; 12, Upper Paleozoic deltaic fan; 13, intraplate magmatism; 14, flood basalt; 15, thrust front,; 16, supposed continuation of Precambrian block boundaries; 17, oceanic floor in the Peri-Caspian depression.
Numbered are: B - Baltic Shield; 1, Murmansk block; 2, Kola block; 3, Pechenga zone; 4, Imandra-Varguza zone; 5, Belomorian (White Sea) block; 6, East Karelian zone; 7, Karelian block; 8, West Karelian zone; 9, Sveco-Fennian block; 10, Gothian volcanic belt; 11, Sveco-Norwegian block.
U - Ukranian Shield; 12, Volyno-Podolian block; 13, Odessa-Belotserkovsky block; 14, Kirovograd block; 15, Krivoy Rog zone; 16, Pridneprovsky block; 17, Orekhovo-Pavlograd zone; 18, Pri-Azov block. V - Voronezh massif; 19, Mikhailovsky zone; 20, Vorontsov zone; V-U - Volgo-Uralian massif.
Aulacogens: 21, Kandalaksha; 22, Mezen; 23, Sredne-Russky; 24, Pachelma; 25, Volyno-Orsha-Krestosovsky; 26, Kaltasinsky; 27, Pripiat-Dneprovsky; 28, Donbass; 29, Karpinsky; 30, South Emba zone; 31, Peri-Caspian depression; 32, Oslo graben.

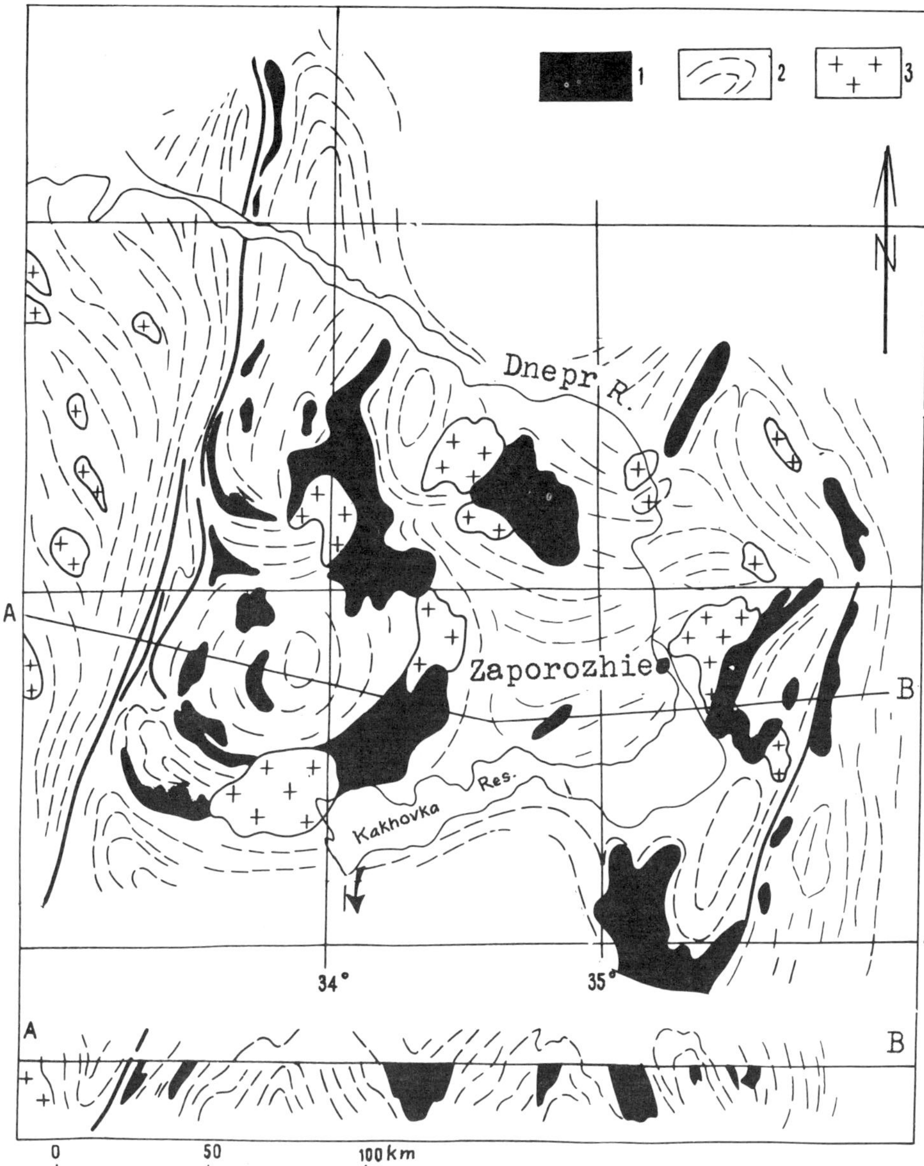

Fig. 3. Geology of the Pridneprovsky block (simplified after Bogolepov and Votakh [1977]). 1, Greenstone belt; 2, structural orientation in gneisso-migmatite complexes; 3, granite.

Podolian block suffered secondary metamorphism 2750 Ma ago [Shcherbak and Bibikova, 1984]. The granites intruded into greenstone belts of the Karelian block (Baltic shield) are 2740 Ma [Shcherbak and Bibikova, 1984].

The primary relationships between greenstone complexes and granite-gneissic cupolas are overprinted by later metamorphism, granitization, and folding. Nevertheless, it may be inferred that the tectonic structure is two-layered. The lower layer consists of the granite-gneissic cupolas, and the upper one comprises the greenstone complexes, which represent intracupola synforms. It may be assumed that prior to the growth of the granite-gneissic cupolas, the greenstone complexes overlay the crystalline basement as a continuous (possibly allochthonous) cover. Mobilization of the crystalline basement, and anatectic melting of the granite material (with metamorphism and migmatization of surrounding sequences, including the overlying greenstone envelope) induced growing of granite-gneissic cupolas and, finally, gave rise to the formation of the two-layered structure that now exists.

In this connection, we may ask whether the volcanic piles of 3250-3000 Ma greenstone belts originally formed on the crystalline basement that now underlies them, or whether they were initially separate and were subsequently stacked tectonically one over another. The available data on the East

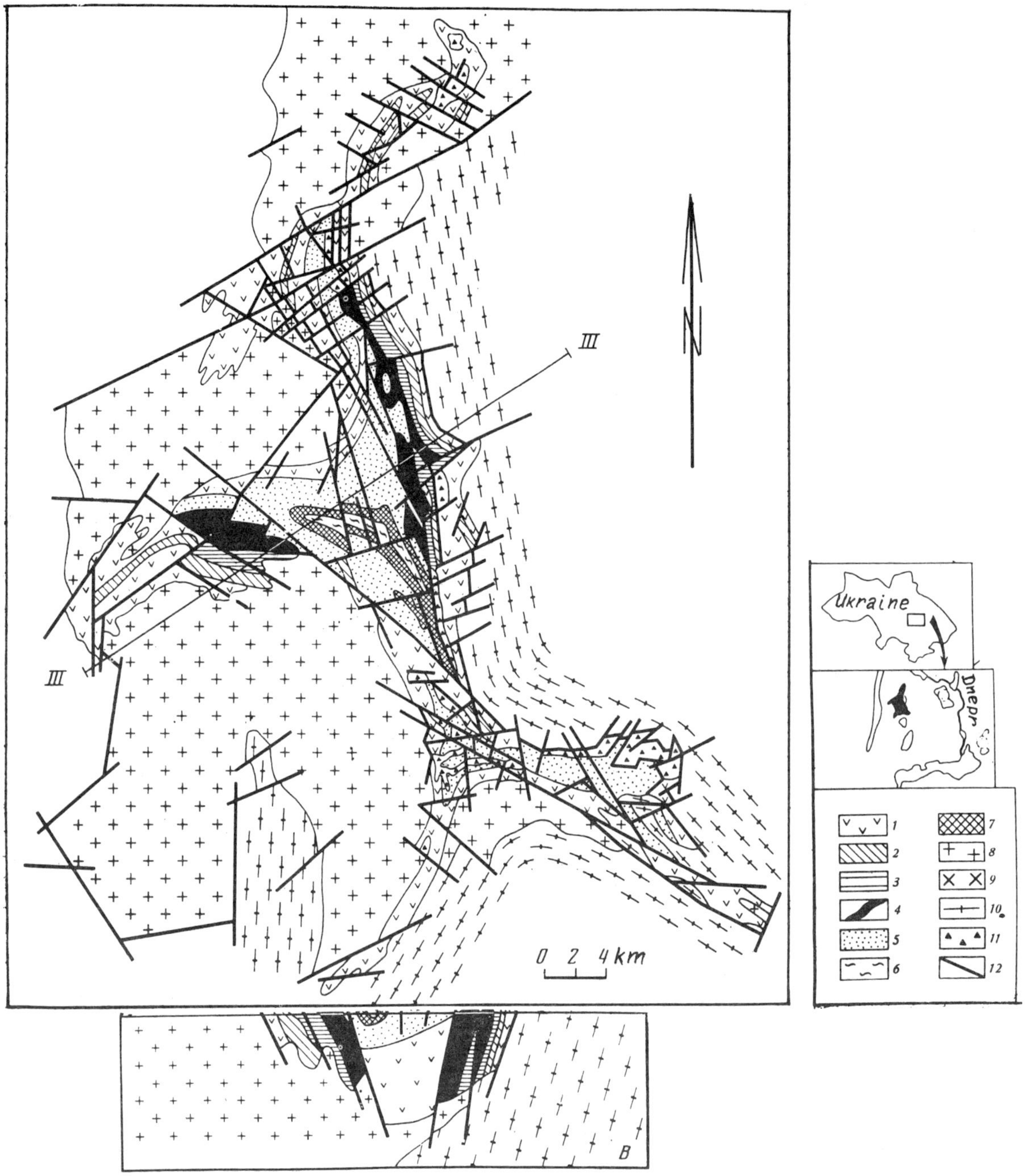

Fig. 4. Geology of the Verkhovtsev greenstone belt (after Sivoronov et al. [1984]).
1, Metamorphosed komatiite and tholeiite; 2, iron formation; 3, meta-dacite and meta-andesite; 4, meta-komatiite; 5, Lower Belozerian member; 6, Upper Belozerian member; 7, Middle Belozerian member; 8, tonalite; 9, potassic granite; 10, gneisso-granite of the old basement; 11, ultramafics; 12, fault.

European platform, particularly those concerning the secondary character of the two-layered structure of granite-greenstone areas, indicate the necessity for large horizontal displacements and underthrusting of some segments of the grey-gneissic (tonalitic) basement beneath the ancient oceanic crust (represented by lower mafic and ultramafic sequences of greenstone belts). If this is so, we can compare these events with Phanerozoic subduction characteristic of active margins. It

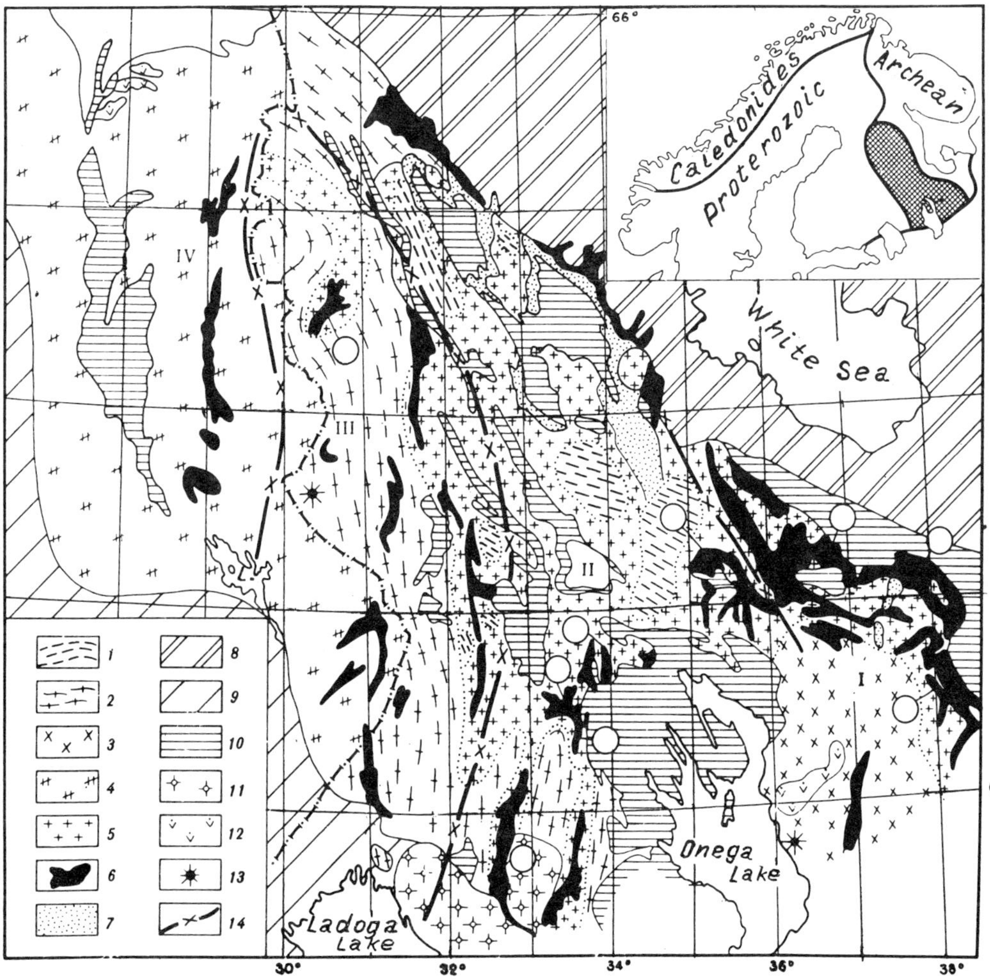

Fig. 5. Structure of the Karelian block (after Krylov et al. [1984]).
From 1 to 5, different types of the gneisso-granitic basement; 6,7, greenstone belts (6-Lopian, 7-Sumian and Sairolian); 8, Belomoride; 9, Svecofennide; 10, Jatulian; 11, rapakivi; 12, mafic intrusion; 13, granulite metamorphism; 14, boundary between different zones.
I, East Karelian zone, II, Central Karelian zone; III, West Karelian zone; IV, East Finland zone.

would seem that the most suitable models are those suggesting that greenstone belts are analogous to marginal seas and are related to subduction zones.

We infer from the age of the latest plutonism and metamorphism in the greenstone belts, that by the end of the Archean, ca. 2600 Ma, the following blocks with continental crust had been created: Murmansk, Kola, Belomorian, Karelian on the Baltic shield and Pridneprovsky, Pri-Azov, Kirovograd, and Volyno-Podolian on the Ukranian shield. As noted above, these blocks are divided by suture zones formed by the end of the Early Proterozoic.

Early Proterozoic Suture Zones

The Krivoy Rog zone is a typical example of a suture zone on the Ukranian shield [Kalyaev et al., 1984]. It is composed of flysch, chert, and subordinate jaspilite (iron formation) and mafic volcanic rocks, with a total thickness of 7-8 km. The clastic rocks contain material from granite-greenstone areas. The age of the Krivoy Rog Formation was determined from authigenic uranium to be 2500 Ma [Shcherbak and Bibikova, 1984]. It is intruded by granites 1880 Ma old. The Krivoy Rog Formation forms a narrow (10 km) synclinorial zone that involves thrusts and isoclinal folds verging eastward, i.e., towards the Pridneprovsky block. The sedimentary facies, particularly its flyschoid character and the widespread cherty rocks, prove that the Krivoy Rog Formation could not have formed initially in the narrow zone that it currently occupies but that its original width was significantly reduced during formation of the present compressional structure.

On the Baltic shield the most prominent suture zone is the East-Karelian lying between the Karelian and Belomorian blocks (Figure 6). It is composed of the Sumii Formation. The age of Sumii silicic volcanics is determined by some researchers to be 2410 Ma [Shcherbak and Bibikova, 1984] and by others 2750 Ma [Krylov et al., 1984]. Pebbles of Sumii acid volcanics contained in the basal Jatulian conglomerates gave an age of 2440 Ma [Shcherbak and Bibikova, 1984].

In many places, an unconformity is observed at the base of the Sumii Formation. Overall, the Sumii Formation is formed by two types of rocks,--volcanogenic (Tunguda Formation) and clastic (Sarioliisky Formation). As in the Ukranian shield, the Early Proterozoic sedimentary rocks underwent intense deformation and in places are metamorphosed in the amphibolite facies; tightly compressed linear folds are typically complicated by small-scale isoclinal folding. The Sumii rocks form part of the tectonites along the East-Karelian shear zone [Novikova, 1975]. The Sumii complex displays a continuous series of volcanics from basalts through andesites to rhyolites, suggesting that they were subduction-related. If so, oceanic crust formerly separating the Karelian from the Belomorsky block was consumed in the East Karelian suture zone.

The Sumii Formation of the Karelian block underwent folding and metamorphism ca. 2200-2300 Ma (Pre-Jatulian, Seletsky orogenic phase) and was intruded by diorites and plagiogranites. The subduction-related volcanism ceased at that time. We tend to connect this folding with the collision of the Belomorsky and Karelian blocks. These motions were reflected in the Belomorsky block by the age of repeated metamorphism of the Inari gneisses, 2200-2300 Ma [Bogolepov and Votakh, 1977].

Proto-Platform Cover. Early Proterozoic, Jatulian

The first platform cover, at least in the Karelian block, formed in Early Proterozoic time, and is known as Jatulian. Facies analogs of the Jatulian are recognized in the Ukranian shield as well. The Jatulian rocks include, on the one hand, clastic sequences--conglomerates, arkoses and quartzites accumulated due to erosion of the Early Precambrian basement--and, on the other hand, piles of mafic and ultramafic lava. From zircons in gabbro-diabases, the age of the Jatulian lava is 2150 Ma and the age of metamorphism is 1850 Ma [Krats et al., 1976]. The Jatulian sediments form a flat-lying cover in the central segment of the Karelian block but strongly deformed tectonic slices within the East Karelian suture zone. The Jatulian platform-like cover is evidence that the Karelian block was already a stable continental craton by the end of the Early Proterozoic.

Suture Zones of the Kola Peninsula

The Pechenga and Imandra-Varguza zones (Figure 7) separate the Kola from the Belomorsky block. The age of the rocks is estimated as 1900-1800 Ma and the age of metamorphism as 1800-1700 Ma, i.e., the same as the Jatulian in Karelia. The section is made up of thick diabase sequences with intervening clastic material, together with gabbro and ultramafic rocks [Novikova, 1975]. Imbricated thrust structures are developed within the zones. According to Novikova [1975], many such rock successions constituting the suture zones are similar to ophiolites, suggesting the presence in Jatulian time of an oceanic basin separating the Kola and Belomorsky blocks.

Svecofennian Schist-Gneissic Block

Khain [1977] emphasized several features of the Svecofennian block. These are: (i) absence of the Archean basement (except in the North Ladoga zone), (ii) predominance of schist and schist-gneissic sequences that suffered strong deformation and intense metamorphism although they are about the same age as much less metamorphosed and less

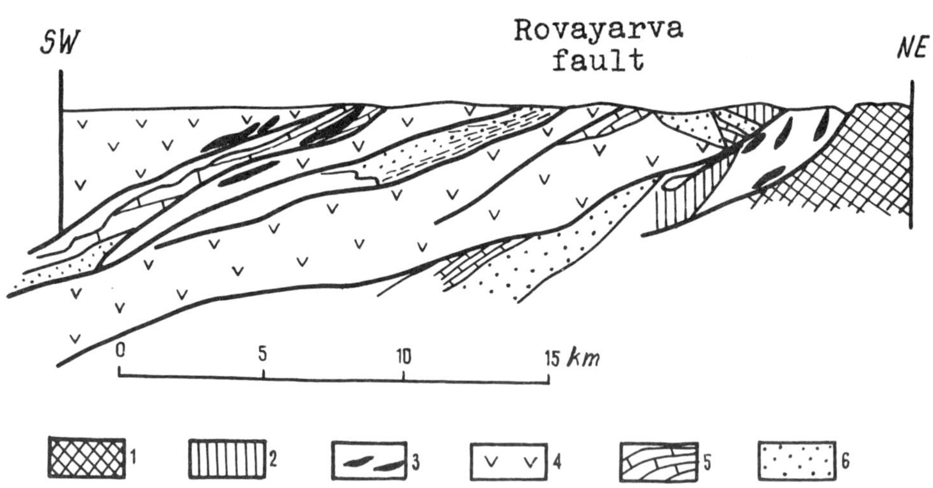

Fig. 6. Cross-section through the East Karelian suture zone 70 km west of Pyaz Lake (after Voitovich [1980] simplified). Great dipping thrusts are clearly seen.
1, Gneisso-granitic complex of Belomoride block; 2, Lopian greenstone belt; 3, ultramafics; 4, greenstone volcanics of Sumian; 5, Jatulian limestone; 6, Jatulian clastics.

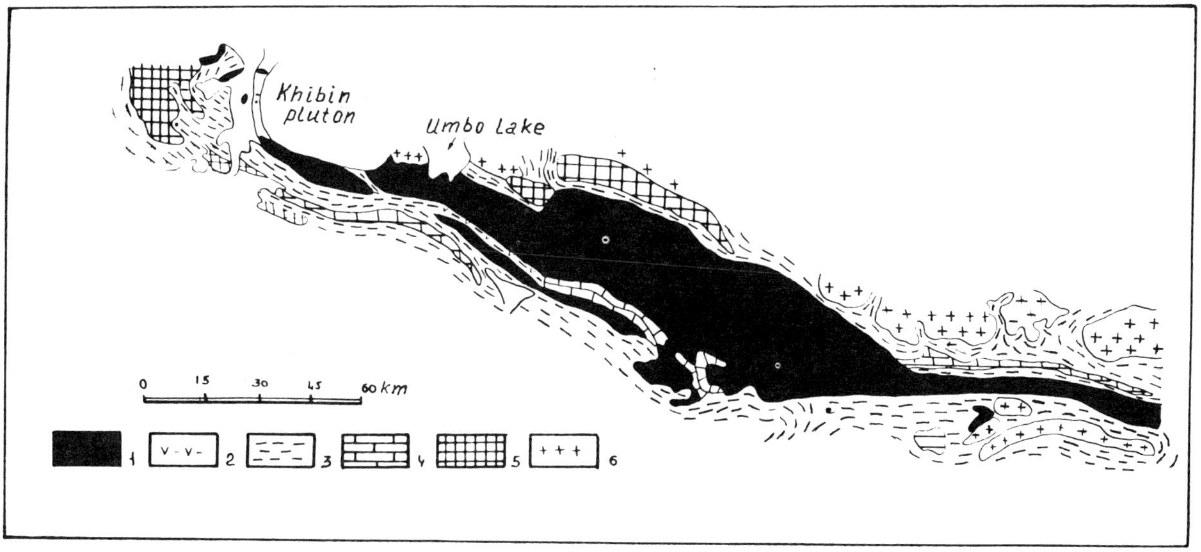

Fig. 7. Structure of the Imandra-Varguza suture zone (after Novikova [1975] simplified).
1, Greenstone formed from lava; 2, greenstone formed from tuffs; 3, metamorphics; 4, limestone; 5, norite+peridotite; 6, alkaline pegmatitic rock.

deformed Lower Proterozoic rocks of the adjacent Karelian block (Sumian, Jatulian, Suisarian and possibly Vepsian rocks); (iii) the important role of volcanic piles, of both mafic and silicic composition (leptites, gelflints); (iv) widespread and huge granitoid plutons, 1950-1820 Ma in age.

The Svecofennian block is separated from more eastern blocks of the Baltic shield by the continuous and very long (ca. 1500 km) West Karelian overthrust zone. As suggested by Voitovich [1980], the West Karelian zone is a deep overthrust along which the schist-gneissic sequences of the Svecofennian block are thrust over the Jatulian platform formations of the Karelian block (Figure 8).

The predominant schist sequences, which include volcanics, of the Svecofennian block are noticeably similar to graywacke-volcanic assemblages of Phanerozoic foldbelts. Borukaev [1985] and Gaal and Gorbatchev [1987] consider this complex to be related to an ancient island-arc-marginal-sea system. If so, the entire Svecofennian block may be interpreted as produced by accretional tectonics. The Karelian craton finally collided with the system of Svecofennian island arcs 1800-1900 Ma ago.

The Gothian Volcanic Belt and Rapakivi Granites

The span 1900-1800 Ma was marked throughout the Precambrian cratons of the East European platform by the first, nearly simultaneous intrusion of potassic (microcline) granites (e.g., Kirovograd-Zhitomir complex in the Ukraine). Their broad occurrence marked the amalgamation of primary continental blocks into the large continental mass that is the East European platform and the melting of the mature continental crust.

The next important stage in the development of the East European platform basement was associated with formation of the Gothian volcano-plutonic belt, which stretches north-south through southern Sweden on the western margin of the

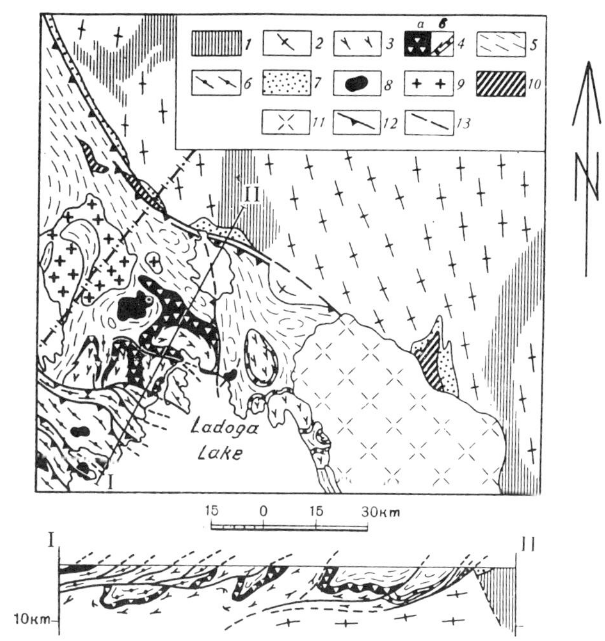

Fig. 8. Tectonic map and cross-section of North Ladoga area showing a wide distribution of thrusts in the West Karelian suture zone (after Martynova [1980]).
1, Archean greenstone belt; 2, plagiogranite and gneisso-granite of the Karelian block; 3, plagiogranite and gneisso-granite of the Sveco-Fennian block; 4, Sartavala series: volcanic succession; 5, Sartavala series: clastic succession; 6, Ladoga series; 7, granitized Ladoga series; 8, Jatulian; 9, gabbro+peridotite; 10, ultramafics within the thrust zone; 11, rapakivi; 12, fault; 13, suspected fault.

Svecofennian block and separates the latter from the Sveconorwegian block. The belt includes terrestrial silicic lavas such as the so-called Dala-porphyries, rhyolites, dacites, ignimbrites, and highly alkalic lavas. The extrusive rocks are associated with granitic batholiths. The age of the lavas and intrusive granites ranges from 1750 to 1650 Ma.

The Proterozoic volcano-plutonic belts, e.g., the Gothian belt on the East-European platform and the Akitkan belt on the Siberian platform, are markedly similar to continental margin belts of the Andean type. Associated with molasse sequences, the lavas coexist with comagmatic plutons. Like the Andes, the Gothian belt probably occupied a continental margin (the East-European) and was formed above the subduction zone which dipped from the ocean under East Europe and consumed the oceanic floor. Thus, the Sveconorwegian block situated to the west should be regarded as part of another continent, which (when this belt started to form) was far from the East-European margin and progressively moved, together with the oceanic part of the plate, towards the subduction zone [Gaal and Gorbatchev, 1987].

After the Gothian volcano-plutonic belt had been formed, by 1650-1550 Ma [Sukhanov, 1988] the entire East European platform was characterized by the intrusion of rapakivi granites, which in some cases succeeded gabbro-anorthosites (e.g., the Korosten pluton on the western part of the Ukranian shield) which are 1740-1890 Ma in age [Sukhanov, 1988]. The first alkaline intrusions originated at the same time, for instance the alkaline and nepheline syenites of the Pri-Azov block. Unfortunately, there is no satisfactory geodynamic interpretation to explain this vast generation of rapakivi granites. Their widespread development within East Europe probably indicates an intraplate origin, and suggests a possible connection with hot spots, or hot fields, in the Earth's mantle.

Jotnian Platform Cover: 1500-1400 Ma

Apparently after the intrusion of the rapakivi granites, the basement of the East-European platform as we now know it was complete. It was then blanketed by sediments of Lower Riphean age, including the Jotnian (Dala) sandstones of the Baltic shield, the Ovruchsky sandstones on the Ukranian shield, and quartzites of the Arlan formation in the Volga-Uralian area. Numerous diabase sills, reminding one of younger trap rocks, are characteristic of the Jotnian cover.

AULACOGENS

The East-European platform basement is interrupted by narrow and deep (more than 3 km) graben-like troughs called aulacogens or failed rifts. Three major epochs of graben formation are distinguished: Riphean, Devonian, and Permian. The Permian grabens, exemplified by the Oslo graben, are not discussed below.

Riphean Aulacogens

The Riphean aulacogens (Figure 9a) form a nearly rectangular network, trending NE and NW and cutting the East-European platform basement into a number of blocks. Aulacogens undoubtedly originated in a period of drastic break-up of the continent. According to Burke's and Wilson's [1977] idea, aulacogens are failed arms of three-arm rift systems. This would imply that where an aulacogen reaches the continental margin (platform edge), two other arms should have existed. Divergence of the continents along these other arms would, if it continued long enough, lead to the formation of oceanic basins. Probably this was just the situation at the eastern terminations of the Sredne-Russky and Kaltasinsky aulacogens as well as at the western termination of the Volyno-Orshano-Krestsovsky aulacogen where they reach the platform border (see Figure 2). It may be concluded that in pre-Riphean time the East-European platform extended appreciably farther east, and in the Middle Riphean time some portion broke off along lines lying at angles from 90° to 120° to the Sredne-Russky and Kaltasinsky aulocogens and then drifted away. As to the Kandalasksha and Mezen grabens, they are parallel to the trend of the continental border which presumably records the divergence. Such a pattern is similar to the distribution of Triassic-Early Jurassic grabens along the Atlantic coast of North America. Thus, we may assume that in Middle-Late Riphean time new oceanic basins came into existence east of the present East-European platform at the site of the Timan and the Urals.

The type Riphean succession in the Western Urals in the Bashkir anticlinorium [Keller at al., 1984] is indirect evidence for the existence of an oceanic area eastward from the platform.

Devonian Aulacogens

The Pripyat-Dneprovksy (Dneprovsko-Donetsky) (Figure 9b) aulacogen and a number of aulacogens in the eastern margin of the platform (Kama-Kinel grabens) originated in Devonian time.

The Pripyat-Dneprovsky aulacogen started to form after the end of the Middle Devonian when the first subsidence occurred, but the main phase took place in the second half of the Frasnian, coinciding with intense basaltic magmatism [Pistrak, Pashkova, 1974].

The position of the Pripyat-Dneprovsky aulacogen on the continuation of the Donbass foldbelt and its connection with the deformed margin of the platform indicates that this aulocagen is a failed arm of a former three-arm rift system. We assume that the oceanic crust preserved near the platform margin in the Peri-Caspian depression has the same Middle-Late Devonian age and hence was produced from one of the active spreading arms of the previous three-arm rift system. Another active arm evidently projected to the south (in present-day coordinates) from the junction of the Dneprovsko-Donetsky aulacogen with the Peri-Caspian arm. Thus up to the end of Devonian time, the East-European continent was larger than now, but in the Devonian a block of uncertain size rifted off and moved laterally toward the SW relative to the main continent.

Peri-Caspian Depression

The central part of the Peri-Caspian depression lacks a granitic layer, which would give seismic P-wave velocities of 6-6.5 km/s. Instead, at 20 to 25 km depth directly under the sedimentary cover, there is a high-velocity layer with seismic velocities from 6.7 to 7.1 km/s (Figure 10) [Fomenko, 1972; Nevolin, 1978]. The deep-drilling and seismic-sounding data provided in Sokolov [1970], Fomenko [1972], Nevolin [1978], Aisenshtadt and Slepakova [1979], Yanshin et al. [1980a] indicate that there is a paleo-oceanic floor formed in the Middle Devonian.

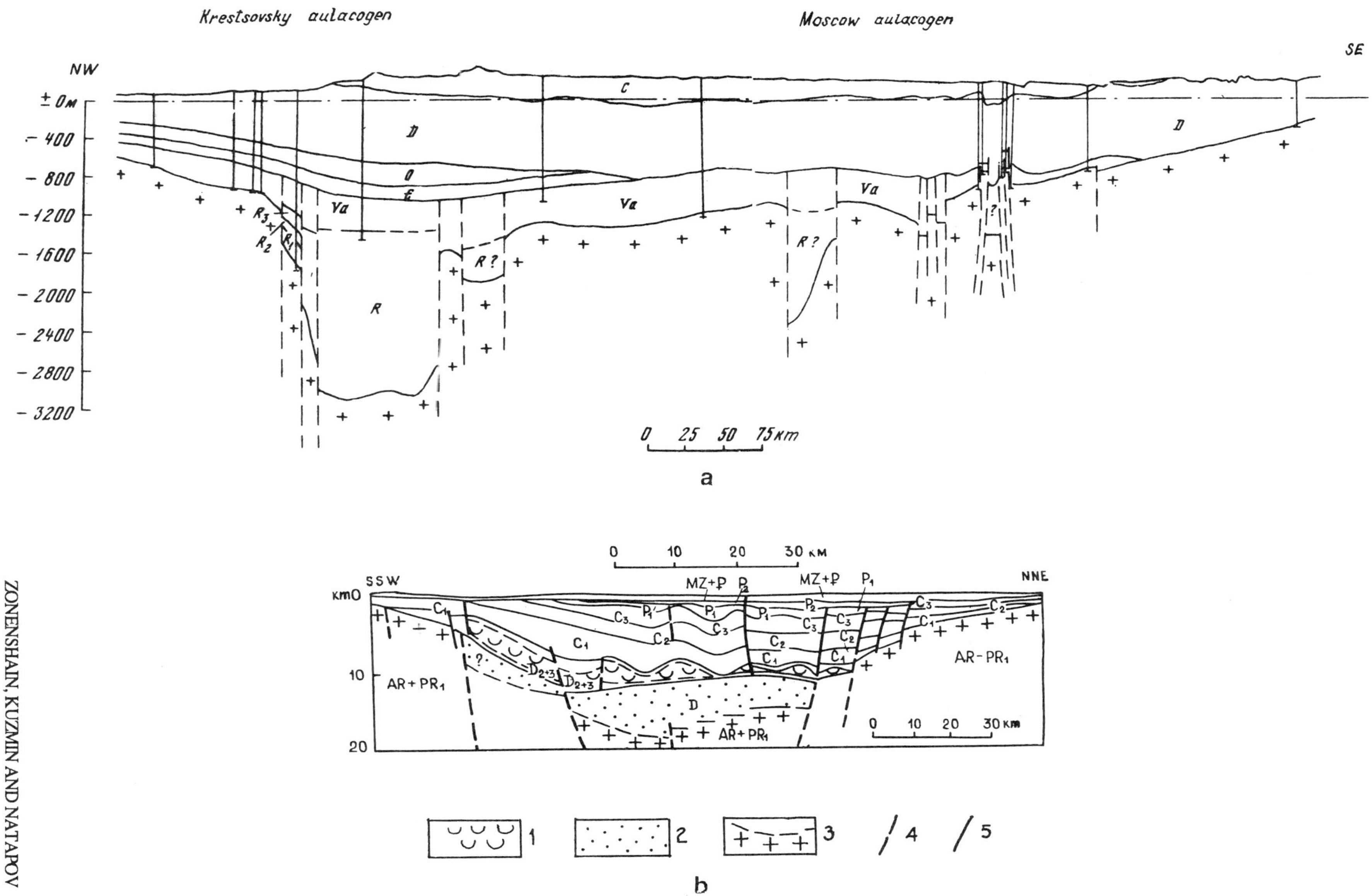

Fig. 9. (a) Cross-section through the Krestsovsky aulacogen (after Peive et al. [1978]). Crosses denote the crystalline basement. Vertical lines show location of boreholes. (b) Cross-section through the Dneprovsky aulacogen.
1, Devonian salt deposits; 2, pre-salt deposits; 3, top of the crystalline basement; 4, faults inferred from geophysical data; 5, faults established from geological data.

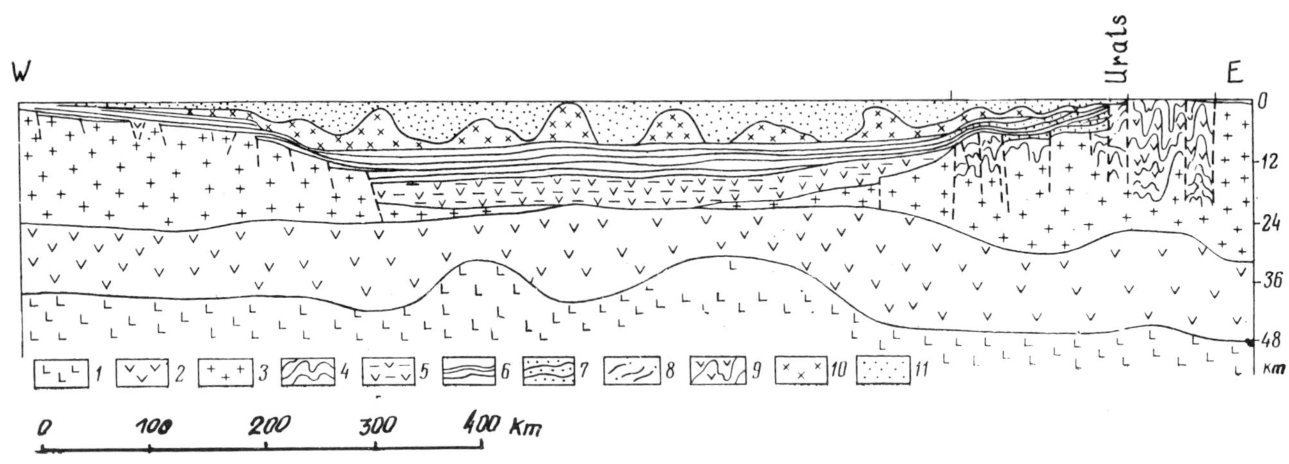

Fig. 10. W-E cross-section through the Peri-Caspian depression along the Kamysh-Emba line (after Nevolin [1978]).
1, Mantle; 2, "basaltic" layer; 3, "granitic" layer; 4, Proterozoic and Lower Paleozoic complexes of the Uralian belt; 5, lower pre-salt unit; 6, upper pre-salt unit in the center of the depression; 7, the same unit on the eastern flank of the depression; 8, Upper Paleozoic of the Pre-Uralian foredeep; 9, Devonian of the South Urals; 10, Kungurian salt; 11, Permian and Mesozoic post-salt deposits.

Facies distribution from the Late Frasnian to the Artinskian, e.g., prior to salt deposition, is very significant (Figure 11). A barrier reef bordered the western and northern slopes of the depression [Grachevsky, 1961; Mirchink, 1973; Kiryukhin, 1982] and sloped down the escarpment toward the center of the depression, where it was replaced by deep-water, thin (a few tens or hundreds of meters thick), carbonate-argillaceous sedimentary sequences of the same age, reflecting the change from shallow-water shelf to deep-sea basin. Such a situation is common on many present passive continental margins.

On the southern side of the Peri-Caspian Depression the pre-salt (Pre-Kungurian) Paleozoic is largely represented by the Late Paleozoic carbonate platform. In two areas, however, thick terrigenous sediments [Yanshin et al., 1980a] constitute deep-water fans or river deltas. The clastic material was supplied from the south (from the uplifted block at the side of Ustyurt and the Paleozoic foldbelt in the Scythian platform basement) rather than from the Urals. These data prove the existence of Late Paleozoic dry land southward (in present coordinates) from the Peri-Caspian depression. This was probably the continental block that moved laterally relative to East Europe when the Peri-Caspian oceanic depression formed.

In Pre-Kungurian time, the Peri-Caspian basin was not isolated from the World Ocean; it filled with normal marine sediments, and reef limestone. However, salt deposition commenced in Kungurian time, and as suggested by Yanshin et al. [1980a], salt accumulated at great depths without any uplifting of the floor. A thick salt layer formed in a short period--about 10 million years--and during subsequent Permian time the basin was rapidly filled with clastic rocks and became the present epicontinental depression. The Peri-Caspian depression has tended to subside until Holocene time, probably for two reasons: increasing density of the cooling oceanic lithosphere and enchanced sedimentary load.

The major cause for the transformation of the Peri-Caspian depression from an open marine oceanic basin to a closed salt-generating basin was its isolation from the open sea. Just at that time, the Ustyurt massif approached the East-European continental margin and blocked the Peri-Caspian depression. The evidence for such an event can be seen in deformation forming foldbelts along the south rim of the Peri-Caspian depression, in particular in the Donetsk-Astrakhan nappe-thrust zone and in the South Embinsky fold zone.

INTRAPLATE MAGMATISM

Intraplate magmatism occurred on the East-European platform in the Early-Middle Riphean, Vendian, Devonian, and Permian.

HISTORY

The development of the East European platform is schematically shown in Figure 12. To restore the former position of the East European continent the authors used the following types of information: (i) for the last 260 Ma, plate-kinematic interpretations in the Atlantic Ocean, paleomagnetic data, and deductions on the movement of plates relative to the hot-spot frame; (ii) for 260-370 Ma, paleomagnetic data and trajectories of the movement of East Europe relative to hot spots; (iii) for more ancient times, paleomagnetic data only. Paleomagnetic poles are listed in Table 1, and the apparent polar wandering path (APWP) is given in Figure 12. The parameters for rotation of the East European continent in the absolute frame since the Early Devonian are given in Table 2. The continental positions on the reconstructions for Vendian and Late Riphean are given latitudinally according to paleomagnetic data only; their longitudinal position is arbitrary.

Figure 12 also shows conventional reconstructions of positions of Precambrian massifs for 2000 and 1700 Ma. No reliable paleomagnetic data for East Europe are on hand.

2000 Ma

This reconstruction approximately corresponds to the time when the Jatulian platform cover formed on the Karelian

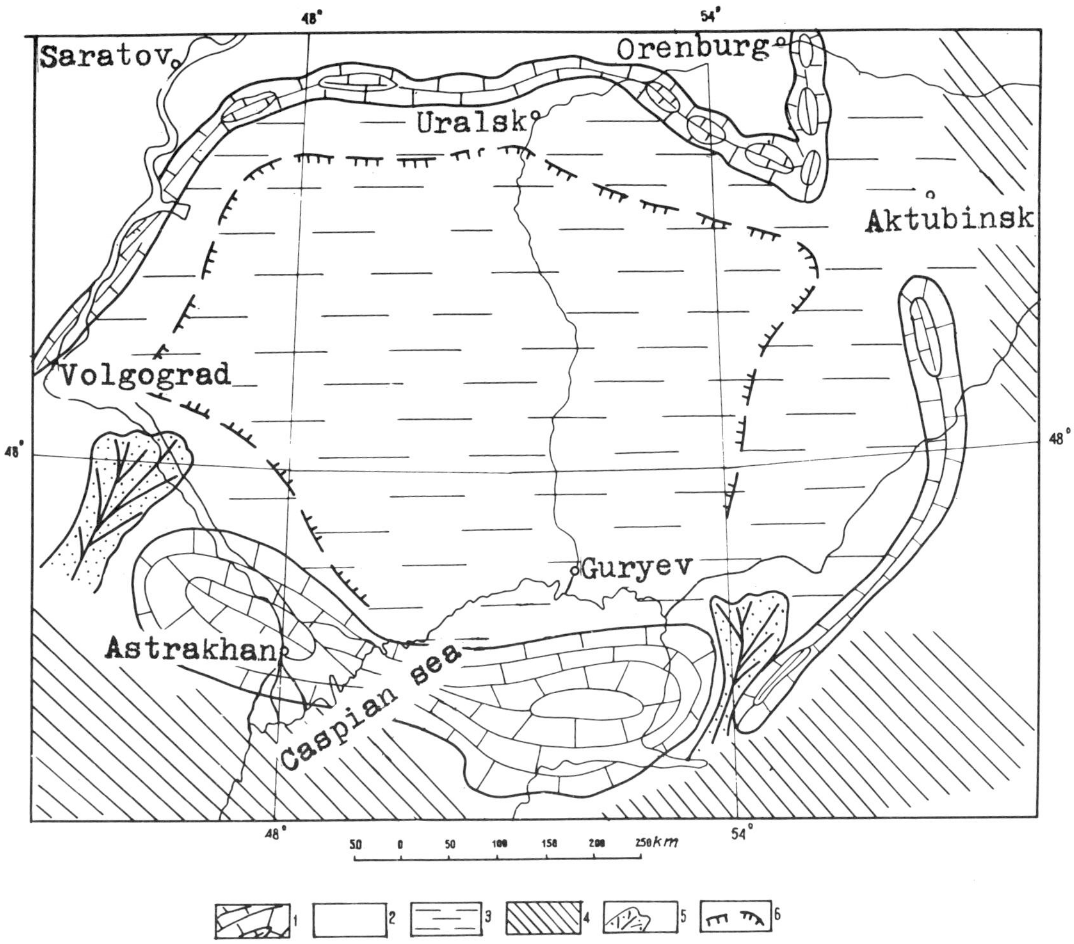

Fig. 11. Facies distribution for Visean deposits within the Peri-Caspian depression (after Kiryukhin [1982] and Yanshin et al., [1980]).
1, Barrier reef and shallow water limestone; 2, other shallow-water sediments; 3, deep-water facies; 4, dry land; 5, fan, (deltaic deposits); 6, approximate limits of oceanic crust (after Fomenko [1972]).

block. The following continental blocks existed at that time: (i) Karelian, including the Karelian block of the Baltic shield as well as a significant portion of the Voronezh massif and evidently the major part of the Ukranian shield--characterized by the development of a Jatulian platform cover; (ii) the Belomorian block, in which Jatulian sediments are not known; (iii) the Kola block, which is markedly different from the Karelian one as it contains widespread high-alumina schists of the Keiva Formation, indicative of a passive margin rather than an island-arc, as in the case of the Sumii volcanics; (iv) the Murmansk granite-gneissic block consisting of a granite-greenstone complex [Bogolepov and Votakh, 1977].

All of these blocks were spatially isolated, but the true distances between them are not known. The Karelian and Belomorian blocks are assumed to have been fairly close in Jatulian time, for the major part of the oceanic lithosphere separating these blocks had already been consumed under the Sumii island arcs. The subduction under these arcs produced a large-scale convergent orogenic phase (Seletsky) at that time and eventually resulted in the nappe-thrust structures of the East Karelian suture zone. The Kola block was separated from the Belomorian and evidently from the Murmansk blocks by appreciably wide oceanic basins, as indicated by greenstone rocks of the Imandra-Varguza and Pechenga Formations. A wide oceanic basin, with a number of island arcs of the Svecofennian zone, existed west of the blocks mentioned above.

1700 Ma

By this time, the formation of the East European craton by convergence and collision of individual blocks was finished. The sequence of events had been as follows: at first the Karelian and Belomorian blocks collided, subsequently the Karelian-Belomorian block collided with the Kola and Murmansk blocks, and eventually the Volga-Uralian block

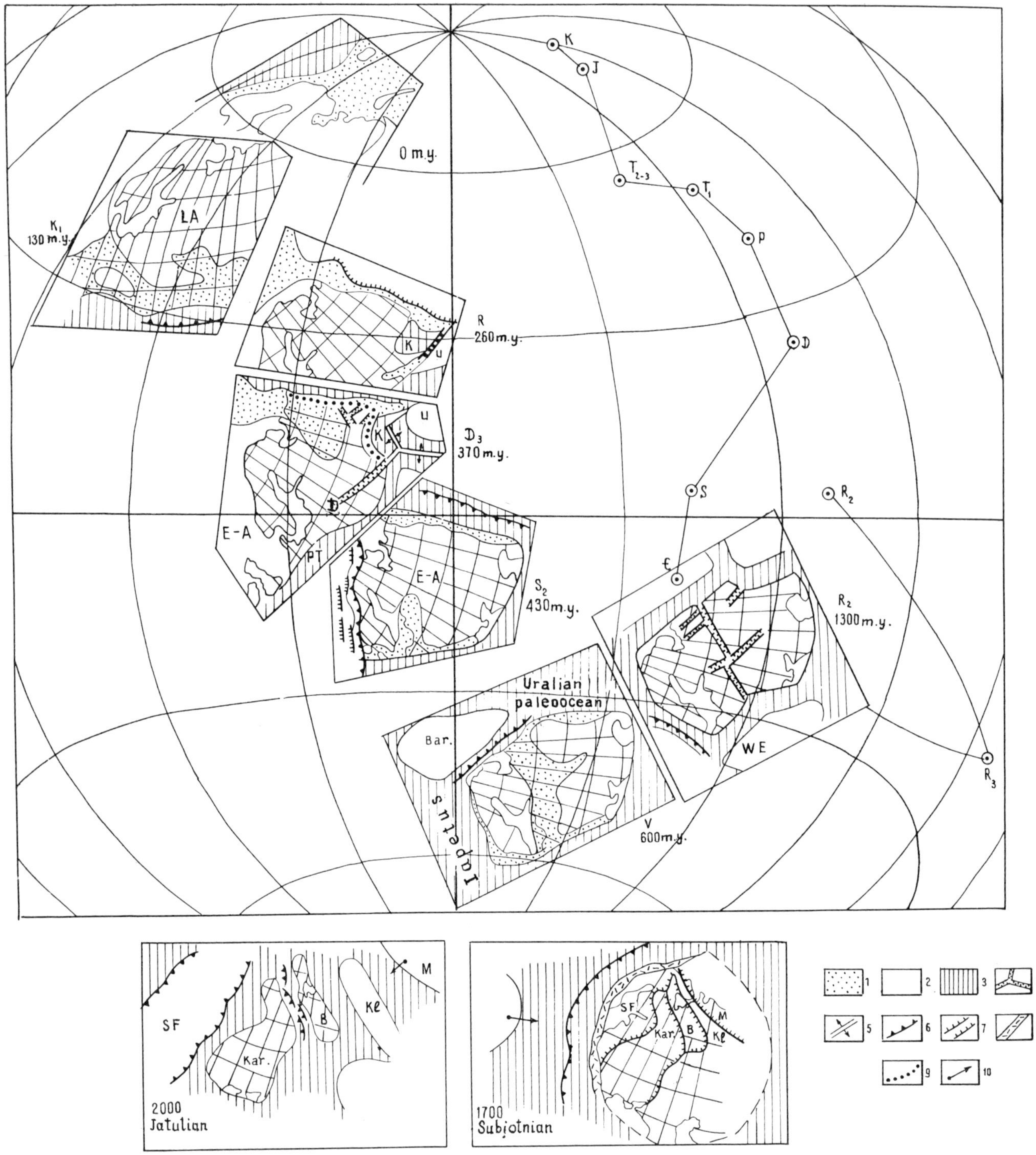

Fig. 12. Palinspastic reconstructions of the East European platform. Apparent polar wandering path is also shown (after Khramov [1982]); encircled points correspond to paleomagnetic pole positions.
1, Shallow sea; 2, dry land; 3, oceanic basin; 4, continental rift; 5, spreading axis; 6, subduction zone; 7, collision zone; 8, continental margin volcanic belt; 9, barrier reef; 10, direction of block motions. B - Belomorian block; Bar - Barentsia; Kl - Kola block, K - Karelian block; M - Murmansk block; SF - Sveco-Fennian block; D - Pripiat-Dneprovsky aulacogen; EA - Euramerica; K - Peri-Caspian depression; LA - Laurasia; PT - Paleo-Tethys; T - Tethys; U - Ustyurt massif; WE - West Europe.

TABLE 1. Paleomagnetic Poles for East Europe (After Khramov [1982])

Age	Degrees N	Degrees E
Cambrian	-13	135
Ordovician	30	143
Lower Silurian	38	155
Lower Devonian	31	161
Lower Carboniferous	46	158
Middle Carboniferous	32	163
Upper Carboniferous	41	169
Lower Permian	40	167
Upper Permian	45	162
Lower Triassic	51	153
Middle-Upper Triassic	55	137
Jurassic	74	167
Cretaceous	77	188

TABLE 2. Parameters for the Rotation of East Europe in Absolute Frame of Reference (After Zonenshain et al. [1985])

	Pole position		
Age (Ma)	Deg. N	Deg. E	Angle of Rotation (Degrees)
0 - 130	20.2	73.1	19.9
0 - 220	53.1	76.9	59.6
0.- 250	55.4	81.3	74.7
0 - 280	50.4	91.7	82.3
0 - 310	47	89.6	87.9
0 - 340	43.9	87.9	93.6
0 - 370	40.1	88.4	98
0 - 400	30.7	97.6	96.5

collided with the main continental mass. This continental mass approached the island arcs of the Svecofennian zone, which were deformed and thrust over the Karelian and Belomorian blocks and accreted to the East European craton on the west.

The most remarkable event of this time is the formation of the Gothian volcano-plutonic belt along the western margin of the East European craton. This belt was undoubtedly associated with a subduction zone upon which the East European craton overrode the oceanic basin to the west, thus converging with some other craton, probably the present Sveconorwegian block.

1300 Ma

The major event of Riphean time was rifting of the East European continent. Presumably, some of the rifting resulted in splitting off and drifting away of fragments, so that the size of the East European continent was significantly reduced. However, some of the rifting was "unsuccessful," and is preserved as aulacogens. On the Scandinavian side, East Europe approached the Sveconorwegian block; the two eventually collided about 1000-850 Ma ago, causing the Dalslandian orogeny.

600 Ma

Well developed passive margins existed in Late Riphean and Vendian times along the Uralian and Baltic-Ukranian sides of the East European continent, indicating the presence of adjacent broad oceans, i.e., Paleo-Uralian and Iapetus (Paleo-Atlantic). In the central part of the continent above the system of Riphean aulacogens in the Vendian, marked subsidence led to the formation of two wide intracratonic basins: the Baltic and the Moscowian. In Vendian time, the margin near the Timan was deformed by convergence and collision with the Barentsia continent.

410 Ma

The Baltic-Ukranian passive margin proceeded to develop throughout the entire Early Paleozoic. The Uralian margin was deformed at the end of Vendian, because of collision with a continental massif. But in the Ordovician, subsidence commenced again on the Uralian passive margin, due to extension and rifting. On the Scandinavian margin, East Europe collided with North America, producing the Caledonian thrust front and forming a single Euramerican continent.

370 Ma

As is well known, in the second half of the Devonian the western portion of East Europe was included in the so-called "Old Red Continent", on which red-colored coarse clastic rocks accumulated because of the erosion of Caledonian mountain chains. The Peri-Uralian and Peri-Caspian margin of the continent was fractured and extended. At this time the Dneprovsky-Donetsky aulacogen, the system of Kama-Kinel grabens, and perhaps the analogous extension structures in the basement of the Pechora lowland were formed. As a consequence, the Peri-Uralian passive margin essentially widened, and the broad eastern part of the platform, adjacent to the Uralian paleo-ocean, was involved in subsidence.

Sialic blocks, currently parts of the basement of the Scythian and Turan lowlands, moved away from the East European continent along two active arms of the Dneprovsky-Donetsky aulacogen. Between them there originated a new oceanic basin; part of its oceanic floor is still preserved in the Peri-Caspian depression.

260 Ma

In Permian time, East Europe was surrounded on all sides by foldbelts: Middle European, Uralian, and Donbass. Only in the Barents Sea and Novaya Zemlya there was no continental collision, and the passive margin sloping down to the adjacent oceanic basin continued to subside.

130 Ma

In the period between Early Permian and Jurassic, East Europe was a part of Pangea and was surrounded on almost all sides by foldbelts. However, on the south it bordered the Tethys Ocean. Thus subsidence was characteristic only of the southern margin of the East European platform. In Cretaceous time Europe occupied approximately the same position on the Earth's sphere as today. The East European platform was little affected during the remainder of the Mesozoic and the Cenozoic.

Chapter III

SIBERIAN PLATFORM

The Siberian platform has an angular shape and is 2500-2750 km across. Figure 13 shows the exposures of Precambrian basement in windows through the platfform cover. The basement crops out mainly on the southern periphery, i.e., in the Aldan shield (Stanovoy Ridge included), the Baikal region, the East Sayan, and the Enisey Range (Figure 13). The Anabar uplift is a large exposure of crystalline rocks of the basement within the center of the platform. The Aldan shield has been an area of continuous uplift, i.e., 'true' shield, since Late Precambrian, and the other massifs became geomorphic highs at different times during the Paleozoic. As compared to the basement of the East European platform, fewer aulacogens are known at present beneath the platform cover. However, two Devonian aulacogens, Viljuy and Ygyatinsky, run NE-SW in the basement of the Viljuy syneclise and have well-defined graben structures. The important Tunguska basin (or syneclise) originated in the Late Paleozoic and filled with the well known Siberian Permian-Triassic flood basalts. Some geophysical data [Shtekh, 1965] provide evidence that under the westernmost downwarped segment of the Tunguska syneclise, and under the eastern part of the Viljuy aulacogen, the continental crust is either considerably thinned or entirely rifted away, giving way to newly formed oceanic crust.

PRECAMBRIAN BASEMENT

A number of blocks are clearly recognized within the Aldan shield where (north of the Stanovoy fault) they have a longitudinal orientation and are divided by meridional faults. From west to east they are the Chara-Olekma, Central Aldan (Iengrsky), Timpton-Uchur, and Batomga blocks. The longitudinal structures of the Aldan shield and the latitudinal structures of the Stanovoy Ridge join like a "T" along the latitudinal Stanovoy fault.

According to Movalev [1981], all blocks of the shield consist of two main rock complexes: older granulitic granite-gneissic basement and younger greenstone belts.

Mafic crystalline schists of the Kurultin-Gonamsky complex [Glukhovsky et al., 1977] appear in some places within the granite-gneissic basement. They include hypersthene and bipyroxene-plagioclase schists with lenses of harzburgites and lherzolites. These rocks are close in composition to ophiolites. Hence, Glukhovsky et al. [1977] suggested that they are the remains of the "Katarchean" oceanic crust. The granite-gneissic basement consists largely of quartzites and high-alumina gneisses (Iengrsky Formation) and charnockites (Timpton-Dzheltulinsky formation). Isotopic ages range from 3650 to 3300 Ma [Bibikova, 1989]. The foregoing rock assemblages are associated with enderbite bodies, characteristic of the grey gneissic complex.

Moralev [1981]suggests that most greenstone belts of the Aldan shield were formed in the period 2600-2100 Ma, but some data indicate that Archean greenstone belts are also present, such as in the Olondin belt in the Olekma River basin [Drugova et al., 1985]. The isotopic age of this belt is 2960 Ma [Babikova, 1989]. Nearly 30 greenstone belts are mapped on the Aldan shield (for example, Figure 14), including the Stanovoy Range. Normally these belts are 3 to 5 km or 10 to 15 km wide and extend a few tens of kilometers, in places 100-150 km. The rock successions of some greenstone belts are not all alike, although it is not known whether they were initially different. Moralev [1981] classifies greenstone belts into volcanic (50% volcanics), terrigenous-volcanic (10-15% volcanics), and terrigenous (less than 10% volcanics). The former are predominantly represented by mafic rocks, and only in one case (Olondin belt) are the volcanics bimodal. Metavolcanics largely correspond to MORB and basaltic komatiites; some lie in the field of peridotitic komatiites and andesites. Moralev came to the conclusion that the volcanics in the greenstone belts formed by remelting of more ancient basaltic crust under high-pressure conditions, perhaps in subduction zones where plagiogranite bodies intruded into the greenstone belts.

Most of the greenstone belts are narrow, tight synclines cut by thrusts. Some are vertically dipping homoclines separating basement granulite-gneissic blocks, and they include zones of blasto-mylonites comprising various crushed rocks, e.g., gabbro, ultramafic rocks, metavolcanics, and metasedimentary rocks. They can be interpreted as collisional sutures between different crustal blocks. Stratigraphic contacts of greenstone belts with surrounding granulite-gneissic complexes were not observed anywhere; the contacts are either tectonic or are obscured by metamorphism around growing granulite-gneissic domes. As in the East European platform, the greenstone belts of the Aldan shield may be considered as tectonic synforms, the remains of an upper tectonic sheet thrust over the granulite-gneissic basement. The sheets were predominantly preserved between the granulite-gneissic domes as

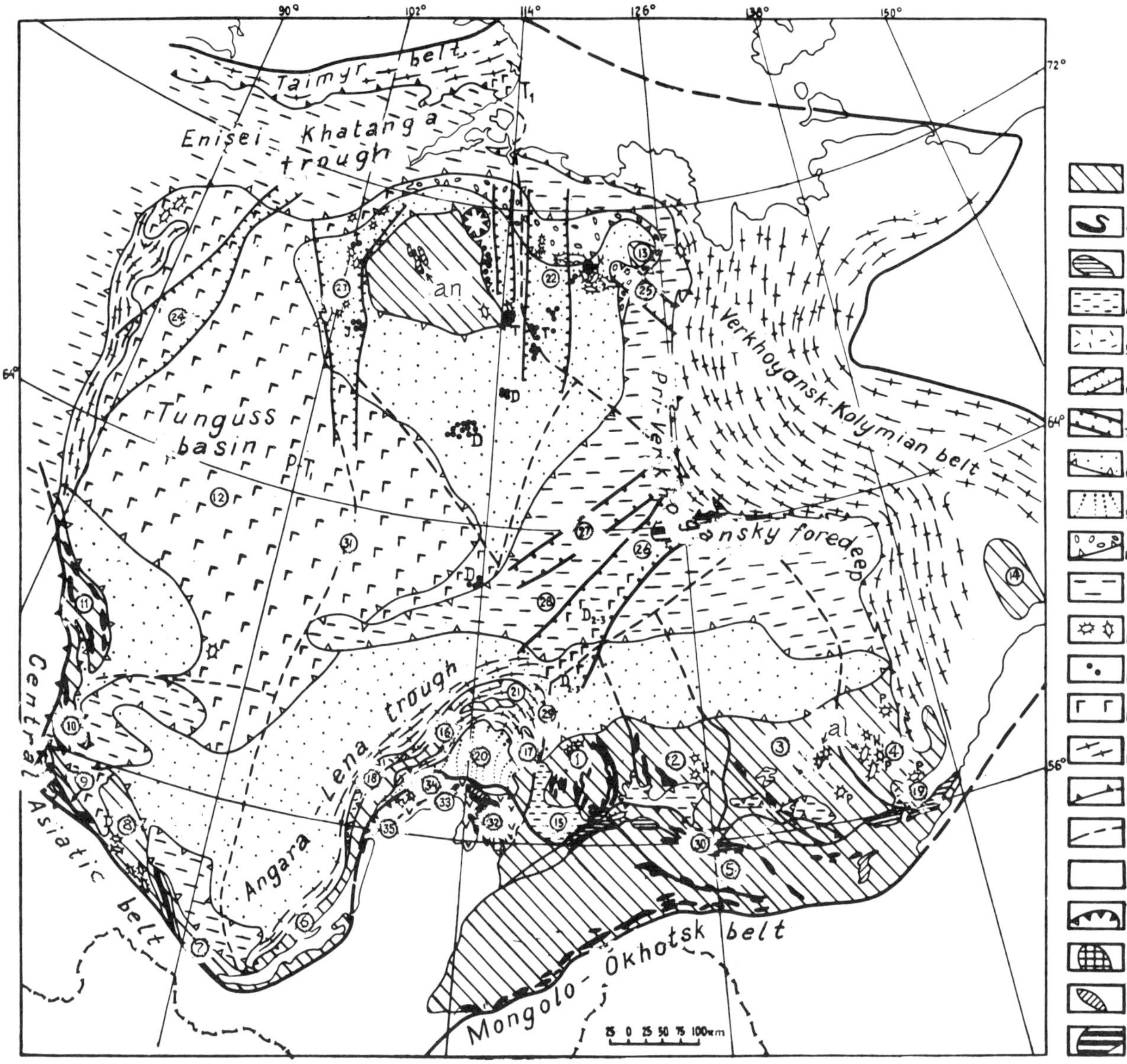

Fig. 13. Tectonics of the Siberian platform.

1, Shield; 2, greenstone belt; 3, anorthosite; 4, Early Proterozoic sedimentary cover; 5, Early Proterozoic volcanic belt; 6, Riphean aulacogen; 7, Devonian aulacogen; 8, Riphean-Paleozoic shelf sediments; 9, Riphean-Paleozoic continental rise sediments; 10, Late Paleozoic and Early Mesozoic sediments; 11, Late Mesozoic and Cenozoic sediments; 12, alkali-ultramafic pluton; 13, kimberlite; 14, flood basalt (trap); 15, folded sedimentary cover; 16, thrust front; 17, boundary of the Pre-Riphean blocks; 18, boundary of the Siberian platform; 19, Popigai astrobleme; 20, Muya block; 21, ophiolites of the Muya belt; 22, possible oceanic crust in Viljui aulocogen.

Numbered are: 1, Chara-Olekma block; 2, Central Aldan block; 3, Timpton-Uchur block; 4, Sutam block, 5, Batomga block; 5, Stanovoi block; 6, Fore-Baikal block; 7, Sharyzhalgai block; 8, Birjusa block; 9, Kan block; 10, Enisei block; 11, Lower Angara block; 12, Tungus block; 13, Oleniok block; 14, Okhotsk massif; 15, Udokan trough; 16, Chuisky uplift; 17, Longdor uplift; 18, Akitkan volcanic belt; 19, Ulkan volcanic belt; 20, Bodaibo trough; 21, Patom uplands; 22, Udzh aulacogen; 23, Maimechin aulacogen; 24, Igaro-Norilsky aulocogen; 25, Kjutingdin aulacogen; 26, Viljuy aulacogen; 27, Ygyatinsky aulacogen; 28, Suntar uplift; 29, Zhuin fault; 30, Stanovoi fault; 31, Taimyr-Sayan fault; 32, Muya block; 33, Muya ophiolitic belt; 34, Muya volcanic belt; 35, Olokit zone. *Al*- Aldan shield; *An* - Anabar massif.

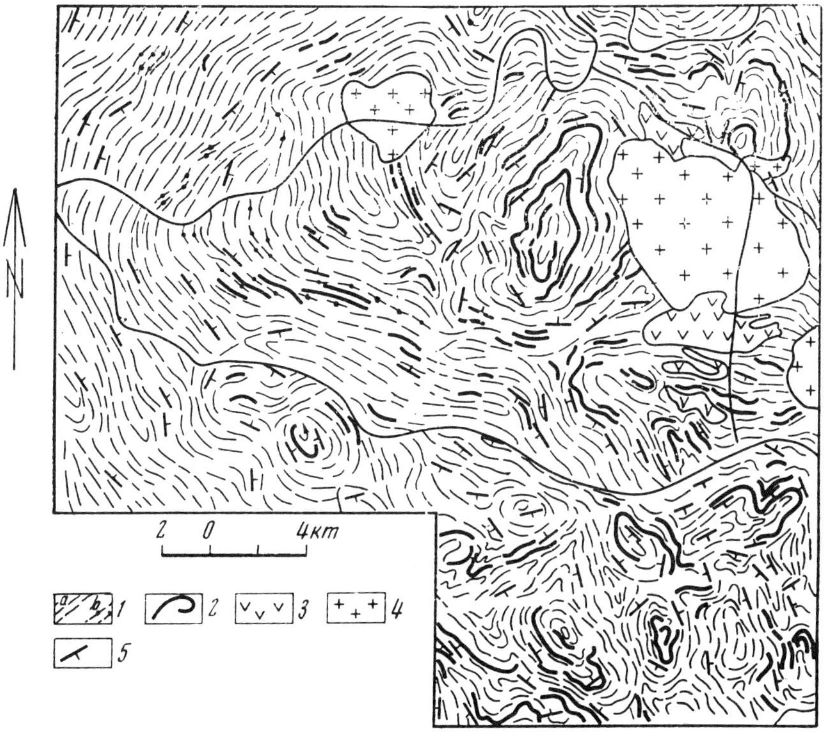

Fig. 14. Geology of the Sutam block within the Aldan shield (after Moralev [1981]).
1, Granulite, a - pyroxzene schist (metavolcanics), b - high-Al schist (metapelite); 2, iron formation; 3, gabbro-anorthosite; 4, Mesozoic granite; 5, dip of foliation.

those domes grew. Again as in the basement of the East European platform, the greenstone belts are evidently crustal remains of oceanic-like basins, judging from the presence of MORB and komatiites. The size and number of these basins are hard to define. The evolving basins closed after convergence of blocks of the granulite-gneissic basement; their crust was subducted, with concurrent formation of calc-alkaline igneous rocks, as the granulite-gneissic blocks approached one another and collided.

A large area on the Aldan shield is occupied by clastic sequences of the Udokan series. They fill the sizable (175 x 75 km) Udokan and Kada troughs situated in the Chara-Olekma block. The Udokan series consists of thick (10-12 km) sequences of clastic rocks that suffered zonal metamorphism in the kyanite-sillimanite facies [Fedorovskiy, 1985]. The metamorphism is reliably dated as Early Proterozoic. The upper age limit of the protolith rocks is defined by metamorphism at 1850-1950 Ma and intrusion of the Kadar rapakivi granites at 2000 Ma. Lavrovich [1970] showed that the Udokan series, except for its upper part, has an oligomictic or monomictic composition. The clastic grains include quartzites, quartz, and very rarely feldspar probably derived from a weathered and peneplaned landmass. Many sediments show evidence of deposition in near-coast shallow-water and delta environments. Both tectonic and paleogeographic settings resemble conditions common to continental margins in the region of large deltas. In Siberia, the Udokan rocks are the first true representatives of mature continental margins. In this respect, the Udokan Formation is the analog of the Jatulian of the Karelian block of the East European platform. Lavrovich [1970] noted a sharp change in composition in the upper section of the Udokan formation where the clastic material becomes polymictic and comprises fragments of cherts, felsic volcanics, plagioclase, and potassic feldspars. Remarkable are ore fragments, e.g., martite- and magnetite-bearing rocks, which resemble fragments of rocks and minerals from iron formation. The change in character of the clastic material preceded deformation and metamorphism of the Udokan series. Fedorovskiy [1985] connects the change in supply of clastic material with growing granite-gneissic domes, which likely originated in the colliding of continental plates. The change of the source seems to mark the time when the Stanovoy block closely approached the Olekma block and adjacent parts of the Aldan shield, providing an opportunity for clastic material to arrive from the south. The collisional suture runs along the Stanovoy fault; the acute "T" junction of the Aldan and Stanovoy Range structures along that fault suggests that the two blocks may have been originally separate.

The Precambrian basement in other parts of the Siberian platform has the same structure as on the Aldan shield. For instance, the Sharyzhalgay, Birjusa, and Kan blocks in the East Sayan, and the Enisey block in the Enisey Range can be dis-

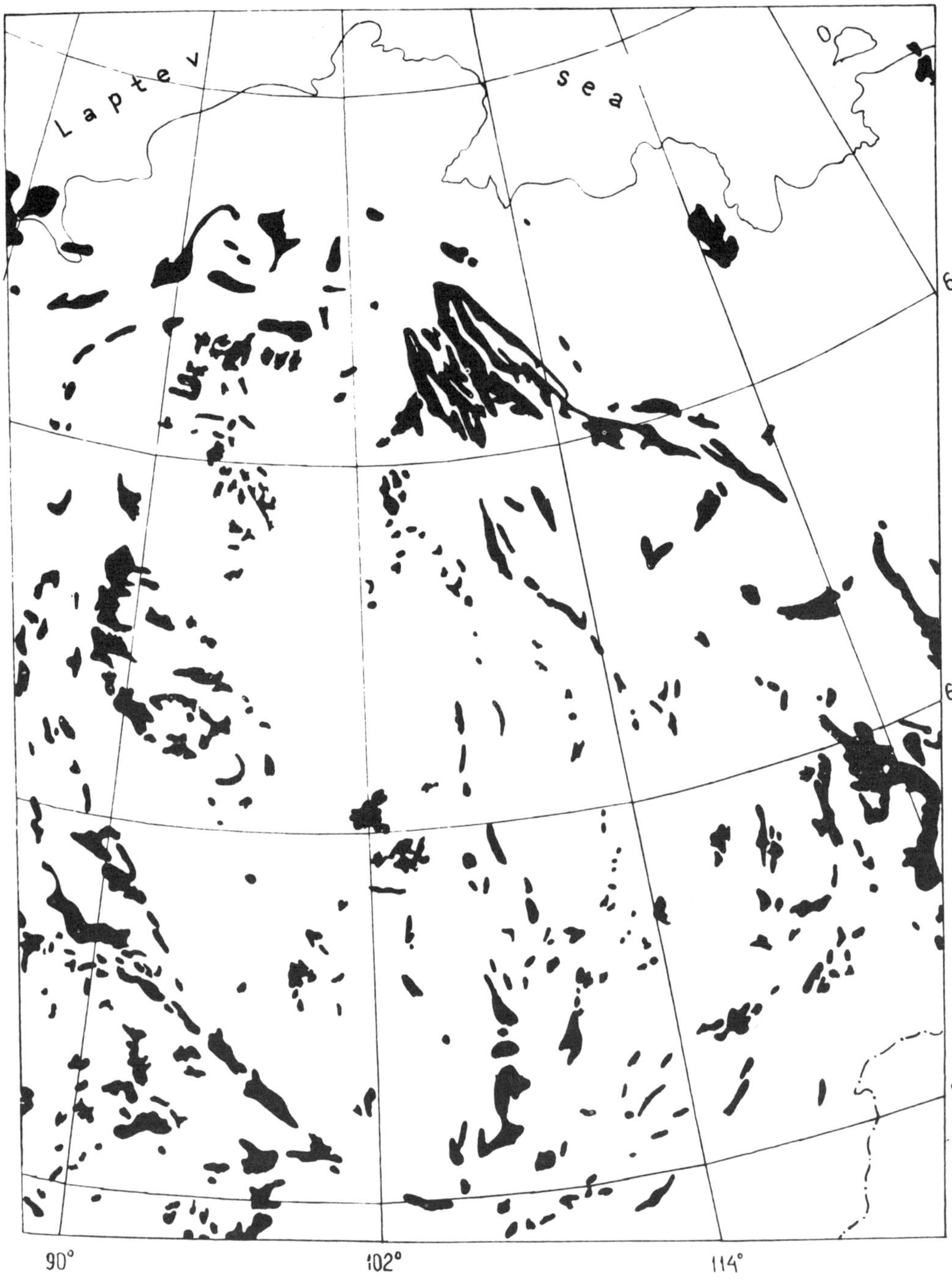

Fig. 15. Aeromagnetic field of Eastern Siberia; positive anomalies in black.

tinguished as individual blocks and include greenstone belts between Precambrian gneissic complexes (Figure 13). The block structure of the basement is well reflected in the magnetic field (Figure 15) [Gafarov et al., 1978].

The Precambrian basement of the Siberian platform also incorporates volcanic belts --the Akitkan, running in the Primorsky Range along the western coast of Lake Baikal, and the Ulkan, on the eastern margin of the Aldan shield. Both belts were formed in the period 1600-1800 Ma. The Akitkan belt [Bukharov et al., 1987] consists of andesites, trachytes, trachyandesites, ignimbrites, porphyrites, and abundant tuffogenic assemblages associated with molasse-like volcanoclastic piles. Volcanics belong to the calc-alkaline series, although in the west of the belt sub-alkalic and latitic lavas are also abundant. The alkali content regularly increases from the east towards the west [Mitrofanov, oral commun., 1986]. Borukaev [1985] correlated these Proterozoic volcanic belts with their Phanerozoic analogs and showed that they mark former Proterozoic active continental margins. The Akitkan belt marks the margin of the Angara-Anabar block; presence of the belt shows that in the Middle Proterozoic this block was isolated from the Aldan shield by an oceanic basin. The remains of this basin were preserved as ophiolites of the Muya belt.

Anorthosites, very abundant in the Stanovoy block, constitute two large massifs--Dzhugzhur in the east and Kadar in the west--and a number of minor massifs in between, stretched along the Stanovoy fault or suture zone. The gabbro-anorthosite massifs [Bogatikov, 1979] result from differentiation of relatively dry magmas of basaltic or andesitic (calc-alkaline) composition that remained for a long time near liquidus [Bogatikov et al., 1984]. Considering the relation of the anorthosites to the zone of collision of the Stanovoy block with the Aldan shield and their Proterozoic age (2200-1800 Ma) [Sukhanov, 1988], therefore, we assume the belt originated above the subduction zone.

AULACOGENS

By 1600 Ma ago, the basement of the Siberian continent was entirely formed and the sedimentary cover began to accumulate. Sedimentation was preceded by fracturing of the platform, and two epochs of aulacogen formation have been recognized: Early Riphean and Devonian.

The Early Riphean aulacogens are discerned where the thickness of the sedimentary cover increases to 7-9 km, e.g., the Udzh aulacogen situated between the Oleniok uplift and the Anabar massif [Mokshantsev et al., 1975]. They display block-faulting and thinning of the continental crust, yielding a vast passive continental margin. As elsewhere, the aulacogens are probably failed arms of three-armed rifts; thus some portions of the continental crust evidently broke off from Siberia and drifted away.

In the Devonian the Viljuy aulacogen was formed, consisting of two grabens--on the south the Viljuy graben proper, on the north the Ygyatinsky graben--separated by the Suntar uplift. In these grabens the basement is 10-12 km deep. The thickness of Upper Devonian and lower Carboniferous assemblages (mostly evaporites) in the Viljuy graben is 7 to 8 km [Gaiduk, 1985]. The graben began to form in the Frasnian, and numerous volcanic domains on the edges of the aulacogen date from this time. The igneous rocks (350-380 Ma) include lava flows, sills, and dikes of tholeiitic and alkaline basalts, monzonite and syenite bodies, and acid lavas and tuffs. Trachyliparitic lavas are also present. Thus, a typical bimodal series formed, as characteristic of continental rift zones. The Viljuy aulacogen runs for about 600 km NE and plunges underneath the Verkhoyansky orogenic belt just at a peculiar knee in the thrust-fold structures. The presence of thick sequences of alkaline basalts in the Devonian in the Sette-Daban area implies that rifting also occurred in the region presently covered by the Verkhoyansk belt. This is the reason that Levashov [1964] suggested that the Viljuy rift system is prolonged northeastward.

Devonian rifts of the Siberian platform appear to be related to several triple junctions, two of them at the east ends of the Viljuy and Kjutingdin aulacogens (failed arms). The active arms caused thinning of the crust of the northeastern (present coordinates) passive continental margin of Siberia, which was subsequently overlain by a thick prism of clastic sediments from the Verkhoyansky orogenic belt. Apparently rifting did not induce rupturing of the continental crust except at the extreme eastern part of the Viljuy aulacogen, where Stekh [1965] using gravity and magnetic data, established that no granitic crust is present in the NE part of the Viljuy graben.

INTRAPLATE MAGMATISM

The Siberian platform was the site of widespread intraplate magmatism. Alkaline, alkaline-ultramafic, and mafic magmatism occurred in the Riphean, Vendian, Cambrian, Devonian, Permain, Triassic, Jurassic, and Cretaceous. It should be emphasized that extrusions of flood basalt and intrusion of kimberlites were among the most remarkable and significant events in the history of the Siberian platform.

As to the flood basalts, the Siberian traps, several points need to be emphasized. First, flood basalts were extruded in an extensional environment, as shown by numerous dike swarms and sheeted dike-like sequences [Kurenkov, in press]. Second, the quiet fissure eruptions were preceded by an episode of explosive activity related to crustal fracturing, which provided subsequent easy eruption of lavas. Third, eruptions occurred within an extremely short period, from 255 to 245 Ma. During these 10 Ma, 1,200,000 km^3 of basalts erupted, a volume estimated as equivalent to the activity of a spreading ridge 1500 km long, assuming the recent rate of accretion of the basaltic layer of the oceanic crust. Fourth, despite the extensional conditions, crustal thinning (8-10 km), and a huge amount of erupted basalts, all these processes did not lead to major subsidence of the Tunguska region except for isostatic subsidence induced by loading. (As is well known, normal processes of rifting cause formation of deep grabens and aulacogens.) Finally, the composition of the volcanic formations of Siberia is identical to that of intraplate basalts of oceanic islands and seamounts. The Siberian traps, like the basalts of other large trappean provinces around the world, are enriched in incompatible rare elements and have high $^{87}Sr/^{86}Sr$ ratios [Al'mukhamedov, 1985]; thus the undepleted mantle was the source of lavas. Inasmuch as these features of trappean magmatism are never associated with plate boundaries, we believe that such magmatism is related to hot spots or hot fields in the Earth's crust.

Kimberlites on the Siberian platform were emplaced during three main epochs: at the end of the Devonian, in the Triassic, and in the Cretaceous. The Devonian and Triassic kimberlites are linked with corresponding epochs of intraplate basalt volcanism. Cretaceous kimberlites are known in the north of the platform on the slopes of the Oleniok uplift, and are confined to the area of Cretaceous intraplate magmatism in the Arctic.

The kimberlitic bodies are distributed randomly on the Siberian platform; within each field or group of bodies, however, networks of fissures preferentially localized the individual kimberlite pipes.

Another manifestation of intraplate magmatism is the Riphean alkaline-ultramafic massifs, often with kimberlites, in the region of the Udzh aulacogen, between the Oleniok uplift and the Anabar massif, and in the east of the Aldan shield on the Konder and Arbarastakh massifs. In southern Siberia there was Middle Paleozoic intraplate magmatism, e.g., bimodal complexes and alkaline granites of the Rybinsk depres sion in the west, ultramafic-alkaline massifs in the north of the East Sayan, and the Synnyrsky complex of alkaline rocks in the northern Baikal region. The Triassic alkaline-ultramafic bodies of the Maimecha-Kotuy province appear westwards of the Anabar shield. In the region of the Aldan shield, Jurassic-Lower Cretaceous alkalic massifs are widespread. The majority of the areas of intraplate magmatism are shown in Figure 13.

HISTORY

Figure 16 demonstrates the path of Siberia throughout the last 2200 Ma. The authors used paleomagnetic data of Khramov et al., [1982] (Table 3) for Paleozoic and Mesozoic times the tracks of plate motions over hot spots were used as well (Table 4) [Zonenshain et al., 1985]. Two main belts of

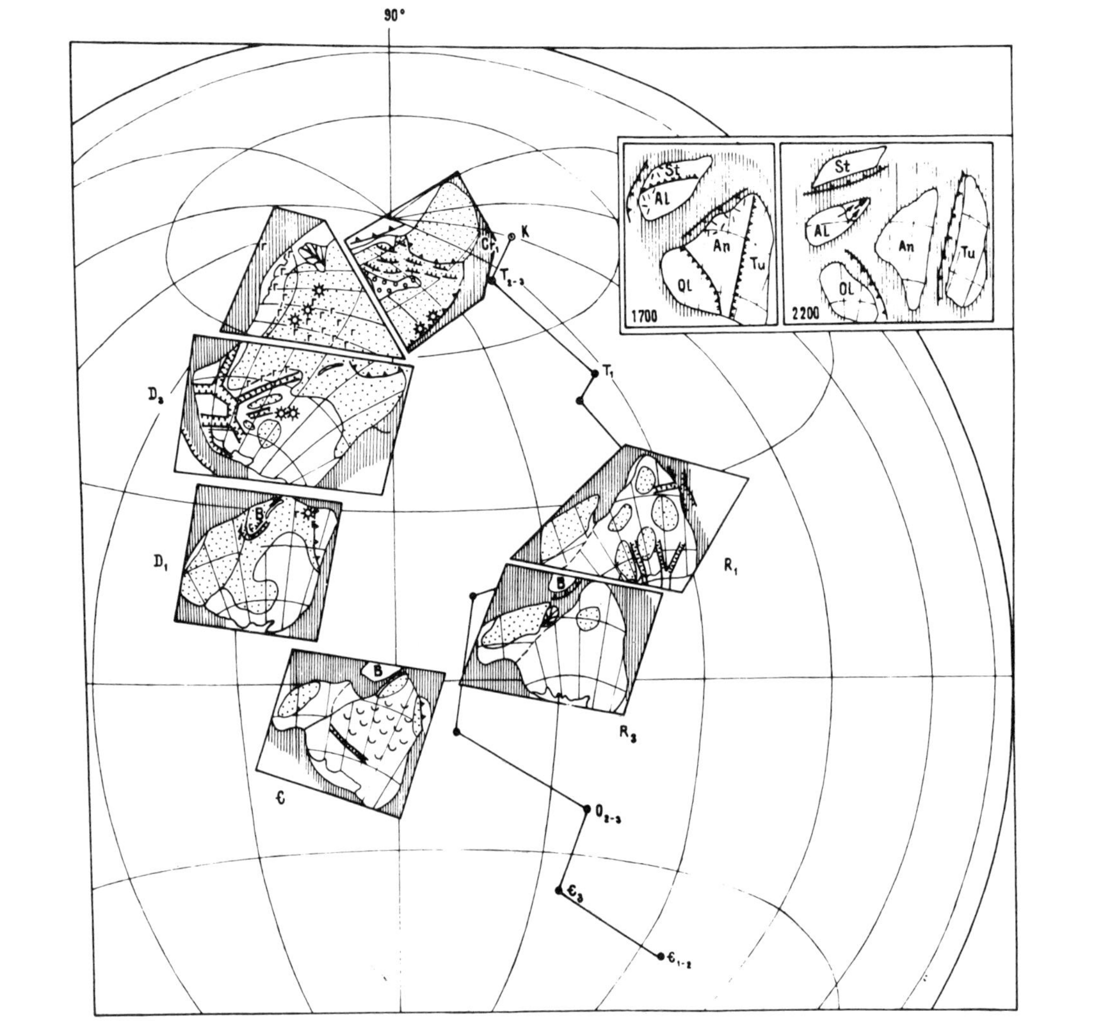

Fig. 16. Palinspastic reconstructions of the Siberian Platform. Inserts show allowable reconstructions for 2200 and 1700 Ma. Apparent polar wandering path is also shown (after Khramov [1982]); encircled points correspond to paleomagnetic pole positions.
1, Shallow sea; 2, dry land; 3, oceanic basin; 4, continental rift; 5, subduction zone; 6, collision zone; 7, flood basalt; 8, intraplate alkalic pluton; 9, evaporite; 10, graben; 11 deltaic fan deposits; 12, continental margin volcanic belt.
Al - Aldan shield; An - Angara shield; B - Barguzin microcontinent; Ol - Oleniok block; St - Stanovoi block; Tu - Tungus block.

TABLE 3. Paleomagnetic Poles for Siberia (after Khramov [1982])

Age	Degrees N	Degrees E
Lower Cambrian, Aldan	-54	108
Middle Cambrian, Aldan	-47	163
Middle Cambrian, Anabar	-44	156
Upper Cambrian, Aldan	-36	126
Upper Cambrian, Anabar	-35	130
Lower Ordovician	-42	127
Middle Ordovician	-24	132
Upper Ordovician	-21	131
Lower Silurian	-10	104
Lower Devonian	15	116
Upper Devonian	10	140
Lower-Middle Carboniferous	10	150
Permian	45	141
Lower Triassic	49	149
Middle-Upper Triassic	59	137
Cretaceous	74	180

TABLE 4. Parameters for the Rotation of Siberia in the Absolute Frame (After Zonenshain et al. [1985])

	Pole Coordinates		
Age (Ma)	Deg. N	Deg. E	Angle of Rotation (Degrees)
0 - 130	20.2	73.1	19.9
0 - 220	53.1	76.9	59.7
0 - 280	57	84.8	89.9
0 - 310	48.1	93.1	93.5
0 - 340	47.5	92.3	105.3
0 - 370	50.4	88.3	120.9
0 - 400	52.6	84.5	136.8
0 - 430	40.6	85.3	156.5
0 - 460	34.8	89.2	151.5
0 - 490	28.7	92.6	146.8
0 - 520	22.4	95.8	142.5

intraplate magmatism are known on the Siberian continent for which regular changes in the age of magmatism have been established along their strike. The North-Mongolian belt (Figure 17) existed in the Late Paleozoic and Early Mesozoic; the South-Siberian belt, in the Middle Paleozoic. Figure 18 shows the correlation between the track of Siberia motion determined from the north-Mongolian belt of intraplate magmatism, and the apparent polar wandering path for the same time, i.e., from the Permian to the end of the Jurassic. It is evident that the two curves show a similar change through time, i.e., the age decreases counter-clockwise relative to Siberia, showing that the movement of Siberia in this time was clockwise. These curves may be referred to as small circles drawn from the same center. This center is located at a point 54° N, 104° E; the rotation angle (both from paleomagnetic data and from the track relative to the hot spot) amounts to 75°.

The South-Siberian belt (Figure 19) stretches from the Minusinsk depression in the west through the Eastern Sayan and the Eastern Trans-Baikal area into the northern regions of the Aldan shield in the east. The age of intraplate magmatism ranges from the Early Devonian in the west to Middle Carboniferous in the east. Combined, the data on paleomagnetism and intraplate magmatism allow us to estimate the position of the pole of rotation of Siberia within the absolute system of coordinates: from Devonian to the Early Carboniferous it was at 70° N and 105° E, and the angle of rotation was 45° clockwise. These parameters reveal the true path of Siberia's motion on the Earth's sphere in the Phanerozoic (Figure 16). More ancient (Pre-Phanerozoic) reconstructions are shown conventionally.

The Early Proterozoic history resembled events taking place at the same time in the East European platform when individual blocks converged and united into greater blocks.

2200 Ma

This period was chosen because the Udokan series was then accumulating, and it is therefore clear that cratonic continental blocks already existed. It may be assumed that at least five individual continental blocks--Angara-Anabar, Tunguska, Oleniok, Aldan, and Stanovoy--were separate. They were from 1000 to 2000 km across and were presumably separated by oceanic basins of unknown width.

1700 Ma

At this time two marginal continental volcano-plutonic belts formed: Akitkan and Ulkan. They were preceded by an important event in the history of the Siberian platform when isolated relatively small blocks amalgamated into two huge blocks or cratons. (The analogous event in the history of the East European platform was approximately contemporaneous.) Great sutures, for example the Stanovoy fault, separating the basement of the Siberian platform into blocks, have commonly been thought to represent the fracturing of a larger terrane into smaller fragments (blocks). Our concept is directly opposite: as in the case of the East European platform, these sutures are regarded as collisional zones.

The Akitkan volcano plutonic belt marks the active margin of the former continent under which the oceanic crust was being subducted. During the formation of this belt the Aldan shield was probably not yet linked with the Angara-Anabar shield.

1600 Ma

By Riphean time the oceanic crust separating the two cratons mentioned above was almost completely consumed under the Akitkan volcanic belt and a single Siberian

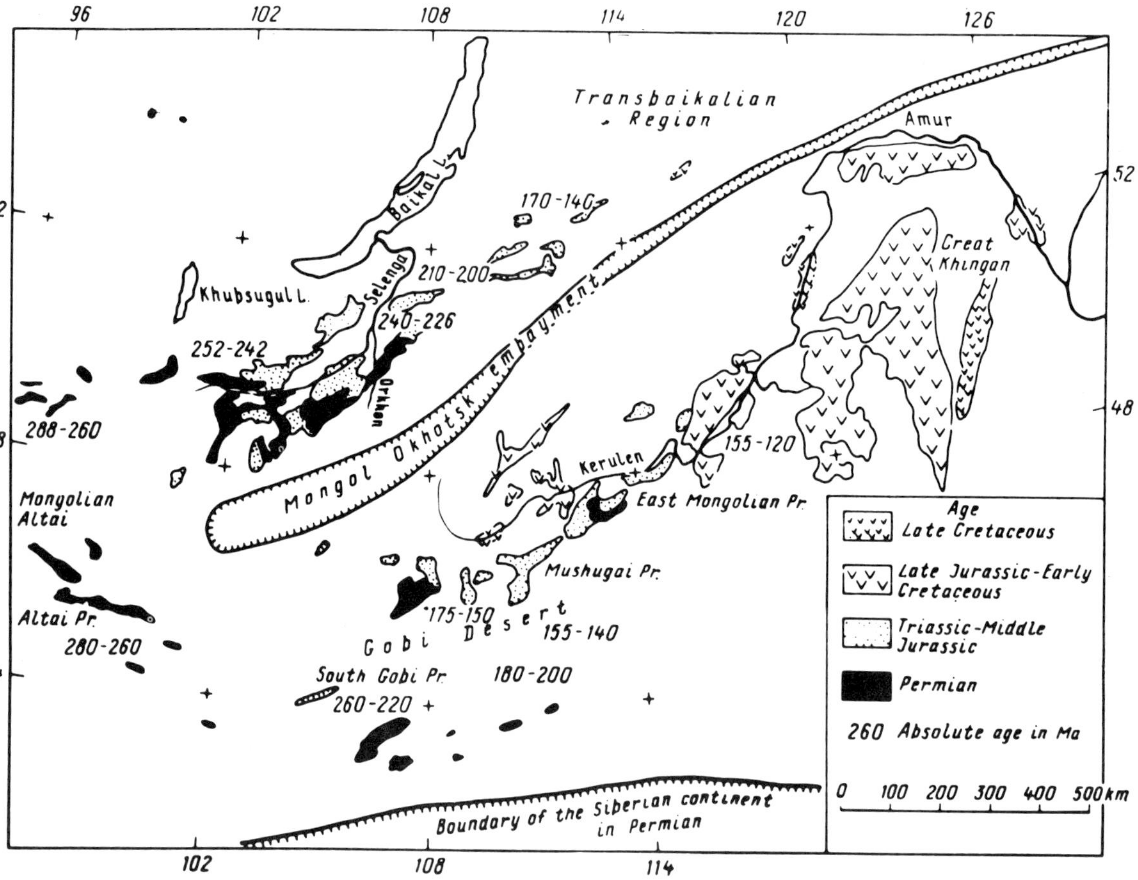

Fig. 17. North Mongolian belt of intraplate magmatism (after Zonenshain et al. [1985]).
1, Lower Cretaceous igneous rocks; 2, Upper Jurassic igneous rocks; 3, Triassic-Middle Jurassic igneous rocks; 4, Permian igneous rocks; 5, isotopic age in Ma.

continent originated. Relative to the Siberian block, the Aldan block lay not less than 400 km towards the north-east as compared to its present position, which results from displacements along the Zhuin strike-slip fault.

1000 Ma

In Riphean time the entire Siberian platform subsided and was covered by a shallow sea; only the Aldan shield and possibly small spots within the platform (Anabar) were above sea level. Apparently the aulacogens formed before the sedimentary cover of the Siberian platform was deposited. Nearly the whole periphery of the platform was a passive continental margin. The sedimentary sequences of the Bodaibo synclinorium were probably formed in a great submarine delta fan.

540 Ma

The Riphean situation persisted into the Cambrian. The Siberian continent was located in low latitudes of the southern hemisphere [Khramov, 1982]; its present northern part (Lena River delta) faced south while its southern part (Baikal) faced north. The (present) southern part of the continent around the Baikal region, East Sayan, and the Enisey Ridge was considerably reconstructed. An angular unconformity beneath the Vendian and Upper Riphean indicates orogeny, perhaps caused by partial collision of minor continental blocks with the Siberian platform or by convergence of the Barguzin microcontinent and Siberia. As a result, a discontinuous barrier of crystalline basement was uplifted, rimming the Siberian continent on the (present) south and separating it from the Paleo-Asiatic ocean. The largest gap in this barrier was in the East Sayan where, judging from Vendian-Lower Cambrian marine sediments of the Mansky trough, the shallow-water Siberian sea was connected to the open ocean. Altogether, the uplift of this barrier gave rise to two events: (i) accumulation of thick clastic red beds (Karagassky and Ushakovsky suites and their analogs) along the southern margin of the platform in front of the barrier and (ii) isolation of the great salt-generating basin over the southern part of the platform (Figure 16). Other remarkable complexes of Siberia are Middle Cambrian reefs and bituminous shales ('Domanikovaya facies') of the lower-Middle Cambrian in the

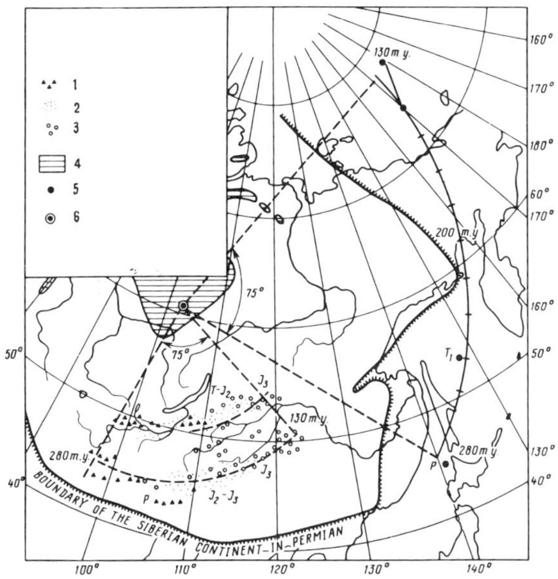

Fig. 18. Comparison between lines of migration of intraplate volcanism and apparent polar wandering paths for Siberia from the Permian to Early Cretaceous (after Zonenshain et al. [1985]).
1, Permian volcanics; 2, Triassic-Middle Jurassic volcanics; 3, Late Jurassic-Early Cretaceous volcanics; 4, flood basalt (Late Permian-Early Triassic); 5, paleomagnetic pole; 6, inferred pole of rotation of Siberia from 280 to 130 Ma.

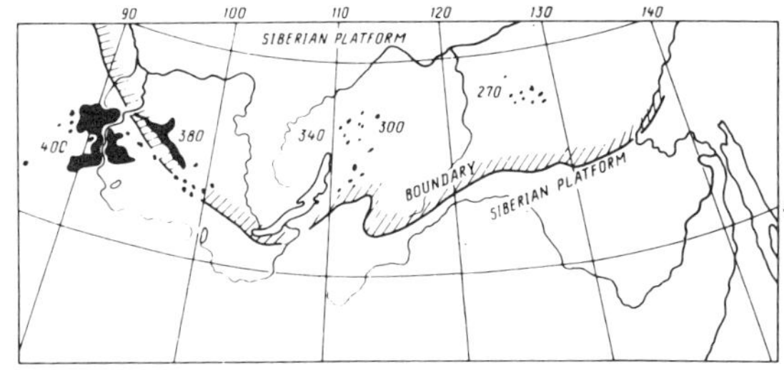

Fig. 19. South Siberian belt of intraplate magmatism; black indicates areas of magmatism. Figures show isotopic ages in Ma (after Zonenshain et al. [1985]).

northeast (in present coordinates) of the platform. These complexes verify the position of Siberia in the tropical zone and most probably in an arid climate.

400 Ma

Prior to the Early Devonian, sedimentation on the Siberian continent was considerably reorganized. The southern (in present coordinates) Siberian margin collided with several microcontinents of the Central-Asiatic foldbelt (see below), producing a vast uplift in the southern part of the platform. Sedimentation proceeded without interruption only in the western (Igaro-Turukhan) zone and northeastern (Verkhoyansk) region. The Siberian continent was still in the arid tropical belt, as shown by the presence of Ordovician-Silurian red beds.

370 Ma

In the Late Devonian, the Siberian continent moved 2500 km to the north, relative to its position in the Early Devonian. In present coordinates, the formerly western part was situated in the area ca 45° N and 60° E. Apparently the continent overlay a hot spot or hot field in the mantle, for this margin of Siberia was intensely fractured and an anastomosing, branching system of aulacogens (Figure 13) came into existence. The paleogeographic pattern of the Siberian platform included a broad area of uplifted dry land on the south (present coordinates) and a vast shallow-water sea on the north. The first intrusion of kimberlites occurred at this time.

220 Ma

From the Late Devonian to the Triassic, the Siberian continent was drifting 3000 km to the north at a velocity of 2 cm/yr, moving from low and moderate latitudes into high latitudes (50°-70° N). Thus, Siberia came into a humid zone; this was possibly a reason for a sudden change in sedimentation, at the end of Early Carboniferous time, from carbonate accumulation, which had persisted through the Early and Middle Paleozoic, to clastic sequences, in particular to coal-bearing sequences in the Late Paleozoic. The Late Paleozoic is also known as the period when the World Ocean level was at its lowest, which also may be responsible for the change from carbonates to clastics. The paleogeographic situation of the Devonian persisted in the Late Paleozoic: uplifts in the south (present coordinates) of the continent, and areas of maximal subsidences on the western and eastern passive margins. The southern uplift, in particular the Aldan shield, was the area from which the clastic material of the Verkhoyansk complex was derived. The lithological data show that material was transferred through large rivers, like the present-day Lena River, which produced extensive deltas and large submarine fans.

By the end of the Paleozoic, the western (present coordinates) Siberian margin reached the area (40° N, 60° E) where the Late Devonian hot spot had been located earlier and where the Viljuy aulacogen was formed. Here intense flood basalt eruptions began toward the end of the Late Permian. Evidently, coincidence of geographical coordinates of the Devonian and Permian-Triassic intraplate magmatism is not accidental; it proves the fixed geographical position of the hot spot or hot field, although available data show that the hot-spot activity was episodic.

This time is also marked by the second episode of kimberlitic volcanism and the intrusion of numerous kimberlitic pipes primarily in the north-east part (present coordinates) of the platform. Evidently the kimberlitic magmatism is also related to the activity of the same hot spot.

130 Ma

Throughout Mesozoic time Siberia rotated progressively clockwise. In this period, the continent was located at the same latitude in approximately the same geographical position as nowadays. The most important events of the Late Mesozoic are undoubtedly (1) the collision of the Omolon and

Chukotka microcontinents with Siberia, which led to the formation of the Verkhoyansk-Kolymian foldbelt, and (2) the collision of the Amur microcontinent with Siberia which resulted in formation of the Mongol-Okhotsk belt. The sedimentary complexes of the former Verkhoyhansk margin were folded and thrust onto the continental margin; the thrust front adapted itself to the outlines of the Devonian aulacogens. The Peri-Verkhoyansk fore-deep originated along the thrust front while the entire platform, formerly flooded by sea, became extensive dry land.

Chapter IV

URALIAN FOLDBELT

GENERAL DESCRIPTION

The Urals are a typical example of a linear collisonal foldbelt. As is well known, the main Uralian structures were formed at the end of the Paleozoic and in the very beginning of the Mesozoic. Throughout the last 20 years, the geology of the Urals has been reconsidered from a new mobilist point of view, first by Kropotkin [1967]. The first schematic plate-tectonic model of Uralian development was provided by Hamilton [1970], showing that the Urals resulted from collision of the East European and Siberian continents, as indeed Kropotkin [1967] had suggested earlier, although not in the context of plate tectonics. Subsequent contributions include those by Peive et al. [1977], Perfiliev [1979], Ruzhentsev [1976], Samygin [1980], Khvorova et al. [1978], Ivanov et al. [1972, 1973, 1974], Puchkov [1975, 1979], Lennykh [1977, 1984], Korinevsky [1984], Koroteev et al. [1979], Kamaletdinov [1974], Maslov [1980], Yudin [1983], Belyakov and Dembovsky [1984], Zhivkovich and Chekhovich [1985], etc. A relatively detailed plate-tectonic pattern of the Uralian foldbelt development applied to the Southern Urals, was suggested by Zonenshain et al. [1984].

The Uralian foldbelt (Figure 20) is divided into externides evolving on the margin of the East European continent (or near it), and internides with widespread Paleozoic oceanic and island arc complexes. Kheraskov and Perfiliev [1963] recognized the Uralides and Pre-Uralides in the internal Uralian zones. The Uralides incorporate complexes formed from the Early Ordovician to the Permian, while Pre-Uralides include more ancient complexes, both Precambrian and Cambrian.

The externides encompass the Peri-Uralian foredeep filled with Permian molasse and the thrust domain situated between the foredeep and the Main Uralian fault. This domain consists mainly of Uralide complexes of two types [Puchkov, 1979]: (i) shelf complexes of the Eletsky zone in the north and of the Belsky zone in the south, and (ii) abyssal and bathyal complexes of the Lemva and Malo-Pechersky zones in the north and of the Sakmara zone in the south. These complexes were strongly deformed, often include nappes, and are overthrust by large allochthonous masses which, together with bathyal complexes of the Lemva type, produce an intermittent chain of marginal allochthons (Sakmara, Kraka, and others) along the entire external Uralian zone. These marginal allochthons frequently have a chaotic inner structure and contain rock complexes characteristic of the internides.

The Uralian internides, situated eastward from the Main Uralian fault, consist of a number of parallel zones aligned along the N-S foldbelt direction. To the east from the Main Uralian fault a series of troughs and synclinoria constitutes the so-called greenstone belt of the Urals, composed of island arc and oceanic rock complexes. The West-Mugodzhar, Magnitogorsk, Tagil, and Shchuchinsky synclinoria are aligned from S to N; they are synforms consisting of a number of nappes stacked one over another. More eastern zones are exposed in the Southern and, partly, the Middle Urals. They include: (i) the East-Uralian anticlinorium, known in the south as the Mugodzhar anticlinorium, including granite-gneissic domes where metamorphic rocks of the Preuralides exposed in the cores are typical for this zone; (ii) the East-Uralian synclinorium, with a possible southward prolongation into the Irgiz synclinorium, consisting of the Uralide complexes; (iii) the Trans-Uralian uplift, in which Cambrian and Ordovician assemblages crop out; (iv) the Oktyabrsko-Denisovsky zone, including well developed Uralide complexes; (v) the Valerianovsky zone--a belt of calc-alkaline volcanics of the lower-Middle Carboniferous, representing an active continental margin of Kazakhstan. Drilling and geophysical data prove that these zones proceed north under the sedimentary cover of the West Siberian lowland. The Pai-Khoy-Novaya Zemlya part of the Uralian belt is made of the externide complex only. Widespread Permian marine sediments are characteristic of this segment of the Uralian belt.

PRE-URALIDES

The Pre-Uralides (Figure 21) exhibit the basements of former continents and microcontinents whose collisions produced the Uralian foldbelt. They can be divided into blocks that became part of the East European continental basement before the Ordovician (these comprise outliers of the platform basement in the externides) and blocks that crop out in uplifted internal parts of the belt and that either belonged to other continents or were detached from East Europe during formation of the Paleo-Uralian ocean and rejoined the East European continent only when this ocean was closed in the Late Paleozoic.

The blocks of the first group, being a part of the East

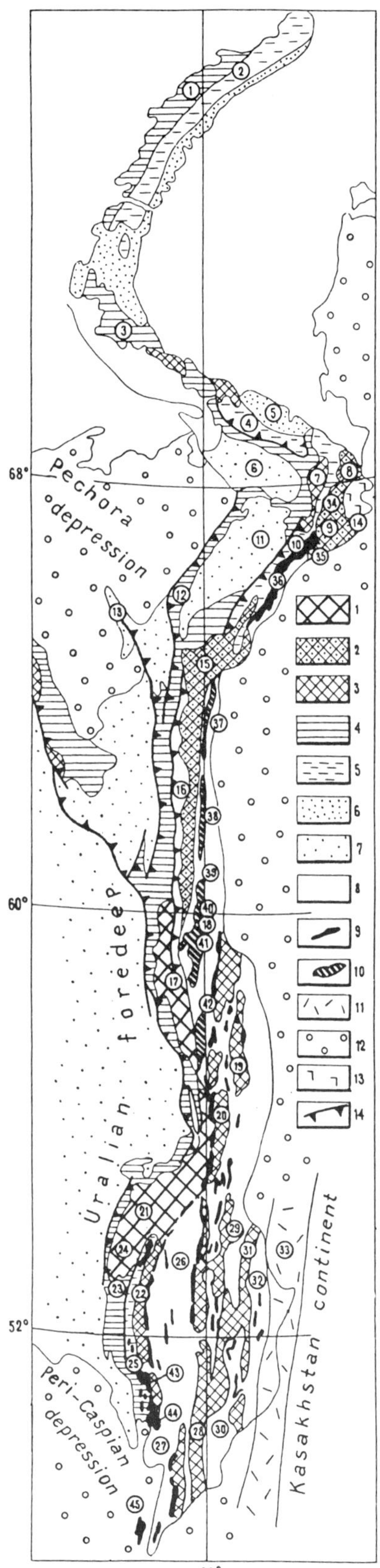

European platform basement, may be classified into two types. The first includes the Riphean sediments that accumulated on the eastern passive continental margin of the Early Precambrian East European continent; they consist of the well known Riphean sequences of the Bashkir anticlinorium and its northern continuation in the Kvarkushsky anticlinorium. The second type includes a Late Precambrian island-arc and various sedimentary complexes (Central-Uralian, Ochenyrd, Kharbei, and Marunkeu anticlinoria) in the Central and Polar Urals (Figure 22) and the Suvanyakhsy and Maksutov complexes of the Ural-tau in the Southern Urals (Figure 23). Late Precambrian granites crop out in the basement of the western Novaya Zemlya externides; they are from 1300 to 680 Ma in age [Korago and Chukhonikna, 1988]. The blocks containing Precambrian complexes were sutured with East Europe at the very end of the Precambrian or even in the beginning of the Paleozoic. According to Lennykh [1984], convergence of these blocks with the East European continent was a cause for widespread folding and metamorphism (Baikalian orogeny) on the eastern margin of the East European continent including the Timan belt. The accretionary foldbelt that appeared as a result became the basement for the sedimentary series of the Uralide externides.

The Pre-Uralide complexes in internal zones of the foldbelt crop out in such uplifted blocks as the East Uralian, Trans-Uralian, and Mugodzhar anticlinoria. Gneiss-magmatitic

Fig. 20. Tectonics of the Urals.
Pre-Uralides (Pre-Ordovician basement): 1, Precambrian basement of the East European platform; 2, Precambrian blocks attached to East Europe in the Late Precambrian and Cambrian; 3, Precambrian blocks attached to East Europe in the Late Paleozoic.
Uralides - Externides. 4, Shelf complexes of the Ordovician-Lower Permian; 5, continental slope and bathyal complexes of the Ordovician-Permian; 6, marine Permian of Novaya Zemlya and Pai-Khoi; 7, Pre-Uralian foredeep.
Uralides - Internides. 8, Oceanic and island arc complexzes, from Ordovician to Lower Carboniferous; 9, ultramafics; 10, platinum-bearing mafic and ultramafic massifs; 11, Early-Middle Carboniferous Valerianovsky volcanic belt.
Younger formations: 12, Meso-Cenozoic sedimentary cover; 13, Triassic floor basalt; 14, thrust.
Numbered are: 1, Western Novaya Zemlya zone; 2, Eastern Novaya Zemlya zone; 3, Vaigach zone; 4, Pai-Khoi allochthon; 5, Kara basin; 6, Korotaikha depression; 7, Ochenyrd anticlinorium; 8, Maruenkeu anticlinorium; 9, Kharbei anticlinorium; 10, Lemva allochthon; 11, Kosyu-Rogov depression; 12, Chernyshev anticlinorium; 13, Pechora-Kozhva swell; 14, Shchuchinsky anticlinorium; 15, Central Uralian anticlinorium; 16, Malopechersky allochthon; 17, Kvarkushsky anticlinorium; 18, Tagil synclinorium; 19, East-Uralian anticlinorium; 20, Ilmensky block; 21, Bashkirsky anticlinorium; 22, Ural-tau anticlinorium; 23, Zilair synclinorium; 24, Kraka allochthon; 25, Sakmara allochthon; 26, Magnitogorsk synclinorium; 27, West Mugodzhar synclinorium; 28, Mugodzhar anticlinorium; 29, East Uralian synclinorium; 30, Irgiz synclinorium; 31, Trans-Uralian anticlinorium; 32, Oktyabrsko-Denisovsky zone; 33, Valerianovsky volcanic belt. Largest ultramafic massifs: 34, Syum-Kieu,; 35, Rai-Iz; 36, Voikaro-Syn'insky; 37, Kharasyursky; 38, Chistopol; 39, Denezhkin Kamen; 40, Kumba; 41, Kytlym; 42, Tagil; 43, Khabarny; 44, Kempirsai; 45, Daul.

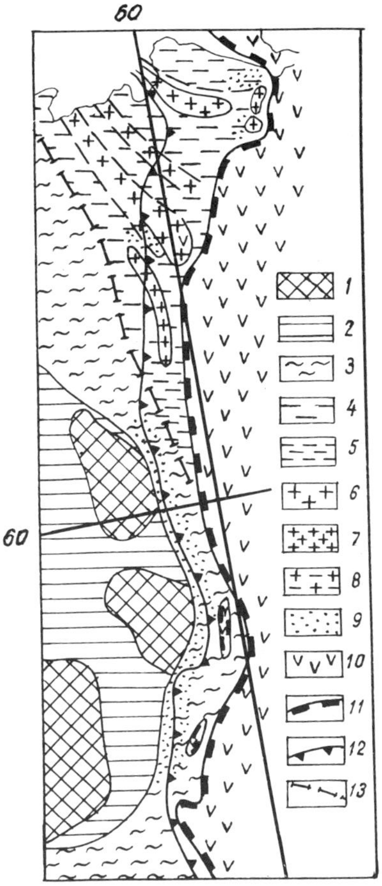

Fig. 21. Relationships between Precambrian and Paleozoic structures of the Pre-Uralide and Uralide internides and externides (after Lennykh [1984]).
1, Archean basement; 2, Karelides; 3, Pre-Uralide externides; 4, Pre-Uralide internides, synclinoria from geophysical data; 5, the same from geological data; 6, gneissic-granitic domes from geophysical data; 7, the same from geological data; 8, uncertain structures; 9, Vendian molasse; 10, Vendian rhyolite; 11, boundary between Uralide externides and internides; 12, Uralide thrust front.

complexes producing granite-gneissic domes are predominent. The age of the metamorphic complexes of the Mugodzhar block was determined by the K-Ar method to be 1100-930 Ma, by the Pb method 880 Ma [Milovsky et al., 1977] and 1300-1500 Ma, and by the Rb-Sr method 580-590 Ma [Kasymov et al., 1980]. The gneiss-amphibolitic sequence in the core of the Ilmen-Vishnevogorsky block in the Miass region is 2100-2300 Ma old.

In addition to the Precambrian, the Pre-Uralides also include Lower Paleozoic sequences, often of unidentified age. They incorporate Cambrian and Ordovician assemblages, in places Archaeocyathid limestones that are exotic for the Uralian and East-European provinces.

According to data by Keilman [1974, 1980], the granite-gneissic domes contain two structural levels. In the lower level, ancient Proterozoic and even Archean rocks are exposed in the cores of the domes. They suffered metasomatic granitization and palingenesis and multi-phase metamorphism. The upper level of the domes consists of the schist envelope surrounding the cores; this envelope is structurally inconsistent with the core and diverse in composition. The envelopes are mostly Paleozoic metamorphic rocks, including ophiolites and numerous serpentinite bodies. The domes grew by uplift of heated material. The central parts of many domes are occupied by granitic masssifs, the products of extreme mobilization and remelting of the crystalline substratum [Fershtatter and Borodina, 1975]. According to Keilman [1975, 1980], the granite-gneissic domes underwent several epochs of metamorphism: 440, 310, and 280 Ma. In the East Mugodzhar, Milovsky et al. [1977] found several stages of activity: (i) 450 Ma,--formation of pegmatites, metasomatism; (ii) 365 Ma,--metamorphism, recrystallization, formation of granites; (iii) 300 Ma,--granite intrusions.

In the granite-gneissic domes, the Pre-Uralides are visible through the cover of the Uralides. As in other foldbelts [Zonenshain et al., 1976], the two levels of the granite-gneissic domes suggest that rocks of the upper level, including the Uralides (oceanic and island-arc Paleozoic complexes) tectonically overlie the Pre-Uralides in the lower level. Formation of the domal structure itself is related to diapiric upwelling of the mobilized sialic basement after the Uralide complex had been thrust over the Pre-Uralides. The envelope of Pre-Uralides also suffered metamorphism, but it decreases in concentric zones towards the periphery of domes and is lacking in the interdomal depressions.

URALIDES: EXTERNAL ZONES

Shelf Complexes

The western Uralian slope consists largely of rock sequences analogous to those developed on adjacent parts of the East European platform but, in contrast to them, the western Uralian sequences underwent thrusting and deformation, often very significant. A wide spectrum of formations, from Ordovician to Late Triassic, are known. The Ordovician formations unconformably overlie the ancient basement consisting of deformed and metamorphosed Vendian and Riphean rocks. Bimodal volcanic complexes in the Ordovician indicate the initial rifting and continental splitting. The basement outliers became morphologically expressed in the topography only at the very end of the Devonian (metamorphic clastics are present in the Zilair suite of the Famennian-Tournaisian) and were vigorously eroded in the Late Paleozoic, supplying clastic material to Permian molasse of the foredeep. Thus, after the Ordovician transgression, the ancient basement was for a long time buried under the Paleozoic sedimentary cover.

Zhivkovich and Chekhovich [1985] propose that formation of the sedimentary basin the Ufa amphitheater of the Central Urals was preceded by formation of a system of listric faults which disrupted the basement into a series of blocks downfaulted to the east (Figure 24). An epicontinental basin was trapped within the continental margin by an uplifted horst block that cut it off from the adjacent oceanic basin. A barrier reef formed on the horst block in the Silurian and Devonian, while stagnant conditions developed in depressions behind and thin carbonate-shale sediments enriched in organics accu-

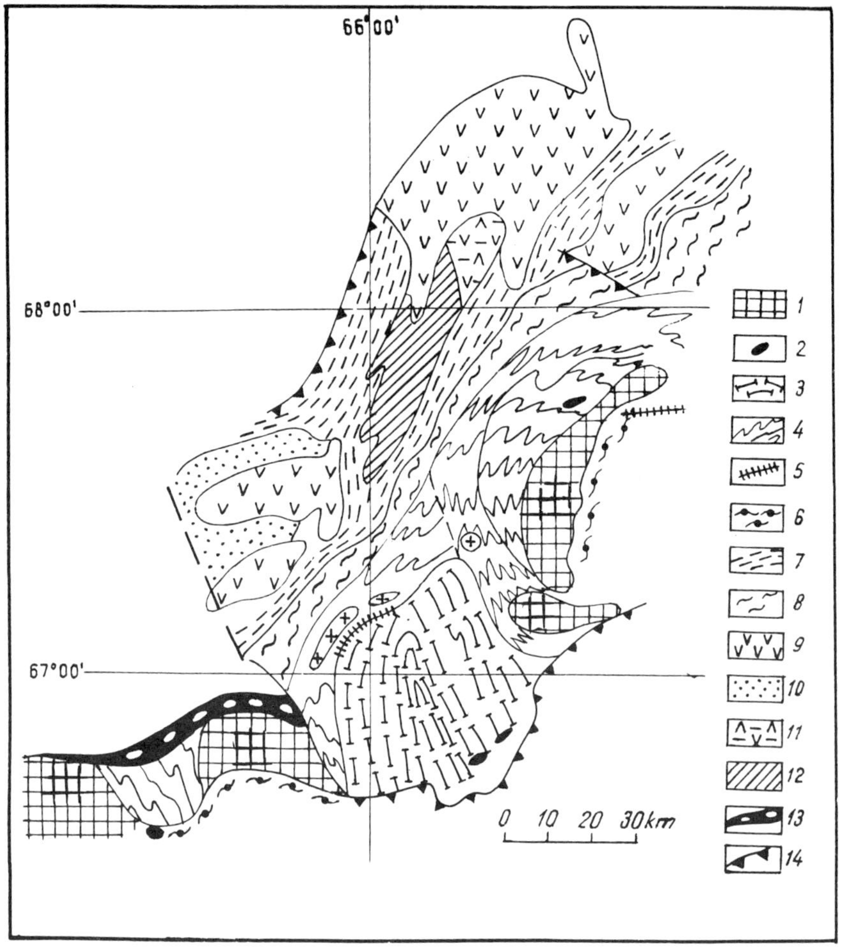

Fig. 22. Geology of the northern part of the Polar Urals (after Lennykh [1984]).
1, Large ultramafic massifs; 2, small ultramafic bodies; 3, high Al schists; 4, metasedimentary rocks; 5, Glaucophane schist; 6, eclogite; 7, early-Middle Ordovician clastics and limestone; 8, Ordovician-Silurian basaltic volcanics; 9, Vendian-Cambrian calc-alkaline volcanics, 10, Vendian-Cambrian mollase; 11, Cambrian-Ordovician bimodal volcanics; 12, Cambrian-Ordovician terrigenous sequence; 13, serpentinite melange; 14, thrust.

mulated there (the so-called black shale 'domanic facies').

To the west on the platform slope, the entire sequence is made up of organogenic limestones from the Silurian up to the Middle Carboniferous, or in places to the Upper Carboniferous and the Lower Permian. Toward the east, the carbonate sediments are followed by flysch, appearing at the extreme east for the first time in the very end of the Devonian, beginning with the Famennian.

The sedimentary cover is strongly deformed, being stacked into several nappes [Kamaletdinov, 1974; Yudin, 1983; Zhivkovich and Chekhovich, 1985]. In some areas, not only the Paleozoic cover, but also Riphean or (more rarely) pre-Riphean sedimentary rocks, and to a lesser degree the crystalline basement, are involved in nappe-fold structures. Deformation commenced in the Late Paleozoic and ceased in the Early Triassic.

In general, the dominantly carbonate Paleozoic sequences of the western Uralian slope may be definitely regarded as the analog of present-day passive margins.

Marginal Allochthons

The marginal allochthons are thrust over the continental margin. They may be divided into individual domains, each considered as an independent zone; from south to north, they are the Sakmara, Kraka, Bardym, Malo-Pechersky, Lemva, and Pai-Khoi zones. The Paleozoic deep-water facies of eastern Novaya Zemlya seems to belong to another of these allochthonous zones. A chaotic mixing of different rock types is characteristic of the marginal allochthons. On the one hand, they include vast allochthonous slabs of ultramafic rocks and other horizons of ophiolite associations; on the

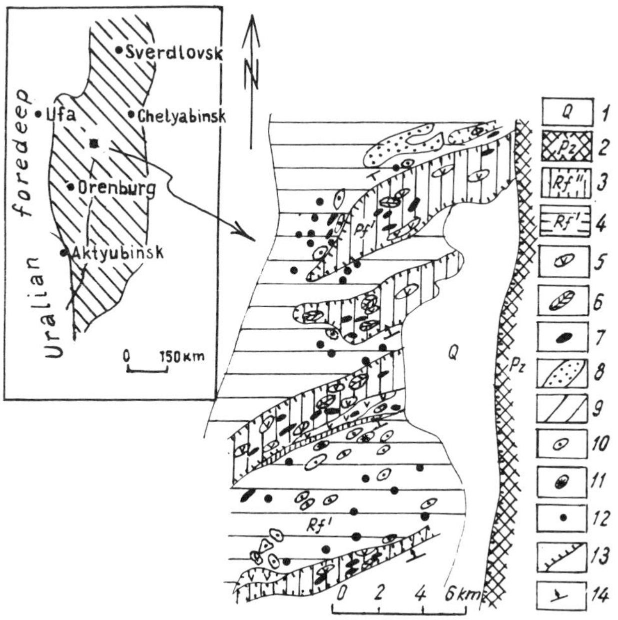

other hand, they display a wide spectrum of facies of the same age. A peculiar feature is a widespread development of deep-water bathyal facies of the Lemva type.

Sakmara Zone (Figure 25)

In the vertical section of the Sakmara zone, Abdullin et al. [1979], Avdeev [1984], and Perfiliev [1979] recognized two structural complexes: autochthonous (or para-authochthonous) and allochthonous. The former contain Precambrian and Lower Ordovician formations exposed in tectonic windows, these formations are analogous to rocks of

Fig. 23. Geology of the Suvonyakhsky and Maksutov complexes on the western slope of the South Urals (after Lennykh [1984]).

1, Quaternary; 2, Paleozoic ophiolite; 3, Upper Maksutov ophiolitic series; 4, Lower Maksutov metamorphic series; 5, greenschist; 6, marble lens; 7, serpentinite; 8, quartzite; 9, mica schist; 10, eclogite; 11, enstatic rocks; 12, glaucophane schist; 13, thrust; 14, dip of foliation. Insert shows location of the region. Hatching shows the Ural belt, thick line is the Main Uralian Fault.

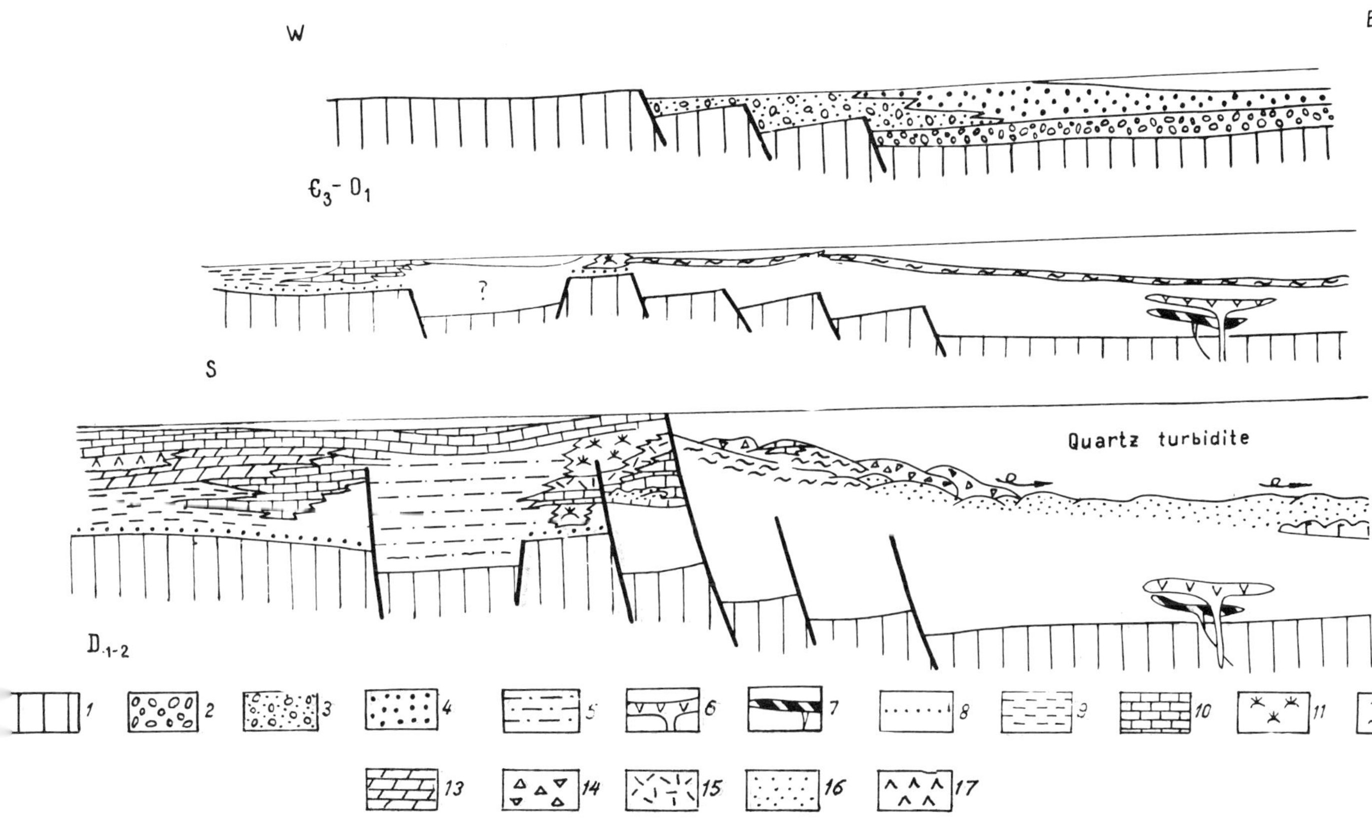

Fig. 24. Cross-sections of facies across the Uralian externides in and east of the Ufa amphitheater (after Zhivkovich and Chekhovich [1985]).

1, Riphean-Vendian basement; 2, coarse clastics; 3, arkosic sandstone; 4, quartz sandstone; 5, shale; 6, basalt; 7, alkaline rhyolite; 8, unconformity; 9, near-shore carbonate; 10, bedded limestone; 11, reef limestone; 12, black shale; 13, dolostone; 14, olistostrome; 15, felsic volcanics; 16, quartz-rich turbidite; 17, evaporite.

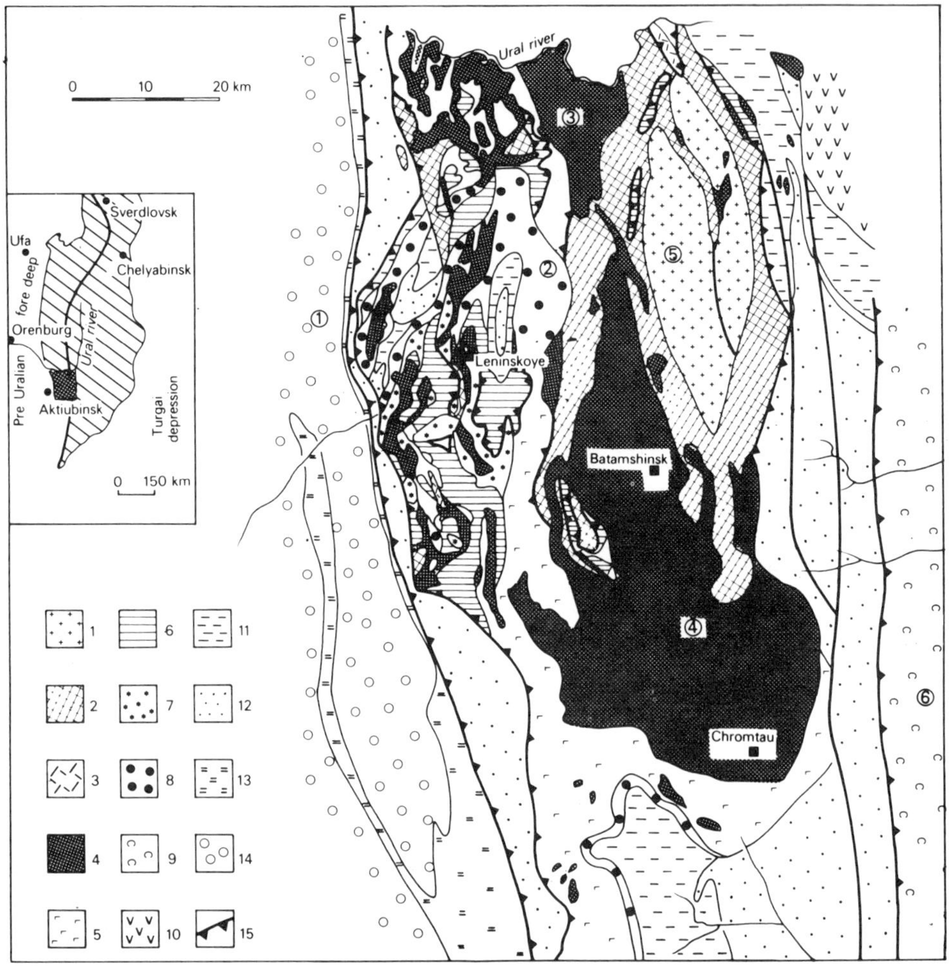

Fig. 25. Geological map of the South Sakmaran zone (after Zonenshain et al. [1984]).
Autochthon: 1, Precambrian rocks; 2, Lower Ordovician rocks; 3, middle-Upper Ordovician rocks.
Allochthon: 4, ultramafics, gabbro, serpentinite melange; 5, Silurian oceanic basalts; 6, Silurian cherts; 7, island-arc assemblages; 8, neo-autochthon; 9, Middle Devonian rocks, oceanic basalts of Mugodjar suite; 10, Middle Devonian island-arc volcanics of Milyashinsky suite; 11, Middle-Upper Devonian rocks; 12, Upper Devonian-Lower Carboniferous rocks; 13, Carboniferous rocks; 14, Permian rocks; 15, main thrusts.
Encircled numbers are: 1, Uralian foredeep; 2, Sakmara zone; 3, Khabarny ultramafic massif; 4, Kempirsai ultramafic massif; 5, Ebetin antiform; 6, West Mugodjar zone. Insert shows location of the area within the Southern Urals. Basement is hatched. Thick line corresponds to the Main Uralian thrust.

the same age distributed in the Sakmara allochthon along the eastern rim of the Zilair synclinorium and in the Ural-tau anticlinorium. The allochthonous complexes mainly include Silurian and Devonian formations including great slabs of ultramafic rocks and gabbro. These complexes compose such massifs as Kempirsai, Khabarny, and Khalilovo. The structure of the Sakmara zone may be divided into two bands. The eastern one, 20-40 km wide, consists of the ultramafic massifs; they are bordered on the west by rocks of other horizons of the ophiolite association: gabbro, sheeted dikes and pillow lavas. A narrower (10-20 km) western band is built up of chaotically arranged tectonic slabs.

The structure of the eastern band was presented by Peive et al. [1977] and Perfiliev [1979] as a recumbent fold overturned westward. Such a recumbent anticlinal fold is reliably established by drilling in the cross-section of the Khabarny massif along the Ural River (Figure 26) and in the Kempirsai massif along the Kuogash River.

The western band of the Sakmara zone has a chaotic pattern and may be compared to a gigantic breccia in which different geological complexes are displaced. The structural relationships and style are very complicated and are not clarified yet. The compositional characteristics of the geological complexes should be underlined. The most ancient sedimentary-volcanogenic piles belong mainly to the Middle Ordovician. They mark the most ancient of the Uralian volcanic island arcs [Korinevsky, 1975]. In younger Silurian-Devonian formations, two lithologically different sequences are known: one

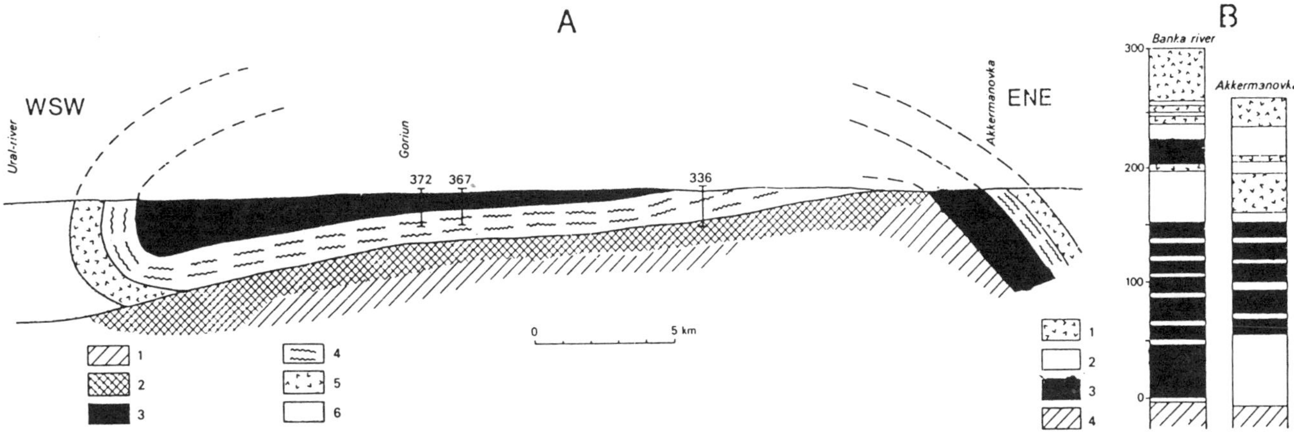

Fig. 26. The Khabarny ulftamafic massif deformed in a recumbent fold (after Tischenko V. T., from Zonenshain et al. [1984]).
A. Geological cross-section: 1, autochthon, Precambrian and Lower Ordovician rocks; 2, mafic metamorphics at the bottom of the ultramafic body; 3, metamorphosed harzburgite; 4, lower part of the layered complex: dunite, wehrlite, lherzolite; 5, upper part of the layered complex: gabbro; 6, basalt.
B. Columnar sections of the layered complex near the Banka River: 1, gabbro; 2, wehrlite and lherzolite; 3, dunite; 4, harzburgite.

with volcanic rocks and hiatuses in the Devonian, and the other lacking hiatuses and consisting of sedimentary rocks only. The Silurian rocks belonging to the first sequences display island-arc and marginal-sea affinities. Commonest among them is the Baiterek type, predominantly composed of volcanics of calc-alkaline composition, i.e., andesites, andesite-basalts, and silicic lavas. The Kosistekh type consists of tuffaceous-clastic rocks deposited by turbid flows; such complexes normally accumulate in a fore-arc basin. The tuffaceous rocks of this type are probably produced by erosion of the island-arc made of lavas of the Baiterek type. Besides the island-arc assemblages, cherty sediments constituting the Sakmara type are remarkable, possibly marking the deepest part of a marginal sea basin.

The Devonian begins with olistostromes (Shandinsky suite) overlain by cherts and pelagic limestones. The sequences encompasse blocks of different rocks: Lower Devonian reef limestones, Silurian cherts and lavas, Ordovician sandstones, and Lower Cambrian Archaeocyathid limestones. This chaotic Devonian sequence clearly indicates that the nappes were moving in the Siegenian-Emsian stages, dating the formation of the first nappes near the end of the Early Devonian. The displacement of nappes and their erosion were both submarine.

The overlying Eifelian sequence resembles the underlying Shandinsky suite but differs in the appearance of peculiar lavas--potassic alkalic basalts or chancharites [Korinevksy, 1975]. Alkalic volcanism is considered to be indicative of intraplate magmatism.

The purely sedimentary Silurian-Devonian sequence of the Sakmara zone is built up of cherts, cherty siltstones, fine-grained sandstones, and pelagic limestones. It involves rocks of Silurian and Devonian age (Frasnian included), but the thickness is extremely small,--from 30 to 40 m,--a typical condensed section representing some portion of an oceanic basin far from sources of clastic material.

Thus the Sakmara zone combines: 1) Lower Ordovician shelf and rift-type facies, 2) Middle Ordovician and Silurian island-arc complexes, 3) Silurian oceanic or marginal basin complexes, 4) Silurian and Devonian continental rise complexes.

Kraka Zone

This zone is somewhat simpler in structure than the Sakmara zone [Kamaletdinov, 1974]. It is a synform with allochthonous ultramafic rocks in the center (Figure 27). Beneath the ultramafics, which are 0.3 to 3 km thick, a widespread serpentinite melange (also allochthonous) encloses blocks, from 40 and 100 m in size, consisting of harzburgites, quartzite sandstones, greywackes, sheeted dikes, pillow lavas, and gabbros. In contrast to recumbent anticlines of the Khabarny and Kempirsai massifs, the Kraka allochthon looks as if it were detached from a flat-lying slab. Structurally beneath the allochthon, the Kraka zone includes: 1) shelf complexes deposited on the East-European continental passive margin, 2) terrigenous sequences of the continental rise, 3) oceanic complexes, i.e., lavas and dikes formed in the axial rift zone of the Uralian paleo-ocean. The Kraka allochthon were emplaced by the Middle Carboniferous.

Bardym Zone of the Ufa Amphitheater

Zhivkovich and Chekhovich [1985] stated that a number of tectonic slabs in this zone incorporate both shelf complexes and overlying allochthonous slabs of the condensed sedimentary section (Lemvinsky type) of Silurian-Devonian age as well as ophiolites building up higher slabs. In cross section, the Bardym zone is a synform incorporating imbricated allochthous (Figure 28). The main nappe in the Bardym zone was emplaced by Kungurian (mid-Permian) time.

Malo-Pechersky Zone

This consists of an allochthon 10-30 km wide and 190 m long in the Northern Urals in the upper reaches of the Pechora River. The allochthon is a synform entirely composed of Ordivician to Devonian cherts in a condensed section of the

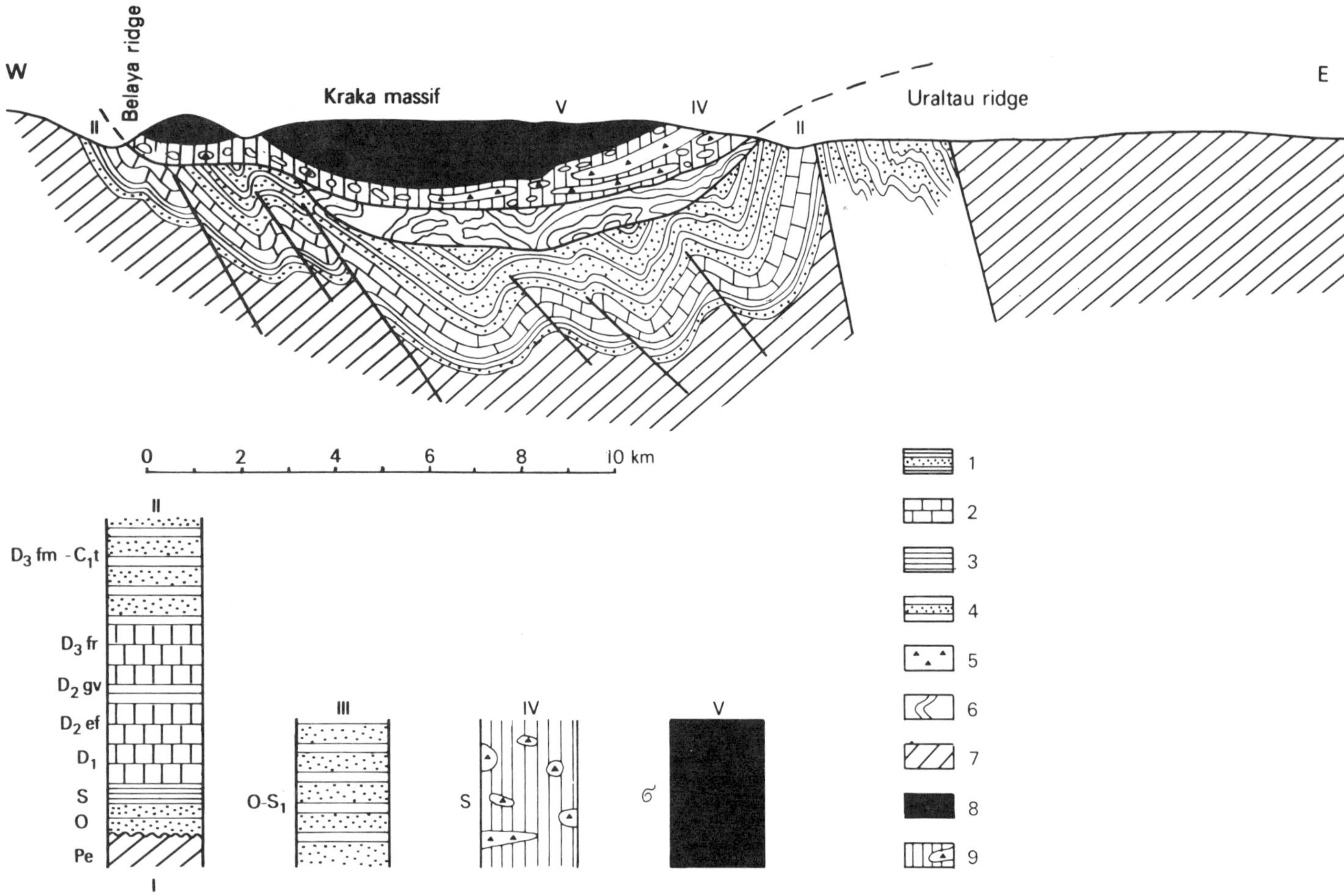

Fig. 27. Geological cross-section across the Kraka zone along the Uzyan River.
1, Zilairian suite, turbidites; 2, Devonian and Siulurian shelf carbonates; 3, Silurian graptolite shale; 4, Ordovician quartzite and shallow-water quartz-sandstone; 5, Silurian pillow lava and sheeted dikes; 6, Ordovician sandstone-shale sequence of the continental rise type; 7, Precambrian Riphean metasediments; 8, ultramafics; 9, melange.

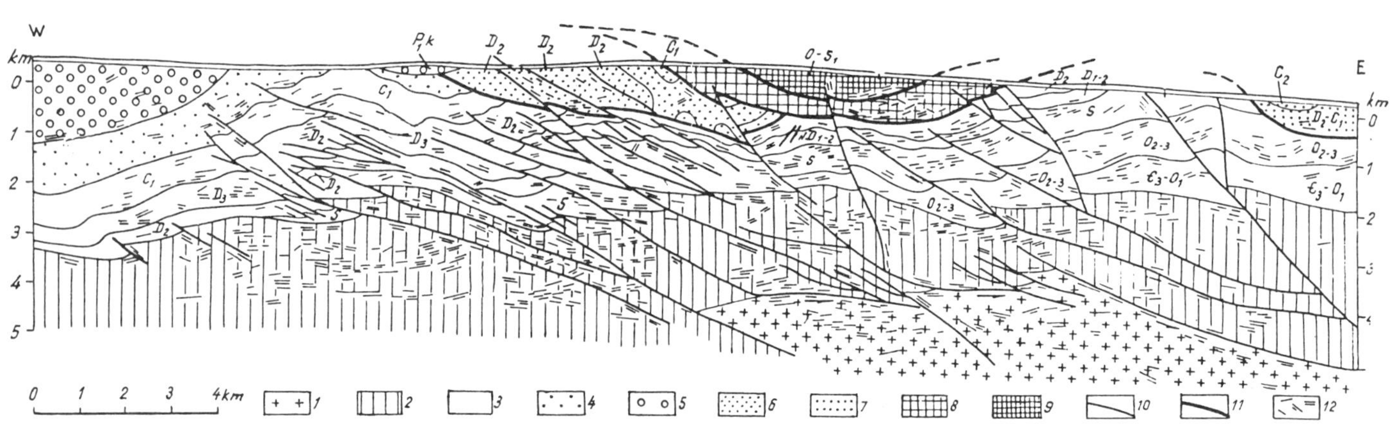

Fig. 28. Cross-section through the Bardym allochthon (after Zhivkovich and Chekhovich [1985]).
1, Basement; 2, Riphean-Vendian metamorphics; 3, Paleozoic autochthon; 4, Middle-Upper Carboniferous flysch; 5, Lower Permian mollase; 6, Kamensko-Demidov nappe; 7, Utkin nappe; 8, nappe consisting of Silurian-Devonian deep-water cherts; 9, melange; 10, fault; 11, main thrust; 12, seismic reflectors.

Lema type; it overlaps the Eletsky-type shelf deposits tectonically.

Lemva Zone

This zone belongs to the Polar Urals. In contrast to analogous zones farther south, the Pre-Uralides are not exposed here betweeen the external allochthons and the Main Uralian fault. The Lemva zone consists of a nappe package [Saveliev and Savelieva, 1977] thrust from east to west over the autochthonous (or para-autochthonous) complex of the Eletsky zone, composed of shelf sequences (Figure 29).

The Lemva package conprises of three nappes (Figure 29). The lowest is made up of a condensed section from Ordovician to Middle Carboniferous, 700 m thick [Puchkow, 1979]. The middle nappe consists mostly of a volcanogenic andesitic-basaltic pile intercalated with cherty shales and graywacke sandstones, their age ranging from Silurian to Lower Devonian. These assemblages are undoubtedly of island-arc origin. Serpentinite melange occupies the bottom of the upper, Maljudoshory nappe; the nappe itself is built up of clastic-chert-spilitic sequences of Ordovician-Devonian age. The major phase of nappe emplacement in the Lemva zone is dated as Late Permian.

Pai-Khoi Zone

Belyakov and Dembovsky [1984] recognized an independent allochthon within the axial part of the Pai-Khoi zone and compared it with the Lemva facies. It is thrust over carbonate shelf formations of the Eletsky complex. The allochthon consists largely of the deep-water Paleozoic Pai-Khoi complex, which is very similar to the Lemva facies of the Polar Urals. The main deformation and displacement of the Pai-Khoi allochthon evidently occurred in the Early Triassic.

It may be assumed that deep-water complexes known in the eastern part of the Novaya Zemiya [Sidorenko, 1970a] are analogous to the Pai-Khoi allochthon. Here, the Ordovician and Silurian are mainly represented by graptolitic shales, while a sizeable part of the Devonian and Carboniferous sequence consists of cherty shales and pelagic limestones. Analysis of the facies reveals a transition from shelf deposits in the southern and western parts of Novaya Zemlya to deep-water sequences developed in the east, indicating that the deep-water sequences belong to the continental-slope and continental-rise facies.

THE MAIN URALIAN FAULT ZONE

This great fault, which was formerly believed to dip steeply and to transect the entire crust down to the mantle, is regarded

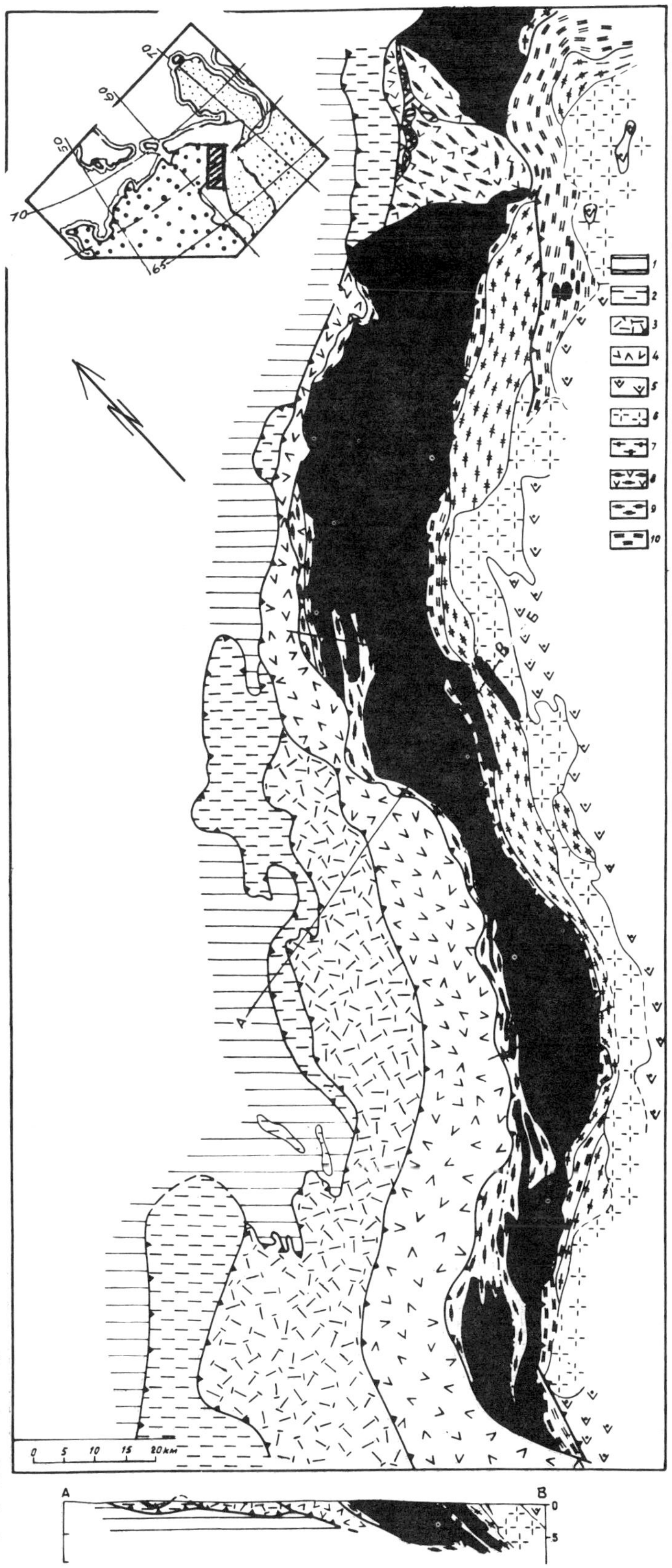

Fig. 29. Geology of the Voikaro-Syn'insky massif in the Lemva zone (after Savel'ev and Savel'eva [1977]).
Ordovician to Devonian of the externides: 1, shelf deposits; 2, chert; 3, clastics; 4, clastics and volcanics.
Silurian to Devonian of the internides: 5, volcanics; 6, diorite; 7, tonalite; 8, metamorphics of epidote-amphibolite facies; 9, garnet amphibolite and glaucophane schist; 10, plagioclase amphibolite. Black - ophiolites of the Voikaro-Syn'insky massif. Toothed lines mark thrusts.

as the boundary between the Internides and Externides in the Urals. In detail, the fault is a swath of tectonic melanges of varying width--from a few to 20 km. This swath incorporates different sized blocks and slabs of dismembered ophiolite association: ultramafic rocks, gabbro, pillow lavas, cherts, and other rocks. In places, it also displays blastomylonites and metamorphic schists, including glaucophane schists and eclogites, i.e., rocks forming under high pressures. The fault often coincides with a major gravitational gradient, to the east of which there is a gravity high. Geological data (including the results of drilling) combined with seismic reflection data [Sokolov, 1985] indicate that the entire tectonic zone dips to the east at angles from 10° to 50° [Puchkov et al., 1985].

The Main Uralian fault zone formed at the same time as the Uralian nappe structures, i.e., toward the end of the Paleozoic, and the fault itself is the frontal zone of the greatest of the nappes.

URALIDES: INTERNAL ZONES

The geological complexes of the Uralian internides are widespread in the western Uralian zones, the Magnitogortsk, Tagil, Shchuchinsky and West Mugodzhar synclinoria. They also compose more eastern zones, the East Uralian trough and the Oktyabrsko-Denisovsky zone.

Volcanic and volcanogenic-clastic sequences predominate at nearly all stratigraphic levels from the Silurian to the Middle Carboniferous. The lower part of the section is composed of ophiolite complexes and the upper one of volcanic rocks of the calc-alkaline series. While the lower complex is oceanic, the upper one is island-arc, showing that island volcanic arcs were emplaced on an oceanic basement.

Oceanic Complexes

Despite their different ages, the oceanic complexes have similar structures and encompass rather similar rocks of the ophiolite association. Like ophiolites in other foldbelts, the ophiolite sequence in the Urals contains metamorphosed harzburgites below, an overlying layered complex, gabbro, sheeted dikes, pillow lavas, and sediments. Unfortunately, there is no place where the entire sequence is exposed in a single undisturbed section. The lower parts of ophiolites--from harzburgites to gabbro--crop out on the margins of the Khabarny, Kempirsai, and Voikaro-Syn'insky massifs. The undisturbed upper parts of the oceanic complex from gabbro to sediments are preserved in the Southern Mugodzhar.

The Khabarny and Kempirsai massifs are predominantly composed of metamorphosed harzburgites, and the overlying units include layered ultramafic and mafic rocks [Varlakov, 1978; Pavlov, Grigorieva-Chuprynina, 1973]. In the Khabarny massif the layered complex is about 300 m thick. In many exposures thin layering is due to change of mineral composition, to enrichment of individual layers with olivine or pyroxene or plagioclase, or in places with chrome spinel. Normally, in this type of layered complexes, there are numerous cross-bedded packages and horizons, about half a meter thick, in which the cross-layers slope at 45° relative to the bottom and top. It should be emphasized that all measurements of cross-bedding indicate a NE dip and it may be concluded that the spreading axis and center of the magma chamber were in that direction.

Greenstones formed from pillow lavas of Silurian-Devonian age are associated with dunite-harzburgites and layered series. These higher horizons of the ophiolite sequence are preserved undisturbed in the Southern Mugodzhar [Ivanov et al., 1973; Korinevsky, 1984; Zonenshain et al., 1984]. Outcrops along the Shuldak River show alternating pillow-lava fields and sheeted-dike swarms (Figure 30). The lavas contain pockets of sedimentary rock. The dike and lava fields are from 3 to 5 km wide.

Vertical or steeply dipping doleritic dikes, averaging 1 m thick, are typical for the sheeted dike swarms; they intrude one another [Didenko et al., 1984]. In petromagnetic parameters the dike dolerites are identical to recent MORB basalts. Reversal of the paleomagnetic field analogous to that in the present oceanic floor was discovered in the above series [Didenko and Pechersky, 1986].

The dike swarms grade into pillow lavas which, as along the axes of present mid-oceanic ridges, represent lava tube flows branching downward. The orientation of flows was used to reconstruct the configuration of volcanic edifices, but in most cases only their western parts were preserved, as might be expected in sea-floor spreading when the basalt floor split along the rift axis and the two halves of volcanic edifices diverged in opposite directions. Here the spreading axis lay towards the east (in present coordinates). Analytical data [Kuzmin and Al'mukhamedov, 1984] definitely show that the Shuldak River basalts are identical to oceanic tholeiites.

Lava vesicularity indicates that the depth of eruption of the Mugodhar lavas was 3 km. Considering the dependence of the titanium content in basalts on spreading rate, the spreading rate was estimated at 5 cm per year [Matveenkov and Khain, 1984]. Conodonts contained in the upper chert layers indicate the Middle Devonian age of the sequence [Korinevsky, 1984].

Volcanic Arc and Associated Complexes

Island-arc complexes are distributed over the entire internal zone of the Urals; they are mainly Silurian-Devonian in age (Figure 31). The best region for understanding their structure is the Magnitogorsk synclinorium [Librovich, 1936; Koroteev et al., 1979; Maslov, 1980; Frolova and Burikova, 1977].

On the west side of the Magnitogorsk synclinorium volcanic complexes are generally described as individual suites: Baimak-Buribaev, Irendyk, and Karamalytash. They represent the Middle Devonian (the Eifelian, and probably the Givetian). The Baimak-Buribaev complex, developed in the west, consists mainly of basalts, although felsic volcanics are also present. The basalts include varieties close to tholeiites; boninites are established, as well. The Irendyk complex occupies the major portion of the western flank in the Magnitogorsk synclinorium, which is characterized by the intercalation of lavas and tuffs of basalt and andesite-basalt composition with numerous volcanogenic-clastic accumulations. Petrochemically, the Irendyk complex belongs to a typical calc-alkaline series and is regarded by geologists as the analog of present island arcs. The Karamalytash complex is located on the eastern margin of the western flank of the Magnitogorsk synclinorium. It is somewhat similar to the Baimak-Buribaev volcanics and consists of basaltic pillow lavas, but subvolcanic bodies of acid composition are present in this complex. The Karamalytash basalts may be attributed

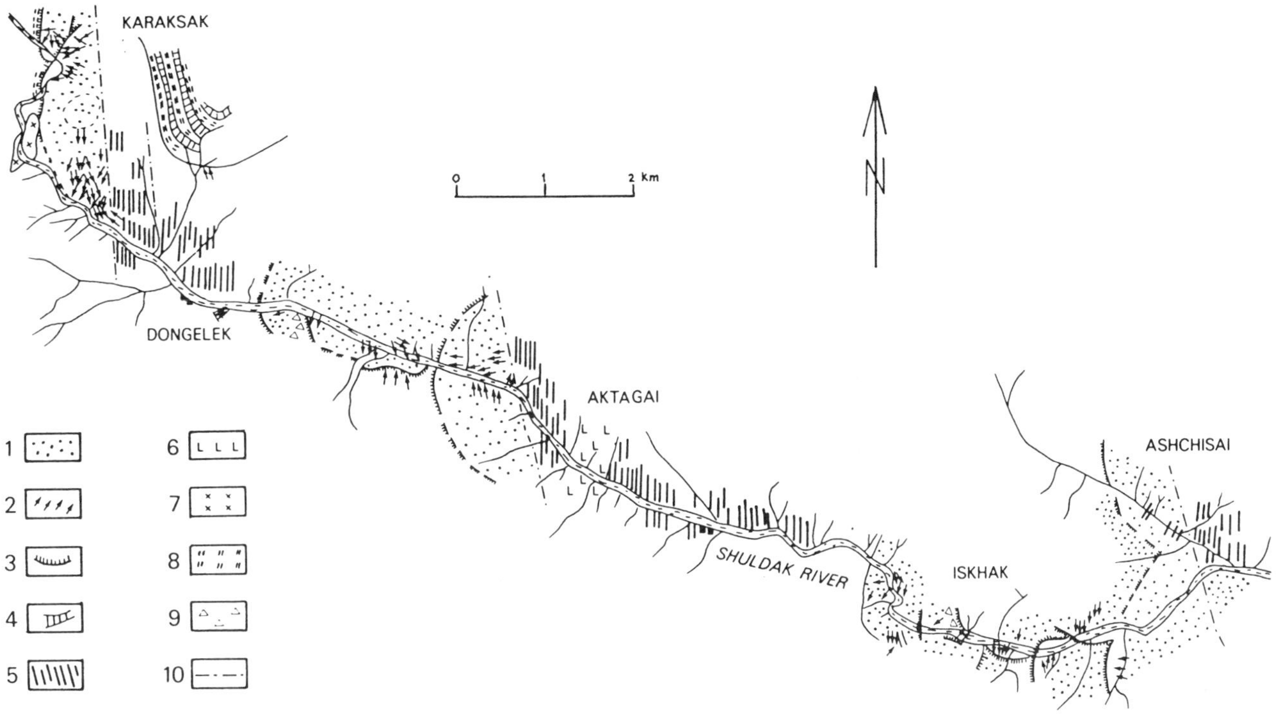

Fig. 30. Geology of the Devonian volcanic pile along Shuldak River, Southern Mugodjar.
1, Pillow lava; 2, flow direction of pillow lava; 3, outline of reconstructed volcanic edifice; 4, dolerite sill; 5, sheeted dikes; 6, gabbro; 7, diorite; 8, chert; 9, hyaloclastics; 10, fault.

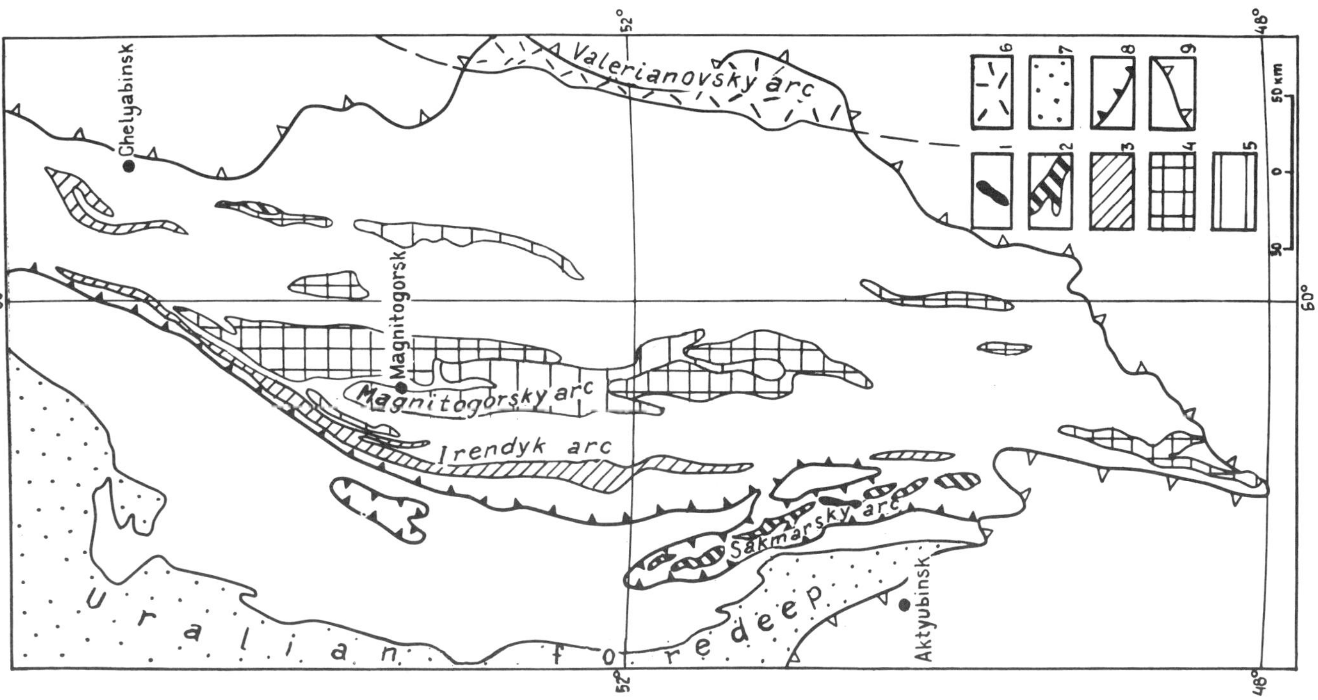

Fig. 31. Distribution of island-arc assemblages in the Southern Urals.
1, Middle Ordovician (Guberlinsky arc); 2, Silurian (Sakmara arc); 3, Lower-Middle Devonian (Irendyk arc); 4, Middle-Upper Devonian (Magnitogorsk arc); 5, Lower Carboniferous (Magnitogorsk arc); 6, Lower-Middle Carboniferous (Valerianovsky arc); 7, Permian of the Pre-Uralian foredeep; 8, boundary between externides and internides; 9, limit of the Meso-Cenozoic sedimentary cover.

to magmas of inter-arc basins. These basalts are definitely of the same age as those of the Mugodzhary suite of the Southern Mugodzhary ophiolite complex.

The volcanic complexes are overlain by Middle and Upper Devonian cherty and tuffaceous-clastic sequences [Maslov, 1980]. The Karamalytash structure shows clearly that volcanism and chert accumulation proceeded contemporaneously. Cherts are dated as Upper Eifelian to Lower Givetian. In the overlying Givetian and Frasnian deposits, fine and coarse sequences of volcano-clastic composition are intercalated. Olistostromes with olistoliths of volcanic rocks and Givetian limestones, some hundred meters thick, appear in the Upper Devonian. Clearly this Middle to Upper Devonian tuffaceous-clastic complex was formed by erosion of the adjacent volcanic island arc. Maslov [1980] provides reliable data indicating that material was carried from the east.

A wide development of Middle to Upper Devonian purely volcanic piles is typical for the eastern flank of the Magnitogorsk synclinorium, in contrast to the western flank.

East of the Magnitogorsk synclinorium the Devonian island-arc complexes are found in several interdomal depressions in the zone of the East Uralian uplift and in the East Uralian trough. In the latter, in addition to Devonian island-arc volcanics, Koroteev et al. [1979] reported Silurian andesites and andesite-basalts, which are underlain by a Lower Silurian oceanic complex.

At the boundary of the Lower and Upper Visean (Early Carboniferous), volcanism ceased in the major portion of the Magnitogorsk synclinorium. Only locally (e.g., the Magnitogorsk region) volcanics occurred throughout the whole Visean. Stratigraphically upwards only sedimentary sequences may be found. Major deformation occurred after the Middle Carboniferous.

The Valerianovsky belt stretches between the Uralian foldbelt and Central Kazakstan [Segedin, 1981]. In the west, this belt is bordered by the Oktyabrsk-Denisovsky zone composed of ophiolites and island-arc complexes of Silurian and Devonian age. To the east of the belt, within the Borovsky zone, there are carbonate and carbonate-terrigenous deposits of the Upper Devonian and Lower Carboniferous, underlain by red beds and volcanic rocks resembling the continental Devonian of western Central Kazakhstan. The Valerianovsky belt consists of calc-alkaline volcanics, andesites, andesite-basalts, and dacites. These are intruded by diorites and granodiorites of the Sokolovsko-Sarbaisky complex. The ages of the volcanic rocks range from the Upper Visean (Early Carboniferous) to the Bashkirian (early Late Carboniferous). Volcanism commenced here just after it ceased in the Magnitogorsk arc. The volcanic belt separates the "Caledonian" (Early Paleozoic) structures of Central Kazakhstan from the "Hercynian" (Late Paleozoic) structures of the Uralian belt. The Valerianovsky belt is comparable to the Andean type of continental margin belt, the Benioff zone dipping underneath the continent, in this case to the east. The Oktyabrsk Denisovsky zone may be interpreted as the remains of a subduction melange formed in front of the Valerianovsky belt.

GENERAL PATTERN OF THE URALS STRUCTURE

The structure of the Urals may be regarded as consisting of two structural complexes (Figure 32): a lower autochthonous and an upper allochthonous complex [Perfiliev, 1979; Kamaletdinov, 1974]. The lower structural complex incorporates the East European platform basement and the Pre-Uralides in the internal part of the belt. The Pre-Uralides may be composed of either the same East European basement or of a microcontinent or microcontinents that were either detached from the platform or approached it. The upper structural complex is made up of allochthous oceanic and island-arc complexes. Synclinoria and anticlinoria, well defined on the surface, are in reality synforms and antiforms. The former contain the allochthonous complex, whereas the cores of the latter expose the (relatively) autochthonous lower complex. Seismic reflection data [Sokolov, 1985] obtained near the Tagil borehole indicate that the top of the crystalline basement of the East European platform dips east to 15 km and can be traced under the overthrust external zones and the Central Uralian uplift. The angle of inclination is approximately 40°. The Main Uralian fault zone also dips gently to the east at angles from 20° to 40° and can be traced to a depth of 7 km. In the eastern part of the profile there are reverse tectonic slabs inclined to the west.

The folded structure of the internal Uralian zones was completed by the end of the Carboniferous or the beginning of the Permian, as shown by the emplacement of many cross-cutting granitic batholiths and the formation of granite-gneissic domes. In the external zone, folding continued up to the Triassic.

HISTORY

Several versions of the structural development of the Urals have been proposed using the mobilistic approach [Hamilton, 1970; Ivanov et al., 1974; Puchkov, 1979; Perfiliev, 1979; Samygin, 1980]. The description given below is after Zonenshain et al. [1984].

Paleomagnetic and Kinematic Data

More or less reliable paleomagnetic data are available for the continents surrounding the Urals, that is, East Europe and Siberia. Figure 33 represents paleolatitudes of Devonian time for East Europe and Siberia, as well as for Central Kazakhstan. The paleolatitudes are entirely compatible with paleoclimatic data. For instance, the Devonian barrier reef of the Western Peri-Uralian region, extending in present coordinates, coincides with the position of the Devonian paleomagnetic equator. In Figure 33 it is evident that the paleolatitudes of Siberia and East Europe are inconsistent. To make them consistent it is necessary to rotate Siberia by 60° anticlockwise relative to East Europe and to move it 3000 km.

Paleomagnetic data for the Uralian internal zones show that the volcanic piles of the Shuldak section, including a series of parallel dikes, formed approximately 500 km north (in past coordinates) of the East European continental margin, within the Paleo-Uralian ocean [Didenko et al., 1984].

Kinematic data also throw light upon mutual displacements of Siberia and East Europe. They were obtained from the absolute reconstructions using hot spot trajectories and apparent polar wandering path data [Zonenshain et al., 1985] (Table 5). The path of Siberia relative to East Europe is shown in Figure 34. In the Devonian, Siberia moved toward East Europe with a velocity of 7 cm/yr. Throughout the Devonian, from 400 to 340 Ma, Siberia traveled about 3000 km in relation to East

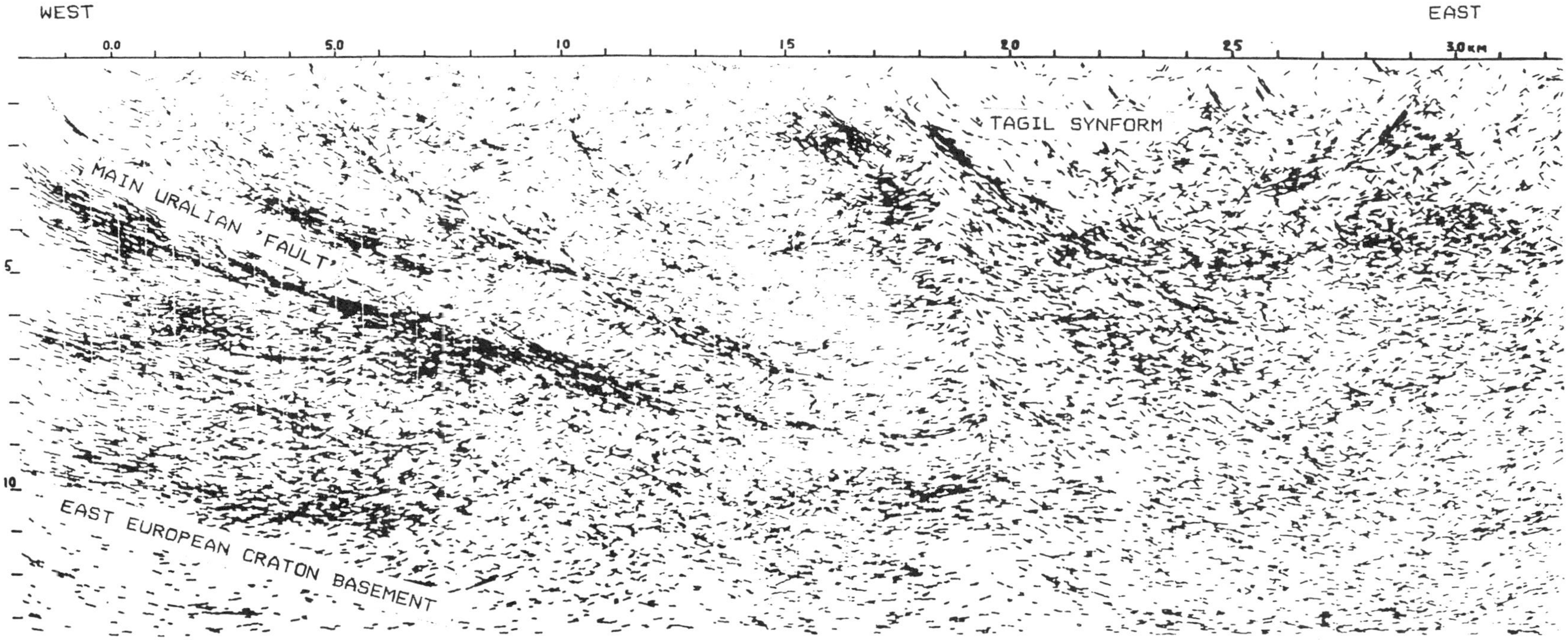

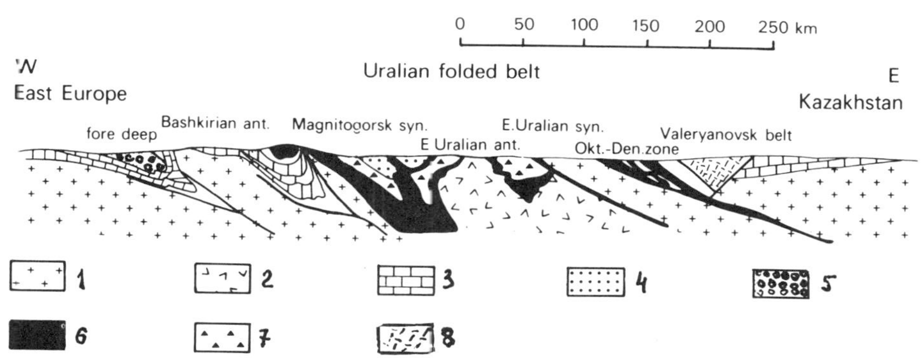

Fig. 32. Cross-sections through the Uralian foldbelt.
a - Multi-channel reflection cross-section near 58° N (after V. S. Sokolov, unpublished material, 1989). Vertical scale - two-way travel time in sec; horizontal scale in km. Sources of energy were explosions in shallow boreholes. Clearly seen are: gently-dipping reflectors in the East European continental crust under the Uralian belt; the basement of the East European continent going down beneath the Urals (10-11 sec depth); the allochtonous, synformal character of the Tagyl synclinorium.
b - Interpretive cross-section through the Uralian foldbelt (after Zonenshain et al. [1984]).
1, Precambrian basement including Riphean metasediments, 2, gneissic-granite dome, 3, platform cover, 4, Devonian-Carboniferous turbidite, 5, Permian molasse, 6, oceanic complexes (Silurian to Middle Devonian), 7, island-arc complexes (Silurian to Carboniferous), 8, Early-Middle Carboniferous Valerianovsky volcanic belt, of continental margin type.

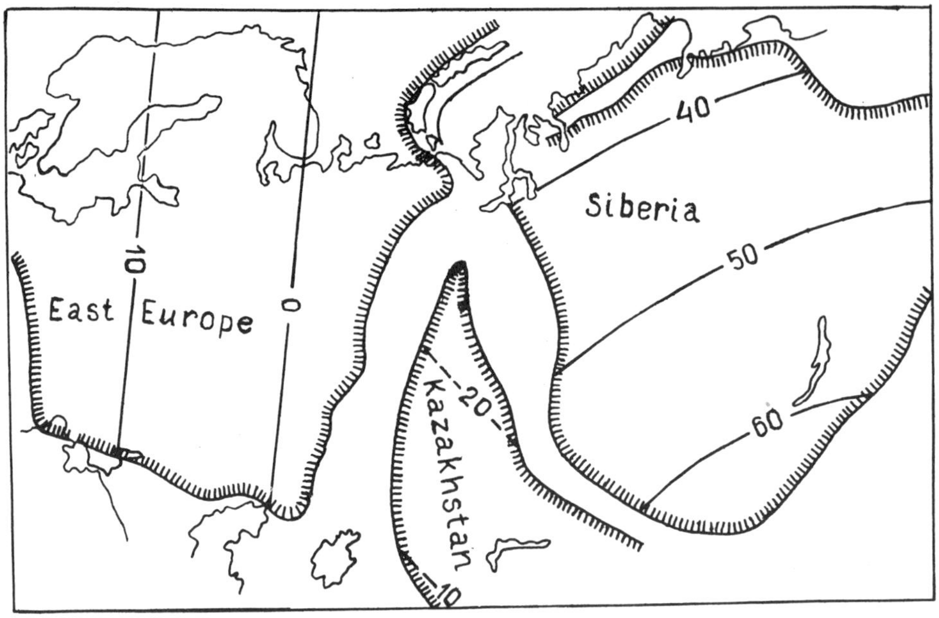

Fig. 33. Comparison of Devonian paleolatitudes for the East European and Siberian platforms. Paleomagnetic data from Khramov et al. [1982].

TABLE 5. Parameters for Differential Motions of the Siberian and East European Continents from 400 to 250 Ma

Age Interval (Ma)	Siberia with Respect to East Europe			East Europe with Respect to Siberia			Motion of Siberia with Respect to East Europe in Points			
	Pole Position		Angle of	Pole Position		Angle of	65°N	60°E	50°N	60°E
	Deg. N (Degrees)	Deg. E Degrees	Rotation Degree	N	E	Rotation Degree	Azimuth (Deg.)	Rate (cm/yr)	Azimuth (Deg.)	(Rate (cm/yr)
280-250	70.24	-119.84	-12.73	70.25	-119.84	12.73	270	3.3	270	4.1
310-280	47.6	-76.26	-12.2	44.88	-84.77	12.2	58	4	61	4.4
340-310	46.01	134.81	-6.72	45.166	137.43	6.72	348	1.7	334	1.85
370-340	57.45	-162.66	-21.62	48	-164.84	21.62	297	6.4	293	7.4
400-370	74.61	96.68	-19.7	85.32	-131.58	19.7	306	2	289	3.5

Europe. In the Carboniferous and Permian, velocity decreased to 2-4 cm/yr due to the initial collision of continents. The direction of movement also changed and became strike slip, left lateral.

Both paleomagnetic and kinematic data allow us to determine the size of the Uralian paleo-ocean. It was found to be 3000 km wide and consequently, it was commensurable with such present-day oceans as the Atlantic Ocean. These data also demonstrate that the Uralian foldbelt itself was indeed formed by collision of continental masses as postulated on geological grounds.

Late Precambrian and Early Cambrian

Many data indicate that in the Late Precambrian the Pre-Uralian paleo-ocean stretched along the eastern (in present

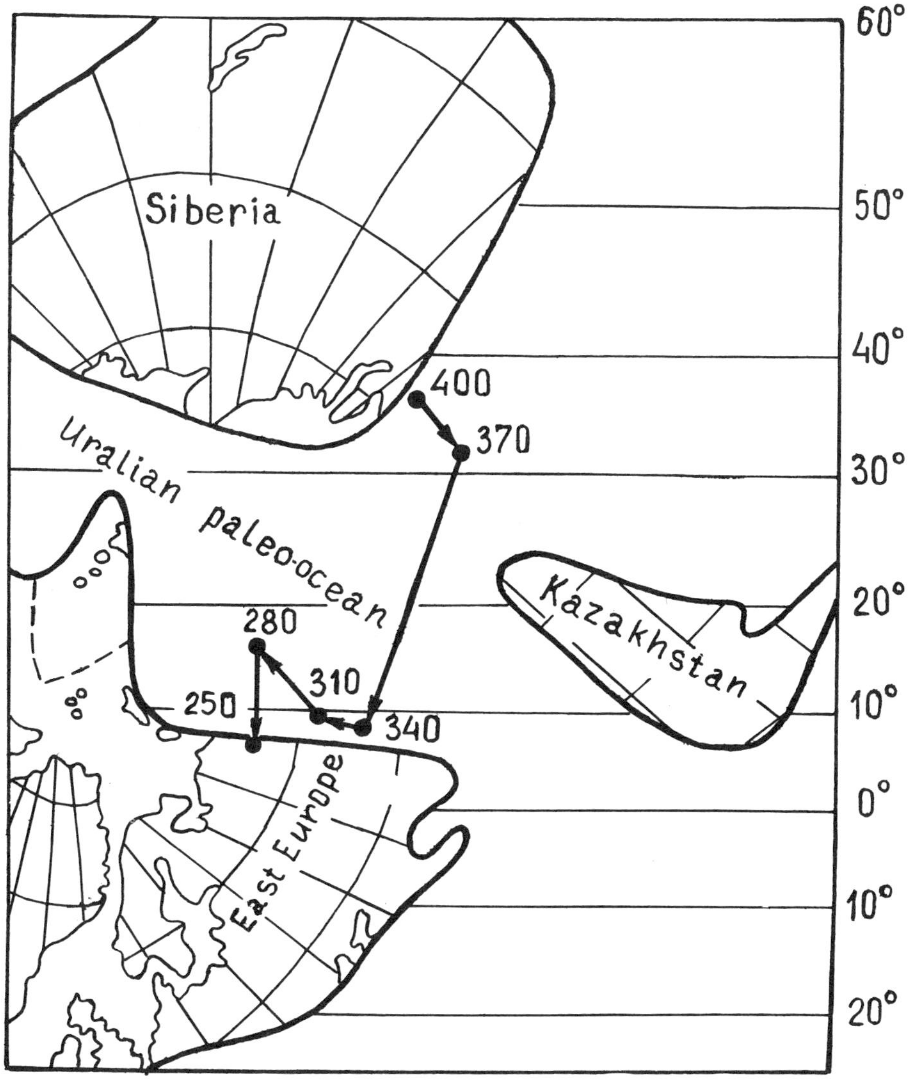

Fig. 34. Motion of Siberia with respect to East Europe from 400 to 250 Ma (after Zonenshain [1984]). Arrows show the Siberian path; figures are ages in Ma.

coordinates) margin of East Europe (Figure 35). This is separated by many island-arc complexes of the Late Riphean, and by thick sedimentary sequences of the Riphean Peri-Uralian area that mark the former passive margin of this ocean. Remnants of the Pre-Uralian ocean floor are perhaps contained in ancient ophiolites of the Maksutov complex in the Southern Urals, and in amphibolites of the Ilmen suite in the Miass region, 700-800 Ma old [Lennykh, 1984]. The ocean began with numerous breakups, as proved by the Riphean bimodal magmatism, the Riphean aulacogens of the East European platform, and the eventual passive margin marked by the Riphean sequences of the Bashkir anticlinorium.

Within the Pre-Uralian ocean there was probably a microcontinent the remains of which now make up the Trans-Uralian rise. Such exotic elements as Archaeocyathid limestones may be related to it. In the Early Cambrian the Trans-Uralian microcontinent was probably a part of eastern areas connected with the Siberian continent.

Pre-Ordovician Continental Collision

Because Ordovician deposits everywhere lie with erosional and angular unconformity on more ancient rocks, orogenic events occurred during a certain span in the Cambrian, probably in its second half. They resulted from collision of island arcs and some microcontinents with East Europe (Figure 36). The whole basement of the Pechora lowland and the Pre-Uralide rocks exposed in the Polar Urals, Pai-Khoi, and Vaigach (in southern Novaya Zemlya) represent accretionary complexes joined to East Europe. In the Southern Urals the Trans-Ural microcontinent probably collided with East Europe in the Cambrian, as suggested by isotopic dating: 590, 570,

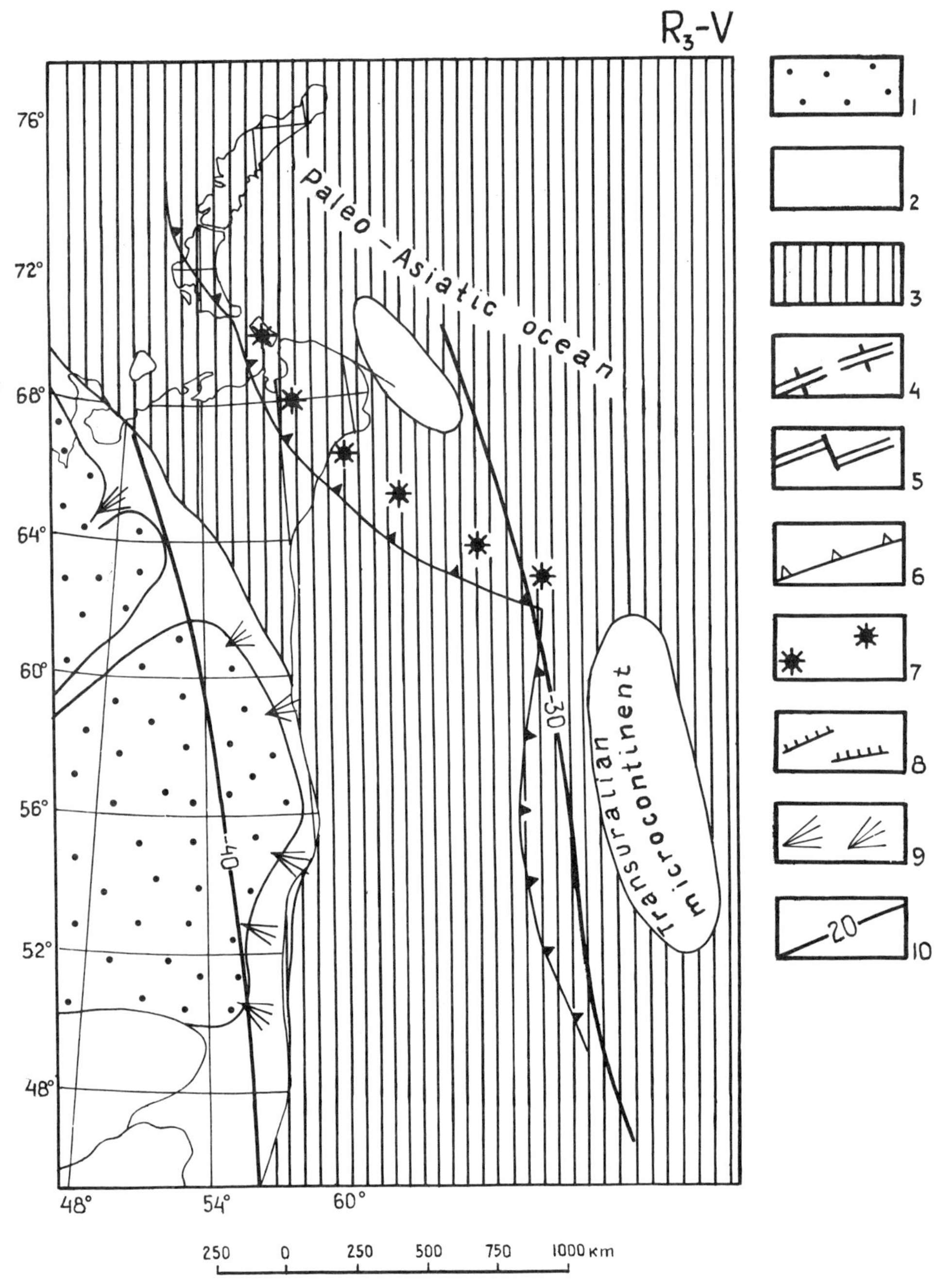

Fig. 35. Palinspastic reconstruction of the Urals for the Latest Precambrian. The reconstruction is made with respect to East Europe which is arbitrarily left fixed. Paleomagnetic data are mostly taken from Khramov et al. [1982] and also from Didenko and Pechersky [1986].
1, Dry land; 2, shallow sea; 3, oceanic floor; 4, continental rift; 5, spreading zone; 6, subduction zone; 7, volcanic arc; 8, folding and thrusting (collision zone); 9, fan; 10, paleolatitude.

510 Ma for the Mugodzhar metamorphics [Milovsky, 1977], 550 and 515 Ma for the Beloretsk metamorphic complex of the Ural-Tau [Lennykh, 1984], and 530-600 Ma for the metamorphic complex of the Polar Urals [Lennykh, 1984].

Early Ordovician, 510-490 Ma (Figure 37)

The Lower Ordovician deposits of the western Urals may definitely be regarded as an indication of continental rifting.

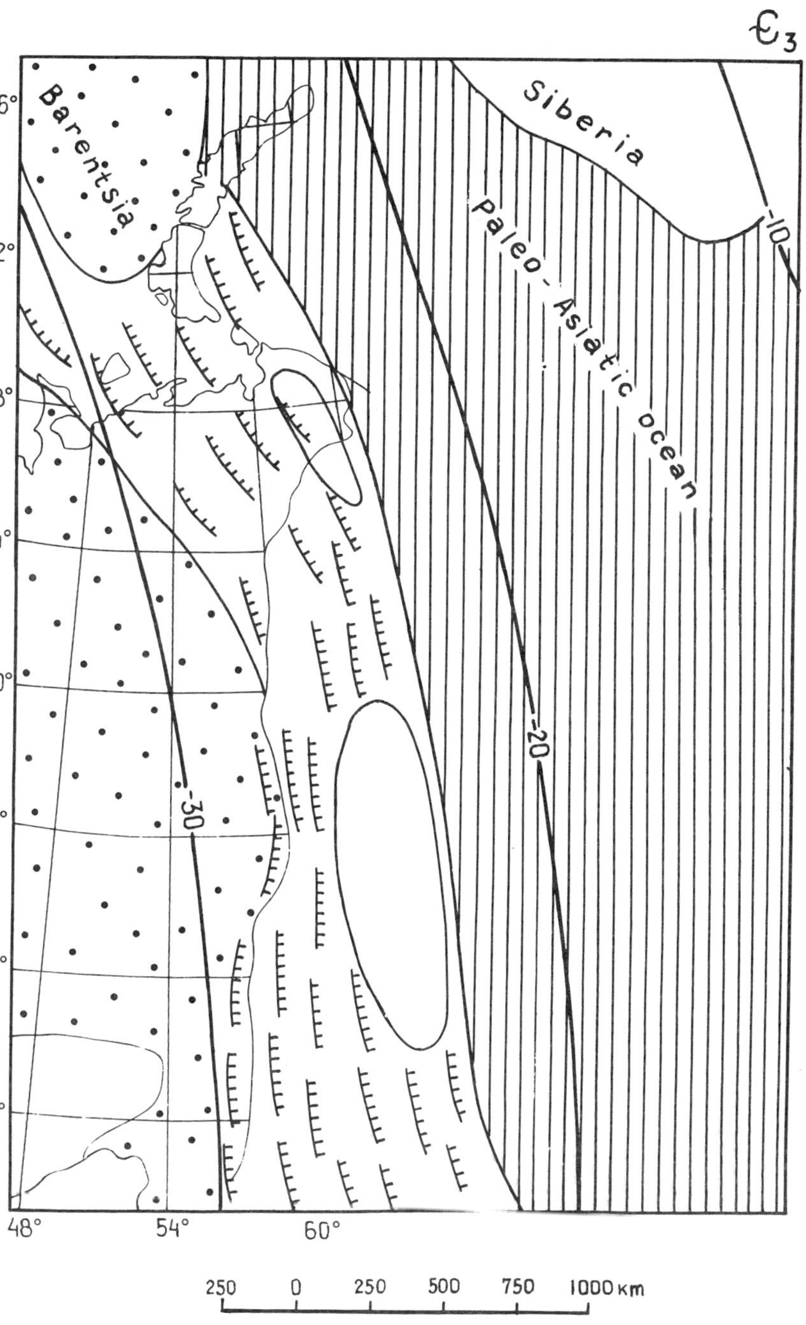

Fig. 36. Palinspastic reconstruction of the Urals for the Late Cambrian. See Figure 35 for explanation.

The breakup occurred through the whole East European continental margin and propagated from north to south.

The Mugodzhar microcontinent was broken off from East Europe. It included parts of the former Trans-Uralian microcontinent and parts of the East European basement. As the Mugodzhar microcontinent drifted away from East Europe, the Uralian paleo-ocean opened. In addition to the Mogodzhar microcontinent, some other smaller massifs were also broken off, their further history being mostly unknown, but two of them are marked by crystalline massifs of the Central Uralian and Ural-Tau uplifts (e.g., the Ebetin antiform of the Sakmara zone, Figure 25).

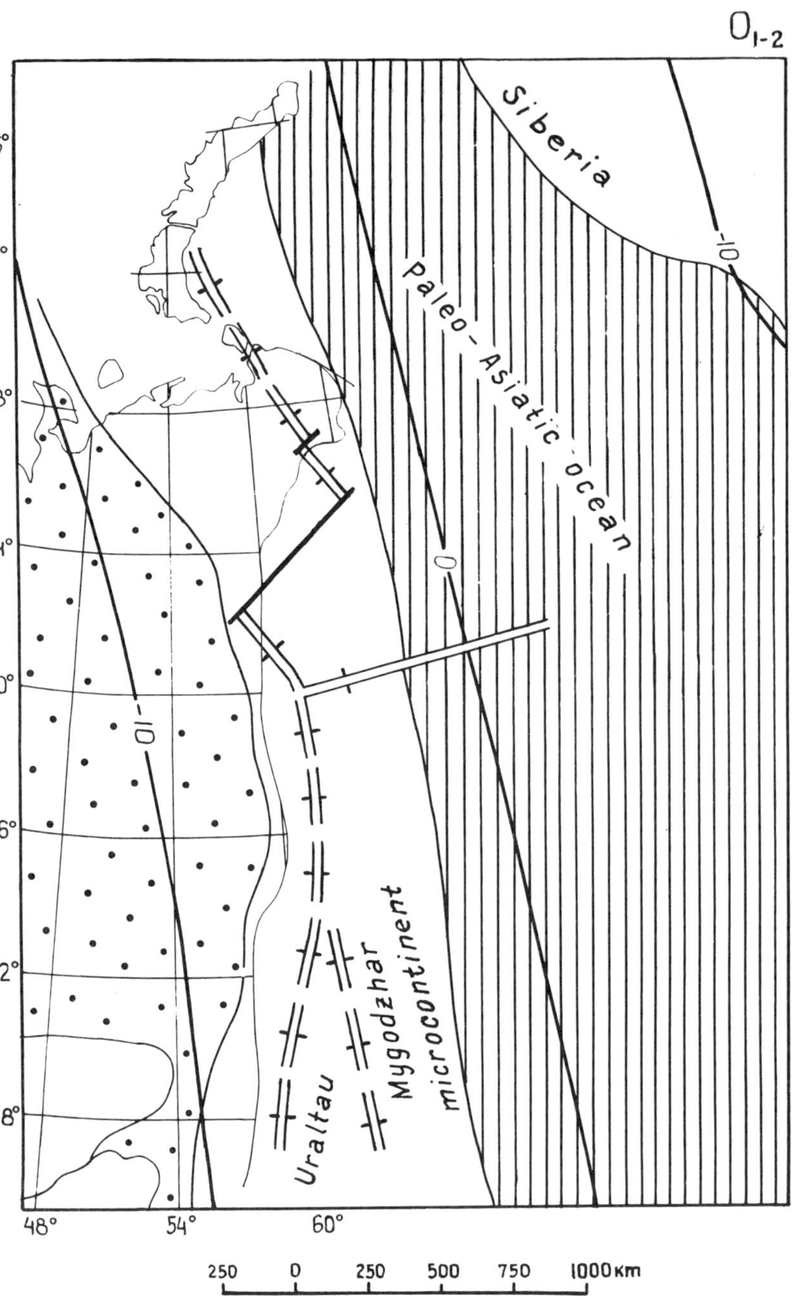

Fig. 37. Palinspastic reconstruction of the Urals for the Early to Middle Ordovician, 510-490 Ma. See Figure 35 for explanation.

Middle Ordovician, 490-450 Ma.(Figure 38)

During the Middle Ordovician, the Uralian paleo-ocean opened rapidly and the East European passive margin was finally shaped. Thus, all along the entire western Uralian slope deep-water (bathyal) cherty shales are widespread within the marginal allochthons. Somewhere in this ocean, an island arc had already formed in the Middle Ordovician. It is inferred from the distribution of calc-alkaline volcanics of the Guberlinsky suite in the Sakmara zone.

Late Ordovician - Early Silurian, 450-430 Ma (Figure 39)

In the Early Silurian, oceanic complexes were emplaced widely throughout the whole Urals. The oceanic tholeiitic basalts of the Surgalinsky and Polyakovsky suites are overlain

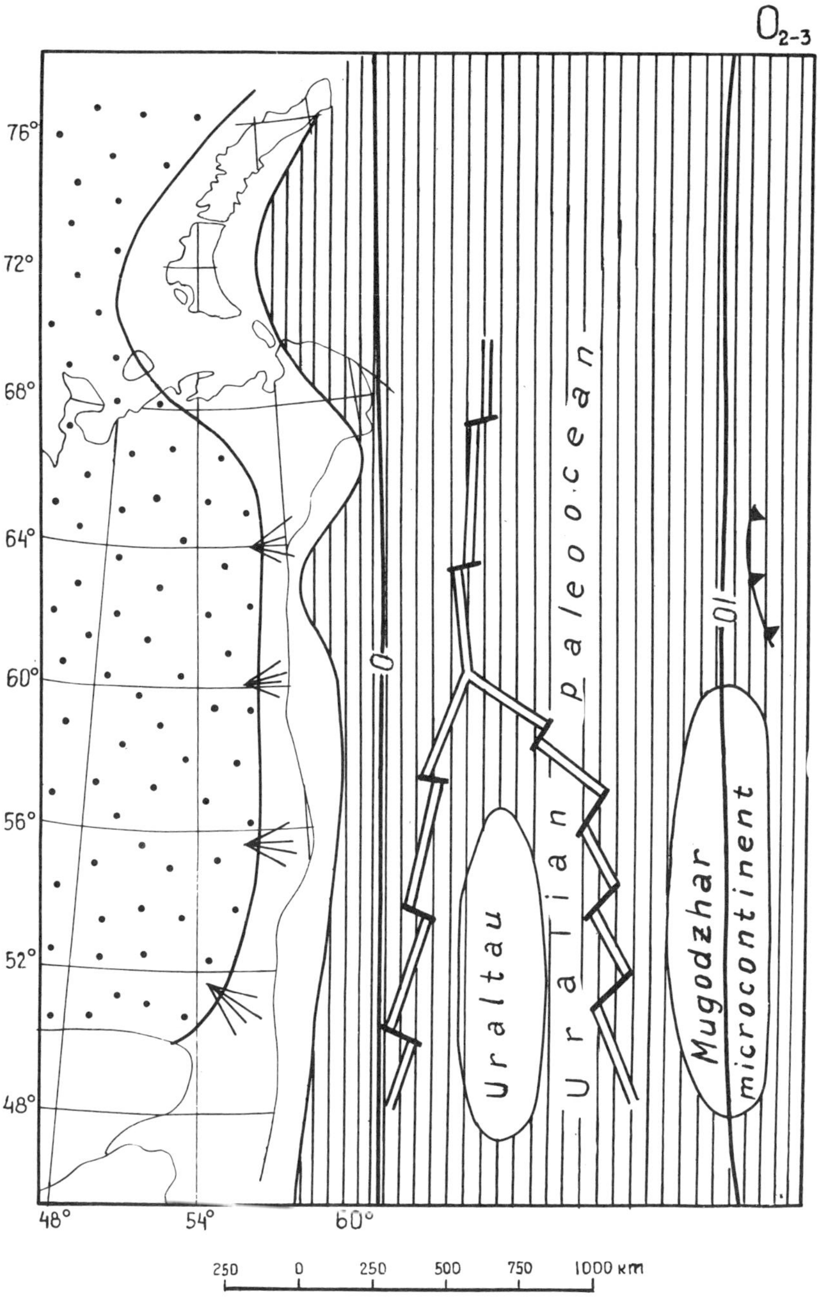

Fig. 38. Palinspastic reconstruction of the Urals for the Middle-Late Ordovician, 490-450 Ma. See Figure 35 for explanation.

by graptolitic cherty shales, proving that the ocean was widening . The spreading axis is supposed to have operated constantly in the central ridge of the paleo-ocean. About 1500 km of the ocean could have been opened if the spreading rate was 2 cm/yr (as in the Atlantic recently) from the end of the Early Ordovician through the Early Silurian i.e., from 490 to 420 Ma. The above figure is accepted as the minimal width of the Uralian paleo-ocean. It should be remembered that according to kinematic data the final width of the ocean was about 3000 km.

Early-Late Silurian, 430-415 Ma (Figure 40)

By the end of the Llandovery, a new island arc--the Sakmara arc--originated within the Uralian paleo-ocean. It is recognized from a typical island-arc volcanic series of the

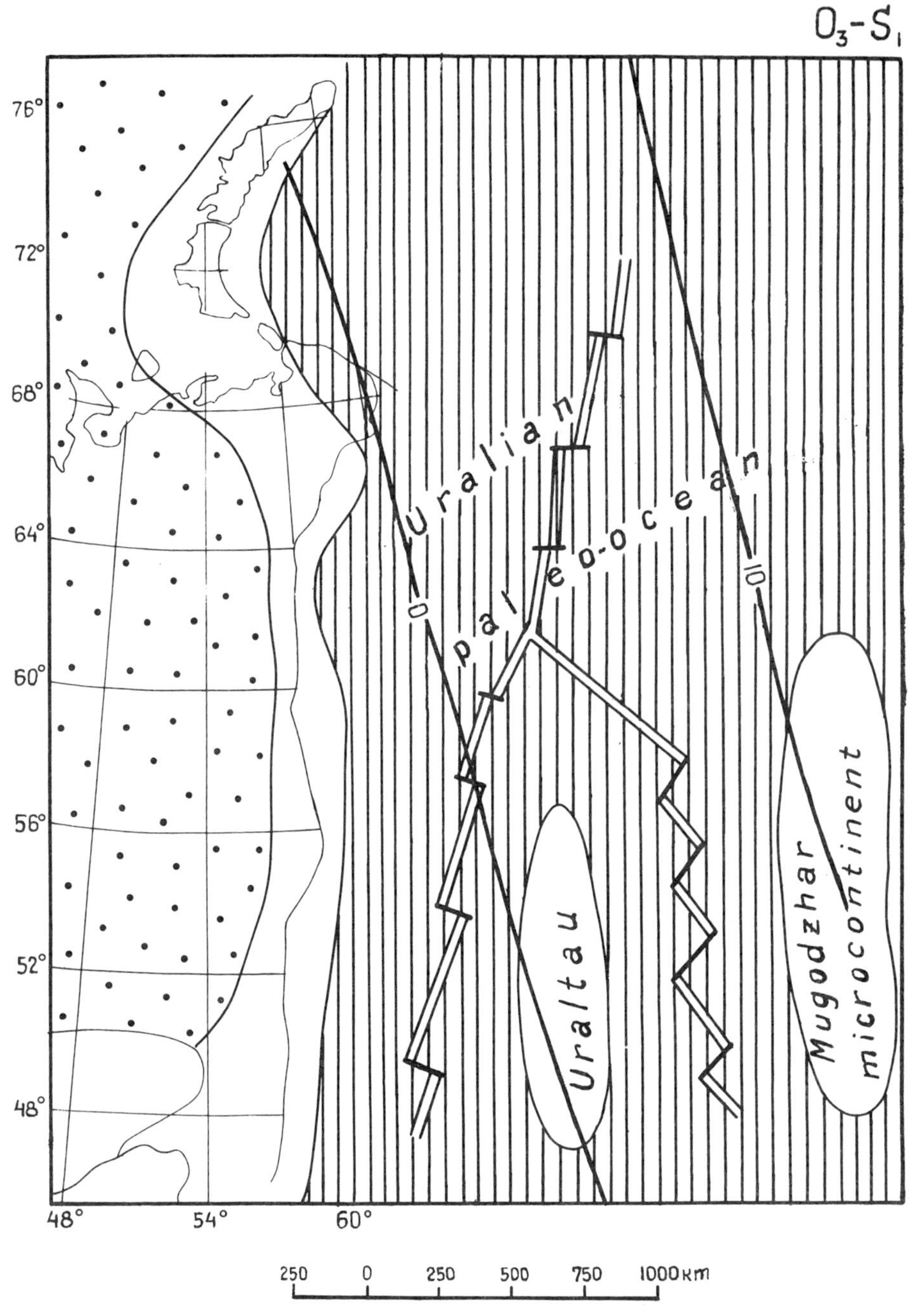

Fig. 39. Palinspastic reconstruction of the Urals for the Early Silurian, 450-430 Ma. See Figure 35 for explanation.

Baiterek sequence of the Sakmara zone and numerous Silurian volcanics of the Tagil and Shchuchinsky synclinoria and Bardym allochthon. Silurian island-arc volcanics are found also in the East Uralian trough. All these occurrences are probably the fragments of the same volcanic arc constructed on an oceanic basement, for calc-alkaline volcanics in places overlie MORB pillow lavas. Consequently, the arc was situated within the Uralian paleo-ocean.

Early Devonian, 405-395 Ma (Figure 41)

Ophiolite was obducted, at least in the southern part of the Urals, in the Early Devonian. The ophiolitic material occurs mainly in the Shandinsky olistostrome but also is known as large slabs. The obduction resulted from the collision of the Sakmara arc with the Ural-Tau microcontinent, the remains of which are present in the Sakmara autochthon, in particular as the Ebetin antiform (Figure 25).

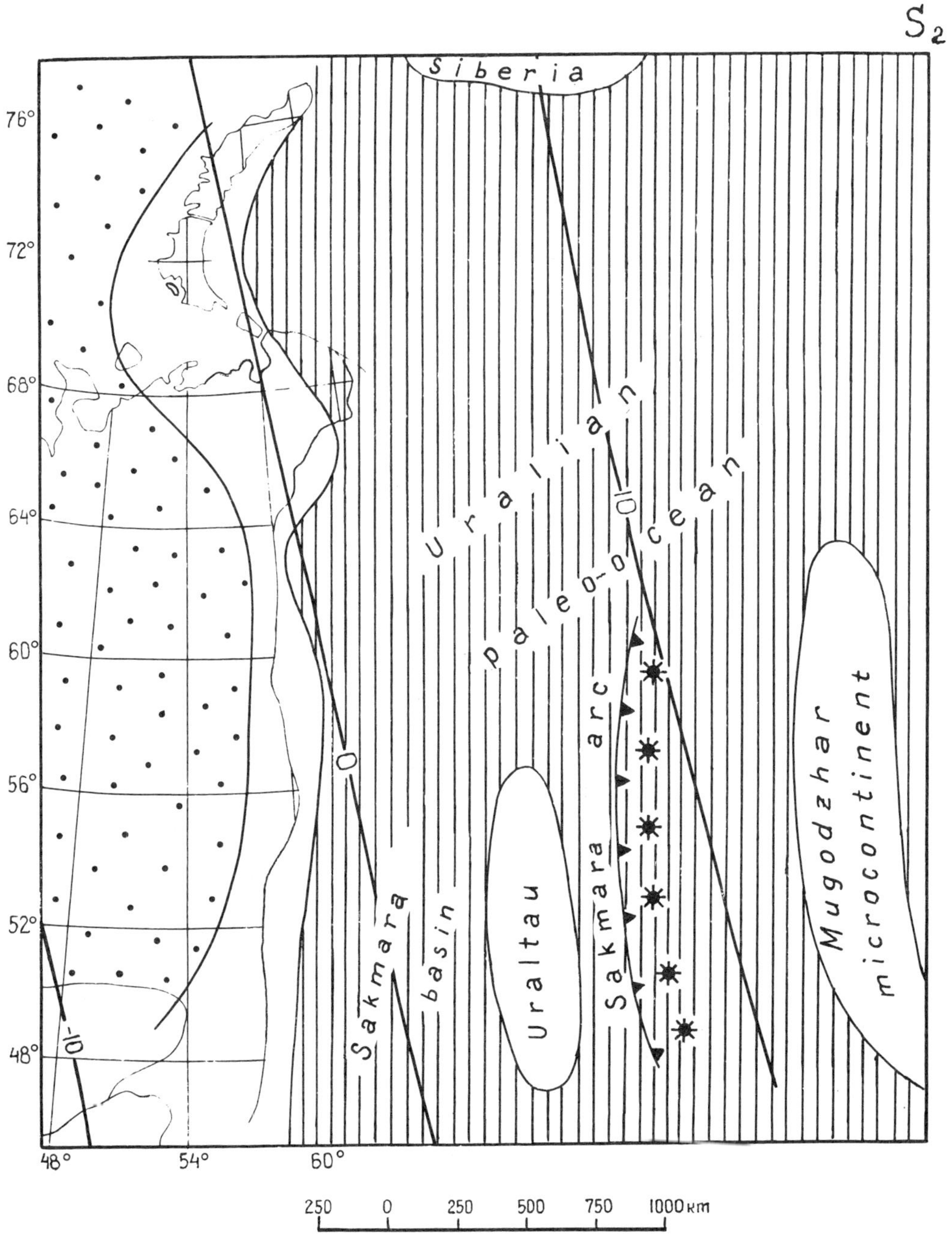

Fig. 40., Palinspastic reconstruction of the Urals for the Late Silurian, 430-415 Ma. See Figure 35 for explanation.

End of Early Devonian - Beginning of Middle Devonian, 395-385 Ma (Figure 42)

The Irendyk arc developed in the second half of the Early Devonian and the beginning of the Middle Devonian. The Baimak-Buribaev volcanic pile west of the Irendyk volcanics may be regarded as the remains of the inter-arc basin, suggesting that the Benioff zone was dipping under the East European continent. If so, the island-arc system within the Uralian paleo-ocean was considerably reorganized at this time,

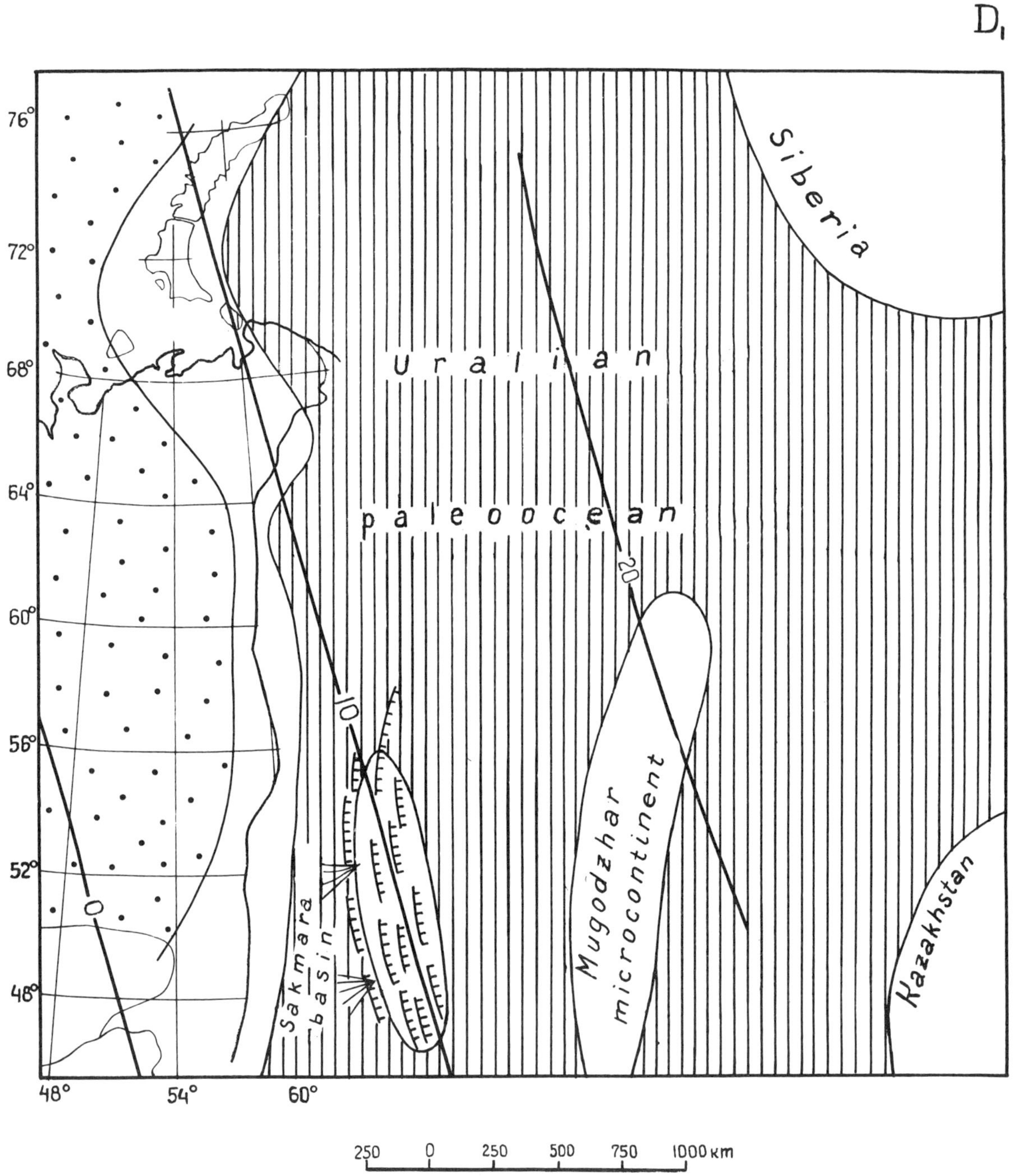

Fig. 41. Palinspastic reconstruction of the Urals for the Early Devonian, 405-395 Ma. See Figure 35 for explanation.

evidently in association with the obduction of the Sakmara ophiolites onto the Ural-Tau microcontinent. Because the arc faced toward the paleo-ocean after the obduction, back-arc basin opening could begin behind the island arc.

Late Eifelian - Early Givetian, 380 Ma (Figure 43)

Throughout a very short span of the Late Eifelian and Early Givetian, not more than 10 Ma, extension and formation of a

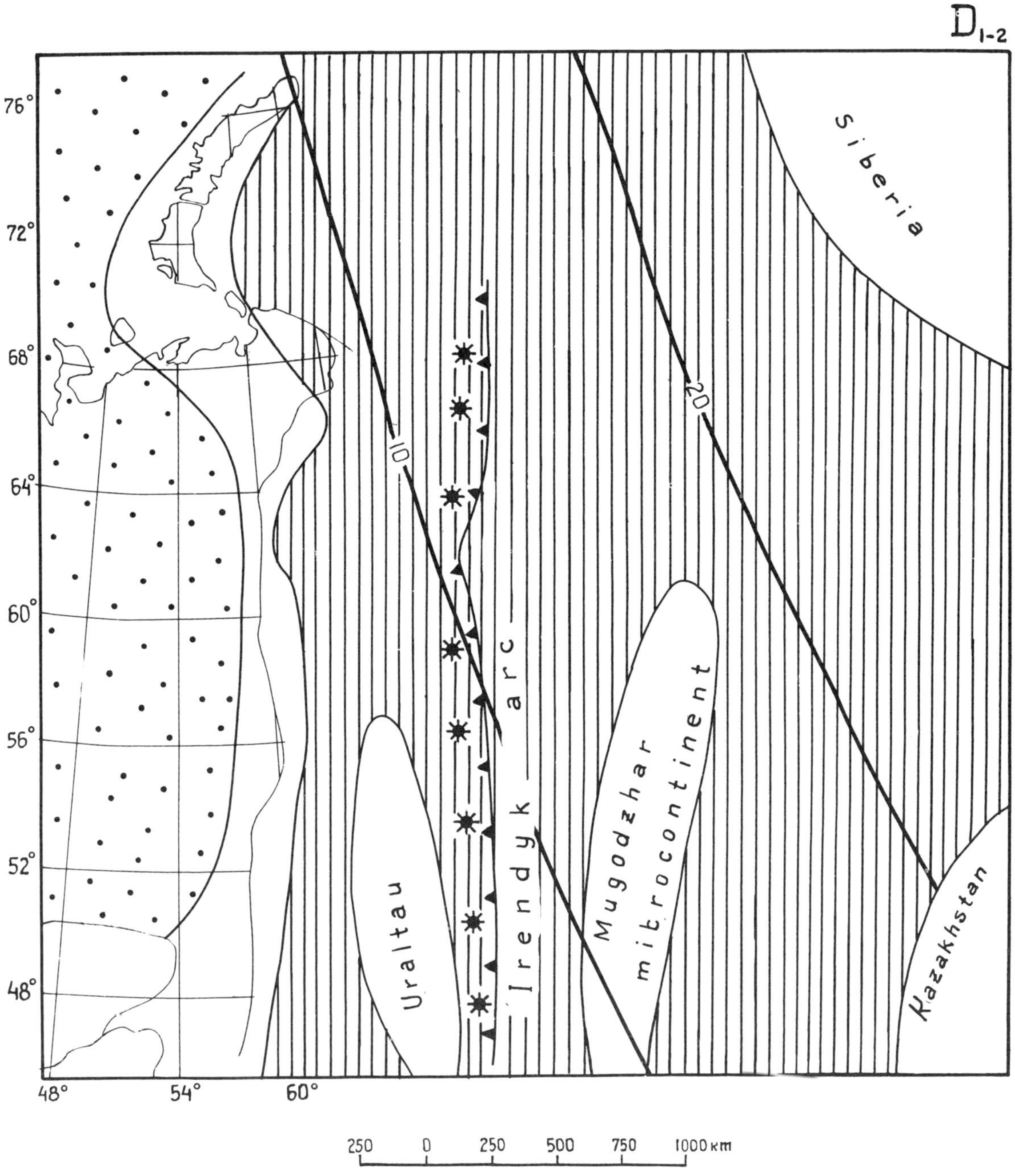

Fig. 42. Palinspastic reconstruction of the Urals for the Early to Middle Devonian, 395-385 Ma. See Figure 35 for explanation.

new marginal sea basin took place. The Irendyk arc ceased to be active. It was active no more than 15 Ma, and in the Late Eifelian it began to subside. During a short time, a few millions of years only, it subsided 2-3 km and deep water cherty sediments began to accumulate on it. Subsidence of the Irendyk arc below the carbonate compensation depth can be best explained if it became a remnant arc when intra-arc splitting and extension formed an inter-arc basin.

A new spreading center generating oceanic crust formed in the Mugodzhar basin. The Shuldak River ophiolites are the

D_2^2

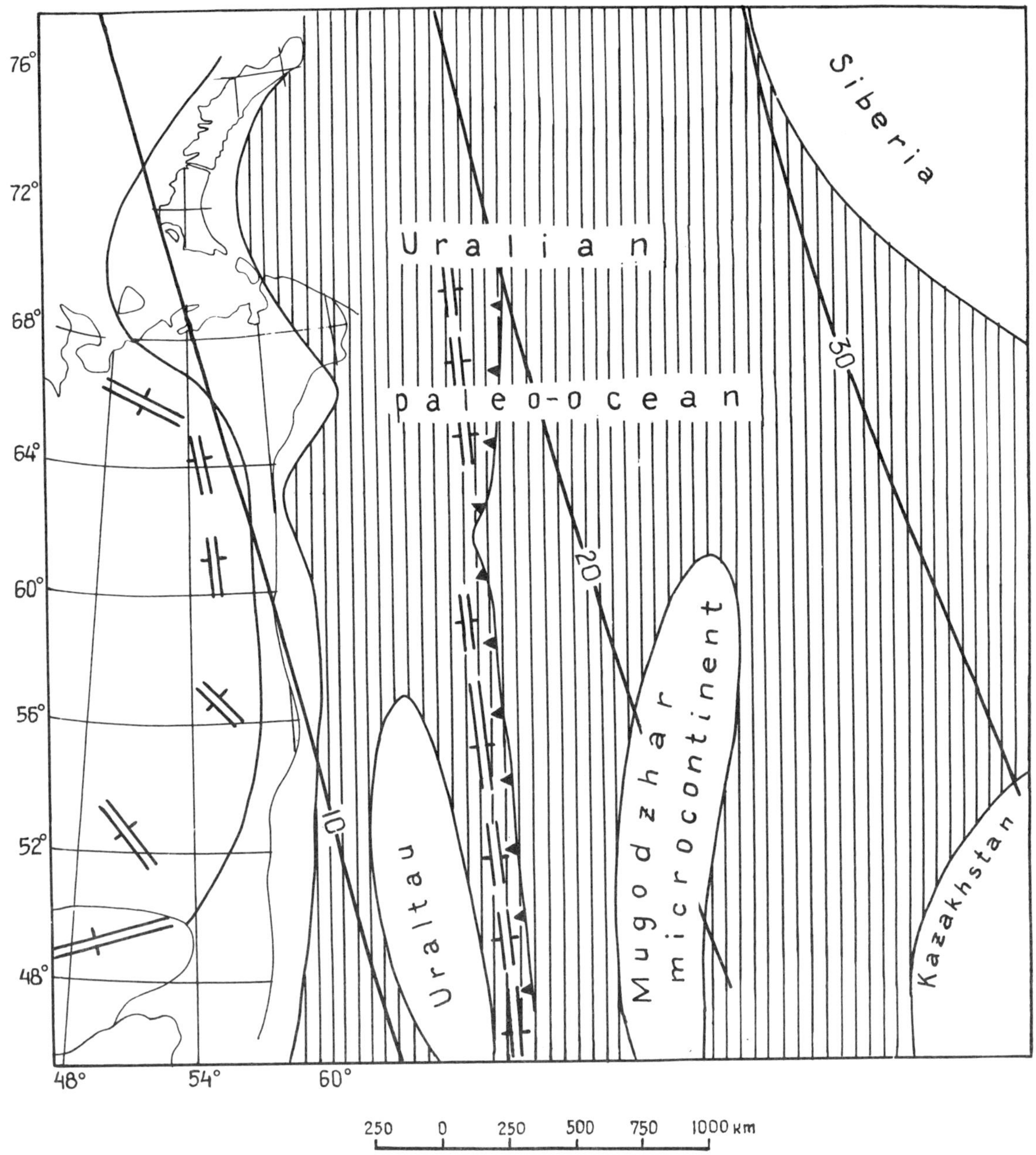

Fig. 43. Palinspastic reconstruction of the Urals for the Givetian, 380 Ma. See Figure 35 for explanation.

remains of the Middle Devonian oceanic floor of this basin. To the north, the newly originated oceanic basin gradually narrowed and wedged out.

According to paleomagnetic data, all paleogeographic elements of the Urals were oriented roughly east-west sublatitudinally, and were situated in the tropical zone of the northern hemisphere.

Late Givetian - Early Frasnian, 380-370 Ma (Figure 44)

Just after the oceanic crust of the Mugodzhar back-arc basin began to form, a new volcanic island arc, the Magnitogorsk arc, became active. The northern part of the arc was active up to the Early Visean, but to the south its activity died out almost immediately, at the end of the Givetian. Such a di-

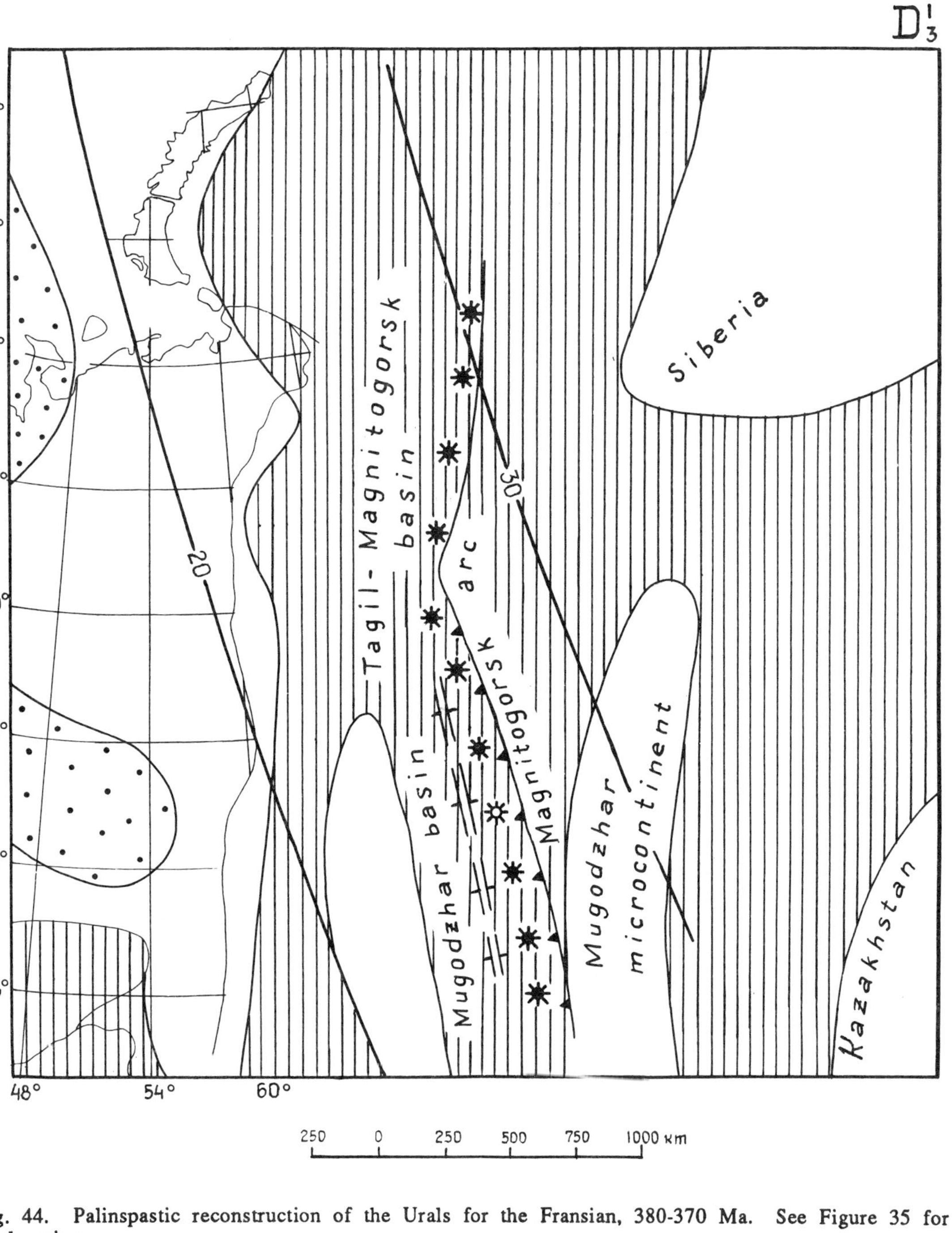

Fig. 44. Palinspastic reconstruction of the Urals for the Fransian, 380-370 Ma. See Figure 35 for explanation.

achroneity of island-arc activity is explained by its oblique convergence and collision with the Mugodzhar microcontinent at different times--earlier in the south, later in the north. A considerable part of the Mugodzhary microcontinent, 50-60 km wide, was overthrust by the island-arc and the arc's underlying oceanic crust. Such large-scale thrusting over continental crust resulted in metamorphism and formation of granite-gneissic domes, 370 Ma old.

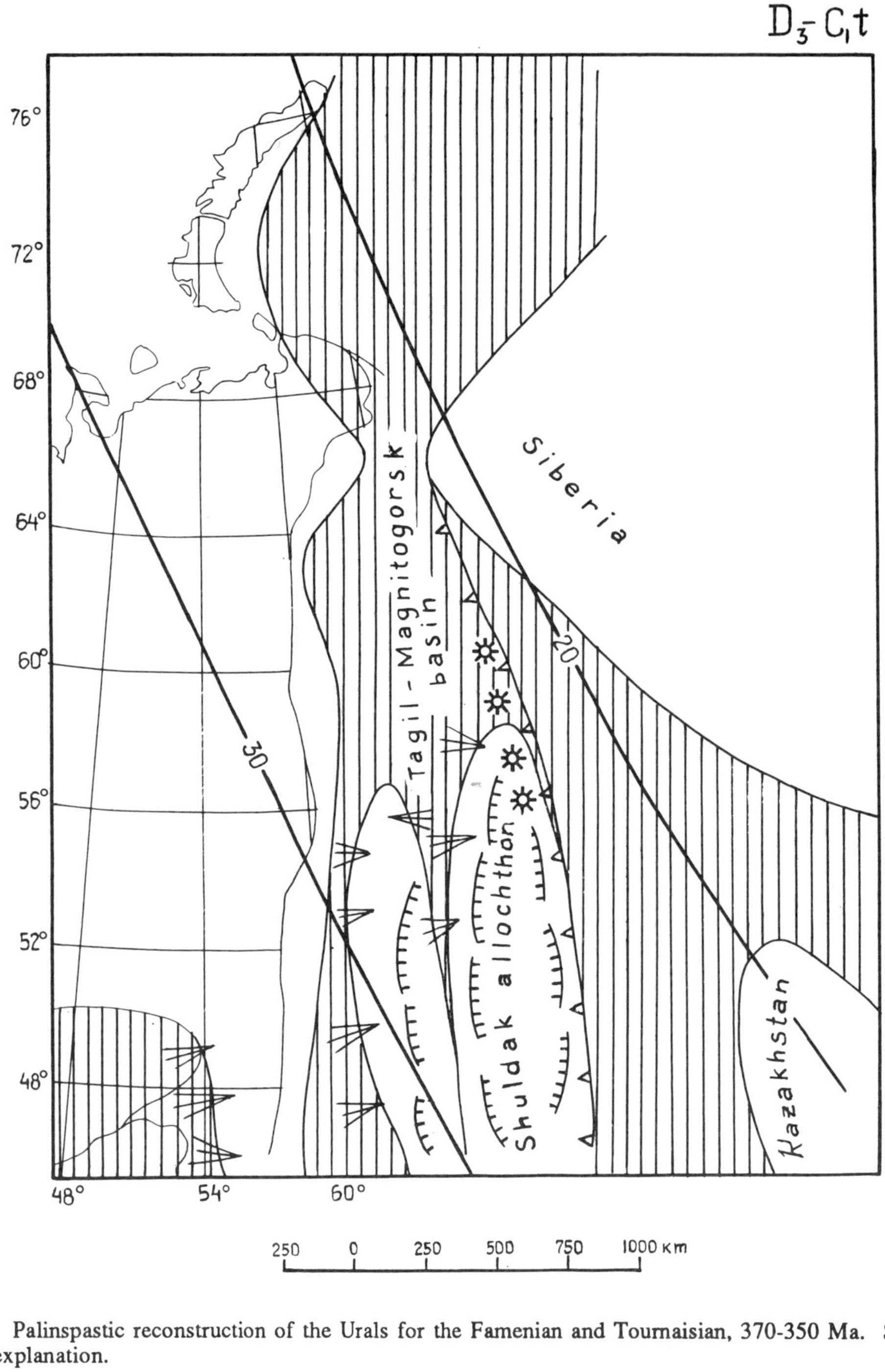

Fig. 45. Palinspastic reconstruction of the Urals for the Famenian and Tournaisian, 370-350 Ma. See Figure 35 for explanation.

Late Devonian - Early Tournaisian, 370-350 Ma (Figure 45)

Throughout the Late Devonian, the Mugodzhar microcontinent's northern (western, in past coordinates) margin approached the Magnitogorsk island arc (Figure 45). This process evidently culminated in the Famennian, when the island arc volcanism ceased almost completely, i.e., the subduction zone was blocked and the arc itself began to rise because a lighter sialic basement was being subducted beneath it. Uplifting and formation of rough topography are implied by the thick clastics of the Zilair suite of Famennian-Early Tournaisian age.

Tournaisian - Early Visean, 350-340 Ma

Abundant felsic volcanics and subalkaline lavas typical for this time interval seem to show the influence of the subducted sialic microcontinental basement. In the middle of the Visean, the subduction zone controlling the Magnitogorsk arc was completely blocked by the subducted microcontinent, allowing the development of granite-gneissic domes under the arc.

Late Visean - Middle Carboniferous, 340-310 Ma (Figure 46)

The northern and southern Urals have different Carboniferous histories. In the southern Urals, the events were connected with the East European/Kazakhstan collision, and in the northern Urals, with the East European/Siberia collison.

In the southern Urals at that time, the oceanic floor was absent between the Magnitogorsk arc and the East European con-

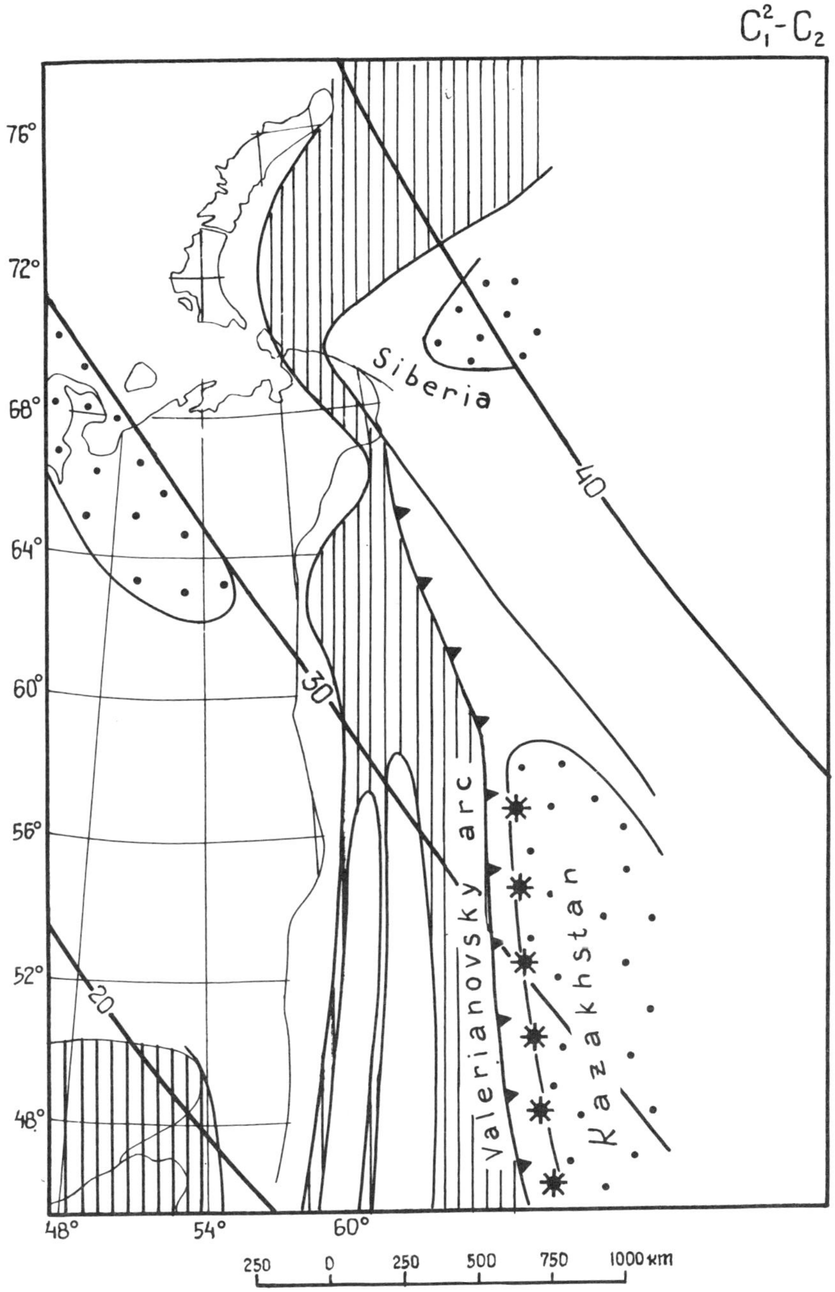

Fig. 46. Palinspastic reconstruction of the Urals for the Early and Middle Carboniferous, 340-310 Ma. See Figure 35 for explanation.

tinent, but an oceanic space nearly 2000 km wide was still left between the arc and the Kazakhstan continent. When subduction underneath the Magnitogorsk arc was blocked, a new zone dipping beneath the Kazakhstan continent originated. During the Visean and Middle Carboniferous, Kazakhstan converged with the East European continental margin. By the end of the Middle Carboniferous the oceanic crust was completely consumed and a direct continental collision began. The Mugodzhar microcontinent with the extinct Magnitogorsk arc above it approached the subduction zone.

In the northern Urals (western, in past coordinates), according to kinematic and paleomagnetic data the distance between East Europe and Siberia was no more than 1000 km, but the oceanic basin endured up to the end of the Carboniferous and beginning of the Permian, as can be inferred from continuation of the deep-water Lemva sequence up to the Early Permian.

Middle Carboniferous - Early Permian, 310-270 Ma (Figure 47)

At this time, the first stage of continental collision was accompanied by folding in the internal zones of the Southern and Middle Urals. This time was also characterized by an intensive growth of granite-gneissic domes, large uplift of the whole territory, and formation of thrust structure analogous to that of recent Himalayas.

Late Permian - Beginning of Early Triassic, 270-220 Ma

For 50 million years the Siberian and Kazakhstan continents were converging with East Europe in the second stage of the continental collision. As Siberia approached East Europe, the collision spread to the Northern Urals and Novaya Zemlya. At that time, a thrust-foldbelt like that of the recent Himalayas came into existence in the Urals. As in the Himalyas, a foredeep was formed in front of the Urals where the products of erosion of the mountains were stored and where the nappes were displaced. The thrusts and nappes of the externides of the western Urals formed during the collision, and individual sheets moved from the Central Uralian and Bashkir uplifts onto clastic sequences of the foredeep and underlying sediments of the former passive continental margin. Longitudinal dextral strike-slip faults appeared in the latest stage of continental collision.

Thus, the Uralian foldbelt appeared in place of the former Uralian ocean whose ocean floor was consumed. The Uralian paleo-ocean had developed at the site of the rifting of the East European continental margin. The background history and the evolution of the Urals from the Vendian to the beginning of the Triassic spanned 430 Ma. The ocean widened during the first 250 Ma, then shortened and closed during the next 90 years. The last 90 years of the Paleozoic was the time of continental collision within the Urals.

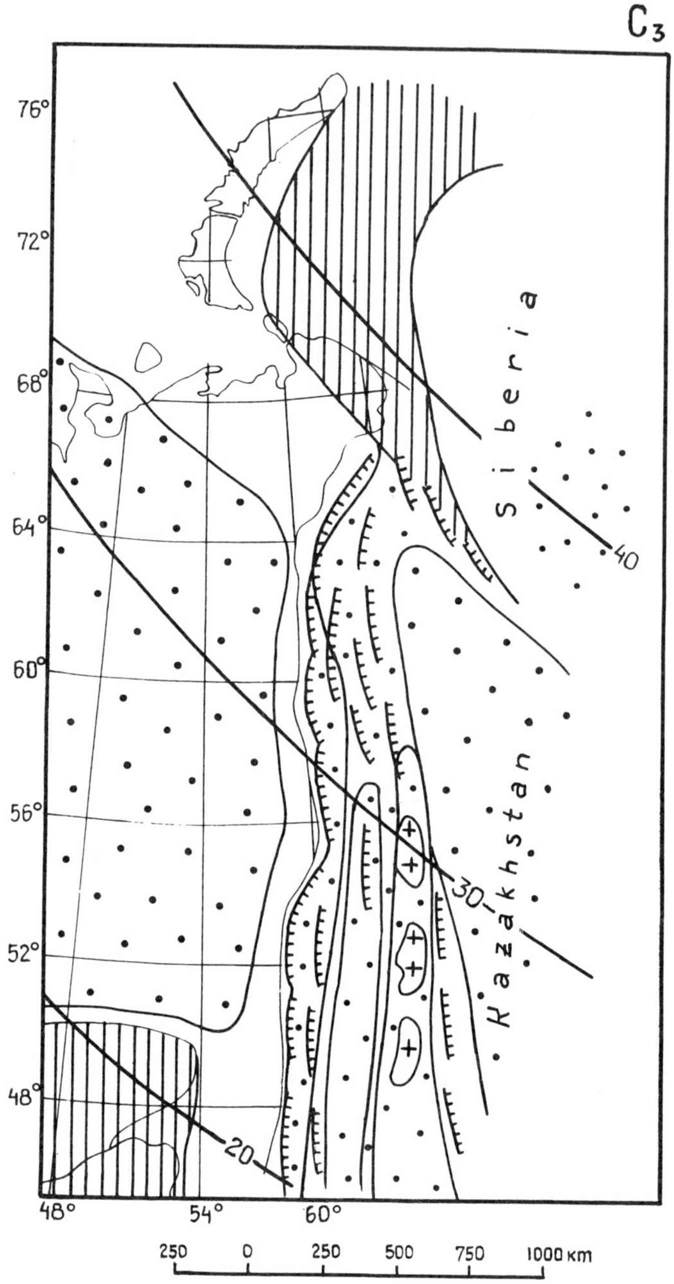

Fig. 47. Palinspastic reconstruction of the Urals for the Late Carboniferous, 310-270 Ma. See Figure 35 for explanation.

Chapter V

CENTRAL ASIAN FOLDBELT, WESTERN PART: CENTRAL KAZAKHSTAN AND TIEN SHAN

BRIEF OVERVIEW

The Central Asian foldbelt as a whole was formed by the progressive convergence and eventual collision of the Siberian continent with the North Chinese, Tarim, Karakum, Tadzhik, and Kazakhstan-North Tien Shan ancient massifs. The foldbelt was completed by the end of the Paleozoic.

Within the terrirory of the USSR, the Central Asian belt has been conventionally subdivided into western and eastern parts, separated by the late Paleozoic Irtysh-Zaisan zone. The western part includes Kazakhstan and the Tien Shan, whereas the eastern part includes the Altay-Sayan area together with North Mongolia. In the discussion below, these two parts of the Central Asian belt are described separately.

Precambrian basement outcrops are known in the western part of Central Kazakhstan (the Kokchetav, Ulyutau, Chu, Kendyktas, Atasu-Mointy massifs) and in the Northern and Middle Tien Shan (the Makbal block and Issykkul massif) (Figure 48). These ancient blocks are regarded as parts of a single Kazakhstan-North Tien Shan Precambrian massif that was reworked by Early Paleozoic deformation. At present, outcrops of the Precambrian massifs form a U-shaped pattern convex to the northwest. The Early Paleozoic ("Caledonian") zones of Kazakhstan together with the Devonian and Late Paleozoic volcanic belts also conform to this shape, running in parallel. The Late Paleozoic ("Hercynian") structures of the Junggar-Balkhash region are located in the center of the U-shaped belt; Precambrian rocks are not found inside the U-shape itself, which instead is typified by late-Early Paleozoic ophiolites and Early Paleozoic island-arc assemblages. The assemblages in the U-shaped region provide good evidence that the Early Paleozoic structures in Kazakhstan formed by accretion of island arcs, and these eventually collided with the Kazakhstan-North Tien Shan massif. Externally, the Precambrian U-shaped zone is rimmed by late Paleozoic zones of the south Tien Shan (and the southern Urals). These late Paleozoic structures are separated from the Kazakhstan Early Paleozoic structures by the Late Paleozoic Beltau-Kurama volcano-plutonic belt, which extends northward to the Valerianovsky belt of the Urals. The Gissar zone of Late Paleozoic ophiolites and island arc complexes separates the Tien Shan from the Pamirs and the Tadzhik depression.

The interpretation of the tectonics of Kazakhstan and the Tien Shan is based on a number of studies including Bogdanov [1965], Antonjuk [1974], Bespalov [1980], Samygin [1974], Markova [1982], Avdeev [1984], Karyaev [1984], Sabdjushev and Usmanov [1971], Shults [1972], Porshnyakov [1973], Burtman [1973, 1976], Khristov and Mikolaichuk [1983], and Makarychev [1978].

LATEST PALEOZOIC-EARLIEST TRIASSIC REDEFORMATION

Both Silurian and Late Paleozoic deformed rock series underwent later deformation immediately following the main phase of Late Paleozoic orogeny, in the Middle Carboniferous and extending into the Triassic. The resultant structures are strike-slip faults, strike-slip shear zones with thrust motion, and a zone of block-faults in the Sarysu-Tengiz uplift.

Strike-slip faults include the Talasso-Fergana, Zhalair-Naiman, Central Kazakhstan, Junggar, and Central Chingiz faults, which are all dextral. The SW flank of the Talasso-Fergana fault [Burtman, 1976] is occupied by the Fergana horizontal flexure, the so-called Fergana sigmoid (Figure 49). All rock series here (up to the Lower Permian) are folded in the horizontal plane (with vertical fold axes) due to N-S compression. Burtman and Gurariy [1973] showed from paleomagnetic investigations of the Fergana sigmoid that it formed after the Early Permian.

Strike-slip shear zones with thrust motion include the Spassky, Uspensky, and Aksoran-Akzhal zones, all with approximately similar structures; these are typified by the Spassky zone (Figure 50). This relatively straight fault, which is a sinistral strike-slip fault, extends east-west within the center of the shear zone and is paralleled on both sides by thrusts.

The block faults of the Sarysu-Tengiz uplift [Tikhomirov, 1975] are concentrated in a zone 150 km wide, which is cut by a system of closely spaced subparallel reverse faults trending approximately east-west into uplifted and subsided blocks, in which the Devonian and Lower Carboniferous sequences are folded into minor disharmonic folds. The compression in this fault zone developed in response to compression of the basement. Estimates, based on unfolding the sequence, show that shortening in the eastern Sarysu-Tengiz uplift was 150-200 km.

The North Balkhash sigmoid in the North Balkhash anticlinorium (Figure 51) reflects the latest Paleozoic transverse compression. The ophiolitic complex of the Itmurunda suite

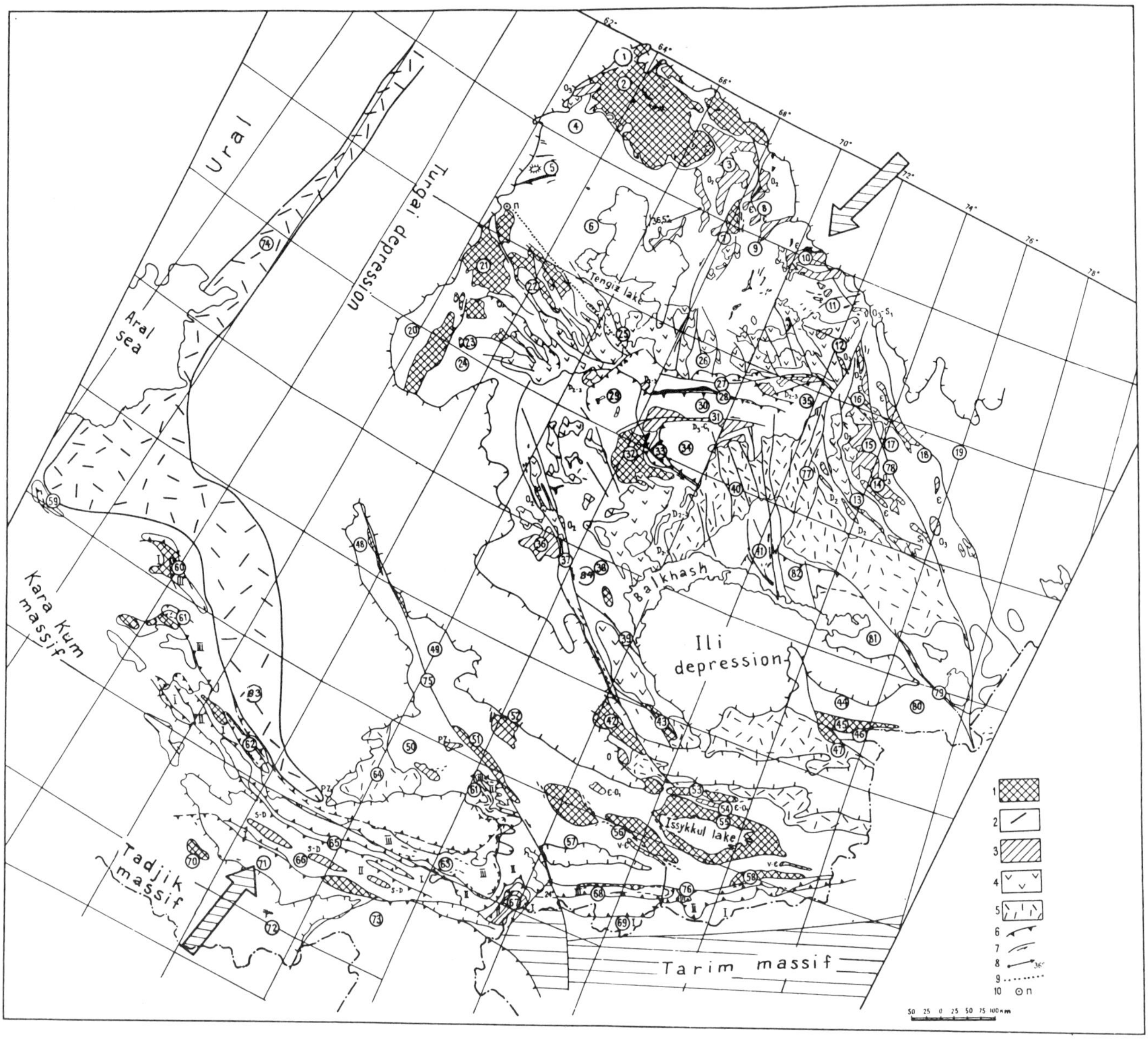

Fig. 48. Tectonics of Central Kazakhstan and Tien Shan.

1, Precambrian massifs; 2, ophiolite; 3, island-arc complex; 4, Devonian continental-margin volcanic belt; 5, Upper Paleozoic continental-margin volcanic belt; 6, thrust; 7, strike-slip fault; 8, Late Paleozoic paleomagnetic vectors (after Khramov [1986]), figure = paleolatitude; 9, approximate boundary between North and South Kazakhstan blocks; (this boundary is the location of data used in the Late Paleozoic reconstruction shown as Figure 53); 10, position of pole of North/South Kazakhstan block rotation. Thin toothed-line marks the limit of the Meso-Cenozoic sedimentary cover. I, II, and II, main nappe units of Tien Shan. Large arrows on the margin of the figure indicate orientation of compression during Late Paleozoic continental collision which joined Europe and Asia.

Numbered are: 1, Mariev synclinorium; 2, Kokchetav massif; 3, Stepnyak synclinorium; 4, Kalmakkol synclinorium; 5, Zharkainagach synclinorium; 6, Tengiz depression; 7, Ishkeolmes anticlinorium; 8, Seleta synclinorium; 9, Erementau anticlinorium; 10, Bozshakol anticlinorium; 11, Maikain-Kyzyltas anticlinorium; 12, Bayanaul zone; 13, Akbastau zone; 14, Akchatau zone; 15, Abrala zone; 16, Central Chingiz zone; 17, Chunai zone; 18, Arkalyk zone; 19, Irtysh-Zaisan zone; 20, Baikonur synclinorium; 21, Ulutay massif; 22, Sarysu-Tengiz uplift; 23, Eskula dome; 24, Dzhezkazgan depression; 25, Marginal Devonian volcanic belt; 26, Karaganda basin; 27, Spassky shear zone and Nura synclinorium; 28, Tekturmas zone; 29, Atasu synclinorium;

30, Sarysu synclinorium; 31, Uspensky shear zone; 32, Atasu-Mointy massif; 33, Aksoran-Akzhal shear zone; 34, Zhaman-Sarysu anticlinorium; 35, Karasor synclinorium; 36, Chu massif; 37, Zhalair-Naiman zone; 38, West Balkhash synclinorium; 39, Sarytum zone; 40, North Balkhash volcanic belt; 41, North Balkhash anticlinorium; 42, Kendyktas zone (Precambrian); 43, Anrakhai zone; 44, North Junggar synclinorium; 45, Central Junggar anticlinorium; 46, Borotala synclinorium; 47, South Junggar synclinorium; 48, Great Karatau zone; 49, Lesser Karatau zone; 50, Chatkalo-Kurama zone; 51, Talas zone; 52, Makbal block; 53, Zaili zone; 54, Kemin zone; 55, Issykkul massif; 56, Terskei zone; 57, Naryn zone; 58, Saryzhas massif; 59, Sultan Uizdag zone; 60, Bukantau zone; 61, Tamdytau zone; 62, Nuratau zone; 63, South Fergana ophiolitic zone; 64, Mailisu zone; 65, Alai zone; 66, Zeravshan-Gissar zone; 67, Terekdavan synform; 68, Atbashin zone; 69, Kokshaal zone; 70, Baisun massif; 71, Gissar zone; 72, Tadzhik depression; 73, Pamirs; 74, Valerianovsky volcanic belt; 75, Talasso-Fergana fault; 76, Zhalair-Naiman fault; 77, Central Kazakhstan fault; 78, Central Chingiz fault; 79, Junggar fault; 80, Tastau synclinorium; 81, Arganat synclinorium; 82, Sayak mulde; 83, Beltau-Kurama fault; 84, Bet-Pak-Dala.

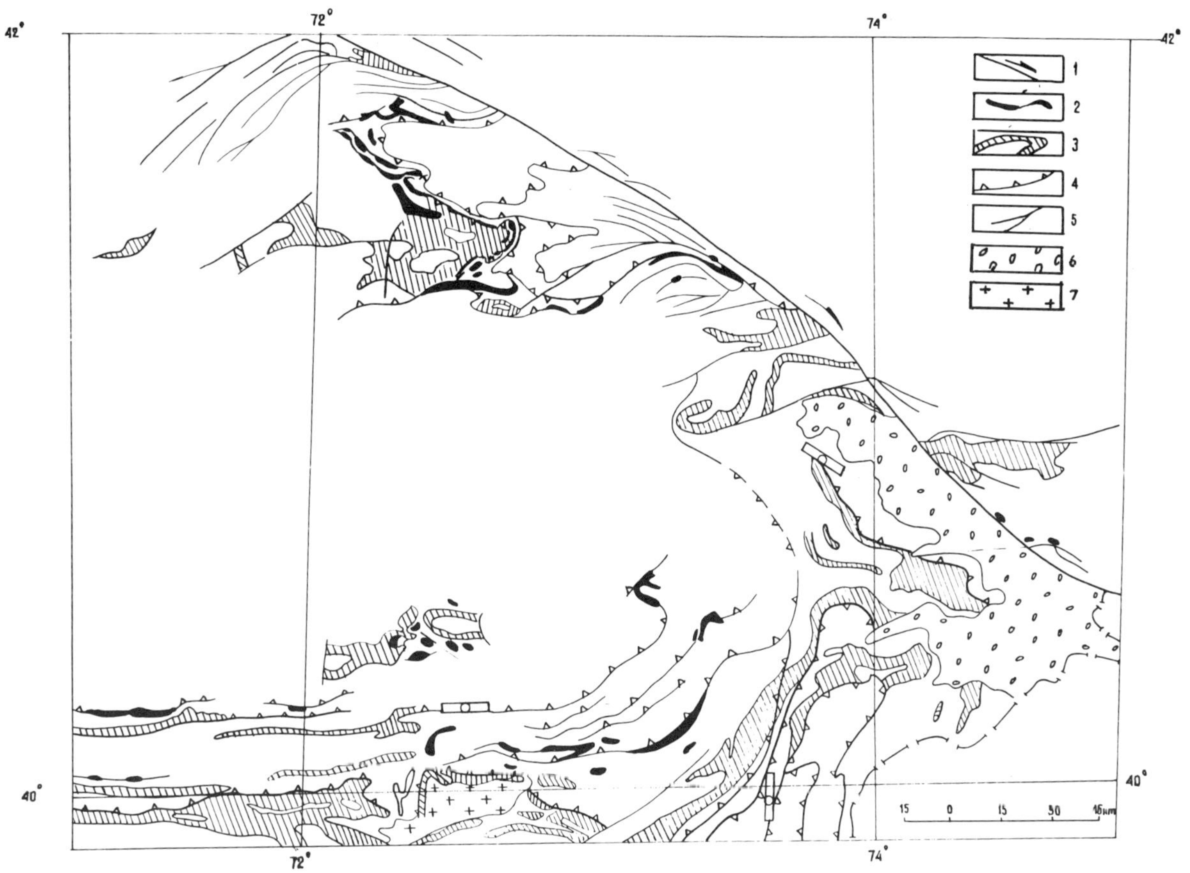

Fig. 49. Map showing structure of the Fergana sigmoid. Compiled from the Geological Map of Kazakhstan [Shlygin, 1981], and partly from Burtman [1976].
1, Talasso-Fergana strike-slip fault; 2, ophiolite; 3, Carboniferous marker beds; 4, thrust; 5, secondary faults branching from the Talasso-Fergana fault; 6, Mesozoic rocks; 7, Permian granite.

(Vendian-Lower Cambrian?) underwent horizontal folding with both an amplitude and a general width of 100 km. In this limited area, shortening is no less than 120 km.

All these latest Paleozoic structures are interrelated in space. The spatial distribution and geometry of the strike-slip faults, the shear zones, and the Sarysu-Tengiz deformational zone are shown on Figure 48. Karyaev [1984] concluded that all these structures formed under NNE-SSW pressure, with vergence from the Siberian continent toward Kazakhstan as the Paleozoic oceans closed and the north Eurasia landmass formed. This NNE-SSW directed major stress induced dextral displacement along northwest trending faults and a sinistral

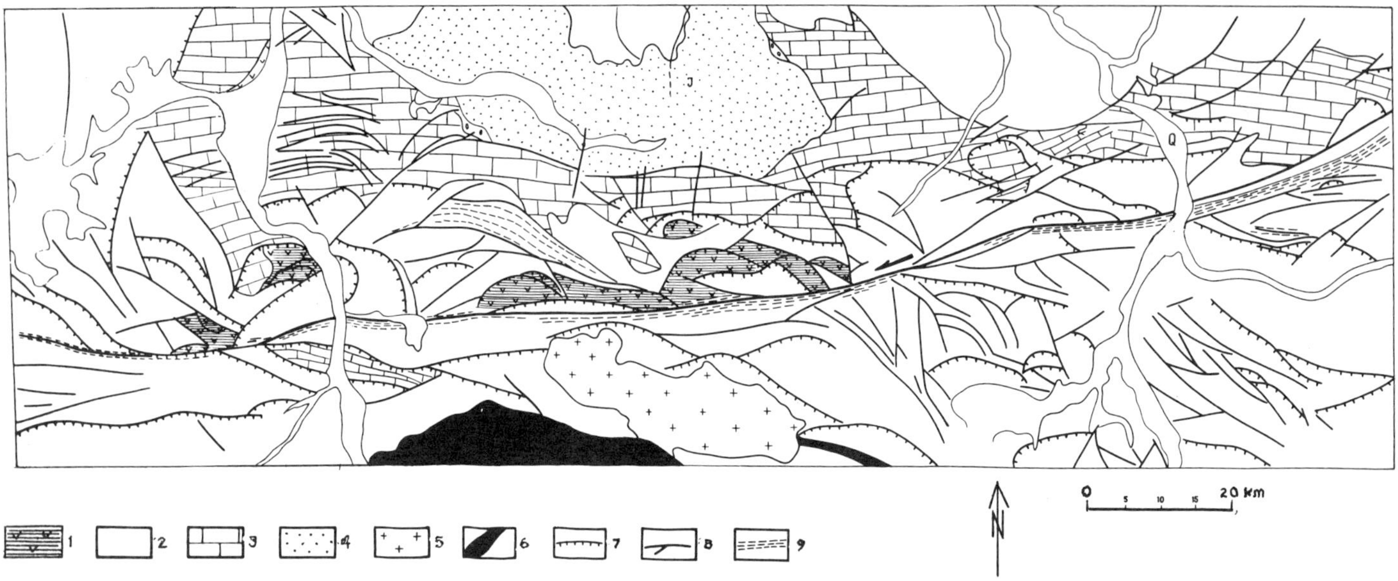

Fig. 50. Map showing structure of the Spassky shear zone, compiled from the Geological Map of Kazakhstan [Shlygin, 1981].
1, Ordovician rocks; 2, Silurian-Devonian rocks; 3, Carboniferous rocks; 4, Mesozoic rocks; 5, Permian granite; 6, ophiolite of the Tekturmas zone; 7, thrust; 8, other faults; 9, shear zone.

component along sublatitudinal faults. Evidently some of the U-shaped structure of Kazakhstan, consisting of several arcs enclosed one in the other, is secondary, resulting from the latest Paleozoic deformations. Therefore, in order to analyse the development of this region in the Paleozoic, it is necessary to rotate the parts of this U-shaped structure to their initial position prior to the final collision of the continents.

The Late Paleozoic reconstruction, corresponding to the late Middle to early Late Carboniferous, is shown in Figure 52, which illustrates the position of all these structures. On the whole, the Ulutau block, together with the Bet-Pak-Dala (e.g., the Chu massif) and the lesser Karatau, is rotated 30° clockwise relative to the Kokchetav block about a point near the northern termination of the Ulutau massif; this caused divergence in the eastern Sarysu-Tengiz uplift. The Atasu-Mointy massif was moved up onto the resultant gap; effects of its Late Paleozoic lateral motion are evident in thrusts of the Aksoran-Akzhal shear zone.

Reconnaissance paleomagnetic data on Late Paleozoic sequences support the 30° rotation of the Ulutau-North Tien Shan block relative to the Kokchetav block (available paleomagnetic vectors are plotted on Figures 48 and 52).

Figure 52 shows that the present U-shaped structure defined by the Precambrian massifs probably did not exist prior to Permian deformation. In the Paleozoic, the Precambrian blocks made up a large Kazakhstan-North Tien Shan massif of nearly meridional strike with respect to the Kokchetav block. The Devonian and Late Paleozoic volcano-plutonic belts were more nearly linear than at present. Figure 52 represents a model that invokes the existence of two Late Paleozoic volcanic belts: North Balkhash and South Junggar. Each of them had an independent Benioff zone.

EARLY PALEOZOIC ("CALEDONIAN") STRUCTURES

The Early Paleozoic structures were commonly called "Caledonian" by fixist geologist, in allusion to the Silurian orogeny in northwestern Europe. However, there is no demonstrable physical connection between the orogenic events of that age in Europe and in Asia.

The Early Paleozoic complexes of Central Kazakhstan have the following spatial distribution (Figure 53). In the west, there is a large (1000 x 700 km) Kazakhstan-North Tien Shan massif and two smaller massifs (250-300 km across): the Kokchetav and the Atasu-Mointy. They include granitoid batholiths, e.g., the Zerendin pluton in the Kokchetav massif. The area east of these ancient massifs, including the North Kazakhstan, Chingiz, Peri-Balkhash and Chu-Iliisky mountains, is occupied by island-arc complexes and ophiolites. Near the ancient massifs, in the Ishkeolmis and Erementau areas, the island-arc complexes have a Middle Ordovician age. In the eastern Bozshakol anticlinorium and in the Chingiz-Tarbagatay area the age ranges from Early Cambrian to Early Silurian. In the extreme south, in the North Tien Shan, several small Precambrian blocks are separated by ophiolite sutures, producing a mosaic structure.

Ancient Massifs

Available data on the ancient massifs prove that they were fragments of former continents containing gneiss domes and greenstone belts (for example, the Efimovsky suite of the Kokchetav massif, and the Karsakpay series of Ulutau). The majority of these sequences range in age from 1800 to 1900 Ma [Avdeev, 1984].

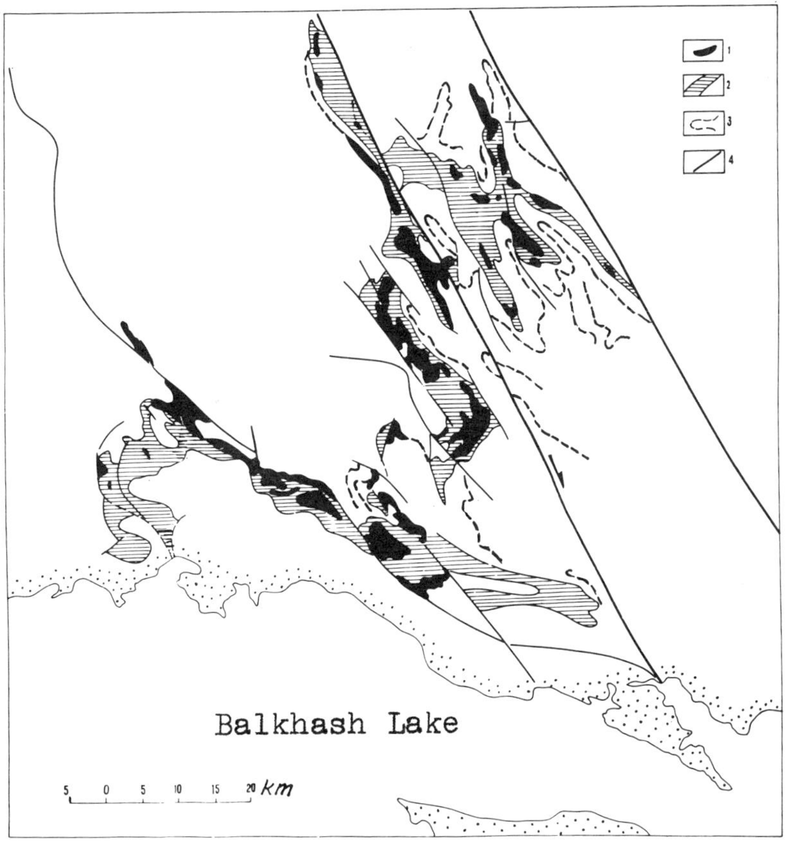

Fig. 51. Map showing structure of the North Balkhash sigmoid. Compiled from the Geological Map of Kazakhstan [Shlygin, 1981].
1, Ultramafics; 2, volcanics of Itmurunda suite (Late Precambrian-Early Paleozoic); 3, structural trends; 4, fault.

A common feature of the whole continental basement of the Kazakhstan-North Tien Shan massif is the presence of widespread pure quartz sandstones transformed into quartzites. The age of this sequence ranges from 1100 to 900 Ma. This massif thus contrasts with the East European continent, where at that time aulacogens were being formed.

Fragmentation

A new stage in the development of the Kazakhstan-North Tien Shan massif--fragmentation--commenced in the Late Riphean, as shown by widespread development of rift-type complexes, in particular bimodal volcanics; an example is the 870-650 Ma Koksui series. Everywhere, except for the Kokchetav massif, the lavas are overlain by a sedimentary sequence devoid of volcanics. These sediments are diverse in composition, varying from mostly clastic to carbonate rocks; their common characteristics include the presence of a coarse-clastic facies at the beginning of the Vendian, the presence of tillites, appearance of phosphate- and vanadium-bearing shales in the Lower Cambrian, including the formation of economically significant phosphate deposits in the Karatau, and the abundance of siliceous shales and cherts. These sedimentary series possess all the properties common to sequences on passive continental margins forming after continental breakup [Markova, 1982].

Oceanic Complexes

Ophiolites occur widely in the Early Paleozoic zones, their age ranging from the Late Riphean to Cambrian (and maybe Ordovician). Moreover, ophiolites of approximately the same age are exposed in the late Paleozoic orogenic belt of Kazkhstan within the Tekturmas and North Balkhash anticlinoria.

In Kazakhstan no complete section of an ophiolite complex is known. Ophiolite associations are usually dismembered, being incorporated in serpentinite melanges. Ophiolites crop out either at the base of tectonic nappes or in so-called suture zones. Bands of melange are found in different regions of Kazakhstan [Antonjuk et al., 1979; Avdeev, 1984; Bespalov, 1980].

As a rule, the ophiolites comprise serpentinized ultramafic rocks, gabbro, and pillow lavas and are evidently of different

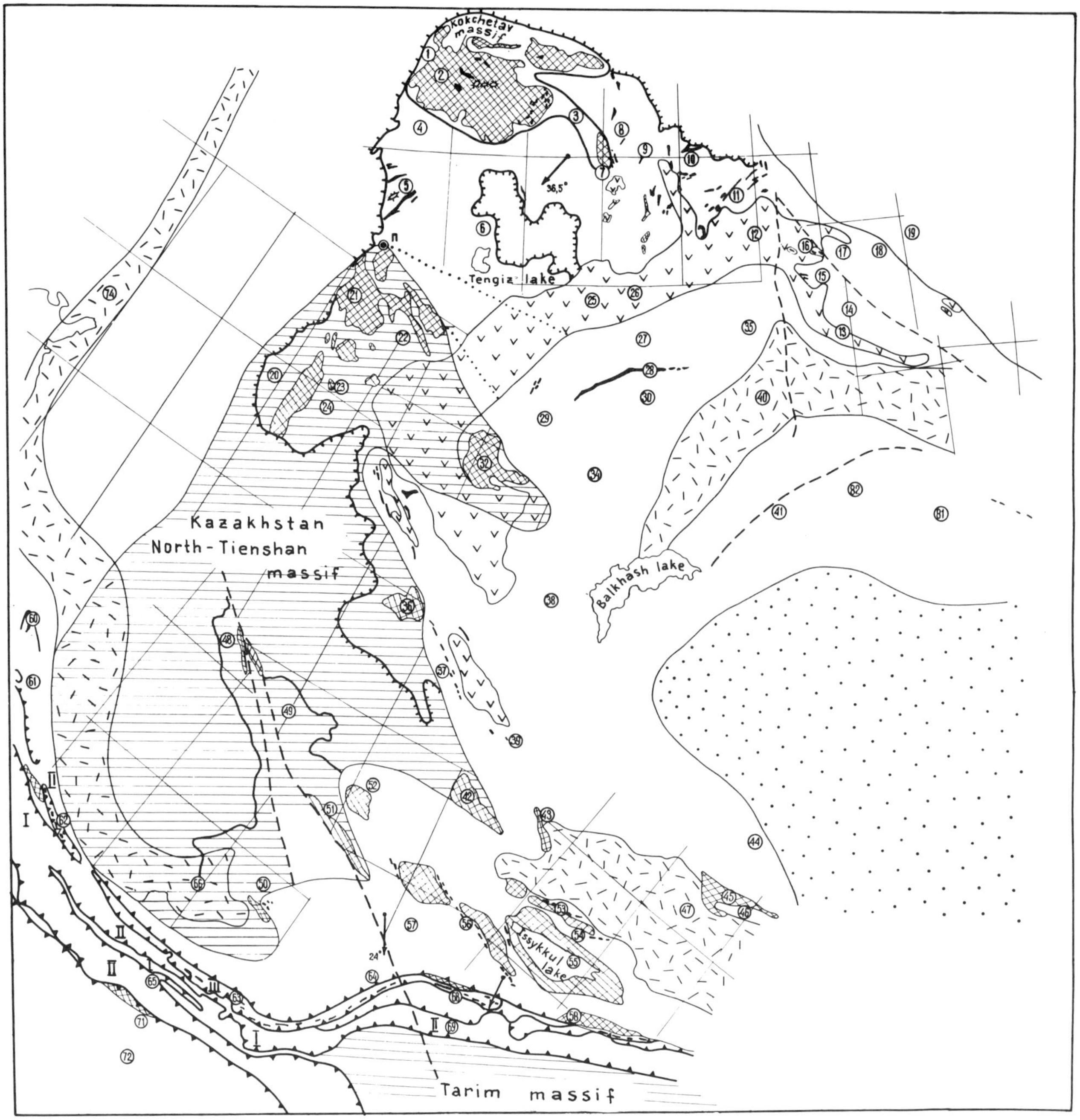

Fig. 52. Late Paleozoic reconstruction of Central Kazakhstan and Tien Shan. The Kokchetav massif is arbitrarily kept fixed. Late Paleozoic displacements along transcurrent strike-slip faults were taken into account. The South Kazakhstan (or Kazakhstan-North Tien Shan) block was rotated 30° clockwise with respect to the Kokchetav massif around a pole at 50.5° N, 62.4° E. The East Tien Shan block was rotated 25° clockwise with respect to the West Tien Shan. Principal Pre-"Caledonian" massifs are horizontally hatched. The inferred Late Paleozoic Junggar-Balkahashian oceanic basin is dotted. Other symbols as in Figure 48.

ages. In the Erementau region, the isotopic ages of lavas were determined to be 1270, 780, and 680 Ma, i.e., Upper Riphean [Antonjuk, 1974]. In Tekturmas, sediments associated with ophiolites contain Early Ordovician radiolaria and Cambrian brachiopods. In the Chingiz region, Upper Riphean diabases are part of an ophiolite section. In the North Peri-Balkhash area, dismembered ophiolite horizons with serpentinized ultramafic rocks are exposed in the core of the Balkhash anticli-

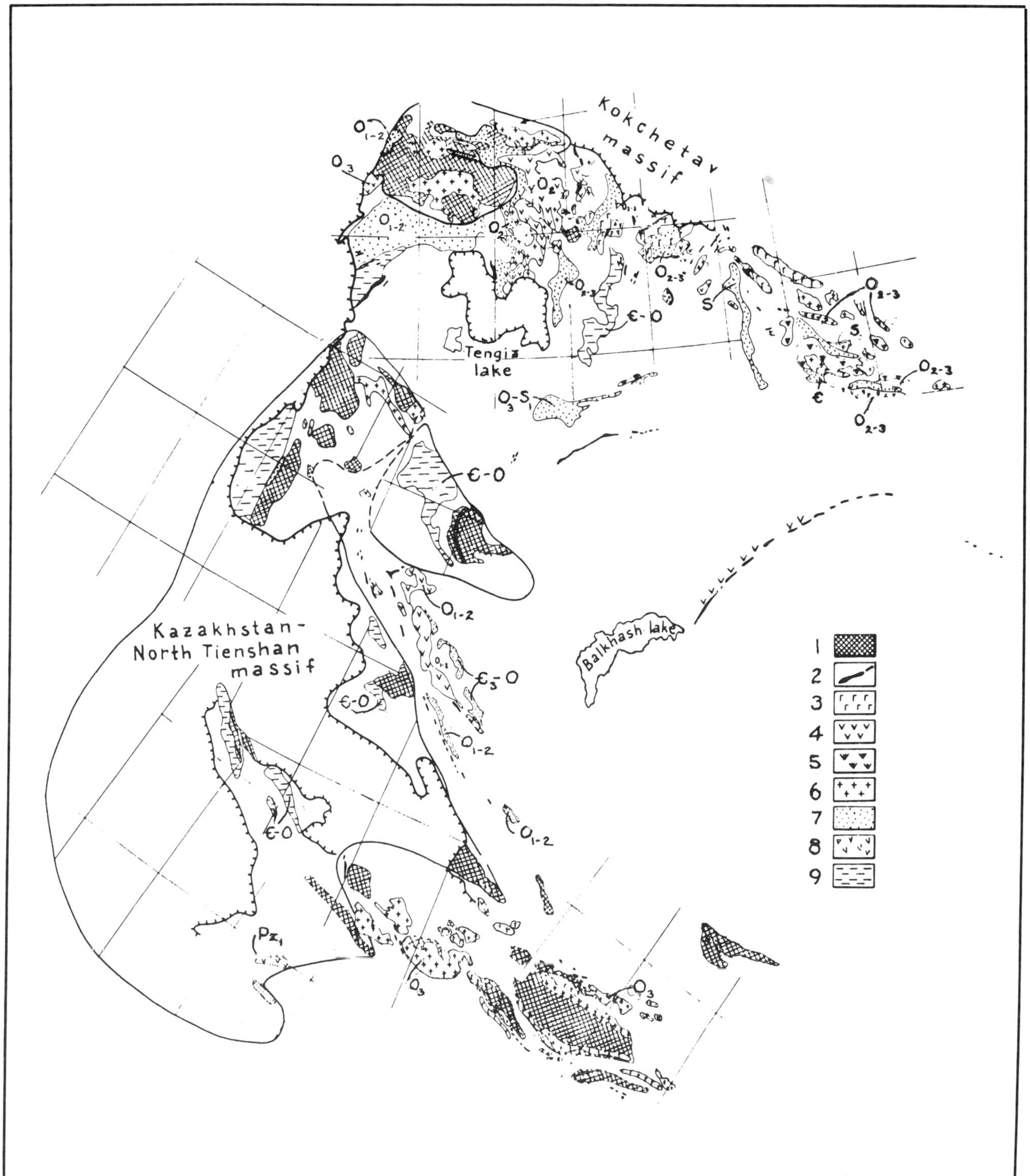

Fig. 53. Distribution of Early Paleozoic ("Caledonian") complexes in Central Kazakhstan and Tien Shan. The base map is the Late Paleozoic reconstruction of Figure 52.
1, Ancient massif; 2, ophiolite; 3, Cambrian island arc volcanics; 4, Ordovician island-arc volcanics; 5, Silurian island-arc volcanics; 6, granite; 7, flysch and olistostrome; 8, bathyal complexes.

norium surrounded by Upper Ordovician and Silurian rocks. They mostly occur in zones of serpentinite melange with large blocks of greenstone-altered diabases (the Itmurunda suite), their isotopic ages ranging from 890 to 904 Ma. They are associated with cherts containing Lower Cambrian radiolaria. The geochemical data of Antonjuk [1974] suggest that spilite-diabase sequences of ophiolite sections are close in composition to oceanic basalts.

From geological data it is evident that the Late Riphean continental breakup occurred at 900-800 Ma and was followed by the separation of continents and formation of oceanic crust. The major portion of this crust was presumably formed at the end of the Riphean, in the Vendian, and in the beginning of the Cambrian, i.e., from 800 to 550 Ma. If the ocean formed continuously for not less than 250 Ma, even at slow spreading rates (for instance, 1 cm/yr), about 2000 km of a new oceanic crust could have been produced by the end of the Cambrian. That a deep oceanic floor existed is substantiated by deep-water cherts of the continental-rise type (Figure 54) that accumulated below the carbonate compensation depth [Markova, 1982]. This oceanic basin was part of the Late Precambrian-Early Paleozoic Paleo-Asiatic ocean [Zonenshain, 1973].

Volcanic Arc Complexes and Paleo-Subduction Zones

Early Paleozoic complexes of calc-alkaline volcanic and plutonic rocks are widespread in the Kazakhstan-Tien Shan region. They are mainly concentrated in three bands: (i) the Chingiz and Tarbagatay ranges; (ii) the Stepnyak - Bet-Pak-Dala (Chu) region; and (iii) the North Tien Shan. In addition, fields of calc-alkaline lavas are found in the Mariev synclinorium west of the Kokchetav massif, in the Spassky zone of Central Kazakhstan, and in the North Balkhash anticlinorium. The magmatic series of each area may be regarded as an indirect result of subduction, and they are related to several separate subduction zones responsible for forming either individual volcanic arcs or active continental margins.

Chingiz Arc

The Chingiz island arc assemblage (Figure 55) represents magmatism during the span from the Middle Cambrian to the Early Silurian. Samygin [in Markova, 1982] indicates that the magmatic complexes belong to a typical sodic calc-alkaline series. Samygin distinguished two stages of the Chingiz arc development: 1) Cambrian to the Late Ordovician, and 2) latest Ordovician-Silurian. The first stage includes magmatic rock series of three age levels: Middle Cambrian, Upper Cambrian-Lower Ordovician, and Middle Ordovician.

According to Samygin, the Chingiz arc in its early stage of development, from the Middle Cambrian to the Middle Ordovician, faced northeastward (in present coordinates), i.e., the forearc basin complexes are developed in the NE Chingiz and the back arc basin SW of the arc in SW central Chingiz. However, judging from widespread olistostromes in sequences of Middle Ordovician age SW of the axis of the arc, one might suggest that at that time the forearc basin was already southwest of the Chingiz arc and the back-arc basin was to the northeast. Samygin also inferred a change in the arc polarity, but only beginning from the Late Ordovician-Early Silurian. He established a distinct pattern of facies distribution: carbonate-shale-terrigenous sequences occur SW of the band of arc volcanics, and farther SW there are cherts accompanied by olistostromes [Rotarash et al., 1980], hence a deep-sea trench was located SW of the Chingiz arc.

Thus, the Chingiz island arc was formed on Late Precambrian-Early Cambrian oceanic crust. As Samygin [in Markova, 1982] pointed out, its polycyclic development lasted 100 m.y. An important feature of its history is the change of polarity. Changes of this kind are usually accompanied by reorganization of the subduction zone due to collision of the arc with a microcontinent or exotic terranes. The time of reorganization and jumping of the subduction zone from NE to SW is inferred to be middle-Late Ordovician.

Considering the duration of the island-arc to be 100 Ma and a minimal subduction rate to be 2 cm/yr, clearly about 2000 km of oceanic crust could have been consumed in the Early Paleozoic.

Stepnyak-Bet-Pak-Dala Arc

This arc is reconstructed from the distribution of island-arc complexes in two separate regions--Stepnyak and Bet-Pak-Dala. Most of the data are available from the Stepnyak region

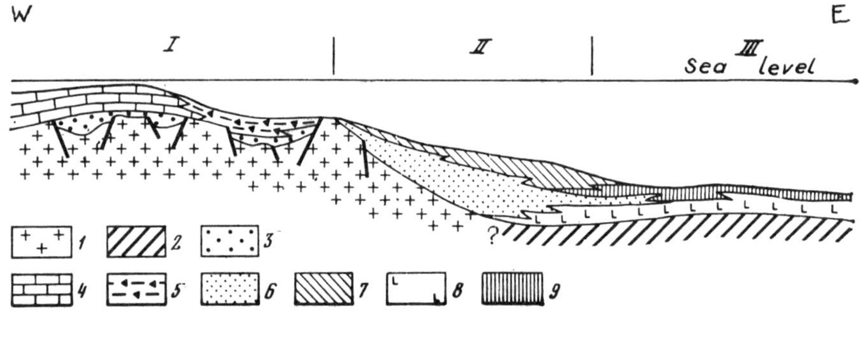

Fig. 54. Paleo-tectonic profile across the eastern margin of the Early Paleozoic ("Caledonian") Kazakhstan continent (after Markova [1982]).
1, Continental crust; 2, oceanic crust; 3, arkose; 4, shelf limestone; 5, chert and black shale; 6, terrigenous sediment with olistostrome; 7, siliceous shale and chert; 8, spilite and jasper; 9, jasper. Tectonic zones: I, Atasu-Mointy, II, Atasu, III, Tekturmas.

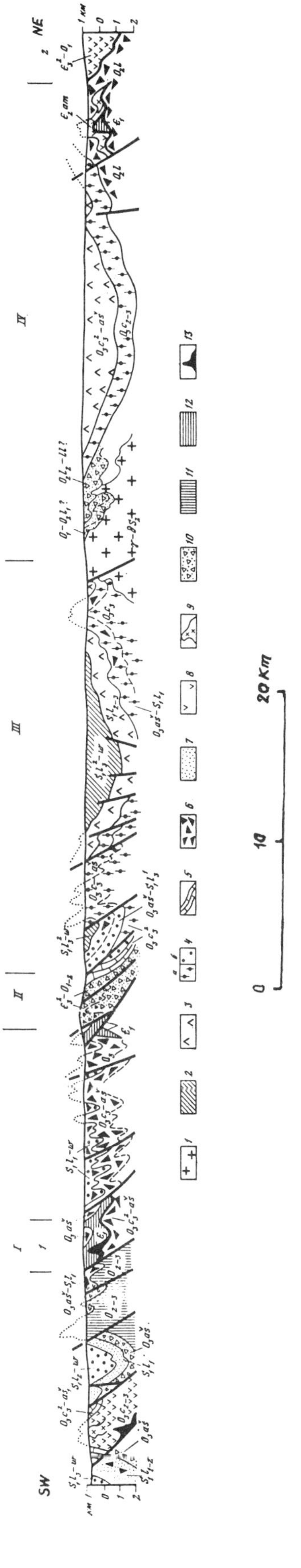

Fig. 55. Geological cross-section through the SW Chingiz Mountains (after Markova [1982]). 1, Granodiorite pluton; 2, basaltic andesite; 3, andesite; 4, shallow marine sediments (a - with tuffaceous material, b - without tuffaceous material); 5, reef limestones; 6, olistostrome; 7, flysch; 8, bimodal volcanics; 9, plagiogranite; 10, volcanics and chert; 11, spilite and diabase; 12, spilite and limestone; 13, ultramafics. Structural zones: I, Akbastau, II, Akchatau, III, Abrala, IV, Central Chingiz. Nappe units: 1, Balbybek; 2, Central Chingiz.

[Spiridonov, 1980], where there is a thick volcanic complex of Middle to Late Ordovician age. The lavas are island-arc andesites or leucobasalts. Spiridonov [1980] noted that the potassium content in the volcanics increases from east to west towards the Kokchetav massif, indicating that the arc faced eastward. The volcanic fields are spatially connected with large massifs of the Krykkuduk granodiorite-tonalitic complex (isotopic age 440-470 Ma). The Stepnyak-Bet-Pak-Dala arc was active for about 40 m.y., hence considerably less than the Chingiz arc, and the Stepnyak-Bet-Pak-Dala region contains the remains of only one paleoarc. Just after activity of the Stepnyak-Bet-Pak-Dala arc ceased, the Benioff zone changed polarity and jumped to the other side of the Kokchetav massif, where the Late Ordovician Mariev arc was formed.

North Tien Shan Arc

This arc is largely represented by the belt of granitoid batholiths confined to the boundary of the early Paleozoic and late Paleozoic complexes of the south Tien Shan. Along this zone are the great calc-alkaline granite-granodiorite batholiths, for example, the Susamyrsky batholith. The North Tien Shan belt of batholiths belongs to the Nevadan type of active continental margin and presumably reflects the intrusive of sialic masses near the edge of the Kazakhstan-Tien Shan continent as it overrode the Paleo-Asiatic ocean.

HISTORY OF THE EARLY PALEOZOIC STRUCTURES

The development of the Kazakhstan and North Tien Shan early Paleozoic orogenic complex is reconstructed in Figure 56. Positions of blocks on the Earth's sphere are given conventionally according to absolute reconstructions [Zonenshain et al., 1985]. For the Devonian, paleomagnetic data [Khramov, 1982] are also taken into account. At that time, Kazakhstan was situated in the northern tropical zone.

It is postulated that prior to the Late Riphean all Precambrian blocks: the Kokchetav, Ulutau, Ishkeolmes, Atasu-Mointy, Chu, North Tien Shan, and Central Junggar blocks, were united in a single Precambrian continent whose internal mosaic and general configuration are impossible to reconstruct.

Early Cambrian, 550 Ma (Figure 56a). The next stage is associated with a breakup of this continental block and the formation of oceanic basins. By the beginning of the Cambrian the ocean had evidently attained a considerable width, about 2000 km.

Late Cambrian, 500 Ma. The Chingiz island arc, situated west (past coordinates) of the Kazakhstan-North Tien Shan and Kokchetav Precambrian massifs, originated in the Cambrian (Figure 56b). Judging from the polarity of the island arc, the Benioff zone dipped to the east, toward the Kokchetav massif. The island arc interrupted part of the Paleo-Asiatic ocean floor, which was about 1000 km wide. The oceanic basin between the Chingiz arc and the Precambrian massif may be considered as a vast marginal sea--the Akdam basin. Its sedimentary fill is preserved as abyssal chert-siliceous shale and turbidite sequences.

Middle Ordovician, 460 Ma. The ensuing history reflects the consumption of the Akdam basin floor by subduction zones situated on both sides, i.e., under the Chingiz arc and under the newly formed Ordovician Stepnyak arc (Figure 56c). A principal reorganization of the geodynamic environment in

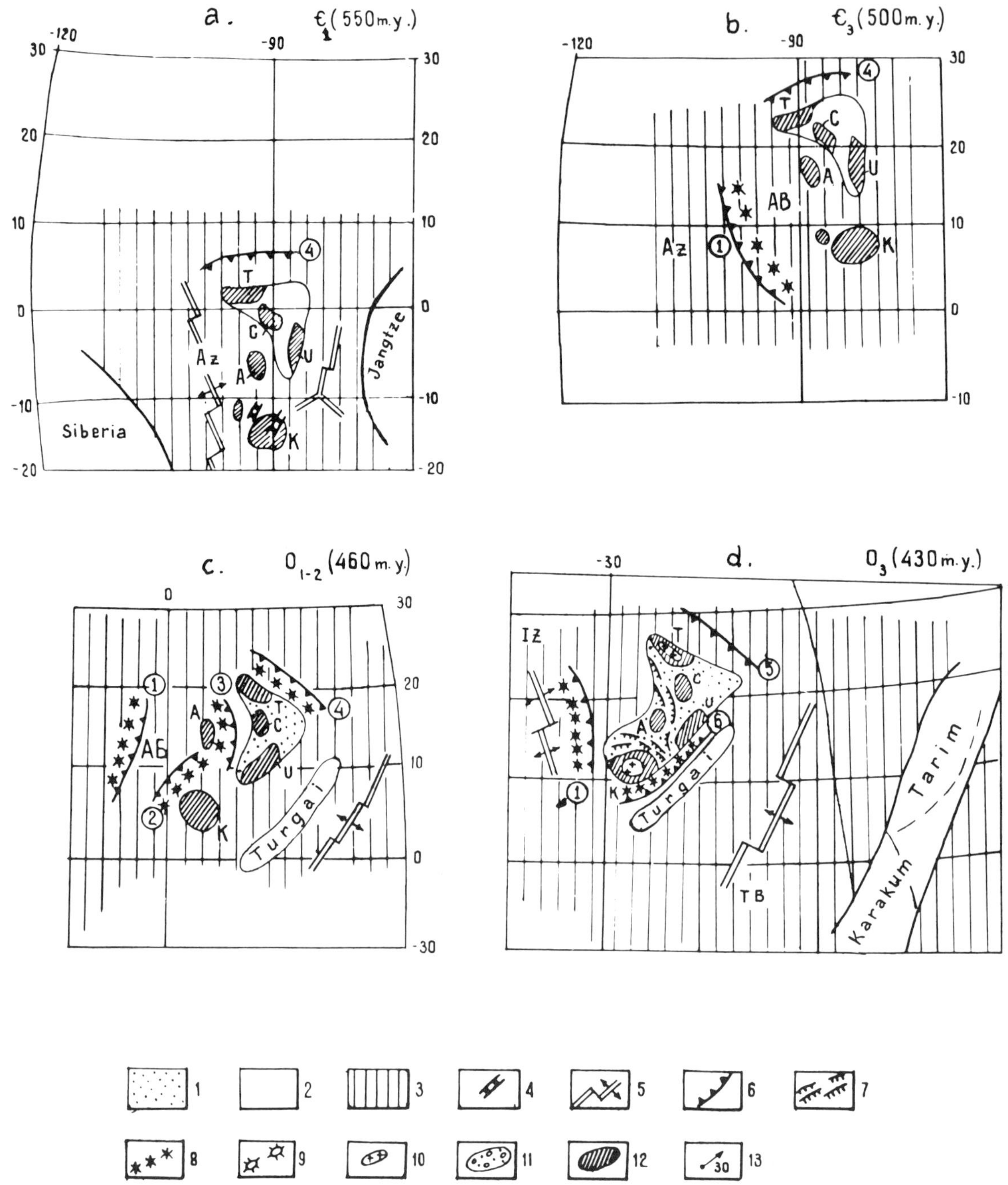

Fig. 56. Paleozoic palinspastic reconstructions for Central Kazakhstan and Tien Shan. Paleomagnetic data after Khramov [1982, 1986].

1, Dry land; 2, shallow sea; 3, oceanic floor; 4, continental rift; 5, spreading axis; 6, subduction zone; 7, collision zone; 8, subduction-related calc-alkaline volcanics; 9, hot-spot-related alkaline volcanics and plutonics; 10, granite and gneissic-granitic dome; 11, molasse; 12, Precambrian massifs; 13, paleomagnetic vector.

Precambrian massifs: A, Atasu-Mointy; B, Baisun; C, Chu; K, Kokchetav; T, North Tien Shan; U, Ulutau.

Oceanic basins: AB, Akdym; Az, Paleo-Asiatic; GB, Gissar; JB, Junggar-Balkashian; PT, Paleo-Tethys; TB, Turkestan.

Volcanic arcs: 1, Chingiz; 2, Stepnyak; 3, Bet-Pak-Dala; 4, North Tien Shan; 5, South Tien Shan; 6, Mariev; 7, Devonian marginal belt; 8, North Balkhash; 9, Balkhash; 10, Gissar; 11, Beltau-Kurama; 12, Valerianovsky; 13, Ili belt.

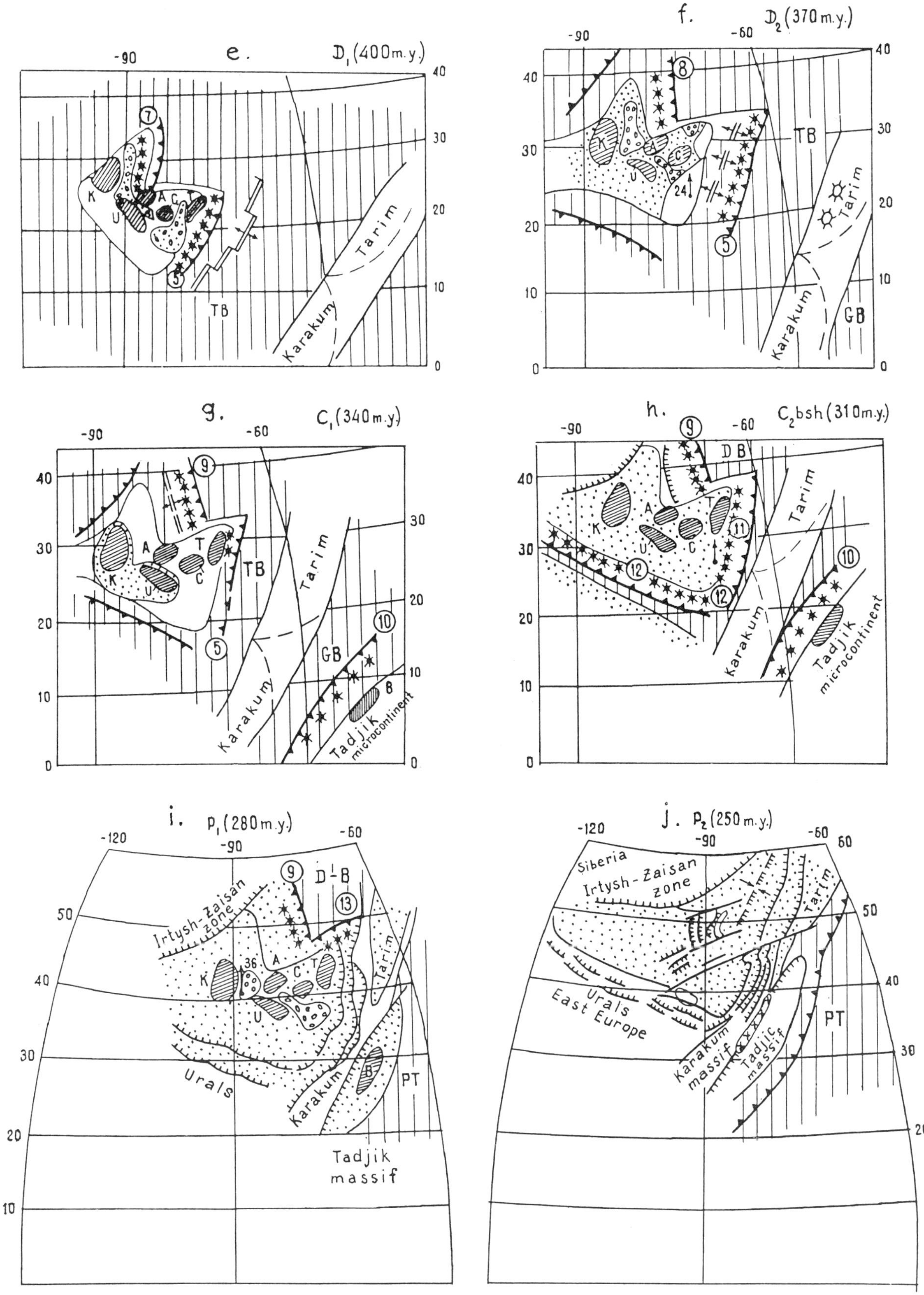
e.
D1 (400 m.y.)
TB
Karakum
Tarim
f.
D2 (370 m.y.)
TB
Tarim
Karakum
GB
g.
C1 (340 m.y.)
TB
Tarim
GB
Karakum
Tadjik microcontinent
h.
C2bsh (310 m.y.)
DB
Tarim
Karakum
Tadjik microcontinent
i.
P1 (280 m.y.)
D-B
Irtysh-Zaisan zone
Tarim
Urals
Karakum
PT
Tadjik massif
j.
P2 (250 m.y.)
Siberia
Irtysh-Zaisan zone
Tarim
Urals
East Europe
Karakum massif
Tadjic massif
PT

the Kazakhstan-Tien Shan region took place in the Middle Ordovician. It resulted in (i) the change of polarity of the Chingiz arc, and consequently the jumping of the subduction zone to the opposite side of the arc, and (ii) formation of the new Stepnyak-Beta-Pak-Dala arc. This reorganization implies general conditions of plate convergence, i.e., compression, following the extensional conditions that dominated prior to Ordovician time.

Early Silurian, 430 Ma. Folding, deformation, and granite emplacement started in the Late Ordovician (Figure 56d). These events may be explained by further convergence and initial collision of the Stepnyak and Chingiz arcs. The extensive North Tien Shan belt of batholiths, formed at this time NE (past coordinates) of the Kazakhstan-Tien Shan continent, was associated with overriding of the continent onto the Paleo-Asiatic ocean. This time is also noted for a wide spread development of olistostromes, zones of serpentinite melange, and tectonic nappes. Collision of arcs ceased by the end of the Silurian, and the subduction zone jumped to a new position on the margin of the newly shaped continental mass of Kazakhstan.

Thus, the early Paleozoic orogenic massif of Central Kazakhstan and North Tien Shan is an accretional structural complex created by collision of Precambrian massifs with the Chingiz and Stepnyak-Beta-Pak-Dala island arcs. Kazkhstan did not exist as a coherent entity until this time.

LATE PALEOZOIC ("HERCYNIAN") STRUCTURES

The term "Hercynian" has been loosely applied to Late Paleozoic structures in the Central Asian Foldbelt in allusion to Late Paleozoic structures and events in central and western Europe. Although there may be a degree of temporal equivalence, there is no demonstrable physical connection. The stable nucleus for the development of the region under study during the Late Paleozoic was represented by the early Paleozoic Kazakhstan-North Tien Shan massif, a small (1000-1500 km long and 1000 km wide) microcontinent. In Late Paleozoic time this massif was surrounded on nearly all sides by subduction zones accompanied by volcanic belts. The spatial distribution of Early Devonian and Early Carboniferous complexes, which are the bases of our interpretations, are shown in Figure 57.

The existence of paleo-subduction zones is implied by late Paleozoic ophiolite sutures, which together with nappe-folded zones constitute belts of collision of the Kazakhstan-North Tien Shan microcontinent with adjacent continental blocks. Two main sutures are the most clearly distinguished: South Tien Shan and North Balkhash. The former marks collision of the Kazakhstan-North Tien Shan massif with the Tarim-Karakum microcontinent. The North Balkash suture divides two areas characterized by essentially different facies of Devonian-Carboniferous deposits. To the north, shallow water terrigenous deposits predominate, whereas to the south in the Arganat zone of North Peri-Balkhash and the Tastau zone of North Junggaria the sediments are of deep water origin. This suture marks the collisional zone of the Kazakhstan-North Tien Shan massif with overriding sialic masses or island arcs that were successively included into the Kazakhstan-North Tien Shan massif. The comparatively small Tekturmass ophiolite zone may be regarded as a suture along which a block coming from the south was accreted to the Kazakhstan margin.

The Late Paleozoic structure of the Kazakhstan-North Tien Shan massif is distinctly divided into two segments coinciding with well known orogenic terrains--Junggar-Balkhash and South Tien Shan, each more than 1000 km long (Figure 57). They are sinistrally displaced 1000 km relative to each other in the zone along the Aksoran-Akzhal and Sarytum faults, called the West Balkhash transform. The knee-like shape of the early Paleozoic Kazakhstan-Tien Shan massif is caused by this 1000 km sinistral displacement.

Junggar-Balkhash Segment

Crossing the Junggar-Balkhash segment from the center (i.e., from North Junggaria and the Tastau synclinorium) NNW toward the early Paleozoic massif, marine sedimentary sequences give way to continental ones, and island-arc series give way to the active continental margin volcanic belt and granites. As Karyaev [1984] noted, the calc-alkaline volcanic belts are almost invariably associated with marine sedimentary sequences in bands parallel to them. These sequences correspond to forearc basins. They became younger from the north to the south. Accordingly, the Junggar-Balkhash segment is divided into two parts: a northern one where continental crust already existed by the end of the Devonian and which was occupied in the Devonian by the marginal volcanic belt, and a southern one covering the North Peri-Balkhash region and North Junggaria where continental basement was absent.

Within the early Paleozoic continental massif, volcanism proceeded on the land only, in a belt stretching for 1200 km along the margin of the massif. The volcanics belong to the calc-alkaline series.

On the continental side the volcanic belt is rimmed by Devonian molasse depressions. In the Famennian, the early Paleozoic continent was peneplained and flooded by a shallow sea that lasted up to the Middle Carboniferous. (In the Early Carboniferous paralic coal had accumulated in the Karaganda, Ekibastuz and other basins.) In the Latest Paleozoic, continental sedimentation was concentrated in two depressions--Dzezkazgan and Tengiz.

On the other (former oceanward) side of the volcanic belt towards Lake Balkhash, notable marine sedimentation was coupled with volcanism. Lower-Middle Devonian sequences fill the Nura and Karasor synclinoria in which clastic material was supplied from both the continental-belt lavas and the continental basement. The sequences of the Nura and Karasor synclinoria may be interpreted as the remains of the forearc basin. The spatial association of tectonic stratigraphic asemblages allows one to reconstruct a subduction zone with a down-going slab dipping underneath the Kazakhstan-North Tien Shan continent to the north (present coordinates).

The southerly Junggar-Balkhash segment proper is characterized by an entire absence of older continental blocks. The most ancient rocks are the Vendian(?) and Cambrian ophiolites, unconformably overlain by the Upper Ordovician sequence. The ophiolites of the Junggar-Balkhash area (Tekturmass and North Peri-Balkhash) strongly resemble accretionary wedges characteristic of outer nonvolcanic arcs. Consequently, all the area south of the marginal volcanic belt toward the North Balkhash suture may be regarded as the forearc domain situated between the volcanic belt and the oceanic basin.

A plausible Middle Paleozoic (Devonian-Lower Carboniferous) open oceanic space may be reconstructed from

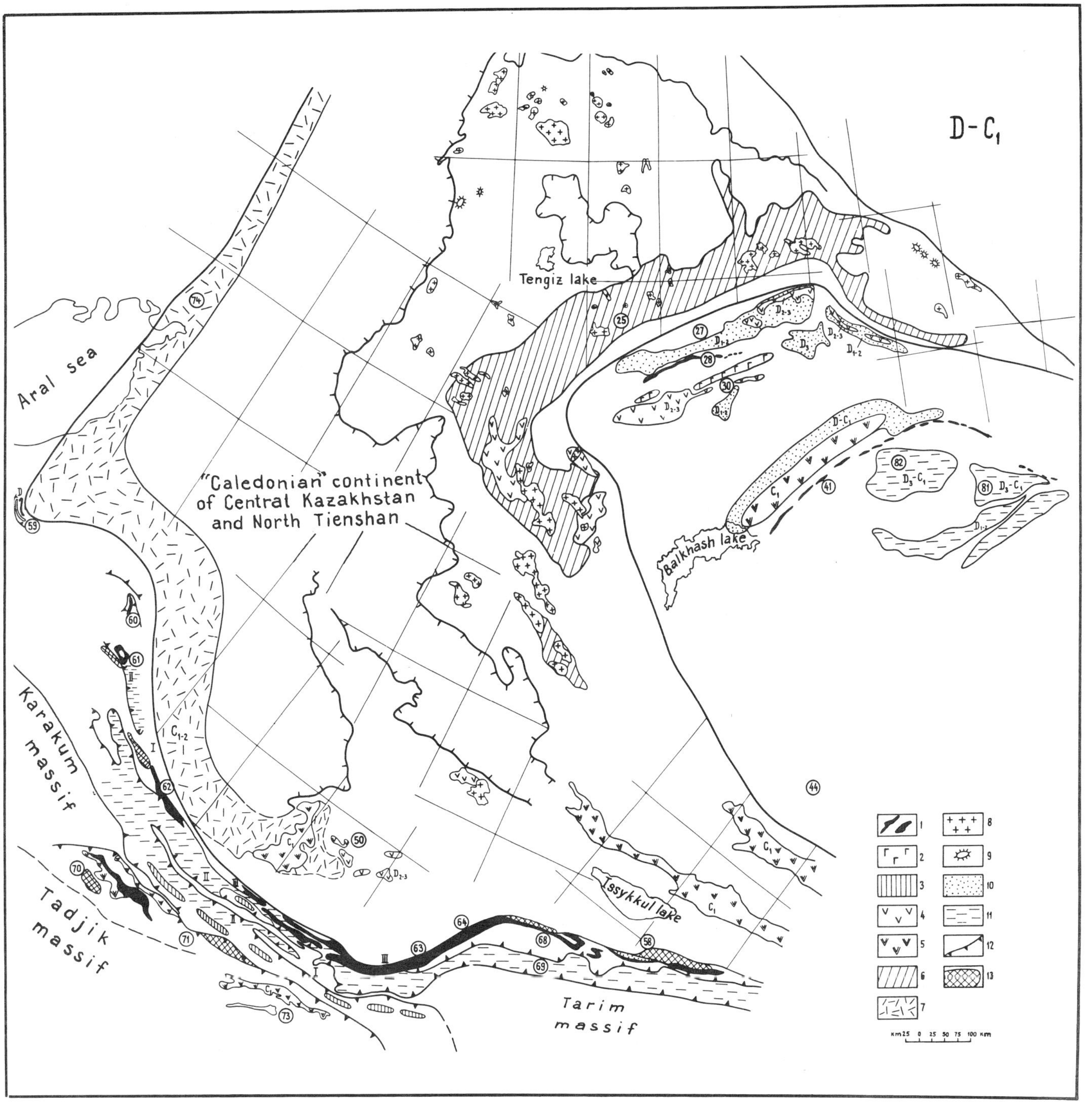

Fig. 57. Distribution of Devonian and Early Carboniferous complexes of Central Kazakhstan and Tien Shan. The base map is the Late Paleozoic reconstruction of Figure 52.
1, Ophiolite; 2, Upper Devonian and Lower Carboniferous back-arc basin volcanics; 4, Middle-Upper Devonia island-arc volcanics; 5, Lower Carboniferous island-arc volcanics; 6, Devonian continental-margin volcanic belt; 7, Lower Carboniferous continental-margin volcanic belt; 8, granite; 9, alkaline rocks; 10, flysch and olistostrome; 11, deep-water complexes of the continental-rise type; 12, nappe units in South Tien Shan; 13, Precambrian basement (shown only in South Tien Shan). Circled figures correspond to structural units in Figure 48. I, II, and III - nappe units of South Tien Shan.

the cherty-terrigenous sequences of North Junggaria and the Arganat zone of the North Peri-Balkhash area. A thick (5 km) sequence of clastic and cherty rocks ranges from Silurian to Lower Carboniferous [Filatova and Bush, 1965]. This sequence is of the continental-rise type, a strikingly principal different from the shallow-water sequences north of the North Balkhash ophiolite zone. Thus, two essentially different facies converged and overlapped tectonically along the North Balkhash suture, most probably in the Middle Carboniferous. The surface trace of the subduction zone during all of the Devonian and Carboniferous, and perhaps even later, may be related to the North Balkhash ophiolite suture, the most outstanding and important structural feature known in this region and the only structure that divides sharply different Devonian and Carboniferous facies zones.

According to Karyaev's [1984] scheme, in the North Peri-Balkhash region one arc replaced another, shifting southward: the Middle-Upper Devonian (Givetian-Frasnian) arc was followed by the Famennian-Early Tournaisian arc. Before the latter arc formed, the forearc area belonging to the Nura trough and Tekturmas was partially deformed (this deformation is often referred as "Early Hercynian"). After the reorganization, the volcanic arc shifted 100 km southward and simultaneously a backarc basin opened in the Uspensky zone, as recognized from comparatively coarse-clastic and cherty sequences of Famennian-Tournaisian age.

The Late Paleozoic continental-margin volcanic-plutonic belt overlies all the structures of the Middle Paleozoic forearc basin. Volcanic activity started within the belt in the middle of the Early Carboniferous and evolved nearly without hiatus to middle Permian time. Volcanic activity renewed again for a short period in the Late Permian [Koshkin, 1974; Kurchavov, 1984].

Kurchavov [1984] studied in detail the zonation of Carboniferous and Permian volcanics in the belt and found an increase of alkalinity and K_2O content in rocks with same SiO_2 content from south to north, i.e., from the margin of the belt toward the interior of the Early Paleozoic continent. These data show that the paleo-trench was south of the belt and the subducting slab dipped northward under the continent. Using K_2O in the andesites of the Kalmakemel suite, it is possible to reconstruct the fossil Benioff zone using the Nivkovich-Hayes diagram; it dipped approximately 35°, and its surface trace must have been 150-160 km south of the present margin of the belt, i.e., 75 km south of the North Balkhash ophiolite zone.

Sediments of the forearc basin in front of the volcanic belt form the purely sedimentary, dominantly clastic section of the Sayak trough [Shlygin, 1981]. The material was derived from the north, from the volcanic belt side.

The Zaili part of the volcanic belt, covering South Junggaria, the SE part of the Chu-Iliisky mountains and the northern slopes of the Zaili Alatau, strongly resembles the northern, Peri-Balkhash branch. As Kurchavov [1984] pointed out, the volcanics of the Zaili belt are characterized by an increase of alkalinity and K_2O content (with the same SiO_2 content) from north to south. Consequently, the related subducting slab dipped southward, i.e., in the opposite direction from the inferred slab dipping under the Peri-Balkhash belt.

South Tien Shan Segment

The present structure of South Tien Shan is divided by the Talasso-Fergana strike slip fault into two parts displaced dextrally nearly 200 km relative to each other. On the Late Paleozoic reconstruction (Figure 52), where this strike-slip displacement has been removed, South Tien Shan forms a narrow 100 km belt stretching 2000 km as an arc, slightly convex to the south through the Kyzyl-Kum desert, the Alay and Turkestan ranges on the west, to the Kokshaal and Atbashin ranges in the east. In the north, this belt is bounded by the early Paleozoic Kazakhstan-North Tien Shan continent; in the south, it borders upon three continental massifs: the Karakum in the west, the Tadzhik in the center, and the Tarim massif in the east. The Karakum and Tarim massifs are parts of the same continental block, the Karakum-Tarim microcontinent, which at the end of the Paleozoic was partially overthrust(?) by the Tadzhik massif and later by the Pamir massif.

In the late Paleozoic orogeny of the South Tien Shan, Burtman [1973, 1976] found three deformation phases: D1, D2, and D3. The D1 deformation is related to thrusting of large nappes at the end of the Middle Carboniferous; D2 deformation caused secondary folding of nappes into a series of synforms and antiforms at the end of the Carboniferous and beginning of the Permian; deformation D3 was characterized by strike-slip faults, such as the Talasso-Fergana fault, and by horizontal flexures like the Fergana sigmoid, at the very end of the Paleozoic. The studies of Porshnyakov [1973], Sabdjushev and Usmanov [1971], Zubtsov et al., [1974], Portnyagin et al. [1973], Burtman [1973, 1976], Akhber and Mushkin [1976], Makarychev [1978], Shults [1972, 1974], Khrisatov and

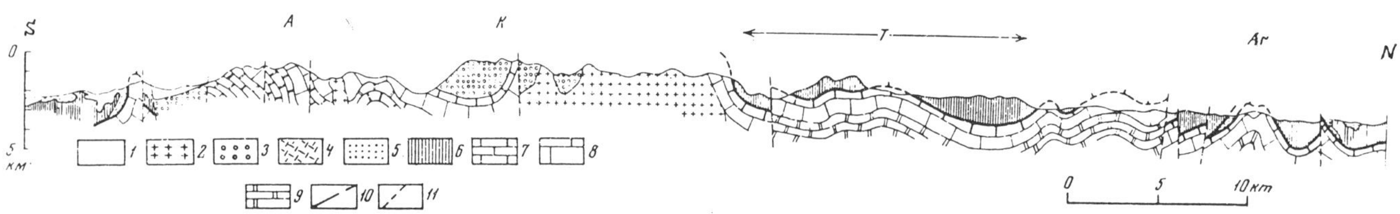

Fig. 58. Geological cross-section through Alai Ridge of South Tien Shan (after Burtman [1976]).
1, Mesozoic and Cenozoic; 2, Permian granite; 3, Upper Paleozoic molasse. Symbols from 4 to 6 correspond to Structural Unit II: 4, Carboniferous limestone; 5, Devonian clastics; 6, Silurian clastics. Symbols from 7 to 9 correspond to Structural Unit I: 7, Middle Carboniferous olistostrome and flysch; 8, Lower Carboniferous limestone; 9, Devonian limestone; 10, thrust; 11, younger fault. A, Alai Ridge; Ar, Arpalyk window; K, Kichikali Ridge; T, Tegermag allochthon.

Mikolaichuk [1983], Brezhnev and Ivanov [1980] clarified the nappe structure of the South Tien Shan.

According to Burtman [1976], the South Tien Shan, including Kyzyl-Kum, is made up of four first-order nappe units. They are all composed of sequences of approximately the same Middle Paleozoic age, from Silurian to Carboniferous (Figures 58 and 59). They are briefly described below.

Structural Unit I

The lowest unit is an autochthon or para-autochthon consisting of sedimentary rocks, mostly of Carboniferous age. Its basement consists of sequences of greenschists, quartzites and marbles (Auminzin suite) overlain by carbonates, cherty rocks and coaly shales (Taskazgan and Bessapan suites), as in the Kyzyl-Kum and Nuratau mountains. Pb-Pb age determinations from schists provide ages of about 370 Ma. However, these sequences must be much older than 370 Ma, because they were metamorphosed and deformed in the Late Riphean-Vendian and they are a part of the Baikalian metamorphic basement on which the Paleozoic sedimentary series of Unit 1 was deposited. In Kyzyl-Kum, the Lower Devonian lies unconformably on this ancient basement. The greater part of Unit I is everywhere composed of shallow-water limestones (commonly of reef origin) of ages spanning the Early Devonian to the Early Moscovian (lowermost stage of the Middle Carboniferous). These limestones belong to the Alay facies. Flysch and olistostrome deposits of the Late Moscovian (in places the lowermost portion of the Upper Carboniferous), overlie the carbonate sequence, marking the beginning of thrusting and movement of nappes over Unit I. The sedimentary sequences of Unit I indicate the presence of a Paleozoic passive continental margin on the Karakum-Tarim microcontinent. Lavas--tholeiites and alkaline basalts--within the limestone sequences in Eastern Tien Shan reflect an intraplate event and probably show that in the Devonian this territory passed over a hot spot.

Structural Unit II

The nappes of Unit II occupy the largest areas and obviously contain several second-order units, as their rock sequences are quite diverse.

A small but particularly characteristic formation is a sequence (the Shalansky suite), no more than few tens of meters thick, of pelagic limestones and colored cherts covering the period from the Middle Devonian to the Bashkirian Stage of the Middle Carboniferous. It was undoubtedly formed in deep water and belongs to the condensed facies characteristic of abyssal depths at a significant distance from sources of clastic material.

The most abundant rocks of Unit II were classified by Burtman [1976] into three major types: carbonate, carbonate-terrigenous, and volcanogenic-terrigenous. The carbonate type embraces the span from the Upper Silurian to the beginning of the Middle Carboniferous. The carbonate-terrigenous type consists of alternating sandstones and siltstones, in places resembling flysch, associated with lava flows of intermediate composition and with cherts and pelagic limestones. In the volcanogenic-terrigenous type, of particular significance are lavas and tuffs of mafic and intermediate composition (andesites and basaltic andesites). They occur at different levels but are particularly abundant in the Silurian and in the lower part of the Devonian. Meta-volcanic rocks that are widespread in the Zeravshan and Gissar Ranges also belong to Unit II. They are often referred to as the Yagnobsky series or 'Yagnobsky schist'. Recently fossils ranging from Middle Ordovician to Ludlovian Stage of the Silurian were found in this sequence. The sequence includes basaltic andesites and other volcanic rocks of the calc-alkaline series [Brezhnev and Ivanov, 1980]. Thus, calc-alkaline volcanic rocks are typical of Unit II; their ages are middle-Upper Ordovician, Silurian, and lower-Middle Devonian.

As in Unit I, the upper part of Unit II consists of flysch and olistostromes of Moscovian (Late Carboniferous) age, proving convergence of these two structural units at that time.

Structural Unit III

This unit is composed of ophiolites. They include igneous rocks of the oceanic crust and their sedimentary cover. The bottom and top of Unit III are formed by ultramafic slabs and by serpentinite melange (e.g., the Kansk melange).

The most complete succession of the Unit III ophiolites is found in the Sartalinsky tectonic sheet (northern slope of the Alay Range near the Fergana Depression) where serpentinized ultramafics are overlain by gabbro-diabases covered by cherts with conglomerate lenses (1-2 m) including pebbles of the same gabbro-diabases and ultramafics [Peive, 1973]. The thickness of the cherts ranges from 10 to 20 m. They are conformably overlain by a sequence 120 m thick of tholeiitic basalts and picrites (picritic komatiites), forming the basal member of a thick volcanic pile. They also contain horizons of cherts with early Middle Ordovician Radiolaria, whereas in higher horizons there are early Middle Devonian Radiolaria. The ophiolitic outcrops along the South Tien Shan mark the suture of the Paleozoic oceanic basin, which Burtman [1976] named the Turkestan ocean.

The geodynamic position of the Unit III lavas is still not clarified. They undoubtedly formed in an oceanic basin, but it remains unclear whether they erupted in a spreading center within the open ocean, in a marginal sea, or within an immature island arc.

Structural Unit IV

This unit occupies limited areas and occurs only where the ophiolites appear. It mainly comprises of metamorphic rocks, and mafic extrusive rocks in the greenschist facies dominate [Makarychev, 1970]. Glaucophane schists are found as well. The age of the metamorphic sequences is problematic. They seem to involve both ancient (Riphean-Vendian) and Lower Paleozoic (up to Silurian) protoliths. In any case, they are older than the Devonian, as the Devonian and overlying Carboniferous rocks rest on the metamorphics with a sharp angular unconformity. The unconformity at the base of the Devonian marks the early Paleozoic deformation specific for Unit IV only. Burtman [1976] believes that the Unit IV rocks are remains of northern sequences detached from the margin of the Early Paleozoic Kazakhstan-North Tien Shan continent. The high-pressure metamorphic assemblages of Unit IV, including glaucophane schists and eclogites were apparently formed in the subduction zone prior to collision.

According to Burtman's [1976] scheme, the successively higher nappes (I to IV) had an increasingly northerly origin, so they probably moved in a southerly direction (Figure 58b).

Thus, in the South Tien Shan foldbelt, complexes belonging to an oceanic basin (Turkestan ocean) and to both its southern and its northern margins are brought together. As mentioned above, Unit I marks the passive continental margin bordering the oceanic basin (Unit III) on the south. Unit II includes bathyal complexes of passive-margin or marginal-sea provenance and calc-alkaline volcanics indicative of former island arcs. Consequently, in the Turkestan oceanic basin, subduction zones must have separated an open oceanic floor from the northern continent. Metamorphics of Unit IV, especially the high P/T assemblages, are evidence for the subduction zone. Thus, the southern margin of the oceanic basin was passive and the northern was active throughout the Ordovician-Middle Devonian. Calc-alkaline volcanics are not found in the Late Devonian or Early Carboniferous, and there is no evidence for a subduction zone at that time. Later on however, a subduction zone along the northern margin of the oceanic basin is well marked by calc-alkaline volcanics with the Serpukhovsky (Bashkirian) fauna, building the very long Beltau-Kurama volcanic belt of island arc origin [Mossakovsky, 1975], and probably continuous with the Valerianovsky belt of the Urals.

The geological information contained in the Unit III sequences indicates that the Turkestan oceanic basin (now partly incorporated in the South Tien Shan foldbelt) developed over a long period of time. Judging from the Sartalinsky ophiolitic succession and overlying cherts, it existed at least from the Early Ordovician to the Middle Devonian. Also, the Sartalinsky succession is important for its evidence of a purely oceanic origin of certain clastic rocks: i.e., deep water deposits of ultramafic and gabbroic rock fragments, perhaps representing deposition in a transform fault zone. As Silurian and Early Devonian lavas overlying these deposits are MORB tholeiites, sea-floor spreading persisted until the Middle Devonian.

The Gissar Zone

In addition to the four nappe units, the Gissar Zone is also recognized as a structural entity in the southern Tien Shan. It is characterized by vestiges of another Paleozoic oceanic basin. The complexes representing this ocean are now closely spaced and spatially overlap the South Tien Shan structural units, although in the past they were significantly farther from them. The Gissar zone can be followed along extensive Lower and Middle Carboniferous lavas that contain ultramafic bodies and are associated with zones of serpentinite melange. The Carboniferous volcanics are present in the Gissar Range around the Permian Gissar batholith and in the southern branches of the range descending to the Tadzhik depression, particularly around the Proterozoic Baisun block. The latter is probably a fragment of the continental basement of the Tadzhik massif. According to Portnyagin et al. [1973], the volcanics consist of two series. The lower one is composed of tholeiitic pillow basalts that underlie a sedimentary series containing the Serpukhovsky (Bashkirian and Late Carboniferous) fauna. The upper volcanic series consists mainly of rocks of andesite-basaltic composition that undoubtedly belong to an island-arc association. Its age is determined as Bashkirian-Early Moscovian (Late Carboniferous). Tournaisian (Early Carboniferous) rocks beneath lie unconformably on the ancient basment and are represented by highly alkalic (up to 7% of alkali content) volcanics of predominantly silicic composition, associated with reef limestones.

The Lower and Middle Carboniferous arc volcanics are a fairly distinctive feature of the Gissar zone. Unfortunately, no reliable data on the volcanic polarity are available. The arc is supposed to face north towards the Tien Shan with which it collided in the beginning of the Permian. The above-mentioned highly alkalic volcanics presumably erupted in a back-arc environment, occur in the Baisun massif, south of the Gissar Range. However, the original spatial relations of some of the rock assemblages is unclear, because the Gissar volcanic belt is cut by the North Gissar thrust.

From the end of the Carboniferous to the end of the Permian, the South Tien Shan and Gissar areas underwent mountain building (often called "Hercynian" in allusion to western Europe) accompanied by emplacement of granites and, in places such as the Beltau-Kurama region, eruption of terrestrial silicic lavas. These events probably indicate continental collision.

Among the Late Paleozoic ("Hercynian") structures, two types have been distinguished. On the one hand, there are narrow troughs initially filled with marine flysch-like and then molasse sequences. To some extent, these structures resemble foredeeps, particularly in the Kokshaal ridge and in the Muzduk trough along the northern margin of the Tarim massif. On the other hand, there are mountainous uplifts with granitic massifs. Small massifs of Late Paleozoic granites are scattered over the whole of the South Tien Shan. The largest, the Gissar batholith, is 250 km long and 180-290 Ma old [Sidorenko, 1972]. Batholith emplacement is believed to have resulted from collision of the Tadzhik microcontinent with the Tien Shan.

HISTORY OF THE LATE PALEOZOIC ("HERCYNIAN") STRUCTURE

Early Devonian, 400 Ma (Figure 56e)

The Kazakhstan-North Tien Shan microcontinent apparently rotated 60° clockwise in the end of the Silurian and the beginning of the Devonian. This rotation was caused by opening of new oceanic basins (Uralian, Turkestan, Irtysh-Zaisan) in the Ordovician and Early Silurian. It follows from paleomagnetic data that in the Devonian the Kazakhstan-North Tien Shan continent was situated between 15° and 40° N [Khramov, 1982]. Geological data, especially the distribution of calc-alkaline volcanic and plutonic rocks, indicate that two subduction zones existed in the Devonian. One of them controlled volcanism of the Central Kazakhstan marginal volcanic belt, an active continental margin of Andean type. Another zone is recognized from the Silurian and Lower-Middle Devonian calc-alcaline volcanics involved in Unit II of South Tien Shan; it may be called the South Tien Shan arc. The subduction zone was also inclined towards the Caledonian continent (to the northwest in the Devonian system of coordinates). In this zone the crust of the Turkestan oceanic basin was consumed. On the other side of this basin was a moderately large continent, the remains of which are preserved as the Tarim and Karakum massifs. It was 3000 km long and several hundred kilometers wide. Judging from the presence of Early-Middle Devonian tholeiitic pillow basalts (Unit III of South Tien Shan), an active spreading center already existed in the Turkestan basin in the beginning of the Devonian. The size of the Turkestan basin is hard to estimate, but we should focus on two aspects. First, the Karakum-Tarim continent is

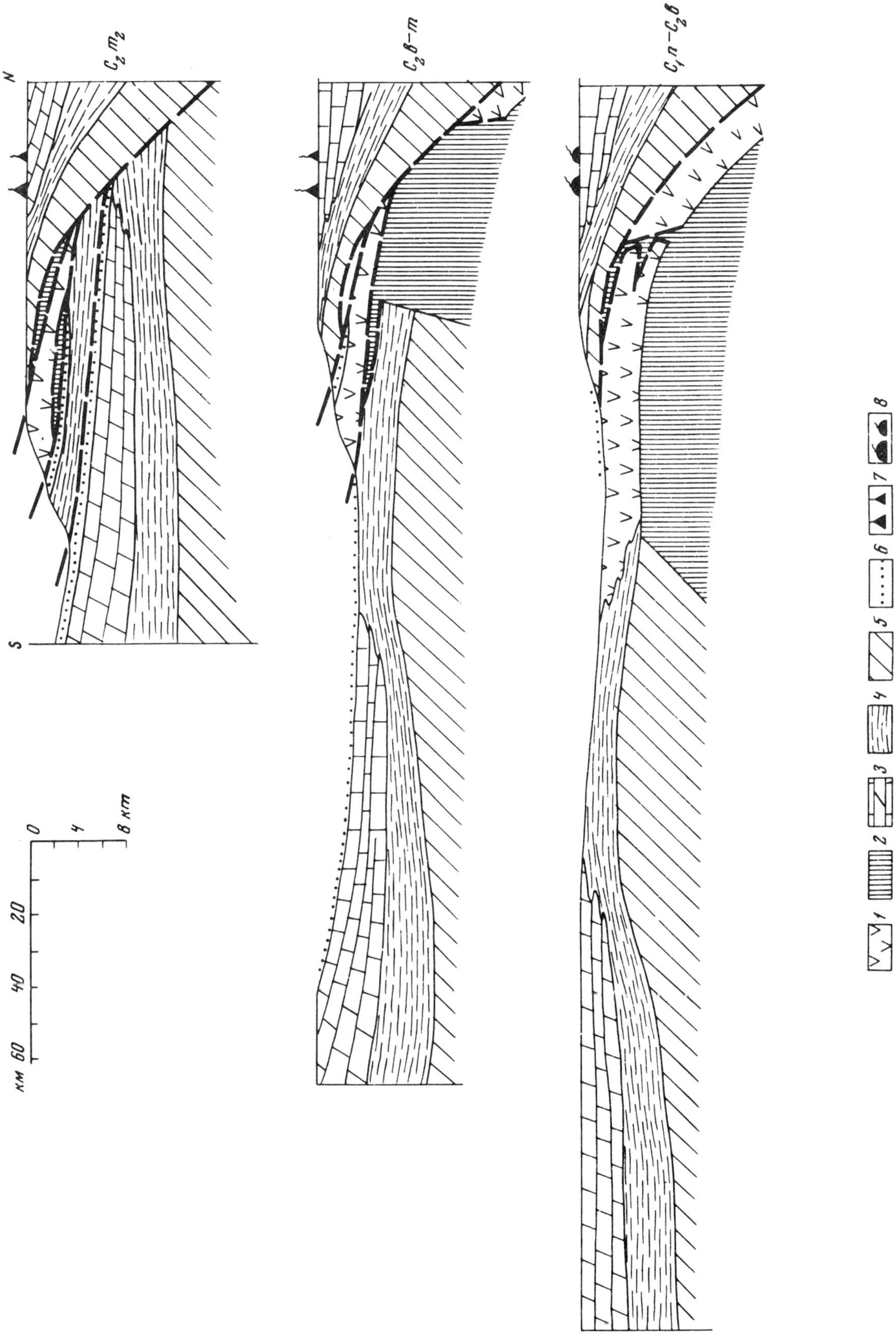

Fig. 59. Evolutionary cross-section of the South Tien Shan region (after Burtman [1976]).
1, Oceanic-type volcanics (remnants of the Turkestan ocean floor); 2, ultramafics and gabbro; 3, Middle Paleozoic limestone; 4, Middle Paleozoic clastics and chert; 5, Lower Paleozoic and Precambrian rocks; 6, flysch and olistostrome; 7, felsic volcanics; 8, intermediate volcanics.

characterized by a development of reef limestones that undoubtedly formed near to the equatorial zone, i.e., they were located ca. 15° from the margin of the Kazakhstan-North Tien Shan continent (about 1600 km). Second, subduction proceeded from the second half of the Ordovician to the Middle Devonian, i.e., around 100 m.y. Even at a minimal subduction rate of 2 cm/yr, about 2000 km of the oceanic crust could have been consumed. Recent paleomagnetic results from Devonian ophiolites of the Tien Shan show that the ophiolites (and therefore a spreading axis) were situated in Devonian time at 22° N while the Alai (Karakum) microcontinent was at 20° paleolatitude either north or south of the equator [Didenko and Pechersky, 1988].

Middle Devonian, 370 Ma (Figure 56f)

Two segments of subduction zones are well recognized for this time. One is the North Balkhash zone, which originated south (present coordinates) of the Central Kazakhstan marginal volcanic belt, and the other is the South Tien Shan zone.

From the Givetian to the Early Carboniferous the Central Kazakhstan volcanic arc moved off the continent towards the ocean, simultaneously with shifting of forearc and backarc sedimentary prisms. Taking into account subsequent folding, it may be estimated that the magmatic front moved nearly 500 km oceanwards. At the beginning of the Late Devonian this migration changed to a backward movement, accompanied by folding in the back-arc region. In the Famennian, on the contrary, the arc again shifted oceanwards and a new back-arc basin opened.

Early Carboniferous, 340 Ma (Figure 56g)

At the end of the Middle Devonian, subduction ceased under the South Tien Shan arc and from that time to the middle of the Early Carboniferous no subduction complexes are known in the South Tien Shan. The only zone of calc-alkaline volcanism of Tournaisian-Visean and partly Late Devonian age belongs to the Zailiisky Alatau and the Ketmen ridge. It is hard to judge what was the environment in the Turkestan basin. Probably, at a certain time the northern and southern margins of the basin were passive and no plate boundaries existed within it. By the middle of the Devonian the width of the Turkestan basin had decreased to one-half its former width in the Late Ordovician.

In the Early Carboniferous, another continent (Tadzhik) was situated east (past coordinates) of the Karakum-Tarim continent; its ancient core is at present exposed as the Baisun massif. It was isolated from the Karakum-Tarim continent, by the Gissar oceanic basin, including the Gissar island arc.

Thus, the Early Carboniferous is characterized by the presence of two continents sequentially approaching the Early/Middle Paleozoic continental margin of Kazakhstan and North Tien Shan.

Middle Carboniferous, 310 Ma (Figure 56h)

The end of the Early and beginning of the Middle Carboniferous (Serpukhovsky - Bashkirian) were noted for an intense calc-alkaline volcanism, concentrated in (i) the long Beltau-Kurama (1500 km) and Valerianovsky (2500 km) volcanic belts on the eastern and southern margins of the Caledonian continent (past coordinates); (ii) the North Balkhash active continental-margin volcanic belt, and (iii) the Gissar volcanic arc. These belts indicate a rapid consumption of oceanic crust and, consequently, a significant convergence of continents. By the Middle Carboniferous (begining of the Moscovian) the major portion of the oceanic crust of the Turkestan and Gissar basins was already consumed, the Karakum-Tarim continent collided with the Kazakhstan-North Tien Shan continent, the Tadzhik continent with the Karakum-Tarim. The start of this collision is clearly reflected in a sharp replacement of carbonate sequences by flysch-olistostrome formations of the Middle-Upper Carboniferous. The Peri-Balkhash region, with a basin 1500 km wide, was the only region where unconsumed oceanic crust was left.

Early Permian, 280 Ma (Figure 56i)

The latest Carboniferous and Early Permian are characterized by a severe continental collision in South Tien Shan. At this time the major nappes of South Tien Shan formed, and oceanic and island-arc complexes of Late Paleozoic age were squeezed out onto the Karakum-Tarim continent. The amount of collision and crustal shortening were maximal in the Alay region, where the Tadzhik continent was involved in collision as well. The whole early Paleozoic Kazakhstan-North Tien Shan continent was squeezed between converging continental and microcontinental blocks. Only the Peri-Balkhash region suffered no collision; instead, oceanic crust was rapidly consumed. Terrestrial calc-alkaline volcanism was concentrated in two belts: North Balkhash and Ili (south of the Ili Depression shown on Figure 48). Volcanic polarity data show that each of these belts was associated with a Benioff zone, one dipping to the west under the North Balkhash belt, and the other to the south under the Ili belt, showing that these blocks were converging. At that time Kazakhstan together with the accreted Karakum-Tarim and Tadzhik continents were merged into the Laurasian supercontinent.

Late Permian, 250 Ma (Figure 56j)

The end of the Permian and evidently the very beginning of the Triassic were marked by fracturing of the newly formed continent and considerable block displacements within it, due to the continued convergence of Siberia and East Europe with Kazakhstan. Deformation bent the Junggar-Balkhash zone into its present U-shape, the oceanic crust of the Junggar-Balkhash basin was almost completely consumed, the system of block-faulted dislocations of the Sarysu-Tengiz uplift formed, blocks were displaced along great strike-slip faults (Talasso-Fergana, Central Kazakhstan, Chingiz), and conjugate shear zones formed (Uspensky, Spassky, Aksoran-Akzhal).

Chapter VI

CENTRAL ASIAN FOLDBELT, EASTERN PART: ALTAY-SAYAN REGION AND BAIKAL UPLAND

GENERAL DESCRIPTION

The Late Precambrian and Paleozoic structures surrounding the Siberian platform on the south and south-east are united into the Altay-Sayan region (Figure 60). It includes East Sayan Mountains, Minusinsk depressions, Kuznetsky Alatau Zone, West Sayan Mountains, Tuva, Altay Mountains, Kuzbass, Salair Range, and Irtysh-Zaisan zone. The structures of the Altay-Sayan region continue in North Mongolia (Mongolian Altay, Great Lakes depression, Khangay upland, Khubsugul area). The Altay-Sayan region is continued to the east as the Baikal upland covering the West Trans-Baikal region and the Vitim-Patom plateau.

This territory has been reinterpreted from a mobilistic point of view by Zonenshain [1972], Berzin [1979], Dergunov and Kheraskov [1982], Mossakovsky and Dergunov [1983], Semenov (unpublished materials), Dobretsov [1985], Gordienko [1984], Kovalev and Koryakin [1975], Rotarash et al. [1982], and Ermolov et al. [1981].

From Figure 60 it is evident that Precambrian massifs are imbedded in the Paleozoic foldbelts. They are concenterated in the eastern part near the Siberian platform, whereas in the West Altay-Sayan region they are extremely scarce and small, and the determinations of old ages for some of them are poorly proved. This part is composed of oceanic, island-arc, and thick terrigenous complexes gathered together as a result of a continuous accretion of material in subduction zones active throughout the Paleozoic. In the East Altay-Sayan region, such complexes are found in narrow belts between ancient massifs, and mark zones of continental collison.

The Altai-Sayan region traditionally has been divided into domains based on ages of deformation. Unfortunately, the scheme is quite imperfect, being influenced by "fixist" concepts and allusions to the classic orogenies of Europe. The following domains were distinguished: (i) domains of early Early Paleozoic deformation, the Salairides, often erroneously called "Early Caledonides", which underwent Early or Middle Cambrian folding, (ii) domains of late Early Paleozoic (post-Middle Cambrian, pre-Devonian) folding, often called "Late Caledonides", (iii) domains of early Late Paleozoic (earliest Carboniferous) folding, often called "Early Hercynides", and (iv) domains of late Late Paleozoic (Permain) folding, often called "Late Hercynides".

The "Early Caledonides" of the region may be subdivided into the Minusinsk zone (East Sayan, Kuznetsky Alatau, the basement of the Minusinsk depressions, Gornaya Shoria), the Tuva zone (East Tuva and North Mongolia), and the Baikal-Vitim (Baikalian) zones. The "Late Caledonides" domain includes the West Sayan and Gorny (Mountain) Altay zones. Recently it was discovered that orogenic deformations of the same age are also characteristic of the eastern part of the East Sayan range, especially of the Oka-Dzhida zone. In addition, the "Late Caledonides" include such large areas as the Chulyshman highland situated at the junction of the West Sayan with Gorny Altay where folding ceased in the Early Ordovician. The "Early Hercynides" domain comprises the Anuy-Chuya and Yustyd (Deljun-Sagsay) zones crossing the Gorny Altay "Caledonides". The "Late Hercynides" include three different zones: Irtysh-Zaisan together with Rudny (Ore) Altay, Salair, and Tom-Kolyvan. These zones were not formed contemporaneously. The earliest orogenic phases occurred in the Rudny Altay. In the Salair there were several phases of folding: pre-Ordovician, Middle Devonian (Givetian), and Early Carboniferous. The Irtysh-Zaisan and Tom-Kolyvan zones underwent folding only in the middle of the Permian.

Obviously, the traditional division of the Altay-Sayan area according to age of folding is unrealistic.

Late Paleozoic Strike-Slip Motions, and Reconstruction of the Middle Paleozoic Structure

The intersecting network of faults is well seen in Figure 60; it indicates post-orogenic motions along strike-slip faults, e.g., East Sayan fault striking NW, Kuznetsky-Alatau fault striking NNW, Kandat fault striking ENE, the sublatitudinal Khangay fault, as well as along the Irtysh and so-called North-Eastern shear zone striking NW. Unfortunately, reliable data on motions along these faults are either scarce or absent. Sinistral displacement is well documented along the Kuznetsky-Alatau fault, with an affect of 120 km (Figure 61). The Kandat fault appears to be sinistral, whereas the East Sayan, Trans-Baikal, and Irtysh faults are probably dextral. A reconstruction after removal of displacements along the strike-slip faults is shown on Figure 62.

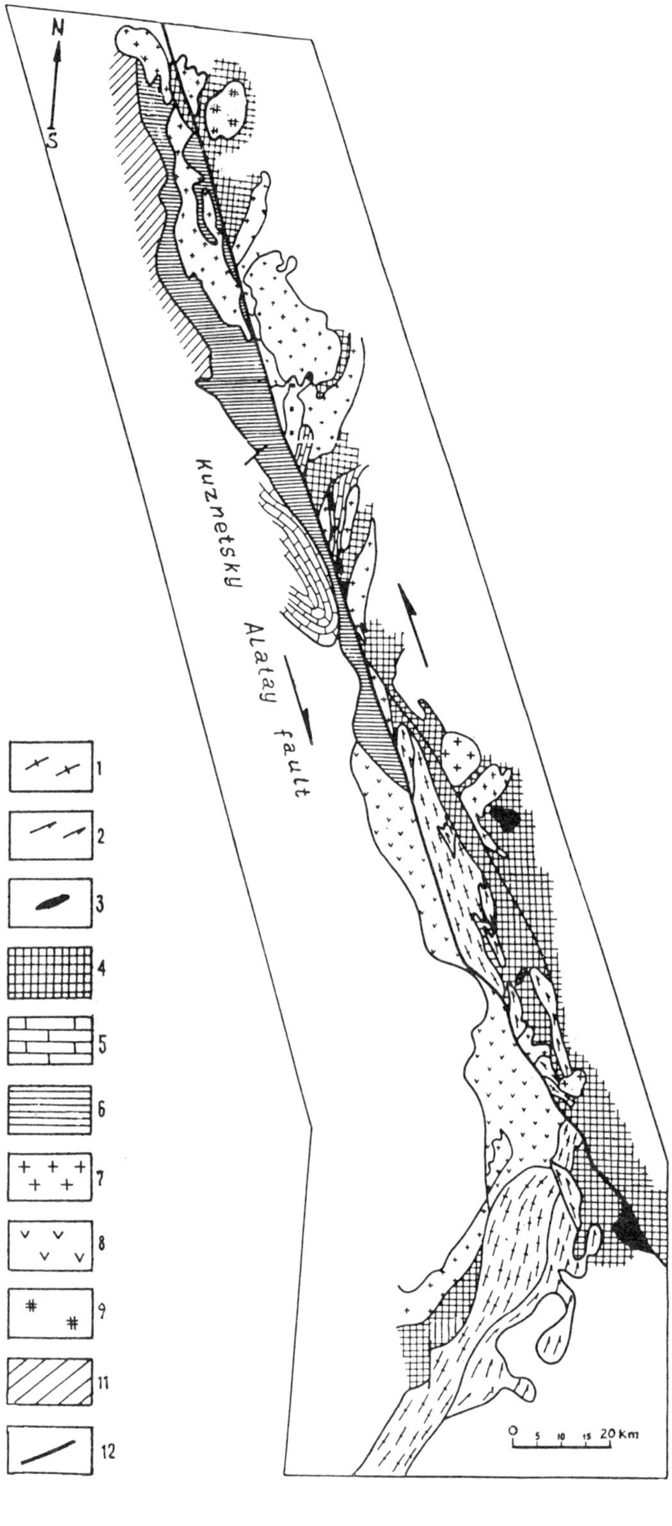

Fig. 61. Geology of the Kuznetsky Alatau fault area.
1, Metamorphics; 2, gabbro; 3, ultramafics; 4, volcanics and chert of the Late Precambrian and Lower Cambrian; 5, carbonates of the Riphean; 6, Ordovician volcanics; 7, Early Paleozoic granite; 8, Lower Devonian continental volcanics; 9, Middle Devonian alkaline granite; 11, sedimentary Devonian sequences; 12, fault.

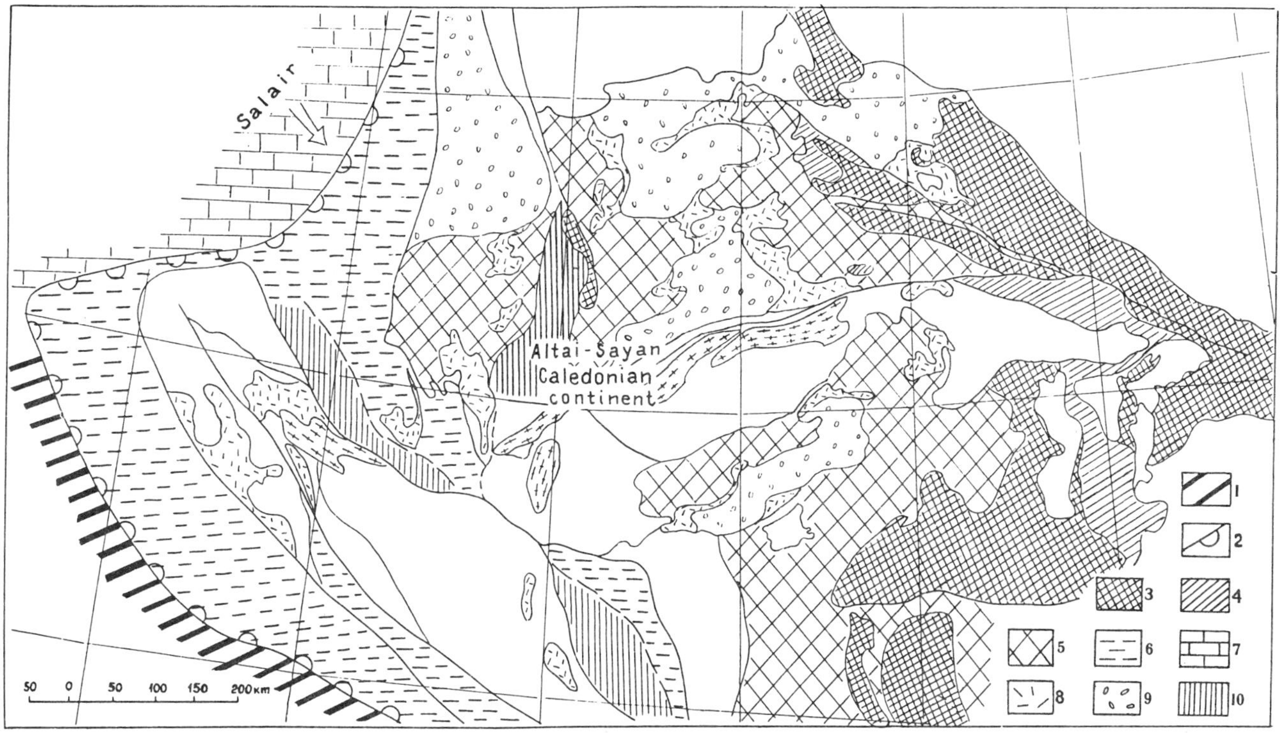

Fig. 62. Late Devonian reconstruction of the Altai-Sayan region.
1, Oceanic crust; 2, subduction zone; 3, Precambrian basement; 4, Late Precambrian and Cambrian cover on the Precambrian basement; 5, Salairide blocks; 6, Devonian marine clastics; 7, Devonian shelf limestone; 8, Devonian subaerial volcanics; 9, Late Devonian molasse; 10, 'gaps' resulting from the reconstruction.

Ancient Massifs

The ancient massifs are divided into (i) inliers of the Siberian platform basement (Birjusa, Kan, Sharyzhalgay, and Baikal massifs), (ii) microcontinental blocks in the eastern part of the Altay-Sayan region near the Siberian platform border (Barguzin, Gargan, Tuva-Mongolian, East Sayan massifs and some others), (iii) outcrops of Precambrian metamorphic rocks in the western part of the Altay Sayan region (the Tomsky, Tersinsky, and Kurchum massifs as well as problematic massifs such as Dzhebash, Chulyshman, Terektin, etc).

Marginal massifs (inliers) of the Siberian platform were involved in a mid-Paleozoic orogeny. These massifs are overlain by the Late Precambrian and Cambrian sedimentary sequence of the Mansky and Mirichunsky troughs [Khomentovsky et al., 1978]. The rocks deposited in these troughs essentially resemble those of the Siberian platform cover, differing however, in the amount of shallow-water limestones in the Vendian and the replacement of shallow-water facies by deep-water facies from NE to SW. In contrast to the platform cover, these sequences underwent intense folding and were later overthrust by flysch and volcanic complexes from the internal segments of the Altay-Sayan area. The sedimentary sequence of the northern flank of the Mansky trough is topped by a series of rhythmically intermittent terrigenous rocks from conglomerates to siltstones (Narva suite). This series is considered to belong to the Upper Cambrian-Lower Ordovician. It is strongly deformed and evidently reflects the last phases of folding and thrusting from the internides of the Altay Sayan area over the platform.

Muya Block and the Barguzin Microcontinent

Internal parts of the Baikal-Vitim Zone are occupied by the vast Barguzin granite batholith (shown on Figure 64). Separate outcrops of Precambrian crystalline rocks are preserved in places within the batholith, exhibiting some kind of metamorphic arc. The Precambrian rocks are unconformably overlain by Vendian-Cambrian coarse-clastic sequences [Klitin et al., 1970; Belichenko, 1982]. Although in the adjacent Siberian platform there is a complete rock succession of upper Riphean, Vendian, and Lower Paleozoic, in the Baikal-Vitim zone the upper Riphean is almost completely absent.

The basement of the metamorphic arc is made up of rock complexes of different ages and different facies; these rocks are presently found to have a nappe structure [Mitrofanov, 1978; Dobretsov et al., 1983; Sizykh, 1985] (Figure 63). The ancient basement of the Muya block consists of gneisses and granite-gneisses probably of Archean age, although isotopic dating has given only 2100 Ma [Mitrofanov, personal communication], surrounded by ophiolites of the Muya or North Baikalian ophiolite belt. In the latter, isotopic age data show that ophiolitic rocks of 2380 to 1750 Ma are present [Sizykh, 1985]. Ophiolites are associated with olistostromes that were

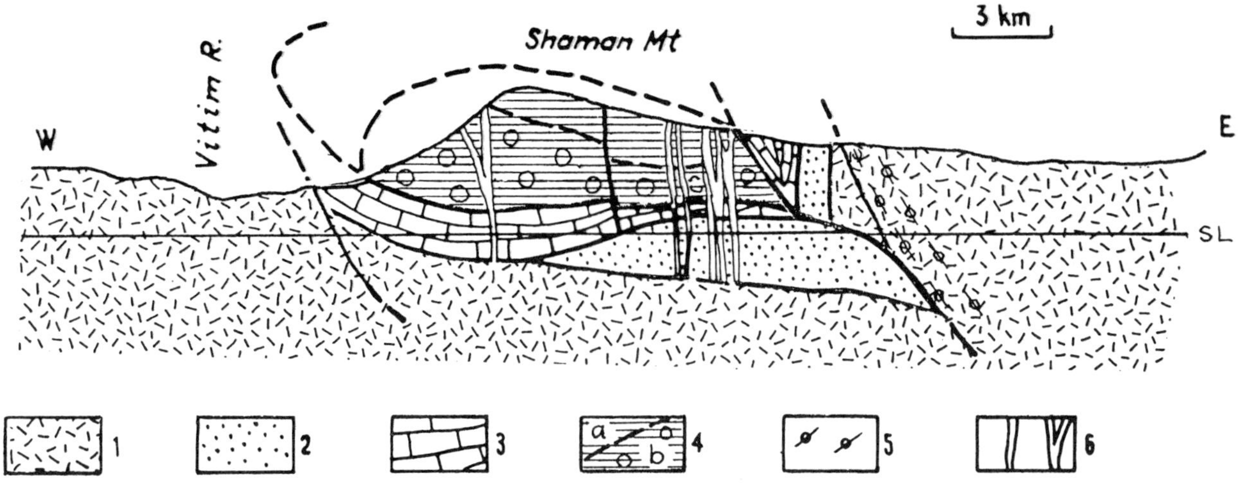

Fig. 63. Geological cross-section across the Muya ophiolite belt, South Muya Ridge, northern Trans-Baikalian region, after Mitrophanov [1978].
Archean-Lower Proterozoic basement: 1, granite-gneiss; 2, schist; 3, marble. Middle Proterozoic ophiolites: 4, ultramafics (a, weakly serpentinized, b, strongly serpentinized); 5, blastomylonite derived from ophiolitic igneous rock; 6, diabase dike.

metamorphosed nearly 1200 Ma ago. Volcanogenic-clastic sequences of Riphean and possibly Lower Proterozoic age are widespread. In the North Baikal region these sequences compose the Olokit zone [Sizykh, 1985] which contains clastic and carbonate sequences below and volcanic rocks above. The volcanics include basalts close in composition to oceanic tholeiites together with andesites and ignimbrites. Isotopic geochronology indicates that the volcanic sequences of the Olokit zone formed from 1500 to 740 Ma [Mitrofanov and Mitrofanova, 1980]. Similar Riphean calc-alkaline volcanics of island-arc affinities are found along the western margin of the Siberian platform within the Enisey ridge where they border upon ophiolites of the Vorogovsky belt.

Deformation and metamorphism terminated before the Vendian, forming the Barguzin continent in this area. In the Early Paleozoic its eroded, smoothed surface was covered by a shallow sea whose floor progressively subsided. The next deformational phase occurred at about the same time as on the passive margin of Siberia, i.e., between Cambrian and Devonian. The Barguzin granitoid batholith was formed at this time. Obviously, all these events--folding, and metamorphism, emplacement of granites--were caused by collision of the Barguzin microcontinent with the Siberian passive margin.

Precambrian Massifs of East Sayan and North Mongolia

A series of ancient massifs south and south-west of Baikal (East Sayan, Tuva-Mongolian, Bilin, Gargan, and Khamar-Daban) were regarded until recently as the continuation of the Siberian platform, but early Paleozoic suture zones separating these massifs from Siberia have now been mapped [Belichenko, 1983, Dobretsov, 1985]. Evidently the massifs do not belong to the Siberian continent but are the remnants of other continent(s) (Figure 64). In fact, the crystalline rocks composing these massifs are considerably different from the basement of the Siberian platform. They include graphitic marbles and quartz-graphitic schists from 1250 to 750 Ma in age [Lepezin, 1978] which are unknown in the Siberian platform basement.

The largest massif, Tuva-Mongolian, is overlain by a sedimentary cover of Upper-Vendian and Cambrian shallow-water and carbonate-clastic sequences with phosphorites. Outliers of Lower Cambrian volcanogenic-sedimentary sequences, commonly altered to greenstone and including bodies of ultramafic rocks, have long been known within the Tuva-Mongolian massif in the Sangilen highland. They were considered [Sidorenko, 1966] to lie unconformably on the Precambrian metamorphics. However, we believe these outliers are synformal remnants of Cambrian ophiolite and island-arc nappes that have been thrust over the Precambrian basement of the Tuva-Mongolian massif.

To the east of East Sayan, adjacent to the Tuva-Mongolian massif, is the Gargan massif, 2300-2400 Ma old [Lepezin, 1978]. Dobretsov [1985] reported remnants of the sedimentary cover in the Gargan massif, represented by limestones and dolomites with rare intercalations of graphitic-cherty schists. This sequence is comparable with the Vendian-Cambrian of the Tuva-Mongolian massif, although some evidence [Belichenko, 1983] indicates the presence of an Ordovician fauna. According to Dobretsov [1985] the Gargan massif crops out in the core of an antiform composed of Lower Paleozoic tectonic nappes.

Thus, these massifs consist of two tectonic units: a lower unit of crystalline Precambrian basement with its overlying non-metamorphosed Vendian-Lower Cambrian cover, and an upper unit formed by nappes of Late Precambrian--Lower Cambrian rocks, ophiolites included. The complicated outlines of the Tuva-Mongolian massif in Tuva, well seen in Figure 60, evidently reflect the tectonic contact of the upper unit with the lower one, rather than the primary boundaries of the massif. We believe that all the massifs are antiforms

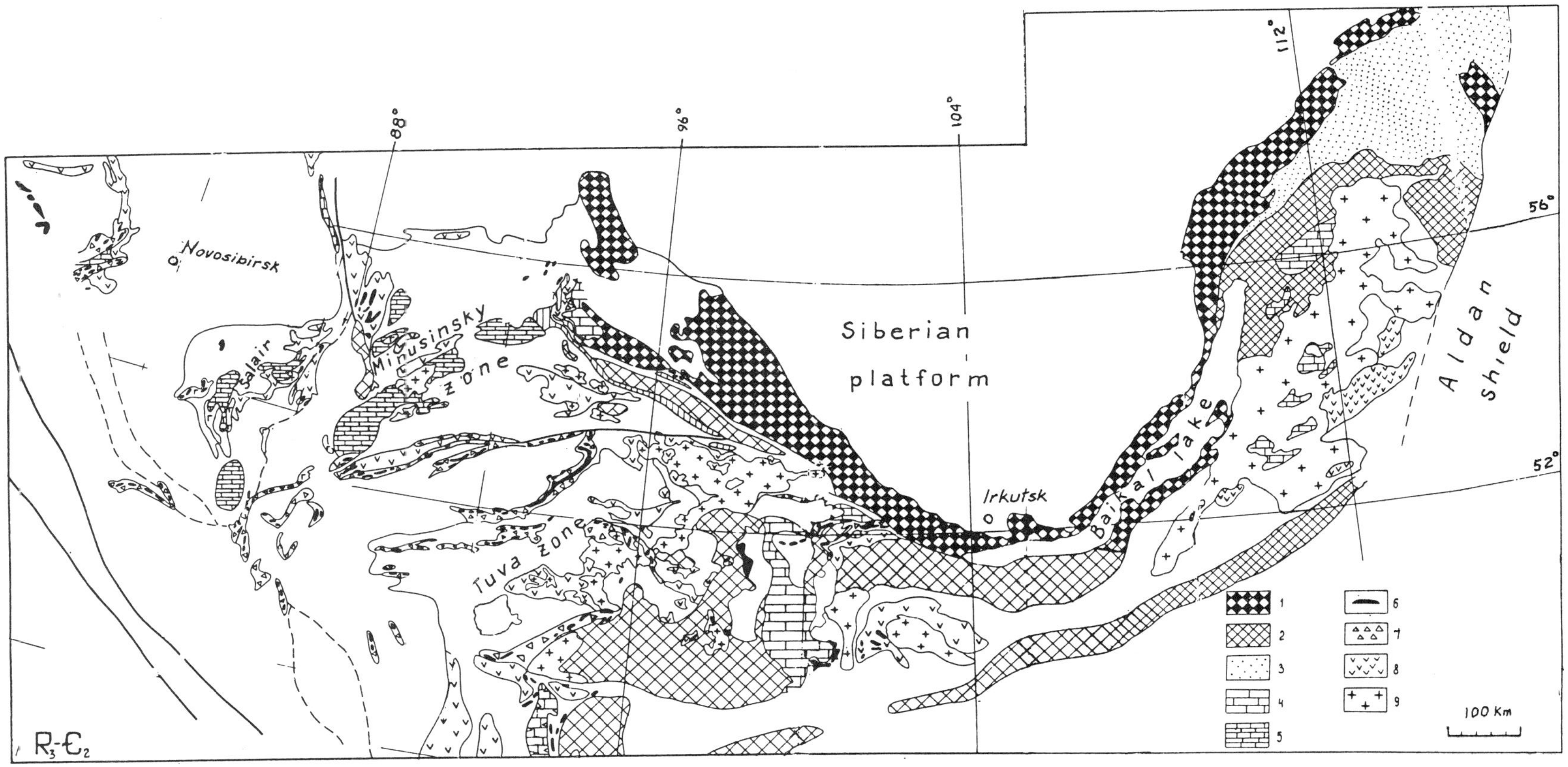

Fig. 64. Distributicn of Late Precambrian and Lower and Middle Cambrian rock complexes within the Altai-Sayan region.
1, Siberian platform basement; 2, ancient massif not belonging to Siberian platform; 3, continental rise deposit; 4, sedimentary (mainly carbonate) cover; 5, limestone; 6, ophiolite; 7, chaotic rocks (olistostrome and serpentinite mélange); 8, island-arc assemblage; 9, granodiorite-tonalite batholith, including Barguzin batholith.

buried under piles of nappes. Rejuvenation of isotopic ages of metamorphic complexes apparently reflects secondary metamorphic processes taking place in connection with uprising of the antiforms.

Massifs of the western Altay-Sayan region are possibly also antiforms overlain by allochthonous Riphean-Low Paleozoic tectonic slices, although some of them may be exotic terranes. These are the Tomsky and Tersin inliers in the Kuznetsky Alatau, their metamorphic rocks being 1470-1880 Ma [Lepezin, 1978]. In the West Sayan and Gorny Altay, metamorphic complexes of the Dzhebash, Chulymshan, Teletsky, and Terektin zones contain metamorphic rocks of epidote-amphibolite and greenschist facies produced from fine-grained sediments and volcanics. Several authors [e.g., Kheraskov, 1979] still doubt the Precambrian age of these complexes, but the isotopic age determinations indicate Precambrian metamorphism from 1120 to 600 Ma [Lepezin, 1978]. A younger age of 520-410 Ma was also determined, obviously showing repeated metamorphism.

The Kurchum massif of the Irtysh-Zaisan zone should be noted as well. The oldest rocks are dated at 600-490 Ma. It can be considered to be an exotic terrane.

Oceanic Complexes

Ophiolites and island-arc complexes of different age are abundant in the Altay-Sayan area. The wide distribution of these complexes allows us to restore a large oceanic basin--part of the Paleo-Asiatic ocean [Zonenshain, 1972].

Comprehensive descriptions of ophiolite assemblages in the Altay-Sayan region are not numerous. They mostly concern the Kurtushibin and Boruss ridges in West Sayan, Chagan-Uzun in Gorny Altay, and the Il'chir complex in the eastern part of East Sayan. Beyond the USSR, in Mongolia the ophiolite complexes are found in the Lakes zone (Khantayshir complex) and in the Khubsugul region.

In the Kurtushibin ridge of West Sayan Dobretsov et al. [1977] reported an ophiolitic sequence (Figure 65) that includes (from bottom to top): a dunite-harzburgite complex 3 km thick; a layered pyroxenite-troctolite-gabbro-norite complex (0.2-0.3 km); taxitic gabbro (1-1.5 km); a gabbro-diabase complex (1.5-2 km). This assemblage is associated with pillow lavas and chert-greywacke sequences of the Chinginsky suite of Vendian-Lower Cambrian age. Gabbro-diabases and pillow lavas of this suite have low-potassium tholeiitic compositions. In places, ophiolites are accompanied by glaucophane schists and compose nappes thrust from south to north over the metamophosed clastic rocks of the internal parts of West Sayan.

The eastern East Sayan ophiolites, as Dobretsov [1985] established, form nappes over metamorphic rocks of the Gargan massif. Here, one can find a full ophiolitic section (from bottom to top): ultramafics, gabbro-pyroxenites, sheeted dikes, and pillow lavas. Widespread boninites are present, along with MORB type tholeiites.

Oceanic rocks are more or less widespread in the lower parts of the Late Precambrian-Cambrian volcanogenic-sedimentary complexes of the Altay-Sayan region. For instance, low-potassic tholeiites close to MORB are known in the Upper Riphean-Vendian Manzherok suite of the Katun zone [Belousov et al., 1969], in the Upper Riphean- Vendian Sugashsky suite, developed in the southern rim of the Terektin uplift of the Gorny Altay [Belousov et al., 1969], in the lower Cambrian Nizhnemonoksky and Vendian-Lower Cambrian Chinginsky suites of West Sayan [Kheraskov, 1979], in the Vendian-Lower Cambrian Kutenbulak suite of the southwestern part of East Sayan, and in the Lower Cambrian Tumattaiga suite of East Tuva.

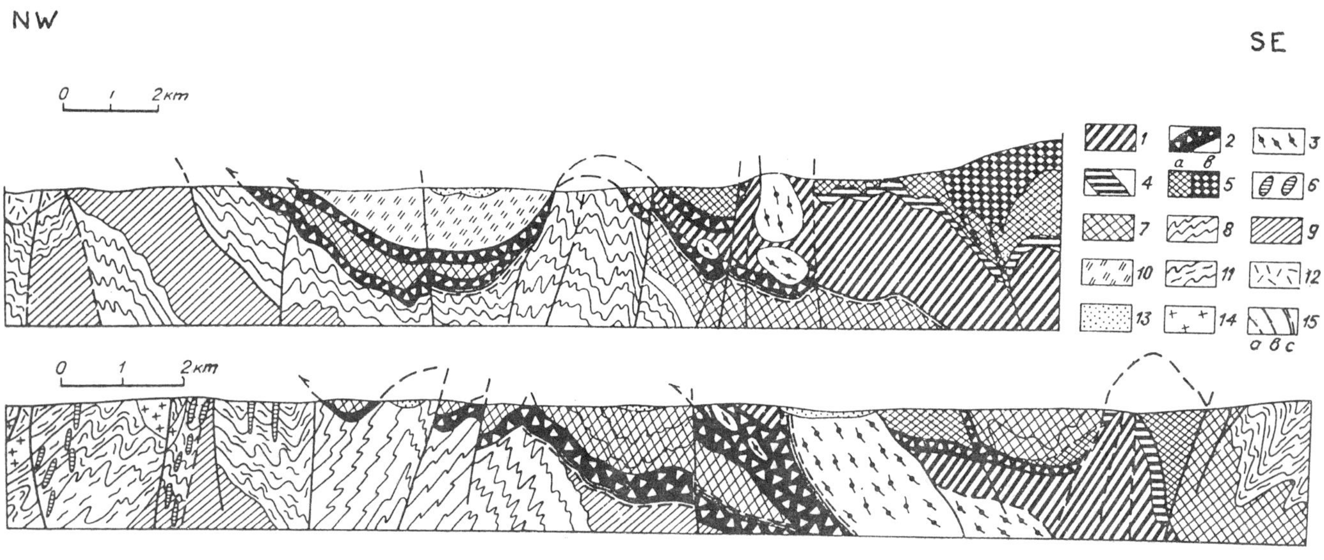

Fig. 65. Cross-section through the Kurtushibinsky ophiolitic belt, after Dobretsov et al. [1977].
1, Serpentinite; 2, melange; 3, peridotite; 4, pyroxenite; 5, (a) gabbro and (b) diabase; 6, chert of Chinginsky suite (Vendian-Lower Cambrian); 7, metamorphics of Dzhebash suite (Riphean?); 8, glaucophane schists of Akkol suite; 9, volcanics of Tereshkina suite (lower Ordovician?); 10, clastics of Ishkin and Alasug suites (Upper Cambrian-Lower Ordovician); 11, Devonian and Carboniferous; 12, Devonian volcanics; 13, Quaternary deposits; 14, granite; 15, (a) thrust, (b) other fault, (c) zone of brecciation.

Thus, the oceanic complexes of the Altay-Sayan region are not younger than Early Cambrian; evidently they started to form in the Late Riphean. This determines the time when ocean floor formed in the Paleo-Asiatic ocean in this area.

EARLY PALEOZOIC ("CALEDONIAN") STRUCTURES

As mentioned earlier, the term "Caledonian" as been applied to Early Paleozoic structures in Asia. This usage may have an approximate temporal connotation but is has no physical basis.

Minusinsk System

The data on the Mansky trough (Figure 66) [Khomentovsky et al., 1978] provide information on the junction of the Minusinsk zone with the Siberian platform. The trough contains three rather diverse domains. From east to west, they are the Solbinsky, Zherzhulsky, and Beretsky domains. The boundaries between them are unusually curvilinear and may be readily interpreted as nappe fronts (Figure 66). The structure described earlier as a simple trough is in reality a package of at least two nappes, Zherzhulsky and Beretsky, thrust over the relative autochthon of the Solbinsky domain, actually the platform cover of Siberia. The sequence in the Zherzhulsky domain is characterized by carbonate rocks and flysch of the same age as in the Solbinsky autochthon, i.e., from the Upper Riphean to the Middle Cambrian, but between there are exotic blocks of Archaeocyathid limestones (Murtuk reef) very similar to those of the internal parts of the Altay-Sayan region and totally different from limestones of the Zherzhulsky domain. Two rock complexes are characteristic of the Beretsky nappe: the first is the Kuvaisky series, a Late Riphean essentially volcanogenic complex, and the second is a flysch complex. The Kuvaisky series incorporates rocks formed in different environments, but island-arc volcanics predominate. Thus, the thrust sheets of the Mansky trough include complexes exhibiting the transition from the continental shelf (autochthonous Solbinsky domain) to the continental slope and rise (limestones and flysch of the Zherzhulsky domain) to a volcanic island arc (Kuvaisky series). The Murtuk reef seems to be an exotic terrane that was incorporated into the arc and then collided (together with the arc) with Siberia. Judging from the Mansky trough, the whole boundary of the Minusinsk system with the Siberian platform should be interpreted as an island-arc-continent collisional boundary.

The Minusinsk system has a mosaic pattern made up of two types of Vendian-Cambrian structural units. These are, on the one hand, blocks basically composed of carbonate and carbonate-chert sequencies and, on the other hand, variously oriented linear zones separating the blocks. These linear zones are largely composed of volcanic complexes having either oceanic or island-arc affinities. Dergunov and Kheraskov [1982] proposed that the carbonate sequences lie unconformably on the metamorphic Precambrian of the Tomsk inlier in the Kuznetsky Alatau.

It seems that the volcanic zones and carbonate sequences are thus separate nappe units brought together tectonically, so that the exposed carbonate blocks, particularly those characterized by granitization, are in antiforms while the volcanic zones are in synforms. As the Middle Cambrian island-arc volcanics are broadly distributed, they may be regarded as neoautochthonous. Emplacement of the volcanic complexes over a carbonate platform must have taken place before the beginning of the Middle Cambrian.

Thus, the Minusinsk zone developed as follows: one or several volcanic island-arcs originated on the oceanic floor of the Paleo-Asiatic ocean in Vendian-Early Cambrian time. The limestone blocks along the Tomsky inlier are considered remnants of a continent that we will call Tomsky, at some distance from the arc. In the Middle Cambrian the Tomsky continent (or its individual fragments) collided with the island arcs, and the ophiolite basement of the arcs and the island arcs themselves were partially thrust over the Tomsky continent. Thus, an accretional mosaic was produced from fragments of continents and island arcs. The Middle Cambrian island arc evidently faced the Siberian continent. Prior to the Late Cambrian, the accretionary mosaic collided with the passive margin of Siberia and was thrust over it. The apparent consequences of this collision were the Salairide folding throughout the entire Minusinsk fold zone, and the emplacement of the Tygertash granitic batholiths.

The Tuva System

The Tuva system comprises the Vendian-Cambrian structures of East Tuva, situated between the Tuva-Mongolian massif and the superimposed Middle-Upper Paleozoic Tuva depression. The Vendian-Cambrian here may be roughly divided into two groups of rock complexes (Figure 64): dominantly volcanic complexes, and chaotic complexes (subduction melanges). Volcanic complexes extend as a band striking NE from the East Tannuola Range to the Khamsara River basin. On both sides, volcanics are bounded by strips of subduction melange. The tonalite-granodiorite batholiths of the Tannuola complex were emplaced along the contact of the Tuva-Mongolian massif within the "early Caledonian" foldbelt. The band of volcanic complexes has an island-arc character and is called the Tannuola (or Tannuola-Khamsara) arc. Available data show that it existed only in the Early Cambrian. The band of chaotic complexes situated east of the volcanic arc separates the latter from the Tuva-Mongolian massif. These chaotics are seen best in the Shurmak zone, which includes blocks of ultramafic rocks, Archaecyathid limestones, and volcanics, and which can be interpreted as a subduction melange arising in the front of the island arc when it overrode the Tuva-Mongolian massif. Another (West Tuva) band of chaotic complexes runs along the north-western margin of the arc volcanic complex. Berzin [1979] proved that this whole band 350 km long and about 50 km wide is a chaotic tectonic melange of differently oriented blocks in a clastic or serpentinite matrix (Figure 67). Olistostromes are closely associated with the tectonic melange.

The following interpretation can be proposed for the evolution of the Tuva system. In the Vendian and Early Cambrian, to the northwest (present coordinates) of the Tuva-Mongolian massif was the Tannuola-Khamsara volcanic island arc, facing toward the massif. In the Early Cambrian the oceanic crust that intervened between the arc and the massif was consumed, producing a subduction melange (Shurmak zone) in the arc front. At the end of the Early Cambrian the arc probably collided with the massif, and subsequently ophiolites and the arc itself were obducted onto the metamorphic basement and sedimentary cover of the massif. The subduction zone was blocked, leading to the formation of a new consumption zone of opposite polarity on the opposite side of the arc. The melange of the West

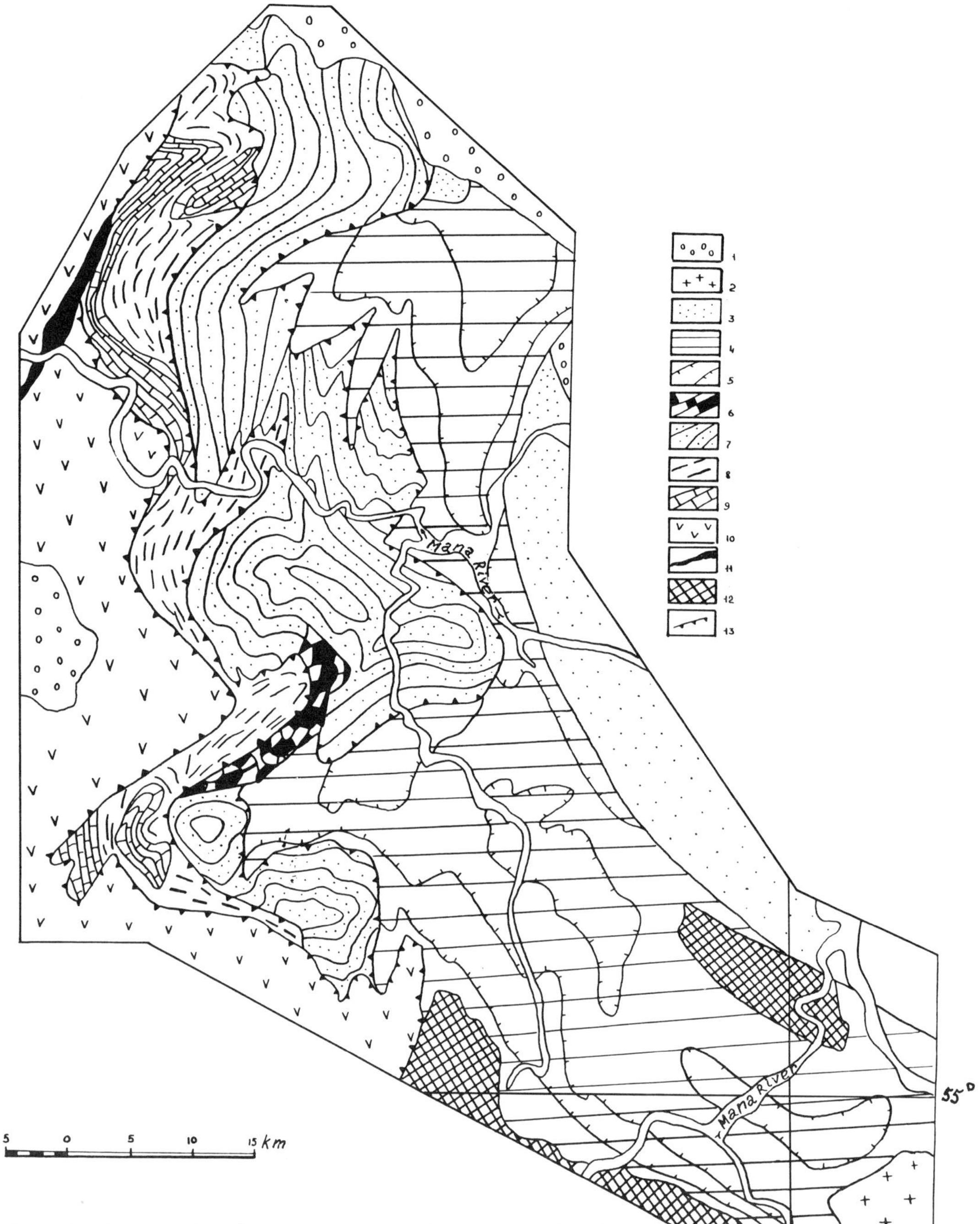

Fig. 66. Geology of the Mansky trough. An interpretation based on the geological map of Khomentovsky et al. [1978].
1, Devonian rocks; 2, Caledonian granite; 3, Upper Cambrian and Lower Ordovician rocks of the Badzhei trough; 4, Upper Riphean and Lower Cambrian rocks of the Solbinsky zone (shelf complex); 5, structural lines within the Solbinsky zone; 6, Murtuk reef (suspect terrane); 7, Upper Riphean and Lower Cambrian of the Zherzhulsky zone (passive margin complex); 8, flysch (continental rise complex); 9, Upper Riphean limestone; 10, Riphean of the Beretsky zone: oceanic and island-arc assemblages; 11, ultramafics; 12, Archean-Proterozoic basement; 13, thrust front.

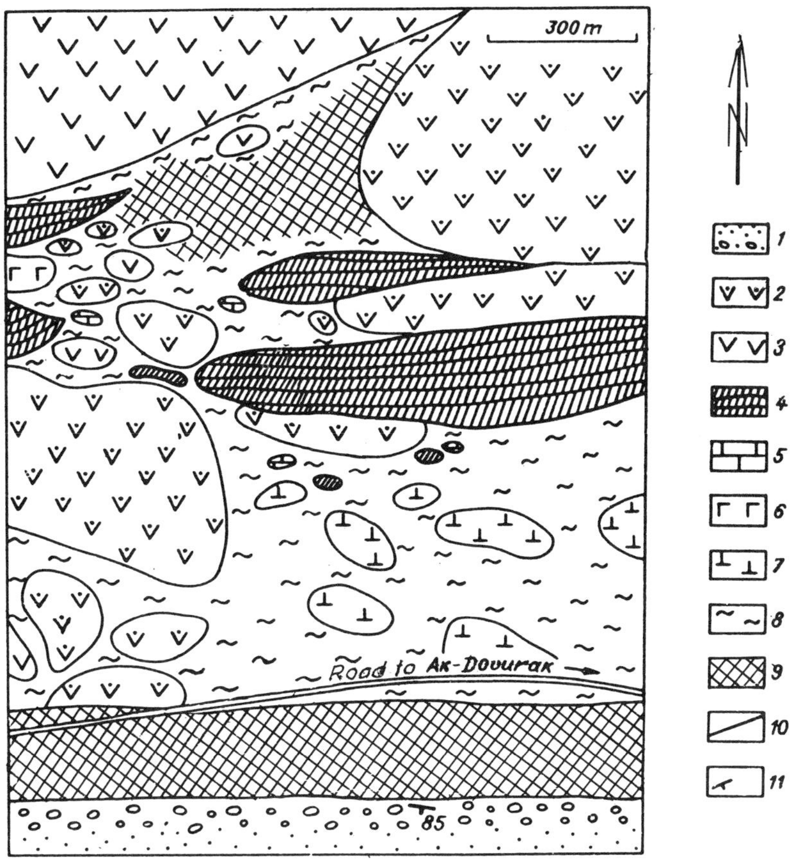

Fig. 67. Detailed geological map of the Lower Cambrian chaotic rocks near Ak Dovurak, Western Tuva, after Berzin [1979].
1, Sandstone and conglomerate (Upper Ordovician); 2, spilite, diabase, basalt; 3, basaltic andesite; 4, chert; 5, limestone; 6, gabbro; 7, massive serpentinite; 8, foliated serpentinite; 9, melange; 10, fault; 11, dip of bedding.

Tuva belt formed in this new subduction zone. As a consequence, a Nevadan-type margin came into existence. Behind the subduction zone, the crystalline basement (buried under nappes) began to rise, and the Tannuola granite batholiths were emplaced.

The Oka-Dzhida System

The Oka-Dzhida system is located between the southern inliers of the Siberian platform and the Tuva-Mongolian massif. According to Dobretsov [1985], this system has a nappe structure (Figure 68). The Gargan massif is autochthonous, and several (at least four) tectonic nappes are thrust over it. The lower nappe consists of ophiolites of the Il'chir complex. The next, the Dabanzhalgin nappe, is made up of the Ordovician-Silurian chert-carbonate sequence of continental rise origin. Large areas are occupied by the Bokson nappe, which consists of shallow-water carbonate deposits of the Bokson series (Vendian-Lower Cambrian). The uppermost nappe unit is composed of calc-alkaline volcanics of the Barungol suite and overlying carbonate-terrigenous deposits of the Toltin suite, evidently of Cambrian-Ordovician age; these rocks formed in island arc environments.

Thick greenschist and flysch complexes, the Okinsky series, are widespread in the eastern part of East Sayan, but their age and origin are unclear. Poor Ordovician fossils [Butov, 1980] and even Silurian fossils [Katyukha and Rogachev, 1983] were found in these rocks. They belong either to the continental rise or to the forearc. We believe that the Okinsky series is an independent nappe unit preserved in several synforms.

The position of the Oka-Dzhida system between the Siberian platform and Tuva-Mongolian massif shows that it originated by the collision of these continental masses. In this small knot (eastern part of East Sayan), several essentially different Early Paleozoic complexes are brought together: shelf, continental rise, island arc, and oceanic volcanic complexes. They are evidence of considerable shortening by consumption of oceanic crust as well as continental collision. Consequently, a subduction zone existed in Cambrian time within the oceanic basin between the Tuva-Mongolian massif and Siberia. The remnants of the volcanic

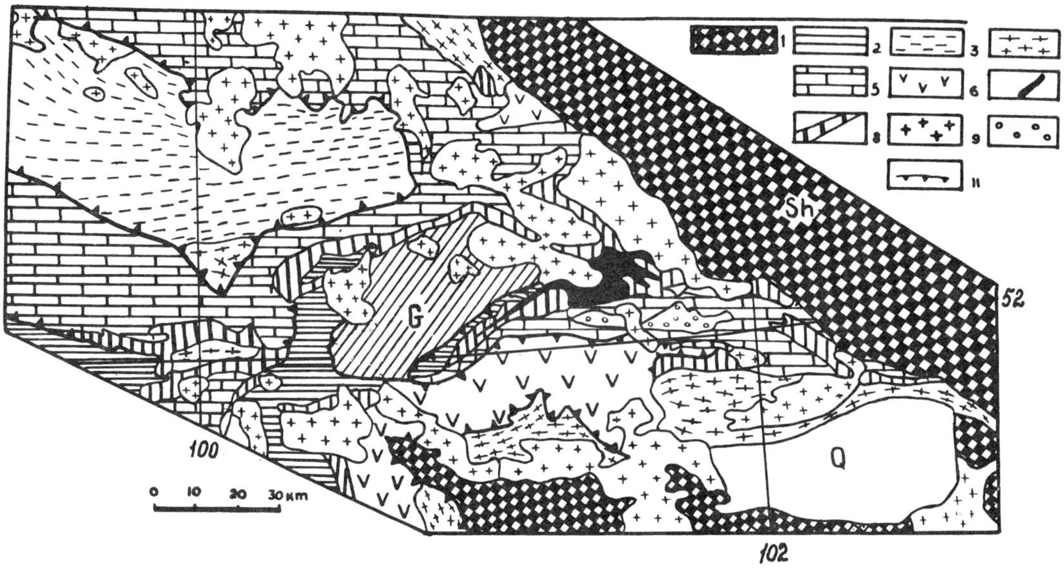

Fig. 68. Main nappe units of the southeastern part of East Sayan, after Dobretsov [1985] simplified. 1, Archean-Proterozoic basement; 2, Riphean-Lower Paleozoic sedimentary cover; 3, Oka nappe; 4, various schists of the Bokson nappe; 5, Vendian-Lower Cambrian limestone of the Bokson nappe; 6, Cambrian-Ordovician island-arc complexes of the Upper nappe, 7, ultramafics; 8, chaotic rocks (melange, olistostrome); 9, Early Paleozoic ("Caledonian") granite; 10, Upper Devonian molasse; 11, thrust fault. G, Gargan massif; Sh, Sharyzhalgay massif; Q, Quaternary deposits

arc above the subduction zone may be seen in the Burungol and Okinsky series. The zone still existed in the Ordovician. Obduction of ophiolites and subsequently continental collision took place along this zone.

The Baikal-Vitim System

The Baikal-Vitim system runs from the Selenga River in the south to the Vitim-Patom plateau in the north, obviously continuing the Oka-Dzhida system. A considerable portion of this system is occupied by the Barguzin batholith (see Figure 64), which has almost totally destroyed the original structure. As stated above, the Barguzin continent, accreted before the Vendian, forms the central part of the Baikal-Vitim system. The two zones on either side of the massif [Belichenko, 1977] are the Okrainny or Marginal zone close to the Siberian platform margin and the more internal Eravnin (Udsko-Vitim) zone. Cambrian sedimentary sequences are typical for the Marginal zone, whereas the Eravnin zone is characterized by a Lower Cambrian andesite-dacitic assemblage undoubtedly of island-arc origin.

From K-Ar data, the Barguzin batholith [Litvinovsky and Zanvilevich, 1976] is from 543 to 350 Ma old. The origin of such a large batholith, 700 km long and discordant to surrounding structures, was evidently related to the collision of continental blocks bounding the Baikal-Vitim system. No reliable data exist for estimating exactly when the strongly arcuate folded structures in the Baikal-Vitim system formed, but apparently the arc developed at some time in the Middle Paleozoic. Soon afterward, all Lower Paleozoic sequences from the Cambrian to the Silurian were folded in front of the arc and were overthrust by the Riphean rocks of the Baikal-Patom highland. Only the Middle Devonian lies unconformably on the older basement and is not deformed [Khain, 1979a].

The history of the Baikal-Vitim system is not easy to reconstruct. Evidently, in the Late Riphean, Vendian, and Early Cambrian the passive continental margin (Marginal zone with carbonate successions) formed over more ancient Riphean accretional complexes, which were part of the Barguzin continent. To the east (present coordinates) this margin faced toward an ocean of unknown size. An island arc was situated within the ocean, and during the early Middle Cambrian, the oceanic floor of this basin was completely consumed in the subduction zone under the arc. Collision of the Barguzin continent with the Siberian (Angara-Anabar) continent apparently began in the Late Cambrian. As a result, melting produced the first granitic portions of the Barguzin batholith, mountainous relief developed in the West Trans-Baikal, and the Istashin molasse formed. However, the marine basin between the Barguzin continent and the Aldan shield was not entirely closed; some oceanic or thinned continental crust remained. The Barguzin continent took its present position relative to Siberia and the Aldan shield not earlier than the Early Devonian.

West Sayan and Gorny Altay

Unfortunately, stratigraphy, petrology, and structure of both West Sayan and Gorny Altay have not been studied sufficiently to give a mobilistic interpretation of their structure and development, so we can provide only a preliminary analysis. A general framework of the structure of West Sayan and Gorny Altay can be seen in Figure 69, which shows the distribution of Ordovician-Silurian complexes within the Altay-Sayan re-

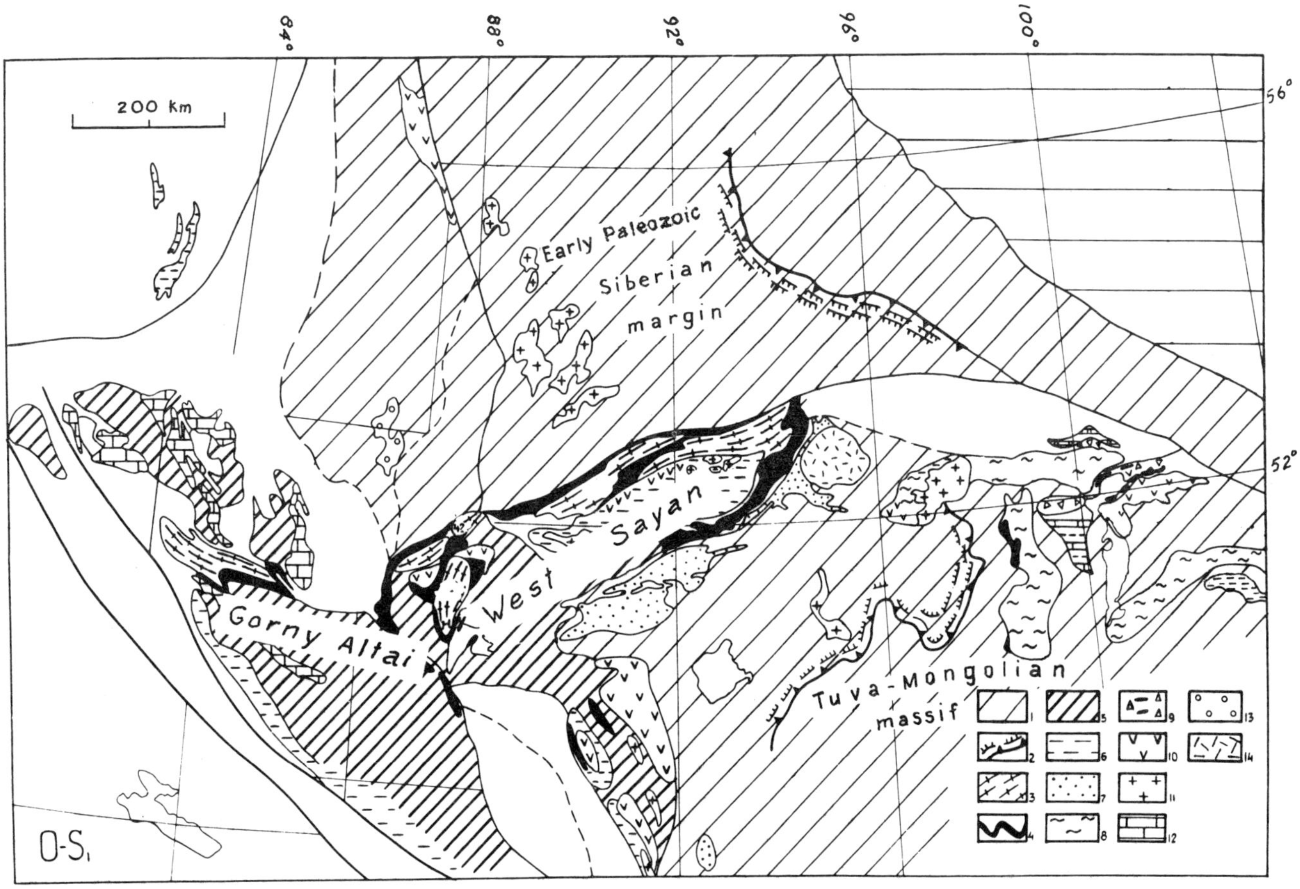

Fig. 69. Distribution of Ordovician-Silurian rock complexes within the Altay-Sayan region. 1, Dry land; 2, Early Paleozoic ("Early Caledonian") thrust front; 3, Dzhebash unit; 4, ophiolite (including melange and olistostrome); 5, Gorny Altai unit (submarine fan); 6, West Sayan unit (flysch); 7, Khemchik-Sistigkhem unit (continental margin clastic wedge); 8-10, Oka-Dzhida unit: 8, greenschist; 9, ophiolite; 10, island-arc assemblage; 11, granodiorite batholith; 12, shelf carbonate; 13, molasse; 14, terrestrial volcancis of intraplate origin. Encircled numbers correspond to numbers in Figure 60.

gion. This system is divided into individual blocks or zones: West Sayan, Khemchik-Sistigkhem , Chulyshman, Kharkharin, Talitsky, Kholzun-Chuisky, and Mongolian Altay.

Several characteristic geological complexes can be recognized within West Sayan and Gorny Altay. Each may be regarded as an independent structural unit, formed under different conditions and possibly far away from adjacent structural units. The following units are distinguished: Dzhebash, an ophiolite unit with olistostromes, Gorny Altay, West Sayan, and Khemchik-Sistigkhem.

The Dzhebash Unit

This unit consists of metamorphic sequences forming the so-called outliers of the basement described above: Dzhebash, Teletsky, Chulyshman and Terektin.

The Ophiolite Unit

This includes oceanic complexes and ultramafic belts of West Sayan and Gorny Altay.

The Gorny Altay Unit

This unit is characterized by a thick, uniform, clastic sequence of flysch-like type (the Gornoaltaisky and Ishkinsky suites) represented by a monotonous sequence of green (rarely violet) sandstones, siltstones, and argillites (phyllites) as well as rare interbeds of coarser clastic rocks, gritstones, and conglomerates. This sequence was deposited by turbiditic currents on the continental rise. The clastic material is uniformly rich in fledspar and quartz. The coarser clastic varieties include fragments of schists, microquartzites, cherts, felsic lavas, pegmatites, and rarely porphyrites. Importantly, amongst the clastic material no oceanic rocks derived from nearby ophiolite melange zones are known. Fragments of limestones common in immediately nearby Cambrian complexes are not present either. Evidently a continental basement made of schists, granites, and felsic lavas was being eroded to produce these sediments. The amount of clastic material in the Gorny Altay unit is no less than 300,000 km^3. In modern times, similar large volumes of clastic material are found only in the submarine fans of the largest rivers (Amazon, Nile, Ganges, Brahmaputra). Clearly, a large continent supplied clastic sediment to the fossil fan in the Gorny Altay unit. It was hardly Siberia, since between it and the Gorny Altay unit is the Minusinsk system, from which products of erosion are absent in the clastic material of the Gorny Altay unit. Therefore, we believe that the Gorny Altay cone formed on the continental rise of some other continent as large as the Siberian continent. Later the cone was separated from the parent continent and moved to the Siberian margin. This whole thick sedimentary sequence suffered strong deformation in the very beginning of the Ordovician.

The history of the different blocks that built the Gorny Altay unit was not similar. Some were covered by a shallow-marine carbonate-clastic series from Middle (Lower?) Ordovician to Eifelian, unconformably overlain by the Eifelian-Givetian volcanic complexes of the Anui-Chuya trough. This shelf complex was deformed mainly in the Early Carboniferous when volcanics of the Anui-Chuya trough were deformed as well.

In the region southwest of the Gorny Altay unit, granitic batholiths of the Altay complex intruded at about the boundary between Devonian and Carboniferous times (isotopic age of granitoids 365-350 Ma [Leont'ev et al., 1981]). A volcanic island arc formed in the Ordovician (and in places in the Silurian) in the northeast Gorny Altay unit.

The West Sayan Unit

This unit includes the Ordovician-Silurian sequences of the West Sayan synclinorium. They are represented by monotonous sandstone-shale successions often resembling flysch. In the northern edge of the West Sayan synclinorium, the flysch incorporates rare andesites and tuffs, plus limestones with an Ordovician fauna; however, these rocks could be olistoliths.

The dominantly flyschoid composition of Ordovician deposits in the internal parts of West Sayan and their supposed conformable deposition over the Gorny Altay series contrast with the closely adjacent Chulyshman highland where molasse and Ordovician island-arc complexes lie unconformably on the folded and metamorphosed Gorny Altay series. Evidently these two units were brought together only after the Early Silurian.

In the West Sayan, the sandstone-shale sequences were deformed in the Late Silurian.

The Khemchik-Sistigkhem Unit

This unit is a body of sediment filling a trough superimposed on the Cambrian (Salairide) basement of the Tuva system. It contains Ordovician coarse clastic deposits and Silurian carbonate and terrigenous rocks. The thickness of the Ordovician-Silurian deposits totals 5000-6000 m. All the rocks accumulated in a shallow-marine environment.

Among the West Sayan and Gorny Altay system structures, only the Khemchik-Sistigkhem trough can be unambiguously interpreted. The trough sediments accumulated as an autochthonous complex on the margin of the Tuva-Mongolian massif. No doubt all the other units are divided by tectonic contacts. Nearly the same structural pattern is observed everywhere: the Dzhebash unit is exposed in the center and is rimmed by narrow bands of ophiolite melange, while in the periphery the Gorny Altay unit appears extensively. Each unit is evidently an independent tectonic nappe, but it is not clear whether the Dzhebash unit crops out in a synform or an antiform. Up to now we have only the evidence of Dobretsov et al. [1977] indicating tectonic overlap of the Kurtushibinsky ophiolite melange on the Dzhebash unit. If this relation is accepted, the Dzhebash unit must be exposed in antiforms, whereas the Gorny Altay unit must compose the upper structural level in synforms.

The tectonic displacement of these units (nappe formation) is thought to have taken place in the Early Ordovician. The presence of melange and products of high-P metamorphism indicate that the nappes formed above subduction zones. However, no data are available on the position of these subduction zones, and no remnants of volcanic complexes associated with them are known either. Thus all these units probably represent an exotic massif which was already tectonically stacked when it arrived.

The flysch complexes of the West Sayan unit accumulated along a volcanic island arc, as is shown by neighboring calc-alkaline volcanics as well as the presence of volcanic fragments in the Ordovician sandstones. The West Sayan unit evidently reflects the convergence of all the previously separated blocks (Tuva, Minusinsk, Gorny Altay) and the closure of the oceanic basins lying between them. Collision was responsible for the Late Silurian folding and emplacement of granodioritic batholiths in the center of the West Sayan Mountains. Apparently collision did not terminate in the Silurian, but continued in the Devonian as well.

Thus, the early Paleozoic Altay-Sayan area is an agglomerate of heterogeneous blocks, originally situated in distant locations within the Paleo-Asian Ocean and aggregated together at different times. It is a typical example of the accretion of exotic blocks.

Salair

The Salair Range is composed of a NE-facing arc thrust over the Kuznetsky basin [Fomichev and Alexeeva, 1961]. The arc was deformed in the late Paleozoic ("Hercynian") and now constitutes of a series of nappes (e.g., Gurjevsky overthrust) with east vergence. However, Late Cambrian and Lower Paleozoic complexes, the analogs of sequences of the same age in the Minusinsk system, are exposed in a series of slices in the core of the Salair. A volcanic island arc of that age can be reconstructed in the Salair; possibly it was originally connected with the Kuznetsky-Alatau (Minusinsk) arc and was only subsequently separated from it.

The Ordovician-Silurian deposits unconformably overlying the Cambrian island arc are dominantly clastic or flysch-like sequences with limestone horizons. The Lower Devonian and Eifelian sequences consist predominantly of limestones; those from the Givetian upward consist of coarse clastic shallow-marine sediments. The columnar section ends with the Lower Carboniferous. Thus, the Salair history is that of an island-arc segment which split off from the Kuznetsky Alatau (Minusinsk) arc, and which for a long time did not collide with Siberia, remaining inside the oceanic basin. The Middle Paleozoic shelf deposits lying above the island-arc rocks show that this ancient island arc subsided calmly until the Carboniferous, when it and its cover were thrust over the margin of Siberia in the Kuzbass region.

LATE PALEOZOIC ("HERCYNIAN") STRUCTURES

As mentioned previously, the Late Paleozoic structures of the Asian regions of the USSR have often been called "Hercynian" despite the lack of demonstrable physical connection with the Hercynian of central and western Europe. The structures of this age in different regions may have very different origins, and this is evident in the eastern and western parts of the Altay-Sayan area.

Eastern Altay-Sayan Region

The mosaic of blocks that was accreted to Siberia prior to the Devonian added to the Siberian continent on the west and southwest. After the Devonian, numerous block-fault structures were superimposed on this margin.

The middle-Upper Paleozoic rocks of the eastern Altay-Sayan area comprise two lithologically different sequences: (i) Early to Middle Devonian, and (ii) all remaining deposits up to the Permian.

The Early to Middle Devonian (Eifelian)was a magmatic epoch when voluminous terrestrial volcanics were erupted and granitoid batholiths were emplaced (Figure 70). Volcanic sequences are the earliest deposits in several depressions: South Minusinsk, North Minusinsk, Rybinsky, Tuva, and Kuznetsky. The volcanism was subaerial and represented a typical bimodal series including alkaline basalts and other alkaline rocks. The vast distribution of the volcanism, its presence on both Early Paleozoic and Precambrian structures, the lack of relation to any plate boundary, and the bimodal and dominantly alkaline composition of the igneous rocks, suggest that the magmatic activity was an intraplate phenomenon related to a hot-spot.

Granitic batholiths are also characteristic of the eastern Altay-Sayan area . They are known in West Sayan, in the southeast part of East Sayan, and in East Tuva. Their composition is close to the collisional type. Granites were emplaced after the main phase of the Early Devonian volcanism but be-

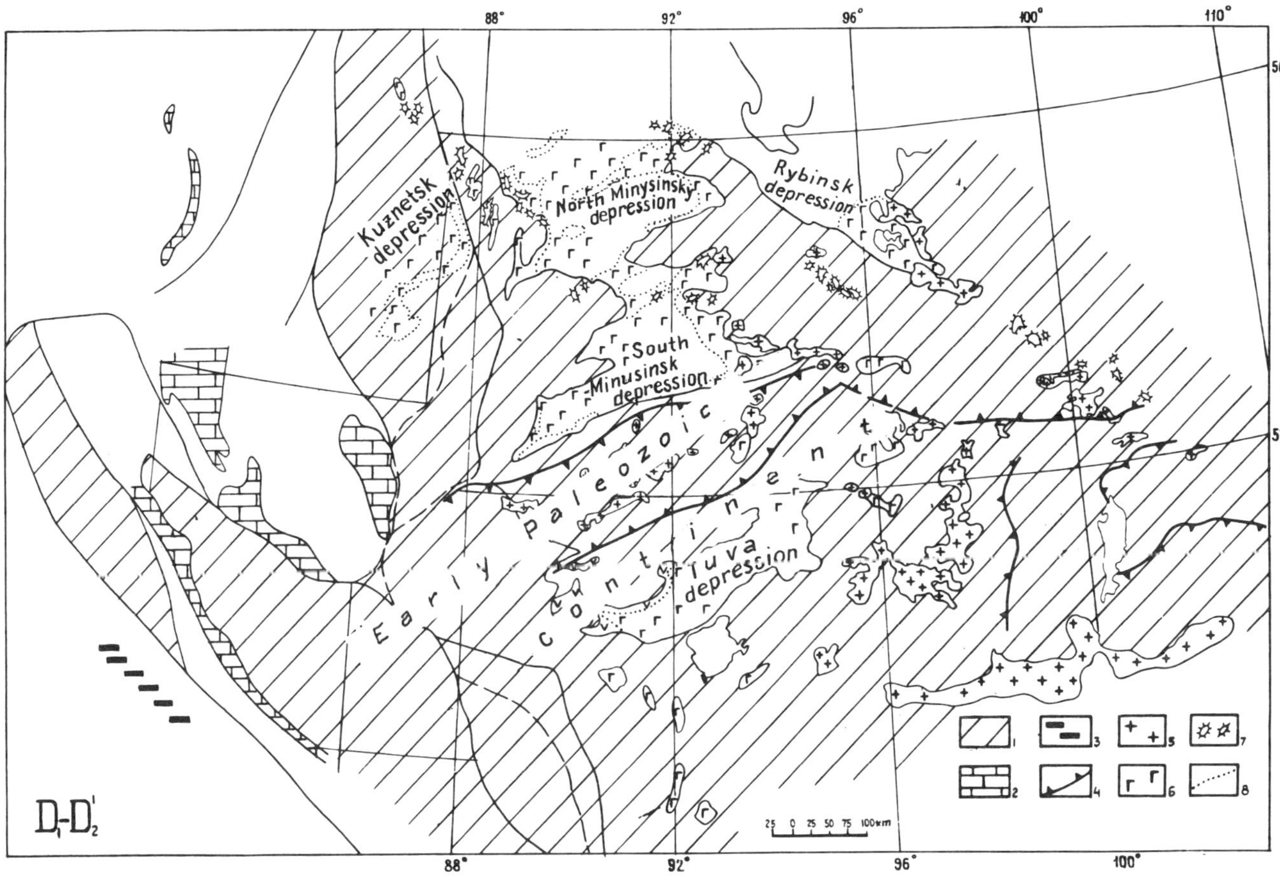

Fig. 70. Distribution of Lower Devonian and Eifelian rock complexes within the Altay-Sayan region.
1, Dry land; 2, shelf carbonates; 3, supposed oceanic floor; 4, Early Paleozoic ("Caledonian") thrust front; 5, granite batholith; 6, intraplate bimodal volcanics; 7, alkaline pluton; 8, limit of outcrop of 6.

fore the deposition of the upper sedimentary sequences. Significantly, the Tuva-Mongolian massif continued to converge with Siberia at that time.

Thus, the Middle-Late Paleozoic structures of the eastern Altay-Sayan area began as a result of intraplate, hot-spot related magmatism. The consequent crustal extension and crustal thinning resulted in the eventual formation and subsidence of sedimentary basins. Simultaneously, the accretion of a mosaic of blocks incremented the Siberian platform, and in some places melting along former collisional zones produced palingenic granites.

Western Altay-Sayan Region

The "Hercynian" structures of the Western Altay-Sayan area mostly originated within, or along the margins of, oceanic basins situated between the Early Paleozoic continents of the Eastern Altay-Sayan area (by then part of Siberia) and Central Kazakhstan.

The Western Altay-Sayan area includes (Figure 60) the Saur, Zharma, Irtysh-Zaisan (with the Chara ophiolite belt), Kalba-Narym, Rudny Altay, South Altay, and Gorny Altay zones. These structural zones are characterized by a vast development of subduction-related rock complexes, including calc-alkaline volcanics, associated flysch-greywacke series, and subduction melanges (Figure 71). Kovalev and Karyakin [1975] proposed the plate-tectonic model of this area as an active margin of the Siberian continent (including trench, volcanic arc, and marginal sea) that collided in the Late Paleozoic with the Kazakhstan continent. This model was later developed in more detail by Rotarash et al. [1982] and Rotarash and Trubnikov [1983].

The Saur Zone

The Saur Zone, which is the southernmost, is composed of thick volcanic sequences of the Middle-Upper Devonian and

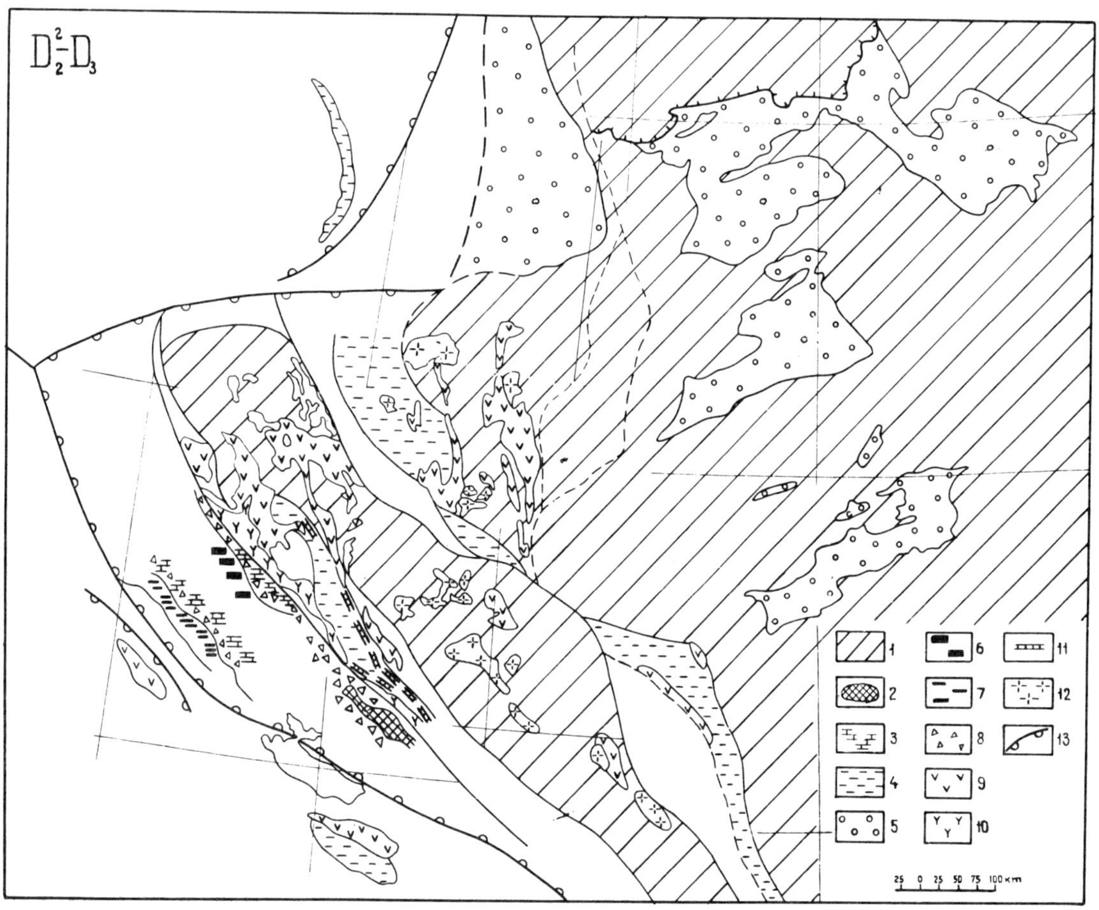

Fig. 71. Distribution of Middle-Upper Devonian rock complexes within the Altay-Sayan region.
1, Dry land; 2, Kurchum exotic terrane; 3, shallow-water limestone, including reefs; 4, marine clastics, including flysch; 5, molasse; 6, oceanic floor; 7, deep-water chert; 8, chaotic rocks; 9, Givetian Frasnian island-arc volcanics; 10, Famenian island-arc volcancis; 11, gabbro-diabase dike; 12, granite; 13, inferred location of subduction zone.

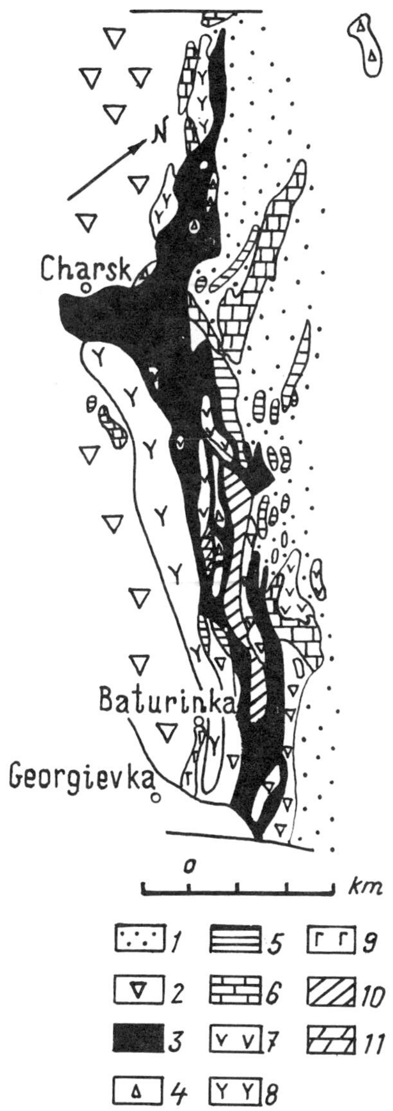

Fig. 72. Tectonics of the Chara ophiolite belt, after Poliansky [1979] simplified.
Autochthon: 1, greywacke; 2, olistostrome. Allochthon: Nappe unit: I: 3, melange; 4, olistostrome (C); 5, chert (S-D); 6, limestone (D_3-C_1); 7, basaltic andesite (C); 8, gabbro-tonalite. Nappe unit II: 9, spilite-diabase and jasper (Pz_1). Nappe unit III: 10, chert and leucobasalt (Pz_1). Nappe unit IV: 11 limestone and chert.

Tournaisian corresponding to an island arc, evidently independent of the Siberian active continental margin.

The Zharma Zone

The Zharma zone is situated between the ophiolite melange of the Chara belt and the Early Paleozoic ("Caledonian") Chingiz-Tarbagatay zone of Central Kazakhstan. Like the Saur zone, volcanic andesite-dacitic island-arc sequences of the Upper Devonian and Lower Carboniferous are widely distributed, but volcanism proceeded throughout the entire Carboniferous time.

The Irtysh-Zaisan Zone

The most remarkable feature of this zone is the Chara ophiolite belt situated near its western margin (Figure 72). The larger eastern part of the Irtysh-Zaisan zone is occupied by monotonous flysch-greywacke sequences and Namurian blue tuffites.

Rotarash and Gredjushko [1974] found that ultramafic rocks of the Chara and other belts are incorporated in serpentinite melanges. They proved that the Chara ophiolite belt has a nappe structure (Figures 73, 74), as was later supported by other investigators [Polyansky et al., 1979; Ermolov et al., 1981, 1983; Belyaev, 1982]. An intensely deformed flysch-greywacke sequence serves as an autochthon, or rather, parautochthon. The overlying allochthonous complexes consist of diverse rock types which occur as blocks in the serpentinite melange and olistostromes. They include ophiolititic rocks, i.e., ultramafics, gabbro, and pillow lavas of spilite-diabase composition, cherts, shales, and jaspers, in places with intercalations of pelagic limestones. Cherts contain Givetian, Frasnian, and Famennian Radiolaria [Ermolov et al., 1981] and can be regarded as pelagic sediments. Metamorphic rocks including glaucophane schists that are 470-545 Ma in age [Polyansky et al., 1979] are also incorporated in the melange. Besides ophiolites and metamorphics, cherty-carbonate rocks of Lower Silurian to Frasnian age, reef limestones of Frasnian, Tournaisian, and Visean age, and andesite-basalt island-arc complexes of Upper Devonian and Visean age occur as blocks in the allochthon. All this chaotic mass is undoubtedly a subduction melange [Rotarash et al., 1982].

The age of deformation and nappe formation is determined by unconformably overlying Middle to Upper Carboniferous andesite-basalt sequences and molasse above the tectonic nappes. The island-arc volcanism proceeded up to the Late Carboniferous. Evidently, the subduction melange, which had been deformed at the end of the Early Carboniferous, was later included in the basement of a new volcanic arc in the Middle to Late Carboniferous.

The Kalba-Narym Zone

The Kalba-Narym zone consists of two subzones [Rotarash et al., 1982]--a western one composed of black shales (Takyr suite) of the Famenian-Tournaisian (latest Devonian and earliest Carboniferous) age, and an eastern one adjacent to the Irtysh shear zone having a chaotic structure and consisting of olistostromes. The chaotic complex includes the Kurchum block of metamorphic rocks. The same subzone incorporates serpentinite melange outcrops with blocks of massive serpentinites, amphibolites, layered gabbro, and diabases, i.e., rocks similar to those in the Chara ophiolitic belt. Rotarash et al. [1982] interpreted this chaotic complex as a subduction melange formed in front of the Rudny Altay volcanic arc. It is assumed that the Kurchum block and also the Middle to Upper Devonian bioherms (of which analogs are not found nearby) are exotic terranes.

A considerable portion of the Kalba-Narym zone is occupied by granitoid batholiths of the Kalbinsky complex, 270-290 Ma old [Leontjev et al., 1981]. The batholiths were em-

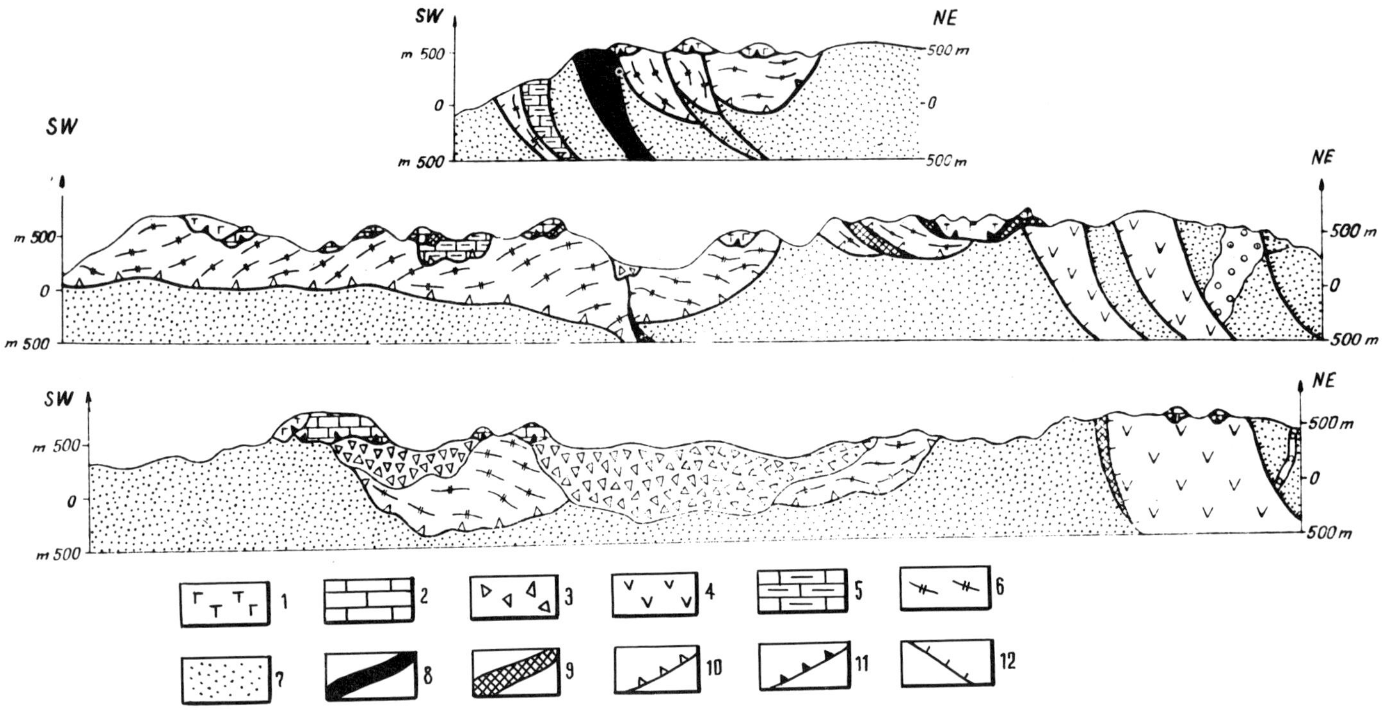

Fig. 73. Cross-sections through the Chara ophiolite belt, after Beliaev [1982].
1, Spilite-diabase and jasper; 2, limestone; 3, olistostrome; 4, basaltic andesite; 5, chert; 6, clastics and olistostrome; 7, clastics; 8, ultramafics; 9, tectonic breccia; 10, base of the Lower nappe; 11, base of the Upper nappe; 12, steeply dipping thrust.

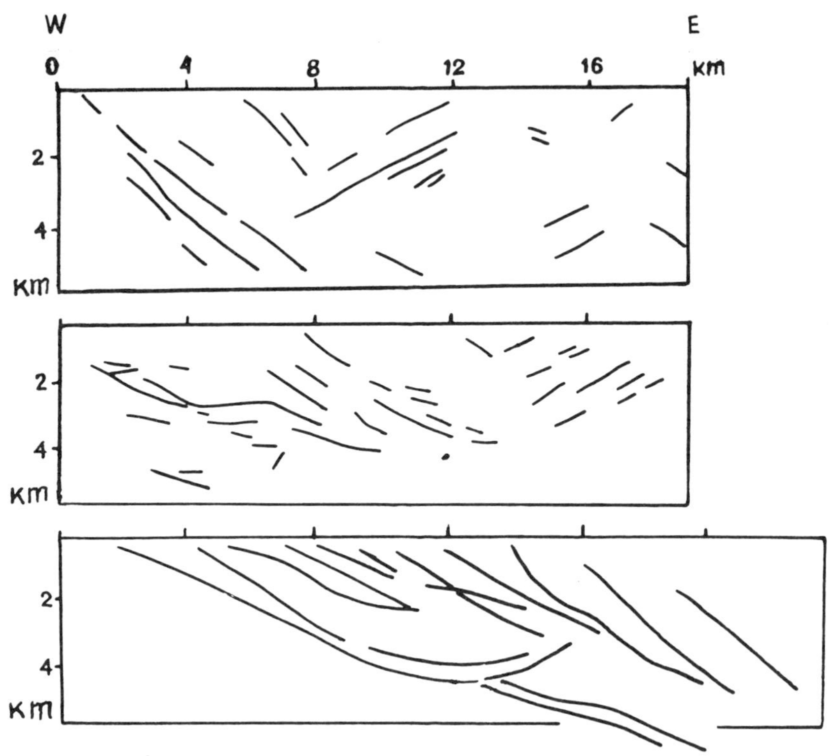

Fig. 74. Seismic reflection structure of the Chara ophiolite belt after Ermolov et al [1981]. Lines correspond to reflectors.

placed after all the island-arc complexes had formed, probably indicating that by that time the Siberian and Kazakhstan continents had started to collide.

The Rudny Altay and South Altay Zones

Between the Irtysh and northeast shear zones (both of which trend NW-SE) are two belts adjoining one another end-to-end: i) the Rudny Altay zone in the northwest, including Devonian volcanics and granitoid batholiths, and ii) the South Altay zone in the southeast with widespread Devonian-Carboniferous marine flysch sequences. In the Rudny Altay zone a thick volcanic complex of Devonian age lies on the Early Paleozoic "Caledonian" basement [Chernov, 1974; Kuzebny, 1975]. It began to develop in the Eifelian Stage and continued with some breaks up to the Famennian. Felsic lavas, with potassium content increasing from west to east, dominate among the volcanics. Rotarash and Trubnikov [1983] interpreted the volcanics as a continental-margin belt related to subduction under the Siberian continent.

The next volcanic epoch occurred in the Famennian and Tournaisian (latest Devonian and earliest Carboniferous). Volcanics of this age (Pikhtovsky suite) have distinct island-arc characteristics. Behind the volcanic belt, near the northeast shear zone, diabase dikes and larger subvolcanic bodies of gabbro-diabase trend NW-SE and record extension. According to Rotarash et al. [1982], the emplacement of the subvolcanic intrusions suggest that a marginal basin formed behind the Rudny Altay island arc. This back-arc basin was filled with clastic sequences of the South Altay zone.

Volcanism ceased during the Late Tournaisian and Early Visean (Early Carboniferous), but it revived in the Late Visean and continued up to the Early Namurian. The Upper Namurian and Middle Carboniferous are poorly represented in the Rudny Altay, but there are some continental and shallow-water marine coarse-clastic deposits of this age. Deformation, in particular folding, took place in the Middle Carboniferous. After folding and the intrusion of Zmeinaya Gora granitoids in Rudny Altay, there were repeated strong, entirely terrestrial, volcanic episodes in the Late Carboniferous and Early Permian, which produced lavas of andesitic through dacitic to rhyolitic composition.

Thus, the Rudny Altay and South Altay zone record the history of a Devonian-Early Carboniferous volcanic arc and marginal sea. The corresponding rock suites were deformed in the Middle Carboniferous and were subsequently incorporated in the basement of the Late Carboniferous to Early Permian volcanic arc. Granitic batholiths record collisional events.

Late Paleozoic Structures of the Gorny Altay Zone

The Gorny Altay suffered significant reorganization in the Lower Eifelian or Givetian Stages of the Middle Devonian,when terrestrial volcanism covered practically the whole Gorny Altay territory. The volcanic rocks are confined to a latitudinal band along the northern margin of Gorny Altay. Andesites and dacites, i.e., rocks typical of the island-arc environment, predominate here. In order to explain the presence of an island-arc volcanic belt along the northern slope of the Gorny Altay, we propose the existence of a subduction zone in the Devonian north (present coordinates) of Gorny Altay, i.e., in the region which is at present buried under the Meso-Cenozoic sedimentary cover of the Barnaul depression of the West Siberian plain. The sublatitudinal belt of ultramafics revealed by drilling in the basement of the Barnaul depression and shown on the Tectonic map of North Eurasia [Peive et al., 1980], possibly represents subduction melanges associated with this zone.

Thick (2000-3000 m) deep-water black shale sequences associated with turbidites are characteristic features of the Upper Givetian and Frasnian of the Anui-Chuya and Yustyd trough occupying the eastern part of the Gorny Altay zone. In accordance with available data, these marine troughs were newly formed in the Devonian and are underlain by the Early Paleozoic ("Caledonian") basement. Presumably, the troughs resulted from extension and the spreading apart of Early Paleozoic blocks.

The Tom-Kolyvan Zone

This zone has a distinct linear structure. It trends SW-NE, cutting across the sublongitudinal Salair and Kuznetsky Alatau zones. It is bounded on the SE by the gently dipping Tom overthrust (nappe), along which Devonian deposits in the frontal part of the zone are thrust southeast over the Late Paleozoic and more ancient deposits of Kuznetsky Alatau, Kuzbass, and Salair (Figure 75). Deep troughs in front of the nappe (the Gorlovsky, Zarubinsky, and Tashlinsky troughs),

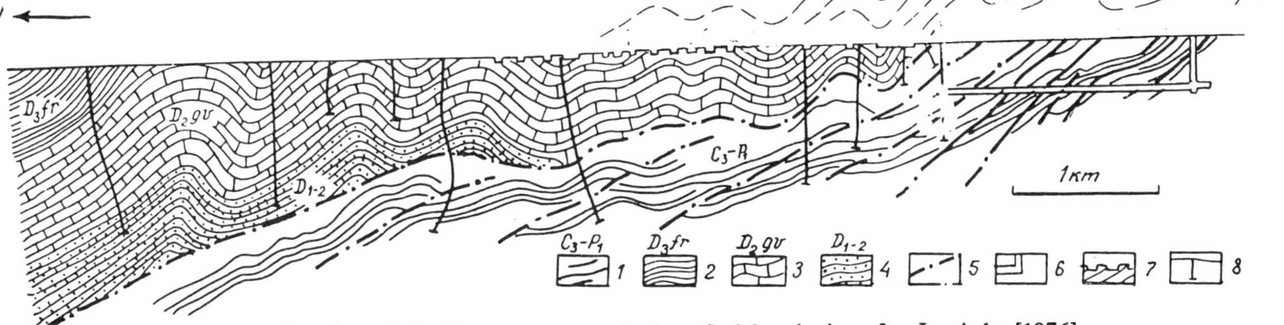

Fig. 75. Cross-section through the Tom nappe near Anzhero-Sudzhensk city, after Juzvitsky [1976]. 1, Permo-Carboniferous coal-bearing deposits; 2, Frasnian sandstone; 3, Givetian limestone; 4, Lower-Middle Devonian redbeds; 5, thrust; 6, mine; 7, artificial outcrops; 8, drill hole.

similar to foredeeps of platforms, are filled with thick sedimentary sequences of Middle Paleozoic up to Permian age. The Tom-Kolyvan zone itself also consists of a series of tectonic slices thrust toward the southeast. The most ancient deposits are Devonian, beginning with Givetian and in places, Eifelian. The Givetian and Frasnian deposits contain volcanic piles composed of diabases, plagioclase porphyrites, albitophyres, and various tuffs--evidently remnants of an island-arc association. The internal part of the Tom Kolyvan zone (Novosibirsk trough) includes a thick sequence (up to 9000 m) of slates and flysch covering the time span from the Upper Givetian to the Namurian. The Tom-Kolyvan zone is part of a considerably larger fold-zone hidden under the West Siberian lowland. The exposed portion displays only small remnants of the Middle Devonian island-arc and the infilling of a great sedimentary basin.

To conclude this consideration of the western part of the Altay-Sayan "Hercynides", it should be emphasized that no subduction complexes older than Middle Devonian are known. However, the melange of the Chara belt contains blocks of rocks of Lower-Middle Paleozoic oceanic floor and oceanic sediments up to Devonian in age. It can be assumed that sea floor spreading was occurring and a passive margin formed in Silurian and Devonian time somewhere between the Early Paleozoic Altay-Sayan margin of Siberia and Central Kazakhstan. In the Middle Devonian, after the Eifelian Stage, several subduction zones (Rudny Altay, North Altay, Saur, Tom-Kolyvan, and possibly Zharma) came into existence during a drastic reorganization.

The history of the Rudny Altay subduction zone can be well reconstructed (Figure 76). The trench was evidently alongside the subduction melange of the Kalba-Narym zone. In the Late Devonian, an exotic Kurchum block of metamorphic rocks (probably together with overlying Middle Devonian reefs) was brought into the subduction zone. This event apparently led to the "blocking" of the latter, creation of chaotic complexes, and jumping of the subduction zone to the west over to the Chara ophiolitic belt. As a result, in the Early Carboniferous we can see the same paleogeographic pattern as in the Devonian, but shifted southwestward.

In the Early Carboniferous, simultaneously with the Rudny Altay volcanic arc, another arc existed in the Zharma zone. Evidently collision of these island-arcs produced the pre-Middle Carboniferous nappes in the Chara belt. However, calc-alkaline magmatism in the Zharma and Rudny Altay zones continued to the Late Carboniferous. Consequently, an ocean basin had to exist up to that time to be consumed in subduction zones. The final closure of the oceanic basin, and collision of Kazakhstan with the "Caledonides" of the Altay-Sayan margin of Siberia, came only in Middle Permian time as shown by emplacement of the Kalbinsky batholith.

A different succession of events is reconstructed for the northern margin of Gorny Altay. A subduction zone dipped under Gorny Altay in the Middle Devonian for a short time. But its activity ceased quickly, for the Salair block moved close to the subduction zone and evidently was thrust over the Gorny Altay margin. Perhaps this push of the Salair block onto Gorny Altay split off the Altay block in the Middle Devonian, forming the Anui-Chuya and Yustyd extensional troughs. During the middle-Late Devonian and Carboniferous, the Tom-Kolyvan arc converged with the Altay-Sayan region until, in the Permian, it too collided with the Altay-Sayan margin of Siberia. As the subduction zone in the Rudny Altay arc dipped under the Altay-Sayan whereas those in the Salair and Tom-Kolyvan arcs dipped away, there must have been a ENE (present coordinates) transform fault between the two arc systems.

HISTORY

Formation of the Altay-Sayan area is shown on palinspastic reconstructions made with respect to a fixed Siberia (Figures 77-84). The paleolatitudes are plotted on reconstructions in accordance with paleomagnetic data on Siberia [Khramov, 1982].

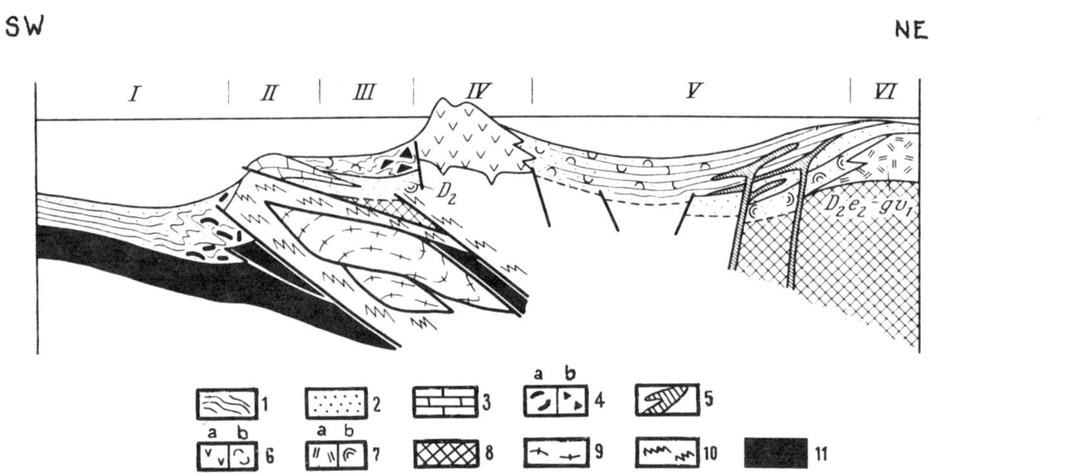

Fig. 76. Paleotectonic cross-section of southwest Altay in the Late Devonian, after Rotarash et al. [1982]. 1, Shale; 2, clastics; 3, limestone; 4, exotics: (a) of various lithologies, (b) of andesite and dacite; 5, diabase and gabbro-diabase; 6, andesite: (a) lava, (b) tuffs; 7, felsic volcanics: (a) lavas, (b) tuffs; 8, Pre-Middle Devonian basement; 9, amphibolite and gneiss; 10, dynamic metamorphic rocks; 11, ophiolite. I - Kalba-Narym zone, II - Irtysh-Zaisan zone, III - Pugachev zone, IV - Rudny-Altai zone, V - South Altai zone, VI - Kholsun-Chuisky zone.

Early Cambrian, 550 Ma (Figure 77)

In the early part of the Cambrian, the Paleo-Asiatic Ocean attained its greatest width, separating the Siberian and North Chinese continents. It is assumed that the Tuva-Mongolian microcontinent was situated within the ocean. Taking into account widespread Lower Cambrian island-arc complexes in the Tuva, Minusinsk, and Salair zones, the subduction zone near the ocean margin can be reconstructed. As these arc complexes are nearly of the same age and have similar compositions, we suggest that they outline a single system of island arcs extending through the whole Paleo-Asiatic Ocean from the North Chinese continent to East Europe. The Tuva, Minusinsk, Salair and Chingiz arcs (the latter was described in the section on Central Kazakhstan) constituted segments of this single Chingiz-Tuva island arc system. This arc is believed to have been associated with a subduction zone dipping from Siberia and the Tuva-Mongolian massif toward, and beneath, the massifs of Central Kazakhstan. Throughout Cambrian and Early Ordovician times, the island arcs of the Chingiz-Tuva system converged with Siberia and the Tuva-Mongolian microcontinent. The hypothetical Tomsky massif covered by a thick blanket of limestones is shown between the arc and the Siberian continent.

Late Cambrian, 500 Ma (Figure 78)

Collision of the Chingiz-Tuva arc with massifs situated between the arc and the Siberian continent was the main event of

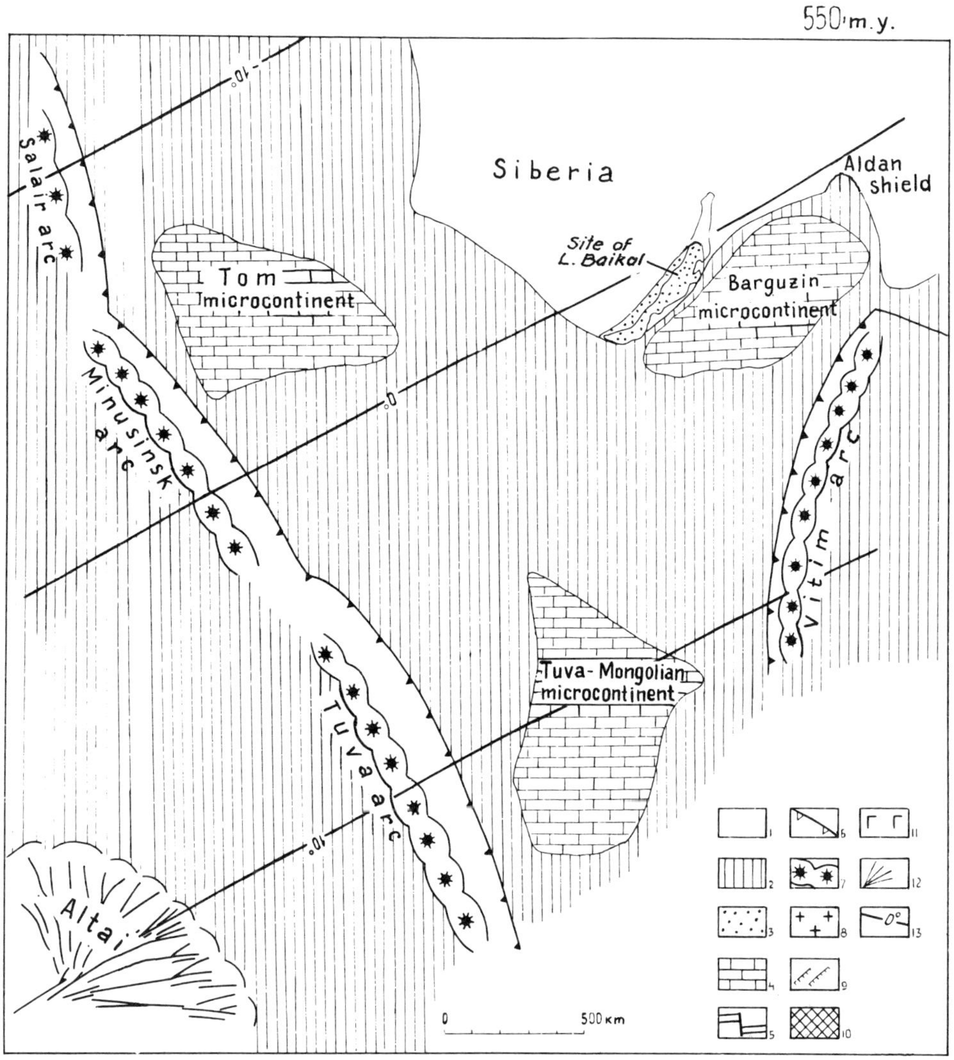

Fig. 77. Palinspastic reconstruction of the Altay-Sayan region for 550 Ma. The reconstruction is made with respect to Siberia, which is arbitrarily left fixed. Paleomagnetic data from Khramov [1982, 1986].
1, Shelf sea; 2, oceanic floor; 3, dry land; 4, carbonate; 5, spreading axis; 6, subduction zone; 7, volcanic arc; 8, granite batholith; 9, collision zone (folding and metamorphism); 10, exotic block; 11, intraplate igneous rocks; 12, fan (deltaic deposit); 13, paleolatitude.

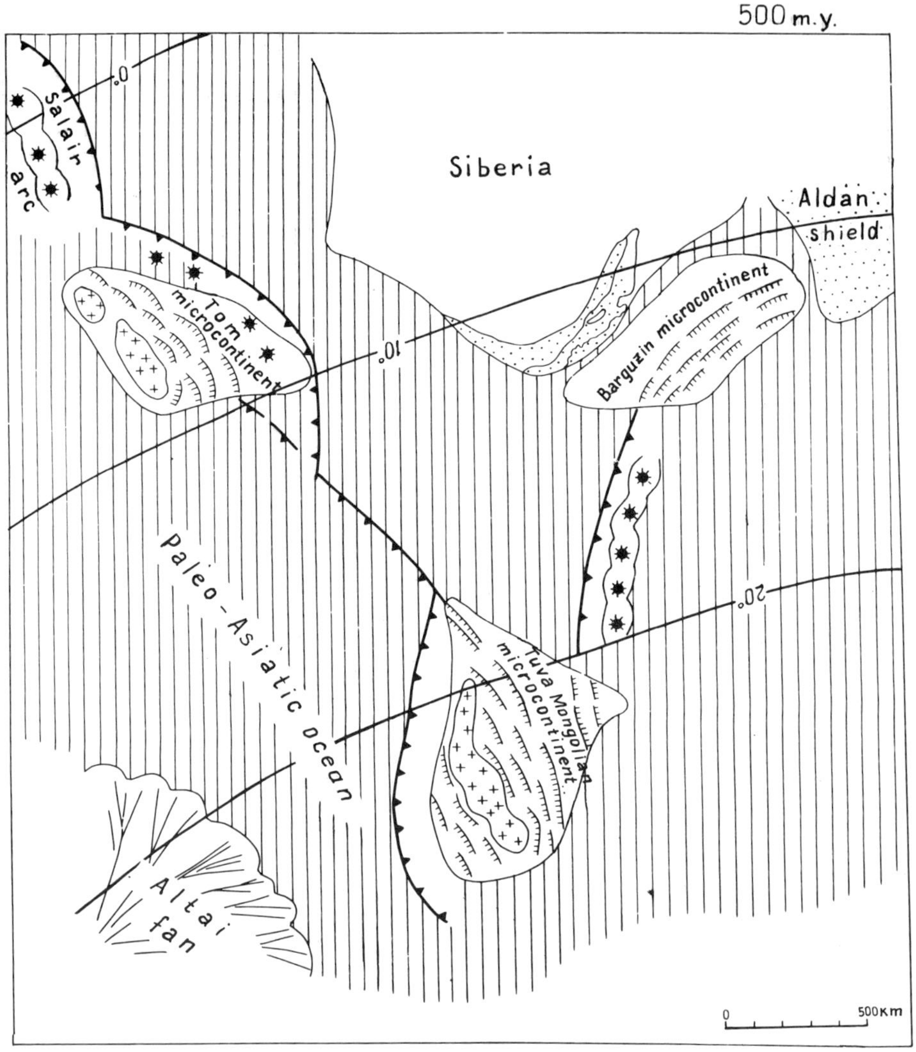

Fig. 78. Palinspastic reconstruction of the Altay-Sayan region for 500 Ma. See Figure 77 for explanation.

the Late Cambrian. Collision with the Tuva-Mongolian massif resulted in the Salairide orogeny, intrusion of the batholith of the Tannuola complex, and growth of granite-gneissic domes. Within the Minusinsk segment, the arc collided with the Tomsk massif, almost entirely overriding it. This event is also associated with batholithic intrusions.

In addition to the segments of the Chingiz-Tuva arc that collided, others, mainly those situated between Siberia and the Tuva-Mongolian massif, escaped collision.

Late Cambrian time was characterized by a maximum supply of clastic material in the Gorny Altay deltaic cone. In the very end of the Cambrian and beginning of the Ordovician, the mass of this cone collided with some island arc. At this very time the Gorny Altay and Dzhebash units were stacked together as nappes separated by ophiolite melange.

Early-Middle Ordovician, 460 Ma (Figure 79)

Plate interactions were obviously reorganized at the beginning of the Ordovician. The single Chingiz-Tuva arc ceased to exist; only the Chingiz segment (in Central Kazakhstan) was active. We assume that back-arc spreading took place in the rear of this arc and that, as a consequence, the Salair block became a remnant arc that progressively subsided and was covered by sediments.

New arcs came into existence. One was a sublongitudinal arc recorded in the island-arc complexes of Kuznetsky Alatau, Chulyshman highland, and Kharkhira zone of Mongolian Altay. Another was a sublatitudinal arc marked by Ordovician volcanics of West Sayan and the Oka-Dzhida system.

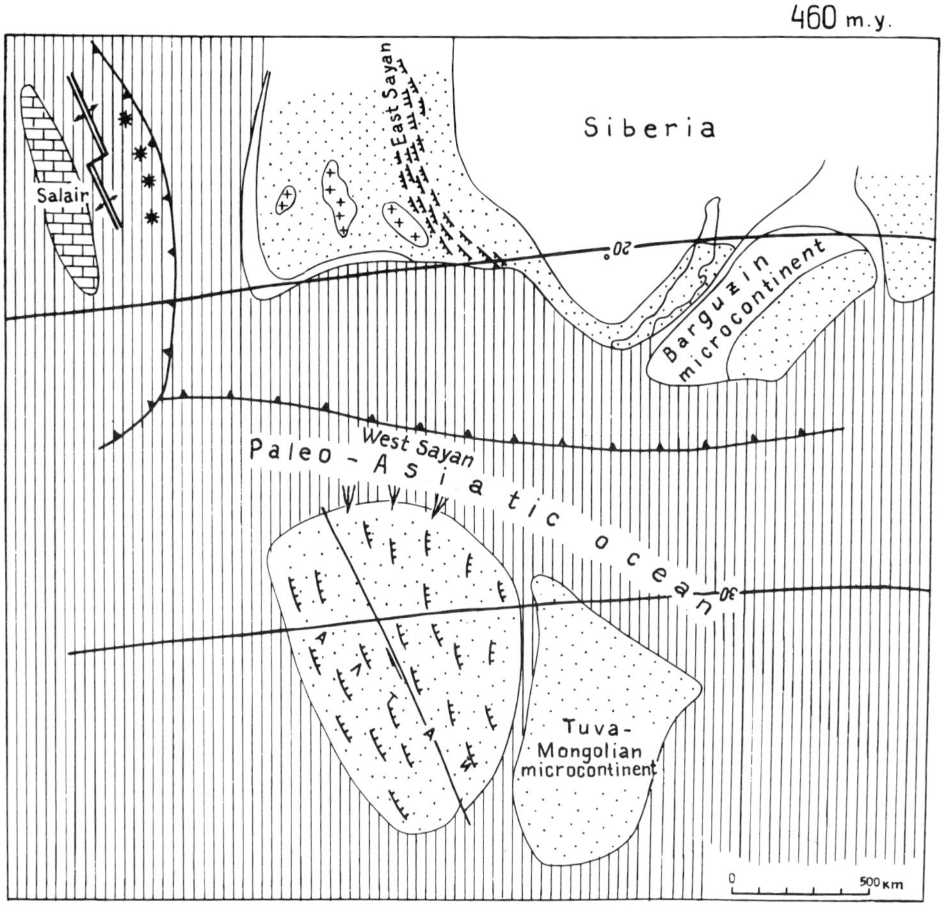

Fig. 79. Palinspastic reconstruction of the Altay-Sayan region for 460 Ma. See Figure 77 for explanation.

Late Ordovician-Early Silurian, 430 Ma (Figure 80)

This was the time of collision of the Tuva-Mongolian and Altay blocks with each other and with Siberia. The first to collide in the Early Silurian were the Altay block and the Kharkhira island arc. They both subsequently collided with the Tuva-Mongolian massif.

Early Devonian, 410-390 Ma (Figures 81 and 82)

At this time, as the Uralian Ocean opened, Siberia moved northwards and collided with the Tuva-Mongolian massif, the Altay microcontinent, and other blocks previously situated north of it, forming the Early Paleozoic ("Caledonian") foldbelt.

Siberia, together with the accreted Altay-Sayan margin, passed over a hot spot. This produced intense intraplate magmatism, extension, block movements, and extensional intermontane depressions.

Middle-Late Devonian, 370 Ma (Figure 83)

From the Middle Devonian, Kazakhstan began to converge with Siberia, consuming the Paleo-Asiatic Ocean floor. As a consequence, a series of subduction zones originated: i) the Rudny Altay zone (together with Gorny Altay) dipping under the Siberian margin, ii) the Tom-Kolyvan zone dipping away from Siberia towards Kazakhstan, and iii) the Saur zone, separating the ancient floor of the Paleo-Asiatic Ocean from the younger Paleo-Tethys. The Salair block converged with the Rudny Altay subduction zone.

Early Carboniferous, 340 Ma (Figure 84)

Kazakhstan and Siberia still continued to converge. Only a narrow strait (500 km wide), remaining from the former Paleo-Asiatic Ocean, connected the Paleo-Tethys and Uralian Oceans. The Salair block, which had for a long time remained in the middle of the oceanic basin, moved into the subduction zone north of the Gorny Altay and was thrust over the Early Paleozoic margin of Siberia.

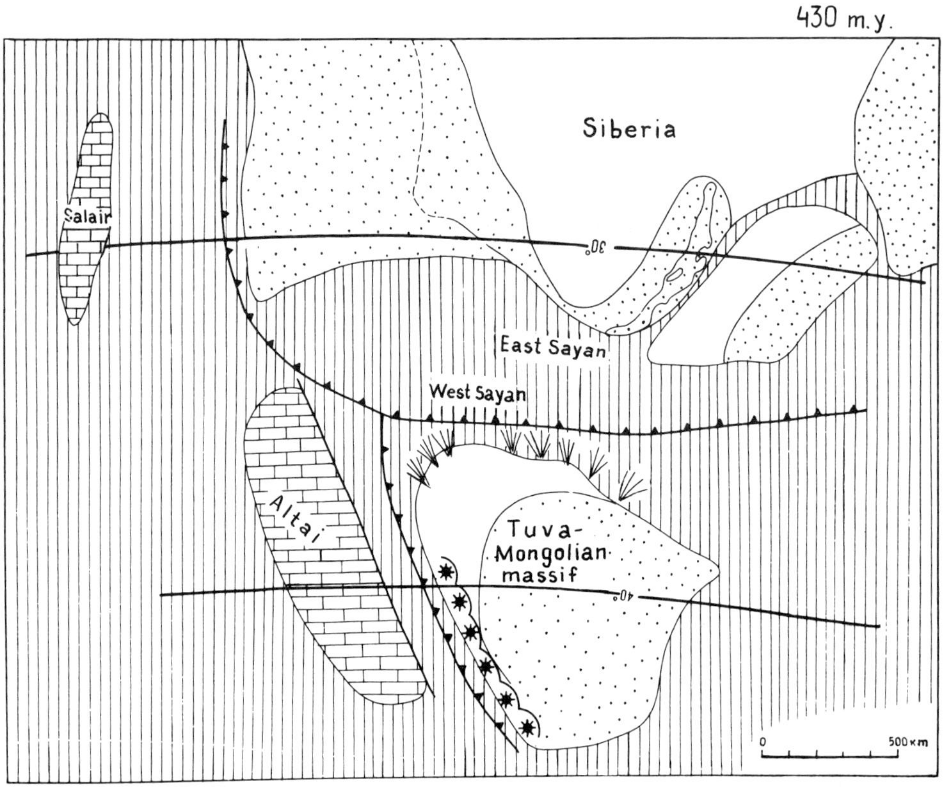

Fig. 80. Palinspastic reconstruction of the Altay-Sayan region for 430 Ma. See Figure 77 for explanation.

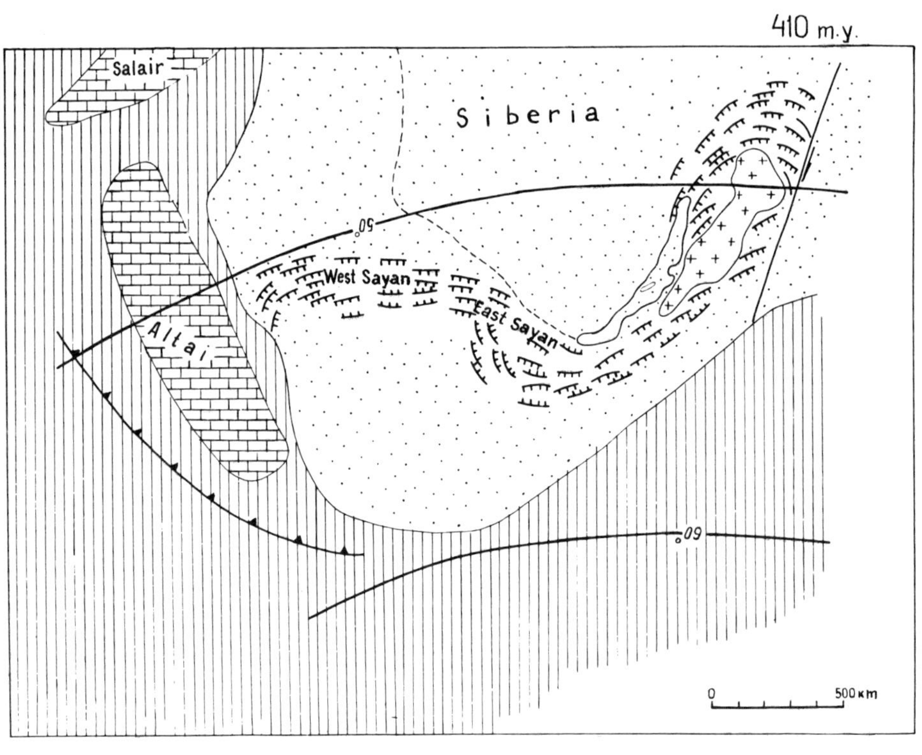

Fig. 81. Palinspastic reconstruction of the Altay-Sayan region for 410 Ma. See Figure 77 for explanation.

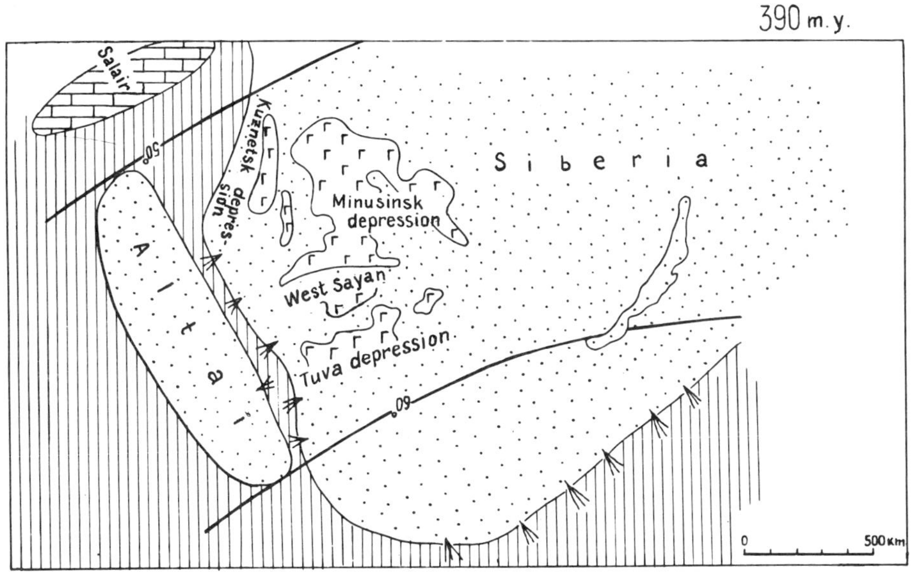

Fig. 82. Palinspastic reconstrucvtion of the Altay-Sayan region for 390 Ma. See Figure 77 for explanation.

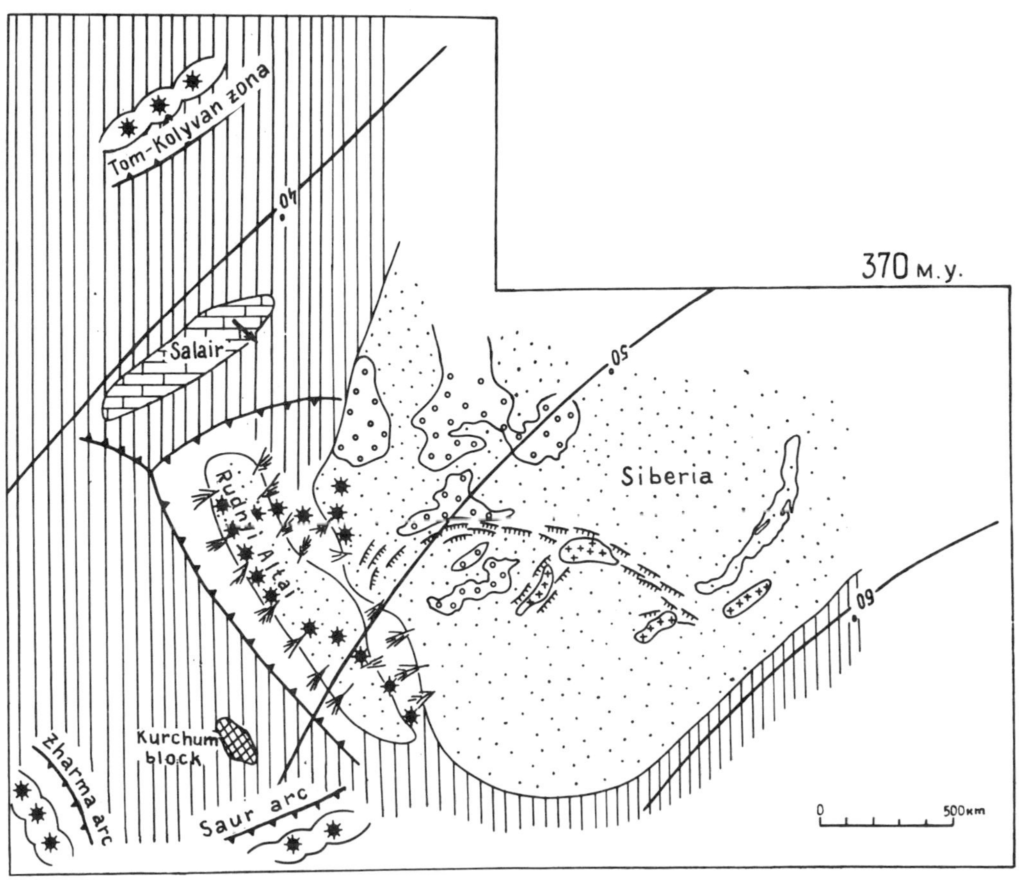

Fig. 83. Palinspastic reconstruction of the Altay-Sayan region for 370 Ma. See Figure 77 for explanation.

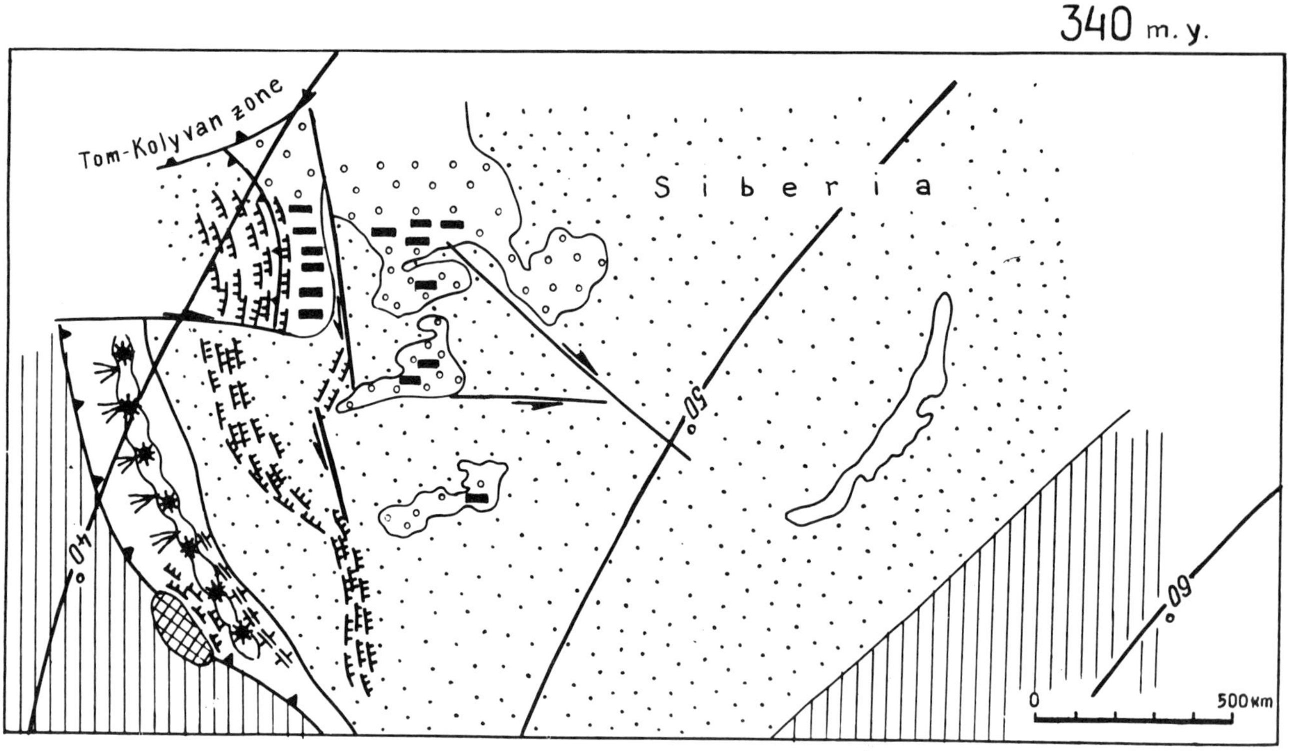

Fig. 84. Palinspastic reconstruction of the Altay-Sayan region for 340 Ma. See Figure 77 for explanation.

Middle Carboniferous, 310 Ma

The main collision of Kazakhstan and Siberia began just at this time. Nappes formed in the Salair front and in the frontal part of the Rudny Altay arc (Char belt). Fragments of the Paleo-Asiatic Ocean crust were still being consumed.

Late Carboniferous-Early Permian, 280 Ma

This time is characterized by continental collision of Kazakhstan and Siberia, resulting in the formation of granitic batholiths, e.g., Kalbinsky.

Late Permian, 250 Ma

The whole Altay-Sayan area was under compression at this time. All oceanic crust was consumed and the area became dry land. Compression-related movements caused nappe thrusting along longitudinal (present coordinates) zones and dextral shearing along latitudinal faults. The Tom-Kolyvan zone was thrust over Kuzbass. At the end of the Permian time all orogenic movements in the Altay-Sayan area ceased and, together with Siberia and Kazakhstan, it became a part of Laurasia.

Thus, the Altay-Sayan Early Paleozoic "Caledonides" are the product of accretion of numerous blocks, while the Late Paleozoic "Hercynian" structures are mainly the result of collision of the Siberian and Kazakhstan continents.

Chapter VII

MONGOL-OKHOTSK FOLDBELT

GENERAL DESCRIPTION

Material published in the late 1970's and early 1980's [Kotlyar and Popeko, 1974; Kirillova and Turbin, 1979; Kuzmin and Filippova, 1979; Parfenov, 1984] allow us to reinterpret the tectonics of the Mongol-Okhotsk belt, which was formerly considered to be a zone of intracontinental reactivation.

The Mongol-Okhotsk belt stretches over 3000 km from east to west, terminating in the center of the present Asian continent in the Khangay highland (Figure 85). It is situated between the Siberian platform and accreted Late Paleozoic blocks of the West Trans-Baikal and North Mongolian areas on the one hand, and ancient Khingan-Bureya, Argun, and Central Mongolian massifs on the other hand. In the Mesozoic, the three massifs situated south of the Mongol-Okhotsk belt made up a single large massif, the Amurian microcontinent. The northern boundary of the belt, separating it from Siberia, is very sharp and is expressed in large part as a distinct lineament referred to as the Mongol-Okhotsk fault. The southern boundary is more diffuse, due to a repeated reworking of the basement of the adjacent ancient massifs.

Typically, marine deposits younger than the Cambrian are almost entirely absent over a great portion of the northern rim. Only in two places on the Siberian margin are there remnants of former Paleozoic continental-margin basins: i) near the Sea of Okhotsk coast (Ayan-Shevla zone) where a full sequence of thick carbonates, including red beds, from the Cambrian to the Devonian is developed, overlapped by coal-bearing deposits of the Middle Carboniferous, and ii) in North Mongolia where Devonian volcanics and marine clastics are overlain by Lower Carboniferous terrigenous sediments. In the Late Paleozoic and Mesozoic the whole northern margin of the Mongol-Okhotsk belt was strongly reworked. Its western part, from the Early Permian to the Late Jurassic and Early Cretaceous, was an area of intraplate magmatism accompanied by graben formation; it includes the Selenga volcanic belt which is the most remarkable feature. The eastern part within the Stanovoy Ridge is occupied by the Uda batholith belt of Late Jurassic-Early Cretaceous age, which formed when Siberia collided with the southern continental massifs.

The Mongol-Okhotsk belt has three particular features. The first feature is widespread development of greenschist sequences and local glaucophane schist. Paleozoic fossils are associated with some of the schists, but metamorphism was found to be Jurassic. The second feature is the presence of linear bodies of gabbro-tonalites of the Pikan and Bereya complexes, associated in places with ultramafics. The Triassic age of these rocks is established in the Trans-Baikal region. The third important feature is the occurrence of numerous granite-gneissic domes, most of which formed in Jurassic time. All of the foregoing features, taken together, can be ascribed to continental collision.

The Mongol-Okhotsk belt is divided into eastern, central, and western segments. In the western and eastern segments the belt is up to 300 km wide, but it wedges out in the central segment where it is represented only by the Mongol-Okhotsk fault.

The eastern segment includes different structural-facies zones (Figure 85). The easternmost Uda-Shantar zone consists of cherty-volcanic and chaotic complexes from the Silurian to the Lower Carboniferous, deformed prior to the Permian. They are overlain by troughs (Torom, Uda) filled with Early Triassic-Late Jurassic marine sequences. Metamorphic schists and granite-gneissic domes are characteristic of the East Dzhagdy zone. The Amgun synclinorium in the south is built of a thick series of marine sandstone and shale from Triassic to Upper Jurassic, strongly deformed and including chaotic complexes. Paleozoic formations, and Triassic flysch and shales are abundant in the Lan and West Dzhagdy zones. Olistostromes are common as well. The Kuturinga zone is known for the gabbro-tonalites of the Pikan complex.

In the central segment the Precambrian complexes of the Stanovoy Ridge (Siberian continent) and Argun massif (Amurian microcontinent) are close to one another. However, there is no break in the Late Triassic-Jurassic flysch sequences preserved in the marine Oldoy trough. The stratigraphic continuity places limits on the mutual approach of the two Precambrian complexes.

In the western segment Devonian and Carboniferous clastic turbidite series are dominant. In the Aga trough marine Triassic and Jurassic series appear above the Paleozoic. As in the eastern segment, the deformation here was Middle Jurassic; however, westward in Mongolia deformation began earlier, occurring as early as the middle of the Permian. Ophiolite zones are aligned along the southern margin of the western segment. The age of the ophiolitic rocks is mostly Late Precambrian, but they were emplaced at the surface only in the Mesozoic. In the East Trans-Baikal region the structures of the western seg-

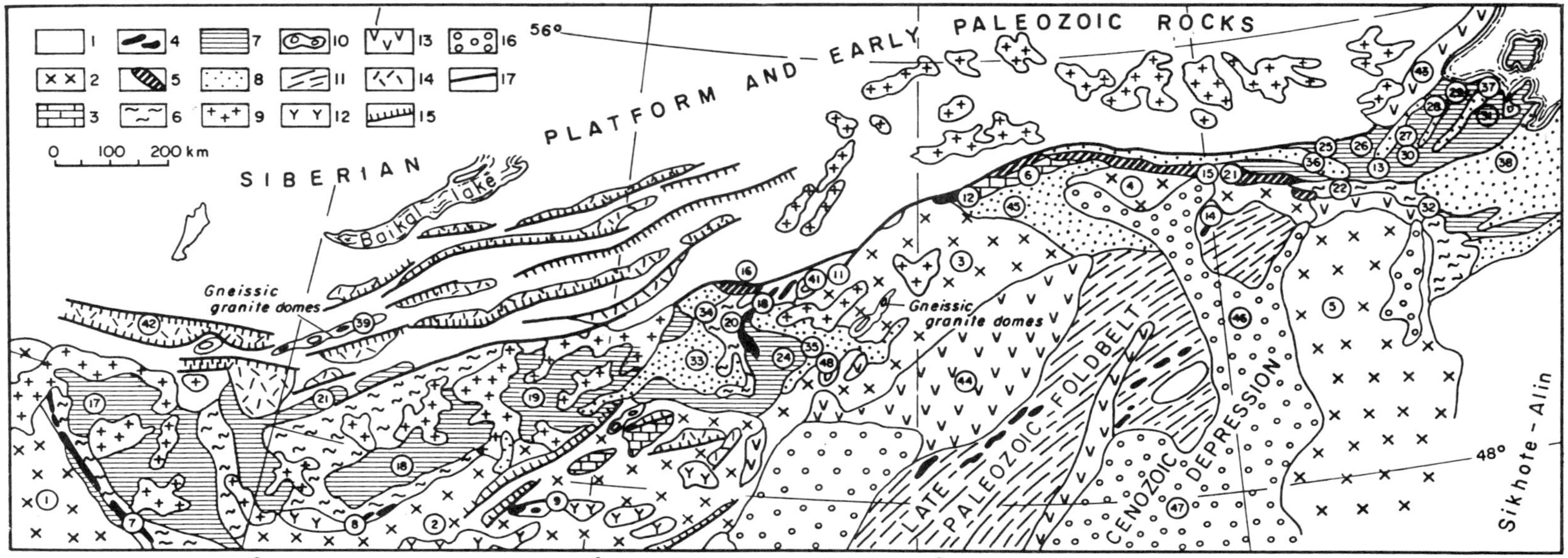

Fig. 85. Tectonics of the Mongol-Okhotsk belt.

Explanation: 1, Precambrian and Lower Paleozoic of the Siberian platform and adjacent Caledonian zones; 2, ancient massif; 3, Middle-Upper Paleozoic sedimentary cover overlying ancient massifs; 4, ophiolite; 5, gabbro-tonalite belt; 6, greenschist; 7, Middle-Upper Paleozoic turbiditic and chaotic complexes; 8, marine Mesozoic (including Upper Permian in Mongolia); 9, granite batholith; 10, gneissic granite dome; 11, Hercynian foldbelt; 12, Permian calc-alkaline volcanics; 13, Mesozoic calc-alkaline volcanics; 14, graben with alkaline volcanics (Permian to Lower Cretaceous); 15, normal fault; 16, Cenozoic depression; 17, Mongol-Okhotsk suture.

Encircled numbers: 1, Baidarag block; 2, Central Mongolian block; 3, Argun massif; 4, Gonzha massif; 5, Khingan-Bureya massif; 6, Upper Amur trough; 7, Bayan-Khongor ophiolitic belt; 8, North Gobi ophiolite belt; 9, Kerulen ophiolitic belt; 10, East Aga ophiolitic belt; 11, Djorol ophiolitic belt; 12, Shilka ophiolitic belt; 13, Shelva ophiolitic belt; 14, Dep ophiolitic belt; 15, Pikan gabbro-tonalite belt; 16, Bereya gabbro-tonalitic belt; 17, Khangai zone; 18, Khentei zone; 19, Daur synclinorium; 20, Onon zone; 21, Tukuringra zone; 22, South Dzhagdy zone; 23, Orkhon trough; 24, Borzya zone; 25, West Lan zone; 26, Lan zone; 27, Ayan-Shelva zone; 28, Uda zone; 29, Tyl synclinorium; 30, Irnimy synclinorium; 31, Uda-Shantar zone; 32, East Dzhagdy zone; 33, Aga trough; 34, Ingoda trough; 35, East Trans-Baikal trough; 36, West Dzhagdy zone; 37, Torom trough; 38, Amgun trough; 39, Butulinnur dome; 40, Tsagan-Oluev dome; 41, Borschevochny dome; 42, Selenga volcanic belt; 43, Uda volcanic belt; 44, Great Khingan volcanic belt; 45, Ol'doy trough; 46, Zeya-Bureya depression; 47, Sunlao basin.

ment are bent into the East Trans-Baikal sigmoid, a secondary dextral bending of all the structural zones by "indentation" of the Argun massif into the Mongol-Okhotsk belt and Stanovoi Ridge. As a result, submeridional Precambrian and Paleozoic rocks of the Argun massif were thrust west along the submeridional thrusts over the Mesozoic clastic series of the Mongol-Okhotsk belt.

Ancient Massifs

The Khingan-Bureya and Argun massifs are located on either side of the Late Paleozoic foldbelt south of the Siberian platform. The foldbelt, a probable continuation of the South Mongolian zone, lies in the Zeya-Selemdzha divide between the two massifs. The Khingan-Bureya massive passes southward into northeastern China, whereas the Argun massive extends southwestward into the Central Mongolian massif.

The Kingan-Bureya Massif

This massif is largely composed of Paleozoic batholiths. The Precambrian basement is preserved in comparatively small areas between the batholiths; it consists of Archean or Lower Proterozoic gneiss-amphiobolite complexes (the Amur series and its analogs) and younger and less metamorphosed Late Precambrian rocks, e.g., variously metamorphosed schists, phyllites, and marbles. The Vendian-Cambrian (Khingan series) lies unconformably upon the metamorphic basement and consists of shale along with widespread carbonate rocks; ore-bearing horizons with Fe-Mn ores and phosphorites are present as well. It is important to emphasize that the metamorphic basement is in places structually overlain by rocks of ophiolitic association (ultramafics, gabbro, lava). Thus, the Khingan -Bureya massif was presumably formed as a stable block after nappe formation, evidently at the end of the Cambrian. The emplacement of large batholiths of biotite granite (Bidzhan complex) at 604 to 483 Ma, is direct evidence that this block became a rigid mass only in the Early Paleozoic [Sidorenko, 1966b]. In these two features, Salairide (Cambrian) orogeny and massive emplacement of granitic batholiths, the Khingan-Bureya massif resembles the Tuva-Mongolian massif of the Altay-Sayan area.

The Argun Massif

In the Argun massif, the Precambrian basement is overlain to a considerable extent by Paleozoic and Mesozoic sedimentary series which, together with the basement, are intruded by Mesozoic granitic batholiths. The Precambrian rocks are visible in individual exposures. Some are strongly metamorphosed rocks, possibly of Archean age, but most are greenschist-facies metamorphics of volcanogenic-terrigenous origin (Borschovochny series) and carbonate-terrigenous sequences (Gazimur series) [Shuldiner et al., 1977]. The metamorphic basement is overlain by the Vendian-Cambrian sedimentary cover [Shuldiner et al., 1977], made up of basal conglomerates and overlying Archaeocyathid limestones alternating with quartzite-like sandstones. The Vendian-Cambrian sequences of the Argun region are similar to deposits of the same age in the Khubsugul and Bokson troughs of East Sayan and North Mongolia.

Judging from geological maps, the Argun massif is the basement for the Upper Amur "trough" composed of a Silurian to Lower Carboniferous carbonate-terrigenous sequence [Sidorenko, 1966b]. Although this structure is called a trough, it is in fact a large tectonic slab whose true relationships with the Precambrian basement remain unclear. The visible section begins with Silurian quartz and arkose sandstones (Omutninsky suite). Upward, there is a full Devonian succession exhibiting alternating limestones, including reef limestones, and sand-siltstone members (Bolsheneversky, Imachinsky, and Oldoy suites). The total thickness of the series is 5000-6000 m. Carboniferous marine sandstones (Tiparinsky suite) lie unconformably upon the Devonian and in places are overlain by Lower Permian continental volcanics. The rock complexes of the Upper Amur trough can be reliably interpreted as belonging to a passive continental margin except that the Permian volcanics show that in the Late Paleozoic this part of the massif became an active continental margin. As mentioned above, in the Early Mesozoic the Argun massif was welded with the Khingan-Bureya massif on which a new passive margin originated, and its remnants are preserved in thick flysch-like Jurassic sequences of the Oldoy trough (Figure 85). Granite-gneissic domes formed in the Middle Jurassic.

Oceanic Complexes

In the Mongol-Okhotsk belt no full sections of the oceanic crust are preserved, but dismembered ophiolites are ubiquitous. They are either contained in narrow ophiolite zones or in blocks within chaotic complexes.

In the western segment of the belt (East Trans-Baikal area, Figure 86), ophiolite zones were recognized by Misnik and Shevchuk [1975, 1980]. These are the East Aga and Djorol zones. Complexes of oceanic type are found at several structural levels. They are evidently present in metamorphic sequences referred to the Late Cambrian-Early Cambrian (Onon suite), represented by thick basalt sequences including low-K oceanic tholeiites [Popeko and Zazulenko, 1975]. Others belong to the Devonian (Undurgin and Ust-Borzja suites) and Early Carboniferous (Urtuy suite). The Devonian lava members, 300-400 m thick, consist of olivine basalts having relatively high TiO_2 (2.5-3.5%) and high alkali contents [Popeko and Zazulenko, 1975], similar to oceanic island basalts. The Early Carboniferous (Urtuy suite) oceanic rocks consist of alkaline basalts associated with cherty siltstones and commonly with reef limestones of the Middle-Upper Visean. They are probably the products of intraplate oceanic magmatism. One cannot exclude the possibility that each of these complexes is an exotic terrane, since they are often associated with reef limestones fitting neither the deep-water environment of ophiolites nor the position of the Siberian margin in Devonian and Carboniferous time in high latitudes of the Northern hemisphere.

The matrix of ophiolite zones, as exemplified in the Djorol zone [Misnik and Shevchuk, 1980], consists of greenschists and gabbro-amphibolites including numerous slabs of serpentinites. Dynamic and thermal metamorphism of ophiolitic rocks in greenschist facies was synchronous with motions and deformations along the Mongolian-Okhotsk suture, so the metamorphism is Mesozoic. At least some tectonic events were later than Early Jurassic, as the Djorol ophiolites are thrust over Lower Mesozoic rocks, while serpentinite melange contains fragments of Jurassic granites.

The eastern segment of the Mongol-Okhotsk belt contains

widespread piles of submarine lavas and deep-water cherts referred to the Silurian, Devonian, and in places Carboniferous. They seem to constitute exotic terranes involved in the Triassic-Jurassic shale matrix. Volcanics are compositionally classified into two groups [Mamontov and Popeka, 1975; Archipov and Panskykh, 1975]: i) basalts close to oceanic tholeiites, and ii) alkaline and high-Ti basalts (schalsteins), i.e., oceanic intraplate volcanics.

Thus, scarce data indicate that remnants of the oceanic floor of two marine basins--Late Precambrian and Middle Paleozoic--are present within the Mongol-Okhotsk belt. The presence of well developed passive margins on the Amurian continent further indicates that oceanic floor existed in both the Late Paleozoic and the Early Mesozoic, but these epochs were times of sedimentation on the ancient oceanic floor. When the Mongol-Okhotsk basin closed, most of the oceanic crust was consumed in subduction zones.

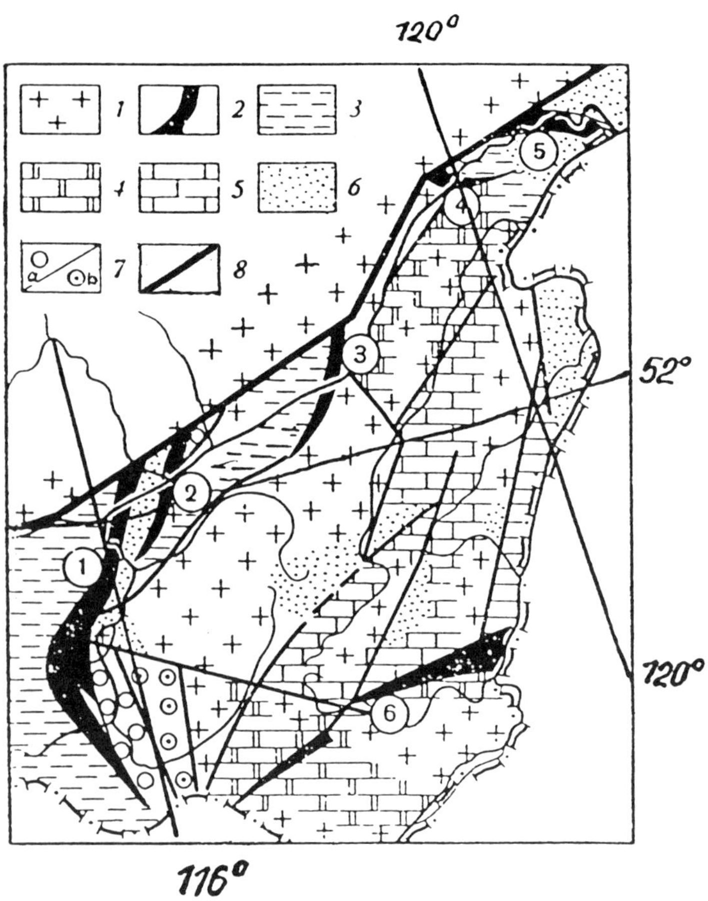

Fig. 86. Ophiolitic zones of the Trans-Baikal region, after Misnik and Shevchuk [1980].
1, Lower Precambrian basement; 2, ophiolitic zone; 3, Upper Precambrian schist; 4, Upper Cambrian clastics and carbonate; 5, clastics and carbonate, Vendian-Lower Paleozoic; 6, Mesozoic clastics; 7, Devonian (a) and Carboniferous (b) ophiolite; 8, suture zone. Ophiolitic zones marked by encircled numbers: 1, East Aga, 2, Urulga; 3, Djorol; 4, Nikolsky; 5, Shilka; 6, Byrka.

Convergence-Related Complexes

In the Mongol-Okhotsk belt and its surroundings, three groups of geological complexes indicating former convergence have been distinguished: i) terrestrial calc-alkaline volcanics on the rim of the belt, ii) tholeiitic and calc-alkaline magmatic series associated with clastic and turbidite sequences within the belt, including gabbro-tonalitic intrusions of the Pikan and Bereya complexes, and iii) chaotic complexes.

Typical subduction-related volcanic complexes are abundant in the Lower Permian south of the Mongol-Okhotsk belt. The Middle-Upper Jurassic volcanics of Bolshoy Khingan and East Trans-Baikal mostly belong to the same type, although they may be of intraplate origin.

North of the Mongol-Okhotsk belt, calc-alkaline volcanics compose the Uda volcanic belt or Uda arc [Parfenov, 1984]. It consists of Late Triassic to Late Jurassic volcanics (K-Ar age, 118-176 Ma), belonging to a typical calc-alkaline series similar in composition to volcanics of the Kurile arc [Parfenov, 1984]. The Jurassic nearshore and marine deposits of the Uda and Torom troughs probably represent the remnants of a fore-arc sedimentary prism, while the paleo-trench lay within the Amgun synclinorium, where Jurassic flysch and chaotic complexes are widespread.

The ribbon-like bodies of gabbro-tonalites of the Pikan and Bereya complexes supposedly formed by subduction. The Pikan intrusion, 300 km long and 15 to 20 km wide, is the largest. It consists of diorites, gabbro-diorites, tonalites related to gabbro, pyroxenites, and in places peridotites. Intrusive contacts of gabbro-tonalites with host rocks are often found, so the intrusive nature of plutons is clear. The gabbro-diorites and tonalites of the Bereya complex are close in composition to oceanic tholeiitic basalts but have higher K_2O content (0.4-0.7%) and somewhat higher Ba, Sr, and Rb [Kuzmin, 1985]. The presence of plagiogranites indicates a calc-alkaline trend of differentiation. The geochemical data on Triassic volcanics in the Aga-Ingoda trough [Barabashev, 1975] are very important for understanding the geodynamic environment within the Mongol-Okhotsk belt. They are basalts and hyaloclastics, similar in composition to low-K (K_2O 0.5%) tholeiitic basalts. According to Rutstein [1970] the Triassic volcanics are comagmatic with gabbro and gabbro-diorites of the Bereya complex; hence, these basalts presumably formed either in the basement of an immature island arc or in a back-arc basin. Both gabbro-tonalites and volcanics are evidence of a subduction zone in the Mongol-Okhotsk basin in the Early Mesozoic.

Chaotic olistostrome complexes are spread over the whole Mongol-Okhotsk belt but are best recorded in the eastern segment [Kirillova and Turbin, 1979]. The sandstone-shale matrix of the olistostromes contains blocks several meters in size, of diverse shapes, and of many different rocks, coarse clastic rocks and reef limestones being especially common. Lower Cambrian Archaeocyathide limestones are present as well. Very important is the presence of olistoliths of Silurian and Devonian reef limestones, since they are exotic considering the high latitudinal position of Siberia in the middle of the Paleozoic. Beside shallow-water sedimentary rocks, there are blocks of various greenstone extrusive rocks (spilites and variolites) and cherts. The chaotic complexes are mainly dated as Devonian; however, some data show that Triassic olistostromes are also present [Kirillova and Turbin, 1979]. The youngest, Jurassic, submarine slump deposits are found in the Amgun synclinorium [Parfenov, 1984].

In the East Trans-Baikal, olistostromes were reported by Belyaev and Chikov [1980] in the Aga synclinorium where sandstone olistoliths are contained in a clay matrix. Reef limestone blocks of middle-Upper Carboniferous and Early Permian age are widespread within Permian and Triassic deposits in the East Trans-Baikal region and adjacent parts of Mongolia, reflecting the wide development of olistostromes and chaotic complexes

Collision Complexes

The tectonic structure of the Mongol-Okhotsk belt was formerly believed to be controlled by long-lived deep-seated faults, including the Mongol-Okhotsk fault. Presently available data show, however, that thrusting and nappe formation were most important (Figure 87), whereas the Mongol-Okhotsk fault is no more than a "scar" remaining from a continent/continent collision. The zones of greenschist metamorphism, granite-gneissic domes, and granitoid batholiths are related to continental collision.

Greenschists in the Mongol-Okhotsk belt appear to have different ages of metamorphism and deformation, and protoliths of various ages and compositions. Most represent ophiolitic rocks and turbidite clastic deposits, i.e., oceanic floor and continental rise rocks. The age of metamorphism is established only in the Western Dzhagdy zone, where two metamorphic phases can be determined: 170-150 and 150-130 Ma [Kirillova and Turbin, 1979]. Metamorphism coincided with the main periods of folding, i.e., with times of continental collision. It is assumed that most of the metamorphic rocks was created when continental blocks first "touched' each other and rock sequences were pushed between the converging plates into the subduction zone to depths of tens of kilometers, where they evidently underwent intense crushing and regional metamorphism.

The granite-gneissic domes have been best studied in East Trans-Baikal [Sinitsa, 1975, 1977]. They are located in the basement of the Argun massif and are surrounded by Jurassic deposits (East Trans-Baikal trough). The cores of the domes consists mostly of gneiss-like granitoids, often intruded by remobilized plutons. The peripheries of the domes are as a rule characterized by blastomylonites and shear zones, whereas the envelope is frequently formed of foliated, metamorphosed, and locally migmatized lower-Middle Jurassic clastic deposits. Sinitsa [1977] suggested that the Jurassic deposits initially lay over the crystalline basement and together with it were involved in formation of the granite-gneissic domes. The domes are Middle Jurassic in age; the isotopic age of displaced granitoids in the core of the domes is 130-140 Ma. Possibly the granite-gneissic domes formed when the basement, together with the sedimentary cover, descended to a considerable depth in a subduction zone. When plate-consumption ceased as a result of continental collision, isotherms rose and the granite-gneissic domes started to grow.

Granitic batholiths are very characteristic features of the Mongol-Okhotsk belt. They include the Uda belt of batholiths north of the Mongol-Okhotsk fault within the adjacent Aldan Shield of Siberia. The plutons are mainly composed of granite and granodiorite. Such granitic batholiths can be postulated to have formed when continental collision resulted in crustal thickening and palingenic granitoid melts. As to the Uda batholiths, their position beyond the belt (100-150 km north of the Mongol-Okhotsk suture), their slab-like shape, and the absence of any Paleozoic-Mesozoic envelope seems to indicate that they are subduction related, although they were emplaced just after the main folding, i.e., when subduction ceased. Probably the Uda batholith belt formed in association with final subduction under Siberia of a sialic slab detached from southern massifs, eventually causing thickening of continental crust and granite melting.

Intraplate Magmatism

There are two discontinuous bands of intraplate magmatism representing north and south of the Mongol-Okhotsk foldbelt--the North Mongolian and South Mongolian belts. Both include intrusive and extrusive alkaline and peralkaline rocks, as well as rare-metal granites. Magmatism was accompanied by

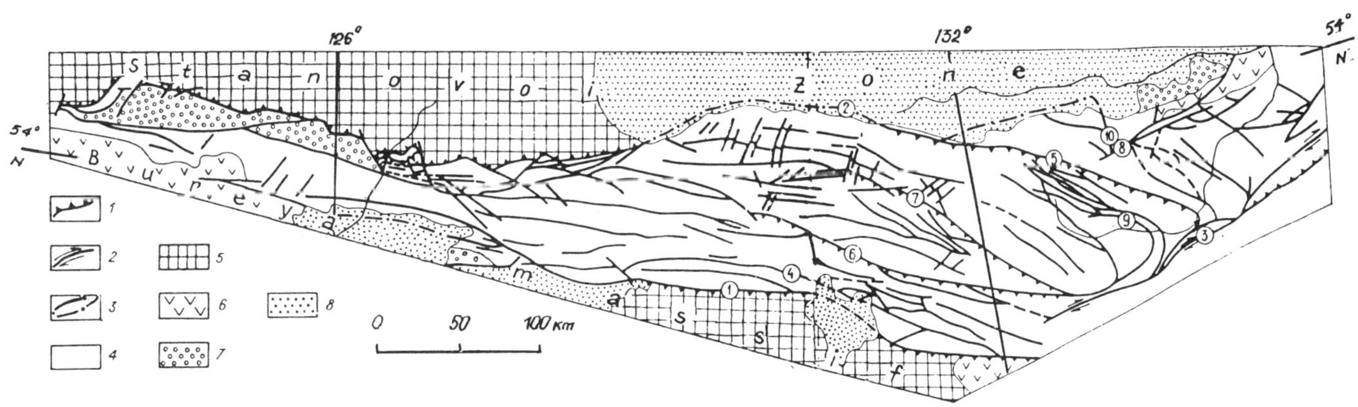

Fig. 87. Fault pattern in the eastern part of the Mongol-Okhotsk belt, after Parfenov [1984].
1, Thrust; 2, strike-slip fault; 3, other faults; 4, Mongol-Okhotsk belt; 5, Lower Precambrian massif; 6, Upper Jurassic-Lower Cretaceous volcanics; 7, Upper Jurassic-Lower Cretaceous sediments; 8, Cenozoic deposits. Faults marked by encircled numbers: 1, Ninni; 2, North Tukuringra; 3, Uligdem; 4, South Tuduringra; 5, Lan; 6, Champulin; 7, Djeltula; 8, Innyakh; 9, Ogdjenon; 10, Koteshtiyak.

formation of grabens filled with molasse, an environment recalling the pattern of Basin and Range tectonics. The area of intraplate magmatism shifted from west to east from Permian to Early Cretaceous time, following eastward migration of folding along the Mongol-Okhotsk belt.

HISTORY

Parfenov [1984] was the first to show that the Mongol-Okhotsk belt formed by collision of the Khingan-Bureya massif with the Siberian continent. Available data do not allow us to reconstruct the earliest history of the belt, so we will begin with mid-Paleozoic events.

Middle Paleozoic

Paleoclimatic data and lithology of Middle Paleozoic (Silurian, Devonian, and Carboniferous) rocks show that in the Middle Paleozoic the Siberian continent and the southern massifs were far from one another. Figure 88 shows the distribution of the Devonian-Carboniferous rocks within the Mongol-Okhotsk belt. On the passive margin of the Siberian platform there are either continental coarse-clastic Devonian sequences (Ayan-Shevla zone) or shelf terrigenous series of the Lower Carboniferous (Orkhon zone of Northern Mongolia).

The Devonian paleolatitudes of Siberia are plotted in Figure 88 after Khramov [1982]. Siberia evidently was inverted to its present orientations, and its edge near the southern tip of Lake Baikal was at 60° N. However, the rocks of the Devonian series in the southern massifs, in particular the Central Mongolian and Argun, are entirely incompatible with this northerly paleo-position. Coral-reef and coral-bryozoan limestones are common for all Devonian sequences in these massifs. If the Siberian latitude grid were extended to the southern massifs, reefs including a barrier reef would lie at 70° N. We must conclude that in the Devonian the Central Mongolian and Argun massifs were in the tropical zone, no less than 3000 km away from Siberia. Figure 89 provides a conventional reconstruction.

For the Devonian and Carboniferous of the Mongol-Okhotsk belt, particularly for its western segment, turbiditic (in places flysch-like) clastic sequences are characteristic. They represent continental rise environment and evidently accumulated near the Siberian continental margin. Their quartz-feldspar composition does not contradict this conclusion, and they completely lack rock types indicating a low-latitude climate, in agreement with their probable accumulation in high latitudes. Sedimentation evidently proceeded on the oceanic crust, since these clastics are interbedded with mafic volcanics and cherts.

Chaotic melanges of the eastern segment of the Mongol-Okhotsk belt contain many voluminous lava and deep-water cherty intercalations and can be regarded as indicating a subduction zone, although the position and polarity of this subduction zone remain unclear.

Thus, the Mongol-Okhotsk belt as such did not exist in the Middle Paleozoic. That is why one should not accept the common but contradictory statements that some present-day

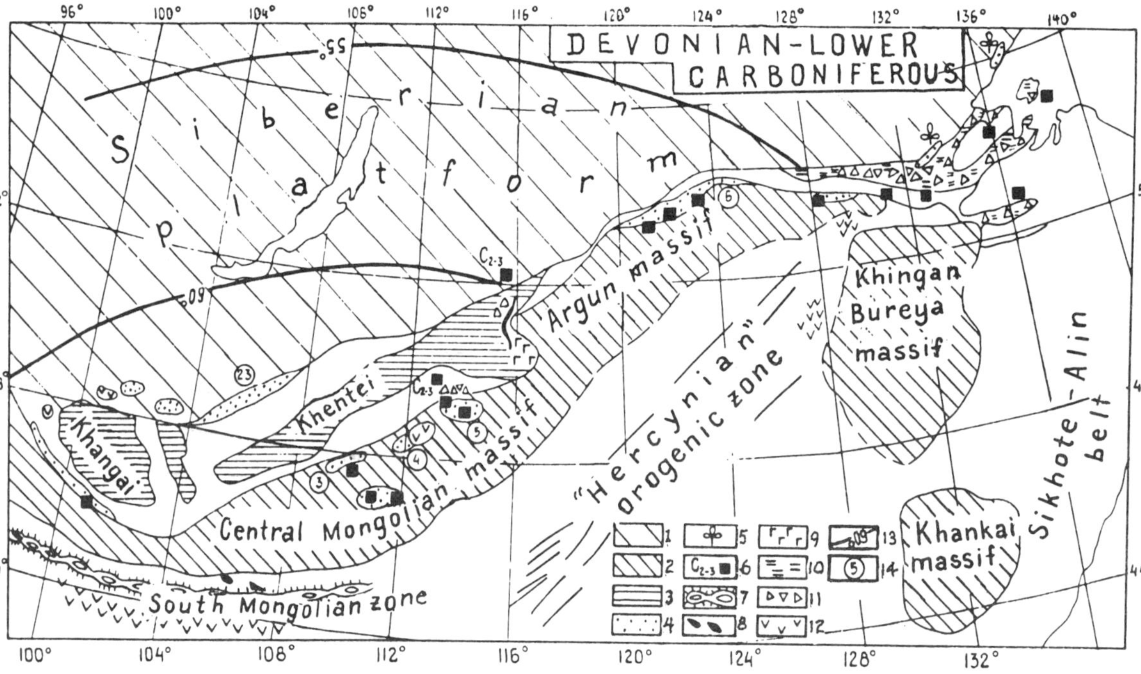

Fig. 88. Distribution of Devonian and Lower Carboniferous rock complexes of the Mongol-Okhotsk belt. 1, Siberian platform; 2, Precambrian massif; 3, continental-rise turbidite; 4, shallow-water clastics; 5, plant remains; 6, reef-limestone olistolith, mainly Devonian; Upper Carboniferous so marked; 7, Middle Devonian barrier reef; 8, ophiolite; 9, pillow lava; 10, chert; 11, chaotic rocks; 12, calc-alkaline volcanics; 13, Devonian paleolatitude for Siberia (after Khramov, [1982]).

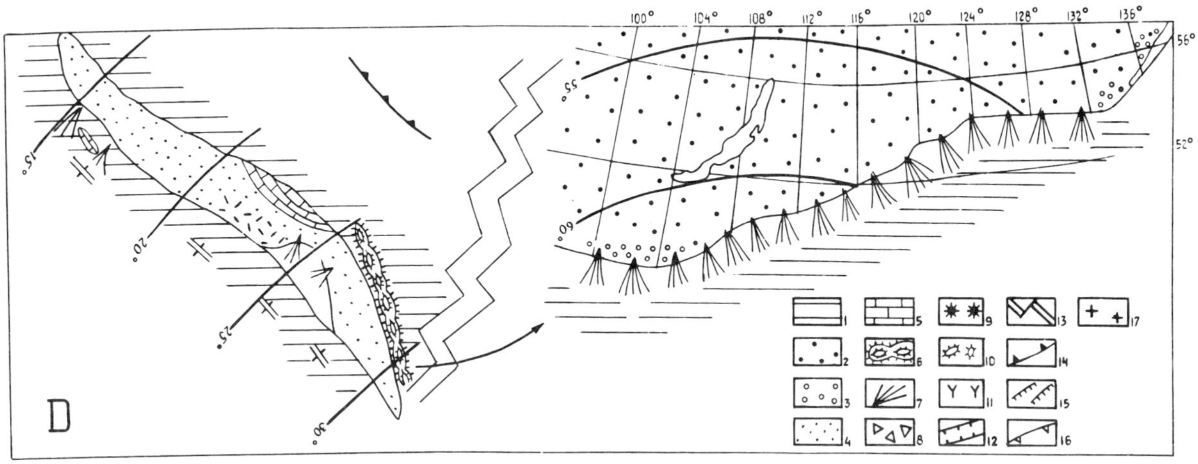

Fig. 89. Palinspastic reconstruction of the Mongol-Okhotsk belt for the Devonian. The reconstruction is made with respect to Siberia which is arbitrarily left fixed. Paleomagnetic data after Khramov [1982].
1, Oceanic floor; 2, dry land; 3, continental depression; 4, shallow-water sediment; 5, limestone; 6, barrier reef; 7, fan deposit; 8, chaotic rocks; 9, calc-alkaline volcanics; 10, intraplate volcanics; 11, Middle-Upper Jurassic latitic series (shown in Figure 96); 12, graben; 13, spreading axis; 14, subduction zone; 15, collision zone; 16, main thrust; 17, granite batholith.

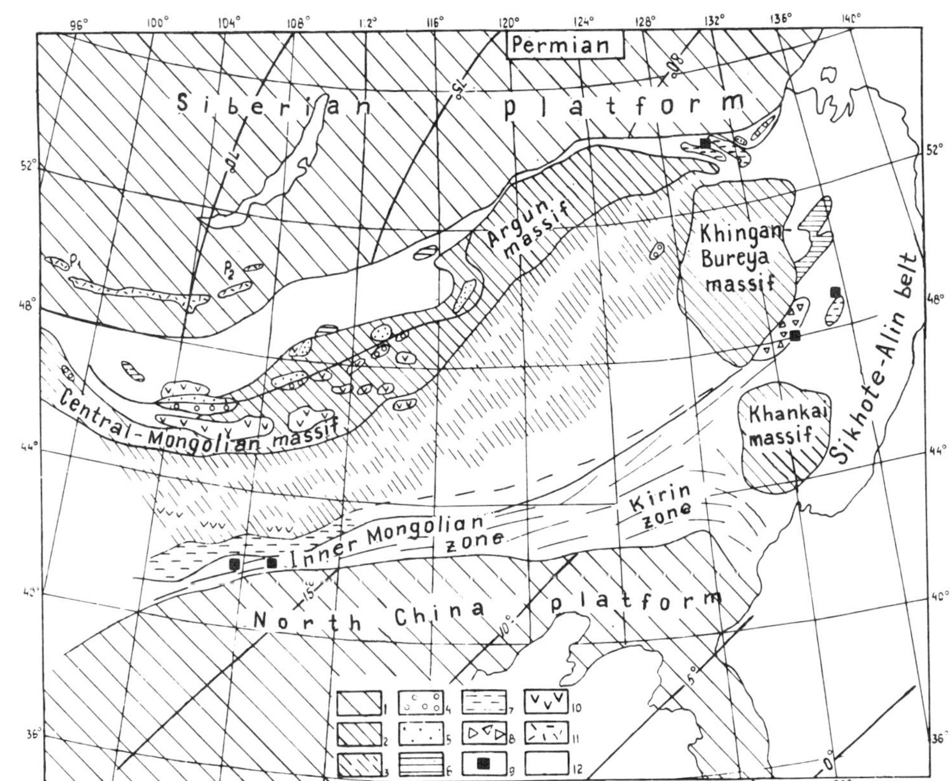

Fig. 90. Distribution of Permian rock complexes within the Mongol-Okhotsk belt.
1, Platform; 2, Precambrian massif; 3, area accreted to continents in the Late Palozoic ("Hercynian orogeny"); 4, molasse; 5, shelf deposit; 6, deep-water complexes; 7, flysch; 8, chaotic rocks; 9, reef limestone, often as olistoliths; 10, calc-alkaline volcanics; 11, alkaline and bimodal volcanics of the Selenga volcanic belt; 12, Permian-Triassic foldzone. Paleolatitudes are shown for the Siberian platform (after Khramov, [1982]) and for the North China platform (from McElhinny et al. [1981]).

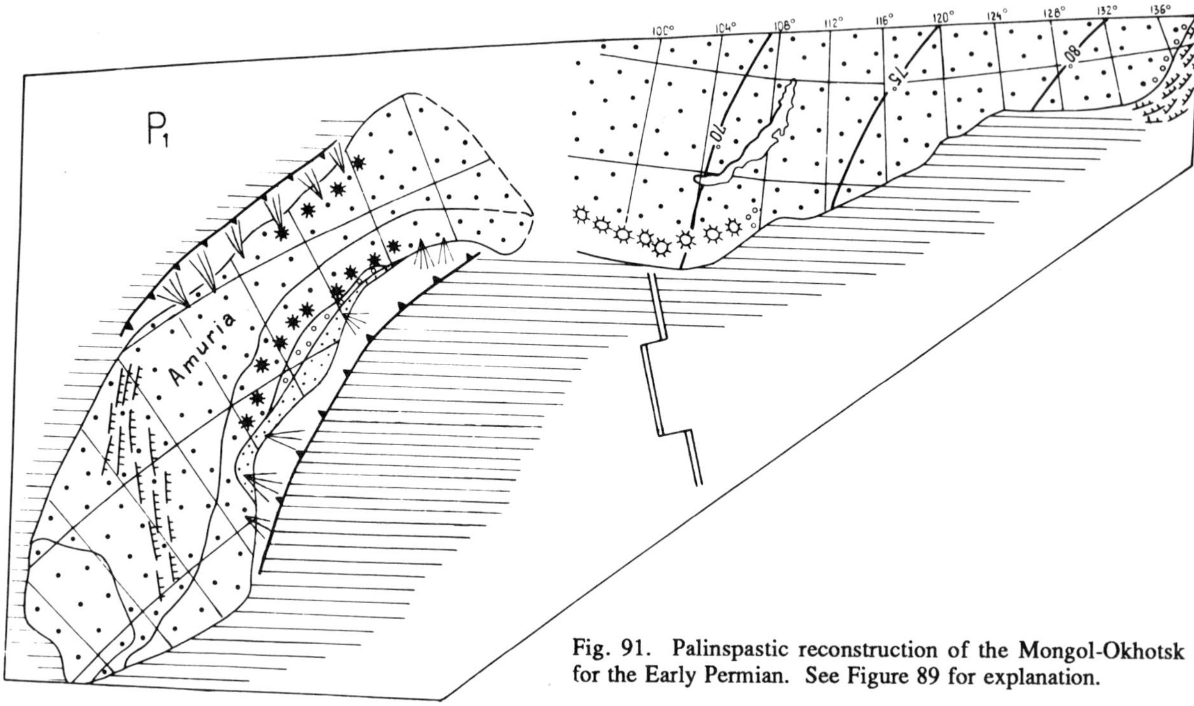

Fig. 91. Palinspastic reconstruction of the Mongol-Okhotsk belt for the Early Permian. See Figure 89 for explanation.

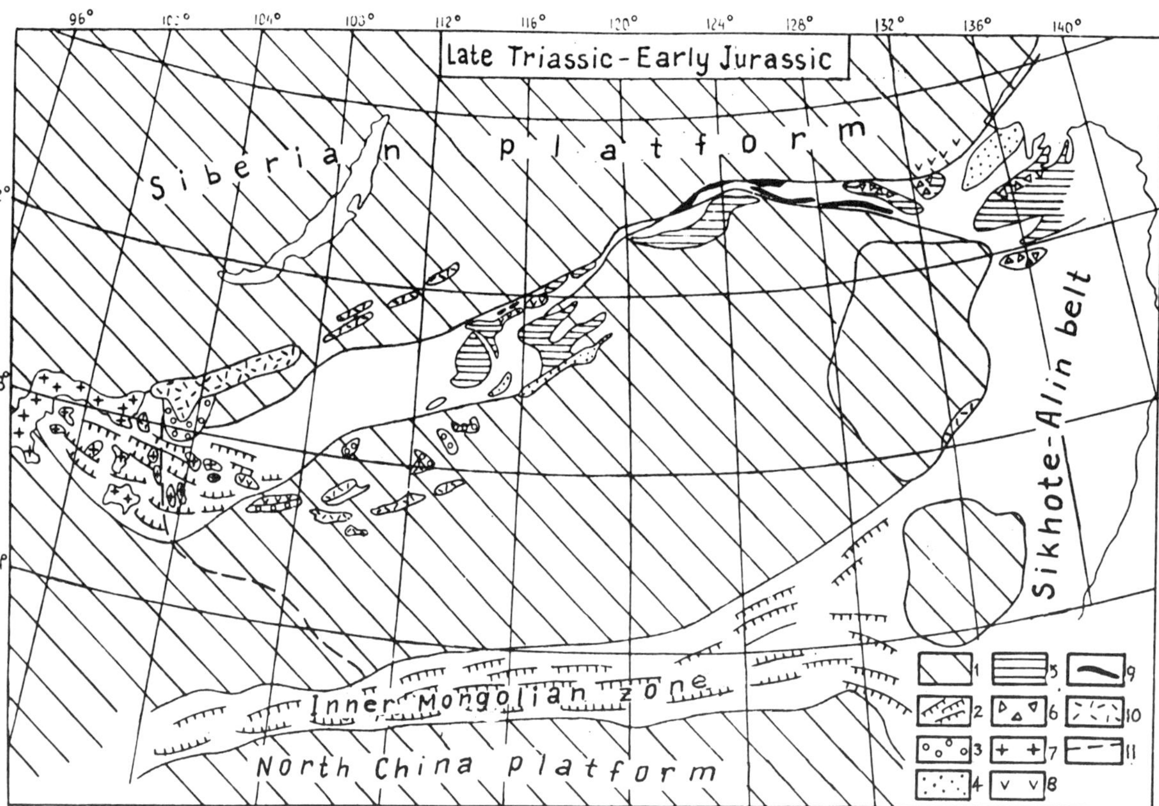

Fig. 92. Distribution of Late Triassic-Early Jurassic rock complexes within the Mongol-Okhotsk belt. 1, Stable area; 2, foldzone; 3, molasse; 4, shelf desposit; 5, deep-water clastics, mostly flysch and turbidite; 6, chaotic rocks; 7, granite batholith, 8, calc-alkaline volcanics; 9, gabbro-tonalite pluton (of the Pikan and Bereya complexes); 10, alkaline and bimodal volcanics; 11, transverse fault zone.

structures of the Mongol-Okhotsk belt are inherited from Precambrian times. In fact, nearly all the different complexes now present within the Mongol-Okhotsk belt were substantially far away from each other in the Middle Paleozoic.

Middle Carboniferous-Early Permian, 310-270 Ma

At the end of the Paleozoic the southern massifs approached Siberia, evidently for the first time, although the basin between Siberia and North China continent remained very wide. The paleolatitudes for Siberia [Khramov, 1982] and the North Chinese continent [McElhinny et al., 1981] are plotted in Figure 90, as well as the distribution of the Lower Permian rock complexes within the Mongol-Okhotsk belt. If this part of Siberia was between 70° and 80° N and North China between 10° and 20° N, not less than 5000 km separated them. This huge space was actually oceanic floor; the microcontinents corresponding to the southern massifs of the Mongol-Okhotsk belt could fill only a small portion. Figure 91 presents an approximate reconstruction of the Mongol-Okhotsk basin for the Early Permian. The Khingan-Bureya, Argun and Central Mongolian massifs were united and formed the single Amurian microcontinent, 2500 km long and 800-900 km wide. In Figure 91 its position relative to Siberia is shown conventionally; the present western end of Amuria joins Siberia in the region of the Khangay upland while the eastern, Khingan-Bureya end is rotated 130° clockwise from Siberia so that Amuria can be placed as far to the south as possible to take into account reef limestones with Tethys fauna, which arrived from the site of Inner Mongolia. The Mongol-Okhotsk basin had the shape of a open wedge or a very large bay, which was bounded by the converging Siberia and Amuria.

In the Early Permian, this basin started to close, as shown by the Early Permian East Mongolian volcanic belt. It is inferred that this belt was associated with a subduction zone dipping, as shown on the reconstruction, from the Mongol-Okhotsk basin beneath Amuria. The Siberia margin southwest of Baikal (present coordinates) happened to be above a hot spot, and here the North Mongolian (Selenga) belt of intraplate magmatism formed.

Thus, only since Permian time has the Mongol-Okhotsk belt be possessed its present features. Its further history is connected with the closure of the Mongol-Okhotsk basin.

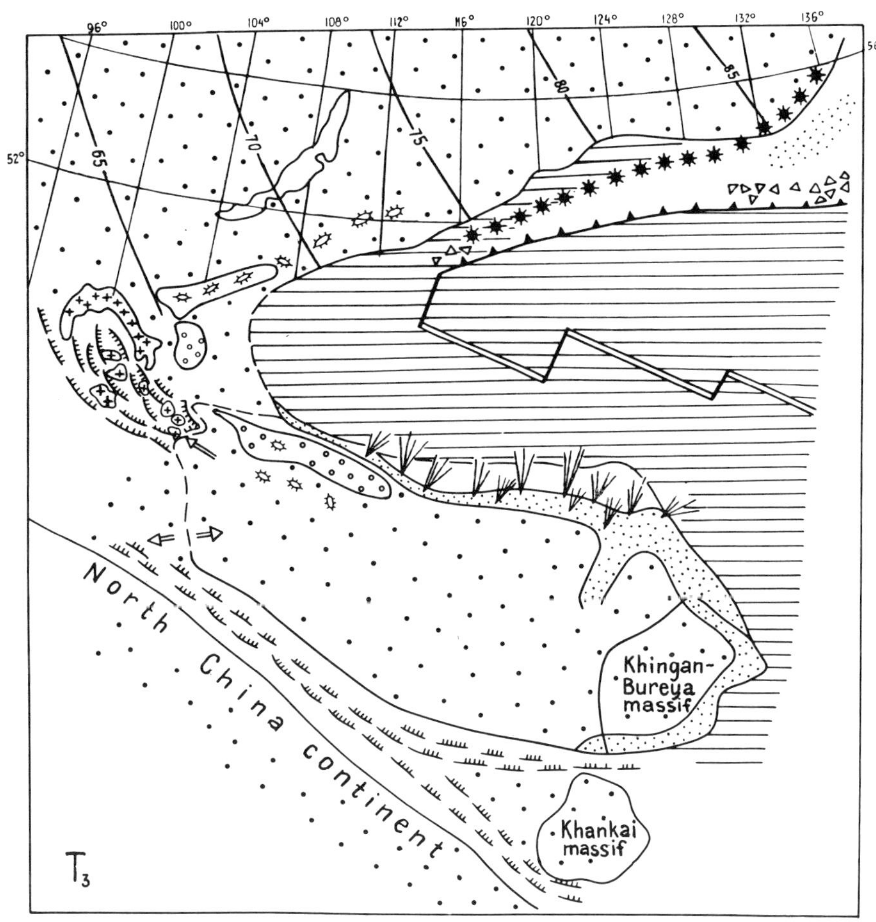

Fig. 93. Palinspastic reconstruction of the Mongol-Okhotsk belt for the Late Triassic. See Figure 89 for explanation.

Late Permian-Middle Triassic, 260-230 Ma

This time span is poorly recorded in the Mongol-Okhotsk belt.

The change in geodynamic conditions, from the Early Permian to the Late Permian, is well recognized in western Amuria where the East Mongolian volcanic belt ceased its activity and was replaced by quiet subsidence and accumulation of clastic sequences in continental and near-shore environments. This change can be easily explained by the cessation of subduction under Amuria. It is not clear whether subduction continued in other places. It should not be excluded that plate intraction relative to Siberia considerably slowed down or completely ceased at that time.

Late Triassic-Early Jurassic, 130-190 Ma (Figure 92)

Collision of western Amuria with Siberia terminated at this time, and a foldbelt formed at the site of the Khangay and western Khentey Mountains (Figure 93). The closure of the eastern Paleo-Tethys Ocean is marked by the occurrence of ophiolites in the Inner Mongolian Zone; it took place in the Late Triassic. As a result of the closure, Amuria was welded to the North China continent, and together they began to converge with Siberia. Therefore, the further history of the Mongol-Okhotsk belt is the history of collision of the North China and Siberian continents. Recent paleomagnetic determinations from the Early Mesozoic rocks of North China [Kent et al., 1987] indicate an inconsistency of paleolatitudes for Siberia and North China of 20-25°; i.e., the width of the Mongol-Okhotsk Ocean basin must have been about 2000 km.

The southern edge of the Mongol-Okhotsk basin was the passive Amurian margin. Thick flysch-like sequences of the East Trans-Baikal and Oldoy troughs are adjuncts of this passive margin.

The northern edge of the basin was rimmed by an active margin. In the distant east, the Uda continental-margin volcanic belt was in existence from Early Jurassic time onward. Parfenov [1984] established lateral zonation for this period (Figure 94). The Uda and Torom troughs which filled with comparatively shallow-water clastic sediments correspond to the fore-arc basin, 120-130 km wide. The Amgum synclinorium with thick shale sequences including chaotic complexes corresponds to the trench infilling.

Westward (in present coordinates) or southwards (in past coordinates) the Uda arc was replaced by an intraoceanic arc which can be reconstructed by the gabbro-tonalitic belt of the Pikan and Bereya complexes and by sporadically developed

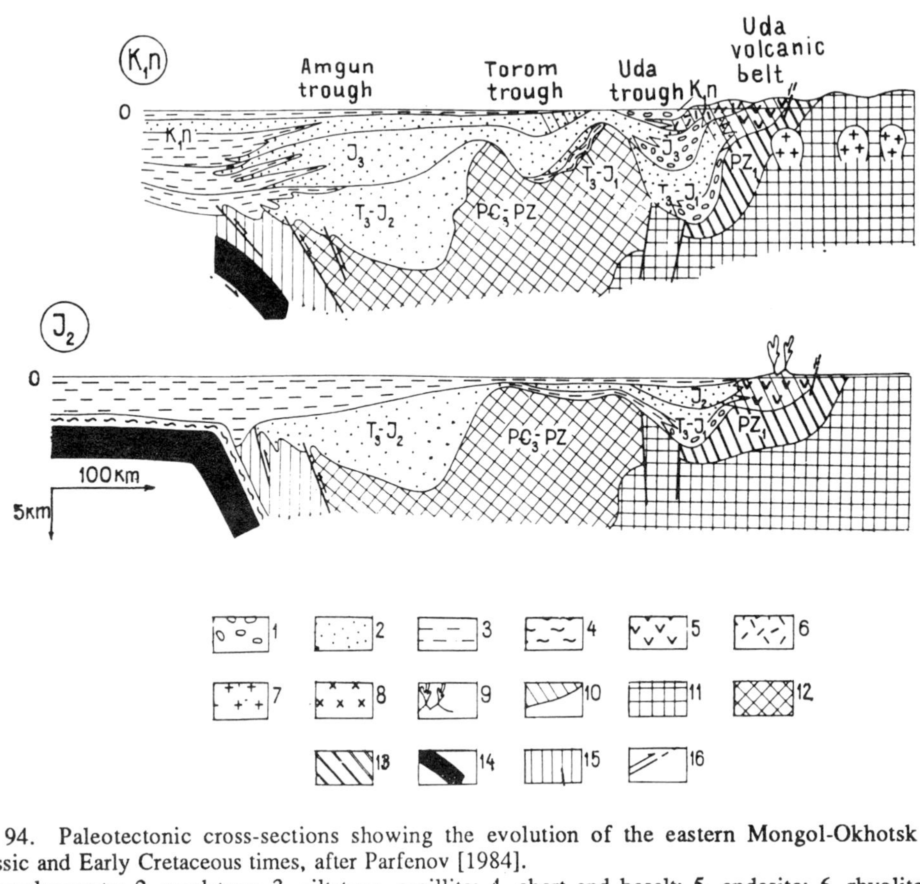

Fig. 94. Paleotectonic cross-sections showing the evolution of the eastern Mongol-Okhotsk belt during Jurassic and Early Cretaceous times, after Parfenov [1984].
1, Conglomerate; 2, sandstone; 3, siltstone, argillite; 4, chert and basalt; 5, andesite; 6, rhyolite; 7, granite; 8, alkaline rocks; 9, active volcanism; 10, continental environment; 11, Siberian platform basement; 12, deformed Late Precambrian and Paleozoic complexes; 13, shelf Paleozoic deposit; 14, ophiolite; 15, greenschist; 16, fault.

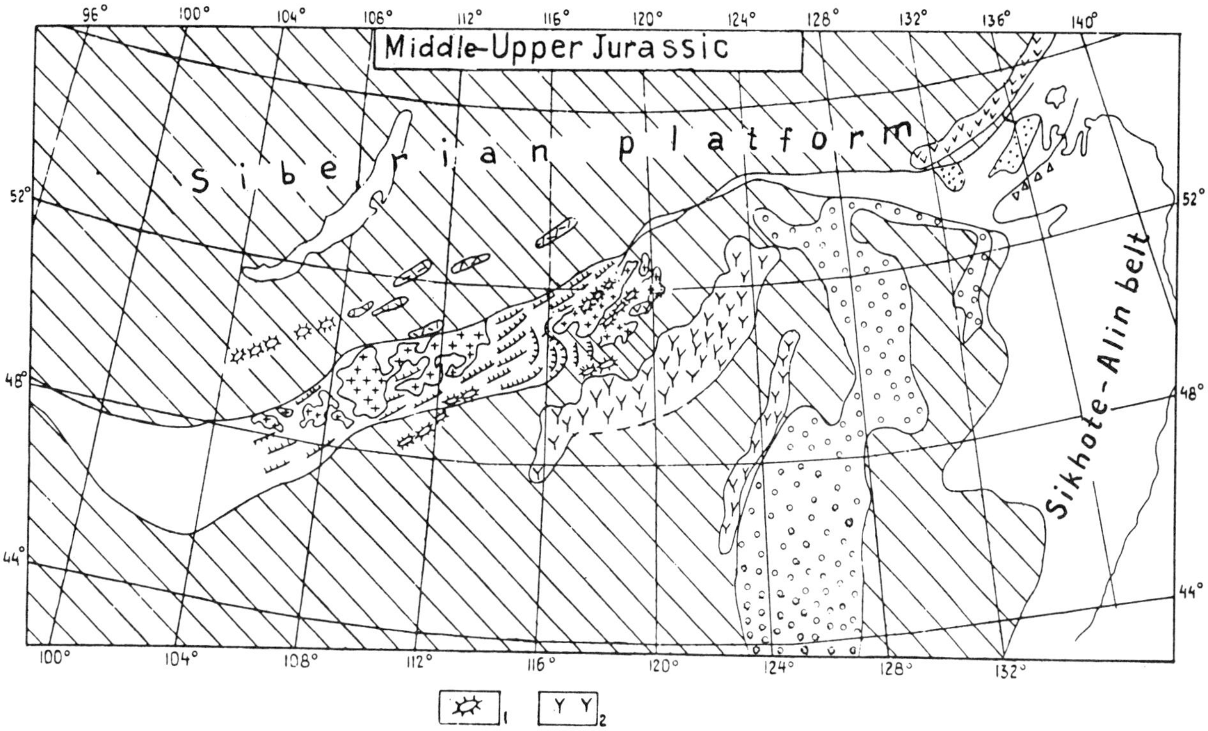

Fig. 95. Distribution of Middle-Upper Jurassic rock complexes within the Mongol-Okhotsk belt. 1, Granite-gneissic dome; 2, latite and trachyandesite. See Figure 92 for explanation of other symbols.

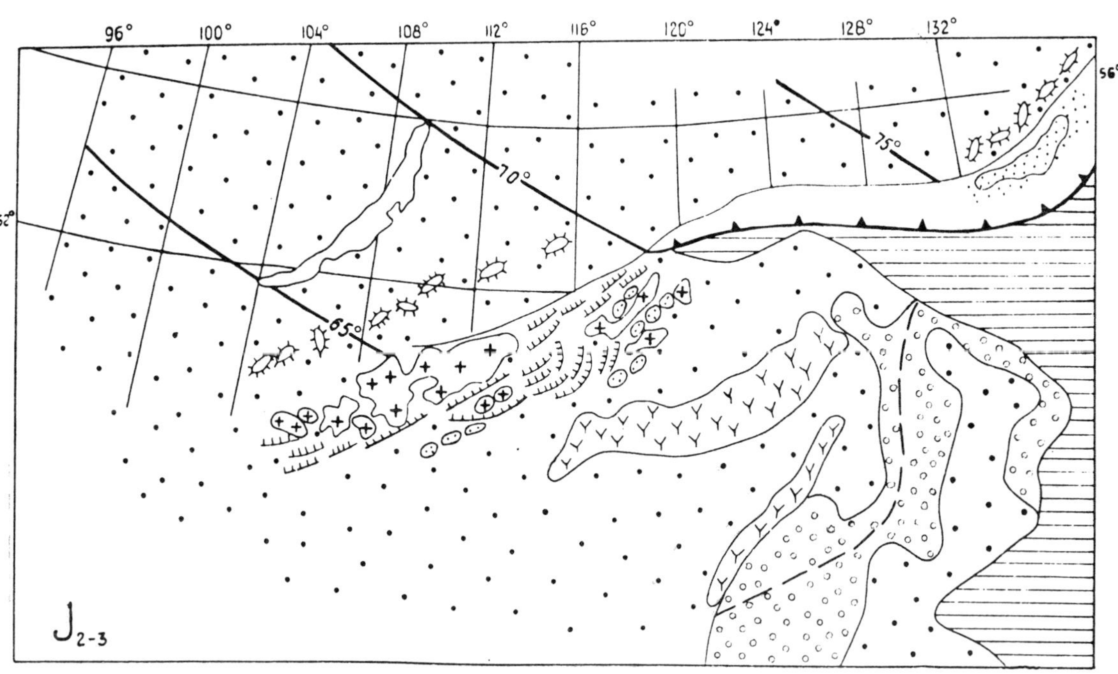

Fig. 96. Palinspastic reconstruction of the Mongol-Okhotsk belt for the Middle-Late Jurassic. See Figure 89 for explanation.

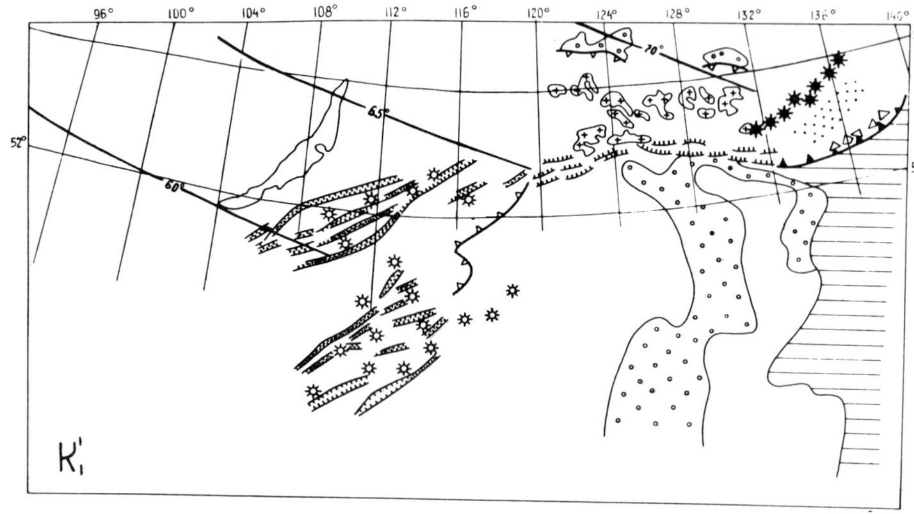

Fig. 97. Palinspastic reconstruction of ther Mongol-Okhotsk belt for the Early Cretaceous. See Figure 89 for explanation.

volcanics (Figures 85 and 92). We suppose that an island-arc back-arc system existed, but that during subsequent subduction and collision, many elements of this system were either consumed or reworked by tectonic and metamorphic events.

Middle-Late Jurassic, 180-150 Ma

It is clear from Figures 95 and 96, showing distribution of the Middle Jurassic rock complexes and a palinspastic reconstruction, that the main portion of the Mongol-Okhotsk basin was closed by the Middle Jurassic. Continental collision terminated in the western segment of the belt where the orogenic system of folds, nappes, granite-gneissic domes, and granitoid batholiths originated. The Argun massif, acted as a promontary indented into Siberia, leading to dextral movement of all structures farther west, and causing their sigmoidal bending.

Only in the east was there a remaining oceanic space, about 300 km wide and still available for subduction.

Early Cretaceous, 140-120 Ma

At this time the Mongol-Okhotsk basin closed entirely (Figure 97). The North China continent collided with and welded to Siberia, and jointly they became part of the Eurasian margin adjacent to the Pacific Ocean. The active Early Cretaceous tectonic belt can be divided along its strike into eastern and western segments. In the former the Uda volcanic arc, under which the Pacific Ocean floor was being consumed, continued its activity. This arc subsequently became a part of the Okhotsk-Chukotka volcanic belt. The western segment from the Dzhagdy ridge to the upper Amur was the zone of continental collision, where continental crust thickened due to partial subduction of the Amuria continent underneath Siberia. As a result, melting produced the granitic batholiths of the Uda belt, and simultaneously the whole Stanovoy Range was uplifted in a broad dome. On the whole, the Mongol-Okhotsk belt is a typical example of an intracontinental fold-belt produced by continental collision.

Chapter VIII

SIKHOTE-ALIN - SAKHALIN FOLDBELT

GENERAL DESCRIPTION

All Mesozoic and Cenozoic structures situated east of the Khingan-Bureya and Khankay massifs are termed the Sikhote-Alin - Sakhalin belt (Figure 98). Recent studies indicate a very widespread development of nappes, ophiolitic melange, and chaotic olistostrome complexes, both in the Sikhote-Alin and Sakhalin regions [Melnikov and Golozubov, 1980; Mazarovich, 1985; Melnikov and Izosov, 1984; Puscharovsky et al., 1983; Rozhdestvensky and Rechkin, 1982; Raznitsin, 1975, 1978; Khanchuk et al, 1988; etc.] The Sikhote-Alin - Sakhalin belt was first interpreted as having an accretionary origin by Parfenov [1984].

The Khingan-Bureya and Khankai massifs represent the western boundary of the belt. By the end of the Paleozoic, their continental crust was formed and they may be considered as microcontinents. They are divided by the Late Paleozoic Kur-Urma, Khabarovsk and Bikin orogenic zones stretching south-west into the Kirin zone of NE China. We regard these Late Paleozoic tectonic zones can be regarded as a collisional belt suturing the Khingan-Bureya and Khankai microcontinents. This is supported by the unpublished data (B. A. Natal'in et al., informal communication, 1989). At the beginning of the Mesozoic they both were joined into a single continental mass called Amuria.

Superimposed Late Paleozoic and Early Mesozoic structures on the eastern margin of Amuria include: the Permian Arseniev volcano-plutonic belt and the Early Mesozoic Suputin trough on the southeastern margin of the Khankai massif, and the Early Mesozoic Bureya trough east of the Khingan-Bureya massif.

Most characteristic for the Sikhote-Alin belt are thick Lower Cretaceous flysch deposits deformed in Late Cretaceous (pre-Senonian) time. Various Paleozoic and Early Mesozoic rock complexes appear between the Khingan-Bureya and Khankay massifs and the Lower Cretaceous flysch in the west and east Sikhote-Alin zones. They mostly include subduction-melange chaotic complexes and also the Sergeev zone in which exotic terranes are widespread, together with nappes, serpentinite melange, and olistostromes.

The Sikhote-Alin can be divided into (i) pre-flysch, and (ii) post-flysch structural units representing two major deformation stages of the Early Cretaceous and the Late Cretaceous (Senonian). All the thrust-nappe structures are overlain by the Sikhote-Alin volcanic belt, which marks an active continental margin of Andean type, from 80 to 50 Ma [Alkhmetov and Filimonova, 1986]. In places, however, older volcanics (Albian to Turonian) appear to be of the same age as flysch with which they are associated, and they evidently are of island-arc origin.

The Sikhote-Alin area is broken by a system of sinistral strike-slip faults, the most significant being the Central Sikhote-Alin fault, with a displacement of about 200 km [Utkin, 1980]. The strike-slip movements were either contemporaneous with, or slightly earlier than, the 80-50 Ma volcanic belt.

The orogenic structures of Sakhalin are separated from the Sikhote-Alin volcanic belt by the Tatar Strait trough and by the West Sakhalin trough composed of thick (about 10 km) clastic series accumulated from the Late Aptian to the Pliocene [Sidorenko, 1970b]. The Miocene deposits of the West Sakhalin trough are characterized by the presence of island-arc volcanics. The main Late Cretaceous deformation typical of the Sikhote-Alin foldbelt did not affect the West Sakhalin trough. The basement of the trough consists of intricately dislocated chaotic complexes of Early Mesozoic age (Daldagan series and its analogs). The rocks of the West Sakhalin trough were deformed in the Late Pliocene and Early Quaternary and were thrust over unconsolidated deposits of the Tym-Poronay depression, situated in the center of Sakhalin Island.

The Central Sakhalin suture, extending longitudinally through the whole island and farther into Hokkaido as the Kamuikotan zone, is the main structural feature of the island. It separates West Sakhalin, where the major structural disconformity is Early Cretaceous (before Aptian), from the East Sakhalin structures.

East of the suture, the basement of the Shmidt Peninsula, East Sakhalin Mountains, Susunai Ridge predominately consists of three rock complexes: metamorphic greenschist and blueschist sequences from 800(?) to 80 Ma old [Sidorenko, 1970b]; chaotic complexes with numerous blocks of Paleozoic and Early Mesozoic rocks; and island-arc or oceanic-type volcanic sequences, mainly of Late Cretaceous age. They were all folded at the very end of the Cretaceous, and the nappes verge westward, toward the continent.

The bottom of the Sea of Okhotsk is morphologically and structurally divided into three parts: (i) the North Okhotsk shelf, (ii) the central subsided shelf, and (iii) the deep South Okhotsk basin. The North Okhotsk shelf is a sedimentary basin filled with a thick (6-7 km) sequence of Late Cretaceous to Recent age. The central subsided shelf is 900-1500 m deep. The basal horizons of its cover are evidently not older than

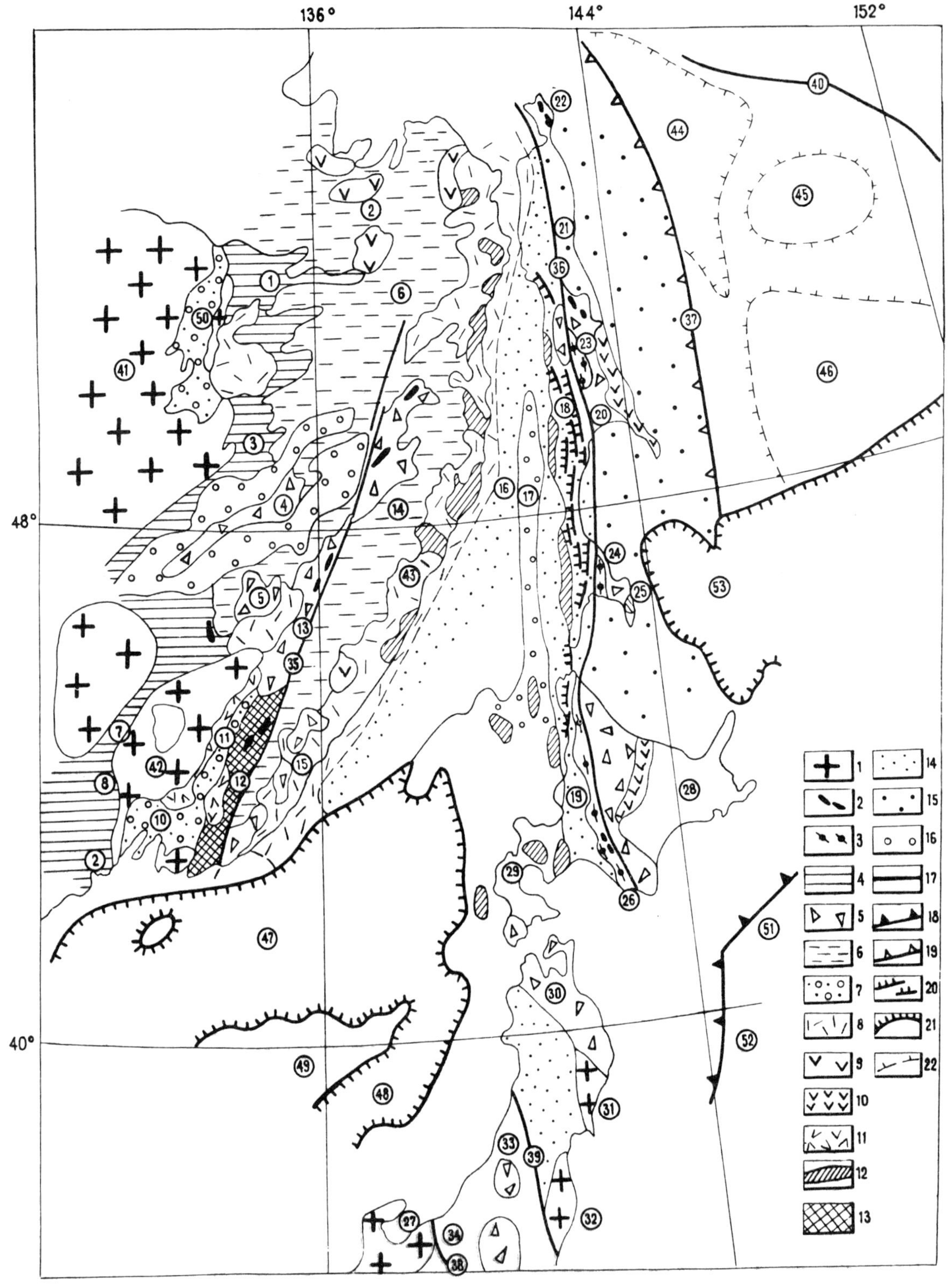

Fig. 98. Tectonics of the Sikhote-Alin - Sakhalin foldbelt.
1, Precambrian basement; 2, ophiolite; 3, glaucophane metamorphisnm; 4, Upper Paleozoic turbidite and chaotic rocks; 5, Mesozoic chaotic rocks; 6, Lower Cretaceous flysch; 7, Early Mesozoic trough; 8, Permian volcanic belt; 9, Lower Cretaceous island-arc volcanics; 10, Upper Cretaceous island-arc volcanics; 11, Upper

Cretaceous-Paleogene volcanic belt; 12, Miocene island-arc volcanics; 13, Sergeev terrane; 14, Upper Cretaceous-Paleogene fore-arc deposits; 15, Neogene fore-arc deposits; 16, Cenozoic basin; 17, strike-slip fault; 18, present subduction zone; 19, fossil subduction zone; 20, Pliocene folding; 21, boundary of oceanic basin; 22, boundary of swells within the Sea of Okhotsk.
Numbered are: 1, Nilan zone; 2, Amgun zone; 3, Kur-Urmy zone; 4, Khabarovsk zone; 5, Bikin zone; 6, West Sikhote Alin zone; 7, Kraev zone; 8, Grodekov zone; 9, Khasan zone; 10, Suputin zone; 11, Arseniev zone; 12, Sergeev zone; 13, Central Sikhote-Alin zone; 14, East Sikhote-Alin zone; 15, Coastal zone; 16, Tatar Strait trough; 17, South Tatar swell; 18, West Sakhalin trough; 19, Issikari-Rumoi trough; 20, Tym-Poronai basin; 21, North Sakhalin trough; 22, Shmidt Peninsula zone; 23, East Sakhalin zone; 24, Susunai zone; 25, Tonino-Aniva zone; 26, West Kamuikotan zone; 27, Hida zone; 28, Hidaka-Nemuro zone; 29, Iwa-Izumi zone; 30, North Kitakami zone; 31, South Kitakami zone; 32, Abukuma zone; 33, Asio zone; 34, Etsu zone; 35, Central Sikhote-Alin fault; 36, Central Sakhalin suture; 37, East Sakhalin suture; 38, Itoigawa, Situoka suture; 39, Tanakura suture; 40, Kashevarov fault zone; 41, Khingan-Bureya massif; 42, Khankai massif; 43, Sikhote-Alin volcanic belt; 44, Derugin basin; 45, Institute of Oceanology rise; 46, USSR Academy of Sciences rise; 47, Japan-Sea basin; 48, Honshu basin; 49, Yamato rise; 50, Bureya trough; 51, Kurile trench; 52, Japan trench; 53, South Okhotsk basin

Late Miocene; before that time, whole block of the subsided shelf was dry land, whereas during the last 10 Ma it subsided to a depth of 1000 m. Island-arc volcanics, plagiogranites, and granodiorites, from 85 to 95 Ma old, have been dredged from relatively high areas [Geodekyan et al., 1976]. The whole block of the subsided shelf can be considered as a large individual terrane, the so-called Sea of Okhotsk massif. The South Okhotsk deep basin (3200 m) is a typical back-arc basin situated behind the Kurile island arc. The time of its opening is conventionally set at Middle Miodcene.

The geologic history of the Sikhote-Alin - Sakhalin belt is usually divided into three unequal stages. The first covers nearly the whole Mesozoic and also the Late Paleozoic. It included the pre-flysch deformations in Early Cretaceous time. The second stage began thereafter and ended when the Sikhote-Alin folded structures had formed, before the Senonian Stage of the Late Cretaceous. The third started at the end of the Late Cretaceous and is still proceeding.

According to kinematic data [Kononov, 1984; Engebretsen et al., 1985; Zonenshain et al., 1987], the Kula-Pacific and Eurasian plates converged throughout the Late Mesozoic and Cenozoic. Therefore, subduction zones had to exist continuously to consume the oceanic lithosphere. Some reorganizations have been recorded during the time intervals 70-60 and 45-40 Ma when plate motions changed. The first was reflected in the very Late Cretaceous orogenic events on East Sakhalin, and the second in the initiation of the present Kurile-Japan island-arc systems.

At present, two seismic belts, the Sakhalin and Kurile belts, occur in the area under consideration. They separate the Okhotsk Sea plate from the Amur plate in the west and from the Pacific plate in the east. The Kurile seismic belt is related to a subduction zone, typical for the West Pacific Ocean, in which the oceanic lithosphere of the Pacific plate is consumed at a rate 10 cm/yr. The Okhotsk Sea plate is rotating clockwise with respect to the Amur plate along the Sakhalin seismic belt [Zonenshain and Savostin, 1979; Savostin et al., 1983].

MIOCENE RECONSTRUCTIONS

The Sakhalin island arc can be accurately restored for Miocene time. Calc-alkaline volcanics (andesite, basaltic andesite) are distributed through the whole Miocene of the West Sakhalin Mountains. According to Baranov [1981], their alkalinity increases from east to west, and the Benioff zone dipped 45° westwards and surfaced at a trench 150 km east of the arc, approximately along the present East Sakhalin fault zone, which separates the structures of the Sikhote-Alin - Sakhalin foldbelt from the Okhotsk Sea plate (in particular from the Derugin depression). The East Sakhalin shelf zone may be regarded as a forearc basin.

Figure 99 is a reconstruction for 15 Ma (Middle Miocene). By this time apparently the South Okhotsk and Sea of Japan deep basins were already opened, and back-arc spreading centers still existed. Extensional conditions are recorded by a vast spread of alkaline rocks in West Sakhalin, on Moneron Island, (just west of the south tip of Sakhalin) and on the eastern margin of Sikhote-Alin.

Simple kinematic considerations show that, as the South Okhotsk and Sea of Japan basins opened, sinistral strike-slip faults must have appeared just in the junction between the basins now occupied by southern Sakhalin and Hokkaido. This motion was opposite to recent dextral movements along the same faults.

The situation was different in the Early Miocene, when the Sakhalin and Kurile arcs began to grow. Figure 100 is a reconstruction for 23 Ma (Oligocene-Miocene boundary), when no deep marginal basins existed. Several available reconstructions show the position of Sakhalin and Japan with respect to Eurasia before the Sea of Japan opened. We follow the Sasajima and Torii [1983] version, based on new paleomagentic data. In that reconstruction of the ancient Khankay and Hida massifs well match, through supposed ancient complexes of the Yamato bank.

It seems that the Sakhalin island arc started operating at this time, and that the Benioff zone occupied the same position as later (15 Ma). The volcanic sequences of this age encompass the volcanics of Sakhalin and the Miocene calc-alkaline lavas (basalts, dacites, dacite-rhyolites) of East Sikhote-Alin [Akhmetjev and Filimonova, 1986].

The Sea of Okhotsk block was situated at some distance from the Benioff zone. At a subduction rate of about 5.0 cm/yr, 350 km of oceanic crust between Sakhalin island and the Sea of Okhotsk block could be entirely consumed by the beginning of the Late Miocene. The reconstruction of the position of the Sea of Okhotsk block is further constrained because by the beginning of the Miocene the block was located between the Kurile subduction zone and the Kashevarov fault zone. The Sea of Okhotsk block moved away from the Kurile zone when the South Okhotsk basin opened. The Kashevarov

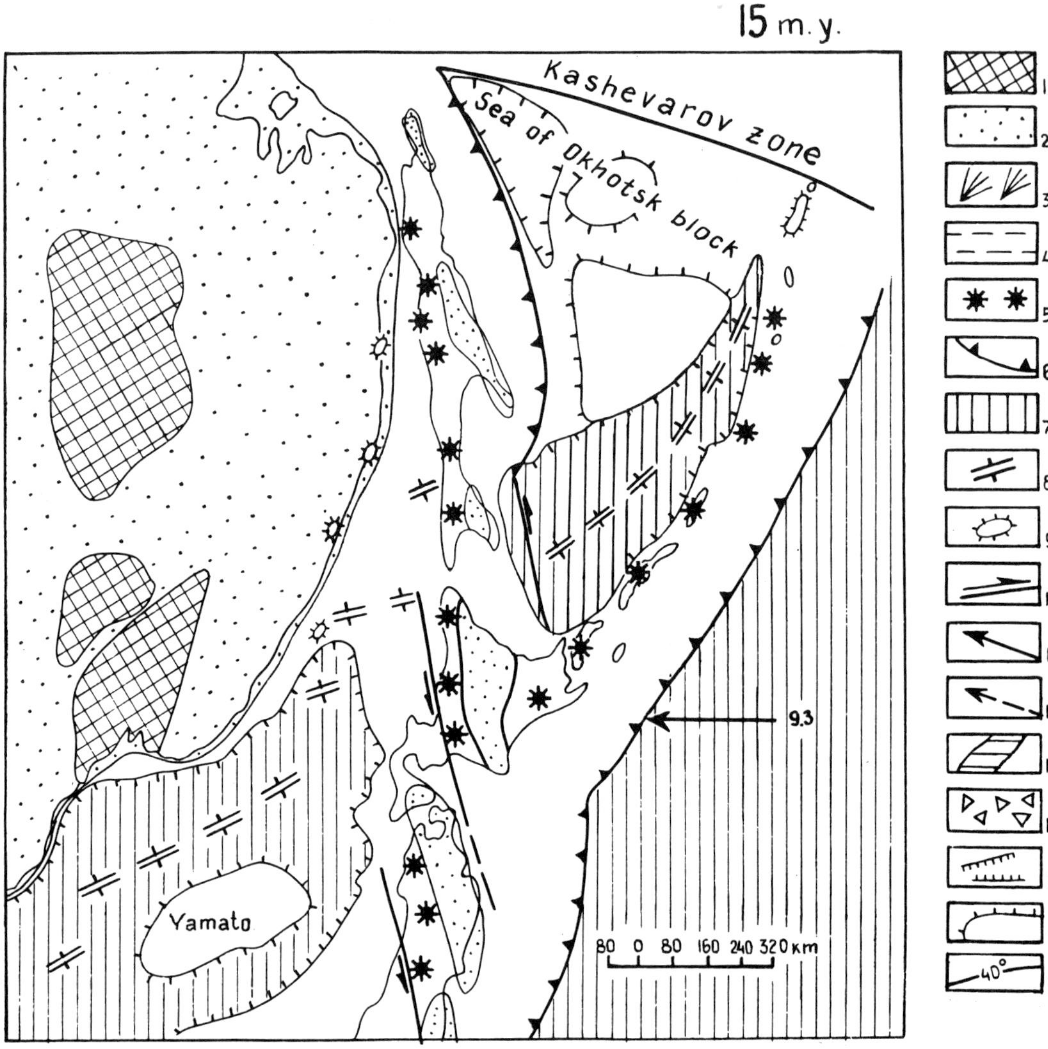

Fig. 99. Palinspastic reconstruction of the Sikhote-Alin - Sakhalin belt for 15 Ma. Siberia is arbitrarily fixed. Explanation for figures 99, 100, 103-108: 1, ancient massif; 2, dry land; 3, fan (deltaic sediment); 4, clastics; 5, calc-alkaline volcanics; 6, subduction zone; 7, oceanic floor; 8, spreading axis; 9, intraplate volcanics; 10, strike-slip fault; 11, direction of relative plate motions, and rate in cm/yr; 12, the same, uncertain; 13, remnants of the Sergeev microcontinent; 14, chaotic rocks; 15, folding; 16, boundary of oceanic basin; 17, paleolatitude.

zone was the transform boundary along which the Sea of Okhotsk block moved relative to the North American plate.

The reconstruction for 23 Ma is taken as a basis for deriving the preceding history of the region.

PRE-NEOGENE BACKGROUND

Ancient Massifs

The Khankai and Bureya massifs represent fragments of a single continent that split in the Early Paleozoic. Ancient metamorphic rocks, 1900 Ma old crop out within the Khankai massif [Melnikov and Izosov, 1984] in the cores of granite-gneissic domes. The dome envelopes and the interdomal synforms are composed of Late Precambrian and Lower-Middle Cambrian deposits, in some places carbonate (with archaeocyathids) and terrigenous-cherty deposits with phosphorites and bauxites, in others volcanic rocks with iron-formation. Upper Cambrian and Ordovician deposits are also found. The basement structures are cut by granitic batholiths 495 Ma old [Melnikov and Izosov, 1984].

A peculiar feature of the Khankai massif, in contrast to the Khingan -Bureya, is the presence along its whole eastern margin of a Permian volcano-plutonic belt (Arseniev belt), an active continental margin of the Andean type. Rich flora contained in the Permian deposits is typically Angarian, indicat-

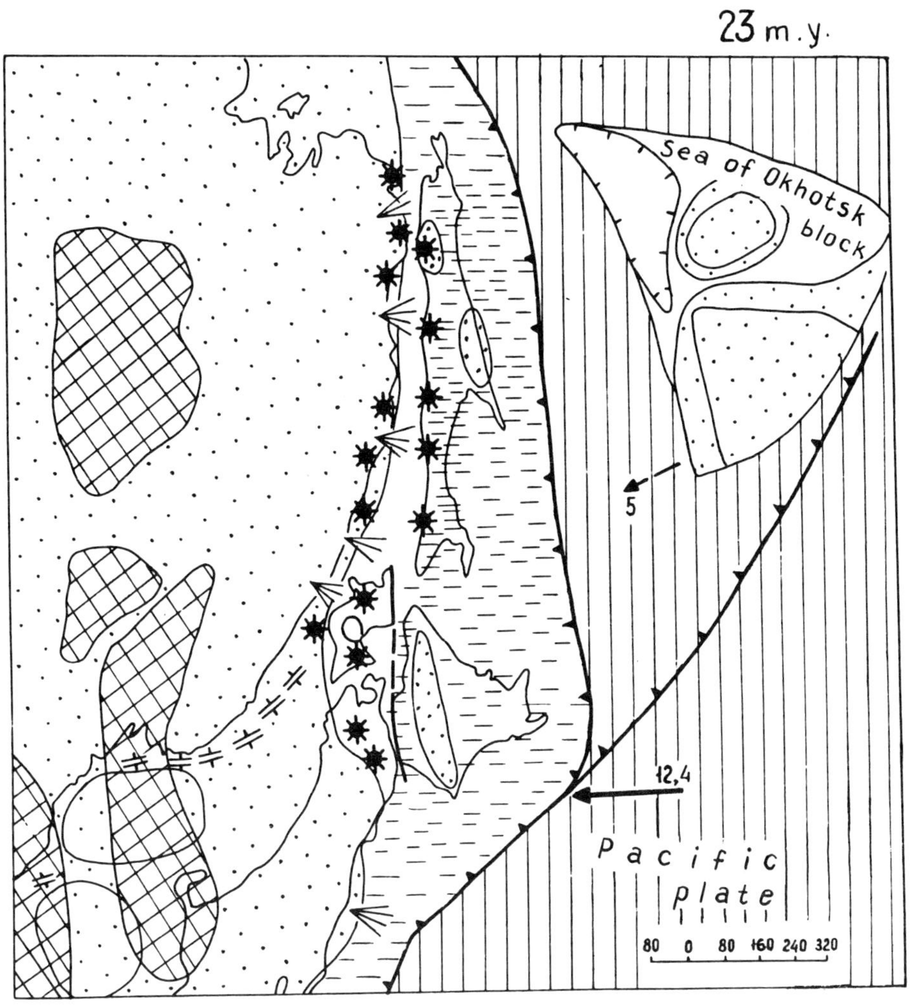

Fig. 100. Palinspastic reconstruction of the Sikhote-Alin - Sakhalin belt for 23 Ma. For explanation, see Figure 99.

ing that in the Permian the Khankai massif was already close to Siberia.

The Early Mesozoic history of the Khankai and Khingan-Bureya massifs is similar. Deep troughs formed along the eastern margin of both massifs on the passive margin of the Amuria continent,--the Bureya trough on the Khingan-Bureya massif and the Suputin trough on the Khankai massif. The available data show that in Jurassic time the Khankai and Hida massifs were close to Siberia. Magmatism was absent throughout the Triassic and Jurassic, also indicating a passive margin on the Amuria continent margin. The Lower Mesozoic sequences were folded in the Early Cretaceous and were intruded by granites. The Lower Cretaceous series lie unconformably on all more ancient formations, evidence for a sharp geodynamic reorganization in the early mid-Cretaceous when the passive Amurian margin collided with the Sikhote-Alin arc and a subduction zone commenced, dipping underneath Amuria and changing the passive margin into an active continental margin.

Exotic Terranes of the Sikhote-Alin Foldbelt

"Suspect terranes" of the foldbelt are shown in Figure 101. According to Melnikov and Golozubov [1980] and Mazarovich [1985], the composition of exotic terranes of the central Sikhote-Alin zone is very diverse. Terranes are often separated by the ophiolite melange. All members of a normal ophiolite association --ultramafics, gabbro, pillow lavas, cherts--are found. The chemical composition and isotopic characteristics show that the lavas are mostly the MORB type [Popeko et al., 1983], but there are also olivine alkaline basalts corresponding to oceanic-island basalt with initial $^{87}Sr/^{86}Sr$ ratios of 0.7046. Judging from Radiolaria, the oceanic complexes are of Upper Triassic-Lower Jurassic age.

Island-arc complexes are another component of Sikhote-Alin exotic blocks. Basalts and tuffs of Middle-Upper Jurassic age have been reported. The presence of Monotis bivalvia is diagnostic of the boreal province.

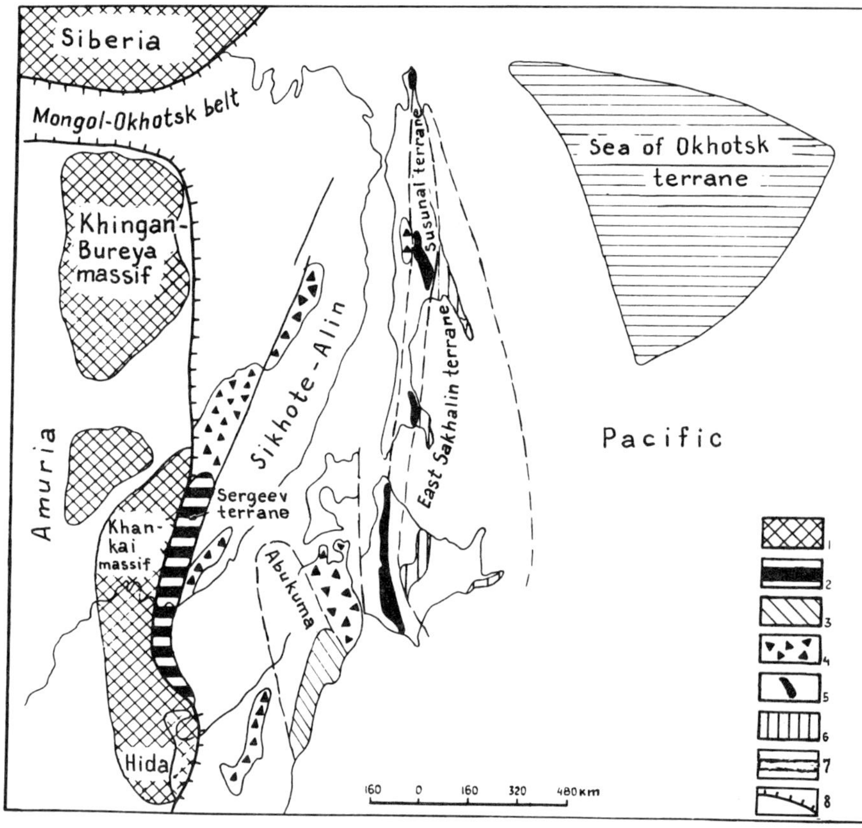

Fig. 101. Exotic terranes within the Sikhote-Alin - Sakhalin belt.
1, Ancient massifs (parts of the Amuria continent); 2, Sergeev; 3, Abukuma; and 4, Sikhote-Alin terranes attached to Amuria in the Late Jurassic and Early Cretaceous; 5, Susunai and 6, East Sakhalin terranes attached to Amuria in the end of the Cretaceous; 7, Sea of Okhotsk terrane attached to Eurasia in the Neogene; 8, continental block boundaries in the Early Mesozoic.

Upper Paleozoic and Triassic rocks of different compositions are very characteristic for exotic blocks of the Central Sikhote-Alin zone. They are involved in olistostrome sequences [Melnikov and Golozubov, 1980]. The blocks contain gabbro, spilites, Triassic radiolarites, Lower Jurassic radiolarites and spilites, Carboniferous and Permian reef limestones. Kanchuk et al [1988] established that many basaltic and other volcanic rocks within exotic blocks as a rule have oceanic island geochemical affinities. This permits the hypothesis that chaotic complexes of the Central Sikhote-Alin belt originated by the accretion of numerous guyots to the Sikhote-Alin subduction zone, mostly in the Early Cretaceous. Warm-water fauna, reef limestones, numerous Permian Fusulinid limestones, and Triassic Tethyan radiolaria are all widely developed here. Low-latitide conodonts of Tethyan provenance were recognized in the North Sikhote-Alin [Dagis, 1984]. The rocks and fauna indicate that in the past these blocks were situated in the tropical belt, i.e., very far from the Amurian margin, which at that time was located in high latitudes of the northern hemisphere. Paleomagnetic data from Permian, Triassic, and Jurassic rocks of exotic blocks within the Mina belt (which surrounds the Hida massif in Japan and is the analog of the Central Sikhote-Alin belt of exotic blocks) indicate the former position of these rocks at latitudes from -5° to +10° [Hiraoka et al., 1983].

The so-called Sergeev block (see Figure 101) is preserved in nappes and olistoliths and is of special interest because of its great bodies of "gabbroids". As Melnikov and Golozubov [1980] indicated, the so-called "gabbroids" include diverse rocks, such as intensely tectonized gabbro, amphibolites, schists, quartzites, and iron-formation (magnetitic gondites). Their K-Ar ages range from 340 to 490 Ma [Mazarovich, 1985] and reflect the time of metamorphism rather than the age of the rocks. The "gabbroid" basement is unconformably overlain by a cover of Permian, Triassic, and Jurassic deposits. The Upper Permian rocks comprise micaceous and arkosic sandstones with detrites from granites and gabbroids and with beds of foraminiferal limestone. The Upper Permian is conformably overlain by siltstones of Triassic through Middle Jurassic age. The whole sequence is only 700-800 m thick. According to Melnikov and Golozubov [1980] the sediments formed in a near-shore environment. The section is topped by a 250 m sequence of middle-upper Jurassic alkaline basalts (Pogsky suite), associated with alkaline and alkaline-ultramafic massifs of the Koksharovsky complex. This alkaline magmatic rock assemblage, as well as its association with a typical platform-like sedimentary sequence, is very indicative of intraplate hot-spot magmatism. The thinness of the sedimentary cover, the abundance of arkose, and the presence of alkaline magmatic rocks are evidence that the Sergeev block was continental in origin and was detached from some greater continent. The composition of the Permian, Triassic, and Jurassic sequences of this microcontinent is markedly different from sequences of the same age of the Khankai massif where the Permian includes a continental-margin volcanic belt (Arseniev zone) and the Triassic is separated from the Permian by a sharp unconformity. Evidently these continental masses were far from each other in Late Paleozoic and Mesozoic time.

The matrix of the exotic Sikhote-Alin blocks and the ophiolite melange is a breccia-like silty mass, often unsorted, sometimes flysch-like. Mazarovich [1985] gave its age as Jurassic.

Sikhote-Alin Flysch Zone

The Sikhote-Alin flysch definitely overlies the chaotic complexes. Valanginian and Hauterivian-Albian flysch complexes are known. Markevich [1978] distinguished between flysch sequences derived from the west and those from the east. The east-derived flysch is enriched in ash and contains layers of volcanoclastic rocks with fragments of bomb size, proving that volcanic activity was simultaneous with flysch accumulation. As the flysch is overlain by volcanics (andesites, liparites, and their tuffs) ranging from Albian to Turonian, the existence of a volcanic arc east of the flysch basin is confirmed.

We accept Parfenov's [1984] conclusion that the Lower Cretaceous flysch zone of the Sikhote-Alin is a filling of a back-arc basin, as proved by the presence of two sources of the sediment and the close association of the eastern flysch with volcanics.

The flysch was deformed prior to the Senonian, before the Sikhote-Alin volcanic belt came into existence.

Sikhote-Alin Volcanic Belt

This belt consists of volcanics of typical calc-alkaline affinites [Fedchin et al., 1982]. Granitic batholiths are closely related to the volcanics. A distinct lateral magmatic zonation is reliably recognized, which shows that the associated subducting slab dipped westward under the continent at about 22° [Zonenshain et al., 1976]. Its surface trace (presumably a trench) was 400 km east of the Sikhote-Alin volcanic belt, in the region of the present East Sakhalin suture, nearly at the same place as the inferred Miocene trench.

The Late Cretaceous-Paleogene sedimentary sequence underlying the Tatar Strait and cropping out in West Sakhalin and Hokkaido belongs to the forearc deposits accumulated in front of the Sikhote-Alin volcanic belt. It is a thick (6500 m) clastic sequence which accumulated continously, beginning with the Albian.

East Sakhalin

East Sakhalin (Figure 102) involves several complexes: metamorphic rocks of the greenschist and blueschist facies (Valzinsky series), chaotic formations (Nabilsky series), and volcanogenic-sedimentary formations of the Upper Cretaceous.

Non-metamorphosed chaotic complexes are unorganized mixtures of blocks and fragments of different composition and size in a breccia-like silt-argillite matrix. The blocks include low-K tholeiitic basalts, alkaline-olivine basalts, sanidine leucobasalts, reef limestones, deep-sea cherts, radiolarities, and coarse-clastic flora-bearing rocks. On all sides the chaotic complexes are associated with tracts of ophiolitic melange, which enclose great blocks of harzburgites, gabbros, pillow lavas, and plagiogranites. The limestone blocks contain remnants of Paleozoic fauna, including Devonian brachiopods [Raznitsin, 1975] and (more abundantly) Upper Paleozoic corals, stromatopora, and fusilinids. Early Mesozoic echinoids are found as well. In other blocks, there are numerous Inoceramus and Radiolaria of Upper Jurassic-Lower Cretaceous age [Sidorenko, 1970b]. The matrix was found to contain Albian-Cenomanian Radiolaria, indicating that the chaotic complexes formed in the mid-Cretaceous, which coincides with the youngest age of metamorphism (90 Ma).

The Upper Cretaceous volcanic complex of East Sakhalin (Rymnik series) has not such chaotic structure as the Nabilsky series and consists of intercalated volcanic rocks and deep-water sediments (cherts, radiolarities, flysch). The composition of the volcanic rocks is diverse [Semenov, 1982]; on the one hand, there are andesites, basaltic andesites, dacites, and liparities of calc-alkaline affinities; on the other hand, there are widespread alkaline basalts and hawaiites. This volcanic complex is considered to be the remnant of a Late Cretaceous island arc.

Folding and thrusting occurred at the Cretaceous-Paleogene boundary. Paleogene strata unconformably overlie the nappes [Sidorenko, 1970; Rozhdestvensky, 1986], and the folded structures are cut by small intrusions of biotitic granites and granodiorites with isotopic ages of 58-66 Ma [Sidorenko, 1970b]. On the whole, much of the East Sakhalin structure can be interpreted as a subduction zone complex that formed or started to form in the early Late Cretaceous.

The timing of the main phase of deformation in East Sakhalin at the Cretaceous-Paleogene boundary indicates the time at which the East Sakhalin island arc arrived at the Sakhalin subduction zone and attached itself to the West Sakhalin forearc.

SYSTEMATIC HISTORY

Triassic-Jurassic

In the Early Mesozoic, the eastern (present coordinates) edge of Amuria was occupied by a passive continental margin (Figure 103). The Sikhote-Alin island arc, as shown on the reconstruction, was at some distance (ca. 1000 km) from the passive margin, judging from paleontologic and paleomagnetic data, the latter being obtained from volcanics contained in chaotic complexes. It is assumed that the subduction zone under the arc at that time dipped westward and that the space between the island arc and Amuria was a marginal oceanic basin. Exotic blocks on the Kula plate (or Izanagi plate) drifted in from the Pacific Ocean into the subduction zone under the Sikhote-Alin arc, and formed accretional complexes in the frontal part of the arc. Evidently the Sergeev block (or microcontinent) was one of the last to arrive in the subduction zone. Its original location is customarily placed at low latitudes, taking into account the above-mentioned Jurassic paleomagnetic data from the Mina belt in Japan. At this time the Sergeev mi-

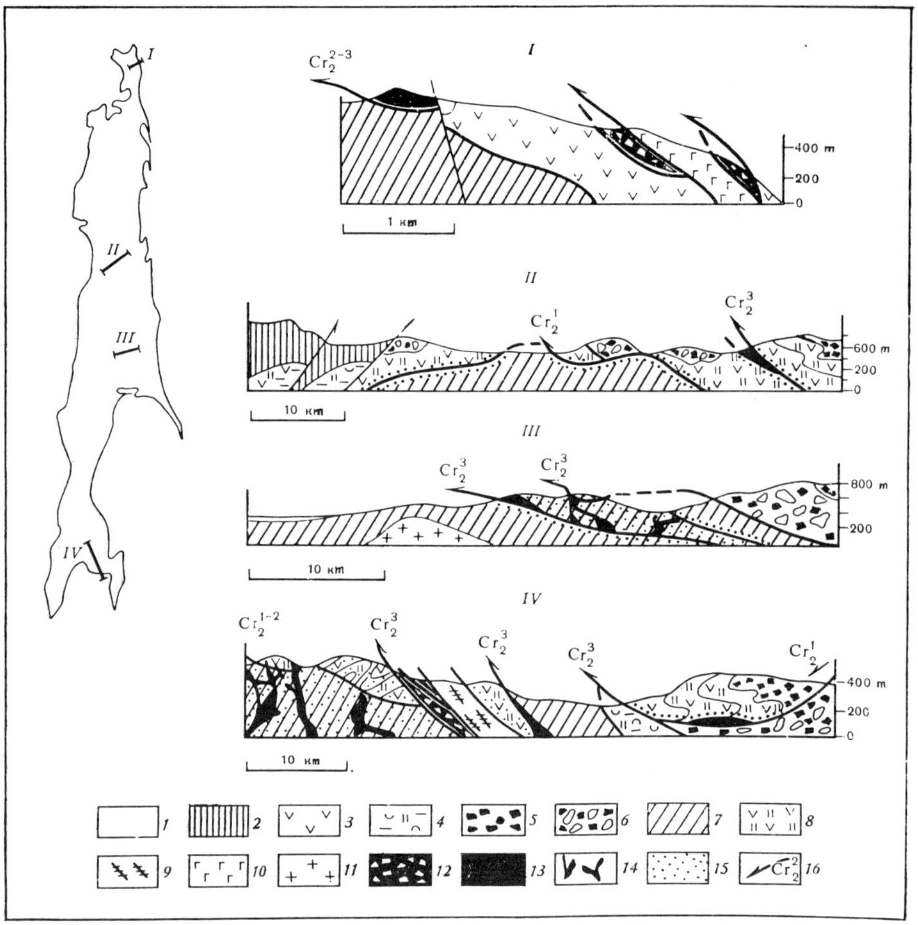

Fig. 102. Nappe structure of Sakhalin, after Pushcharovsky et al. [1983]. Insert shows location of cross-sections.
, Neogene-Quaternary deposits; 2, Upper Cretaceous rocks of the West Sakhalin trough; 3, Upper Cretaceous mafic volcanics; 4, Cretaceous tuffaceous deposits; 5, Upper Cretaceous coarse clastics; 6, olistostrome; 7, Mesozoic volcanogenic-siliceous complex; 8, Triassic-Jurassic chert and volcanics; 9, amphibolite; 10, gabbro; 11, granodiorite; 12, melange; 13, serpentinite; 14, serpentinite protrusion; 15, metamorphism; 16, thrust (age indicated).

crocontinent presumably passed over a hot spot; alkaline basalts (of the Pogsky suite) were erupted and the alkaline-ultramafic Koksharovsky massif was intruded.

Early Cretaceous, 130 Ma (Figure 104)

At this time the Sergeev microcontinent as well as other blocks collided with the Sikhote-Alin arc, forming a comparatively large accretionary island mass, not less than 1500 m long. Formation of this mass was accompanied by folding. It was located where the Sikhote-Alin volcanic arc had been earlier and was separated from Amuria by the same marginal basin as in the Early Mesozoic.

Early Late Cretaceous, 95 Ma (Figure 105)

In the very beginning of the Late Cretaceous, the Sikhote-Alin arc started to converge with Amuria, originating a subduction zone dipping under the Amuria margin, as shown by calc-alkaline volcanics developed on the margin of the Khankai massif. Simultaneously, the Sikhote-Alin marginal basin was closing.

The Early Late Cretaceous deformation and metamorphism in East Sakhalin indicate that at this time the Kula plate brought exotic terranes into the subduction zone under the East Sakhalin block. Obviously, this zone separated the Sikhote-Alin arc from the Kula plate, and an ancient, easily subducted part of the Kula plate was trapped between the two arcs.

Middle of the Late Cretaceous, 85 Ma (Figure 106)

By the beginning of the Senonian, the Sikhote-Alin arc collided with the margin of Amuria. Collision resulted in intense pre-Senonian folding, thrusting, and formation of mountainous topography. Folding was followed by sinistral strike-slip faulting, evidently due to oblique NW-SE convergence.

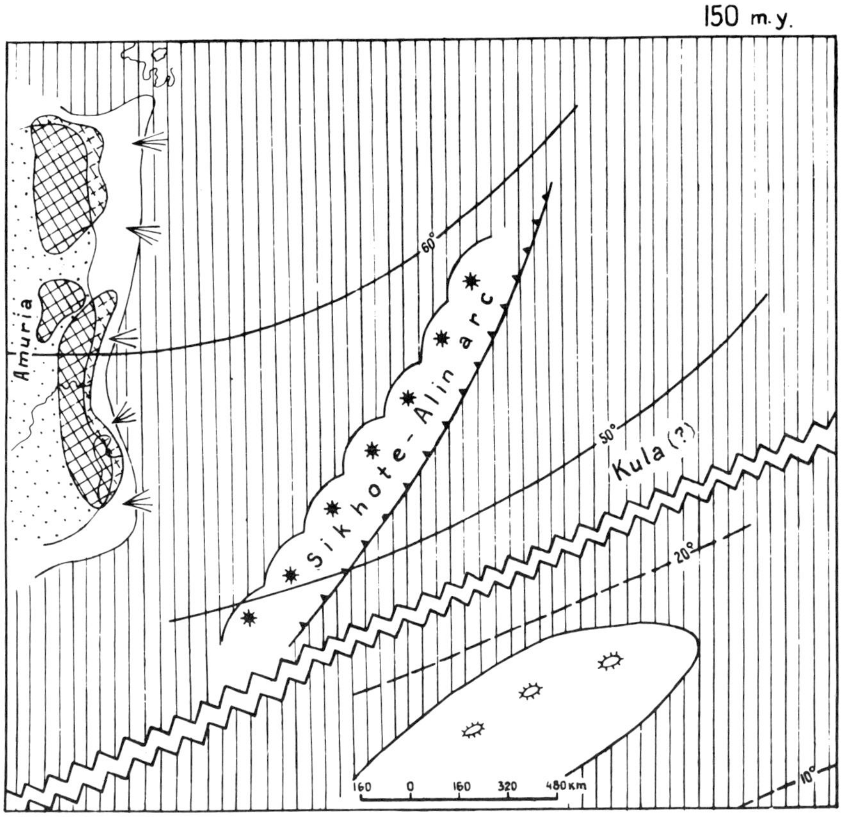

Fig. 103. Palinspastic reconstruction of the Sikhote-Alin - Sakhalin belt for 150 Ma. For explanation, see Figure 99.

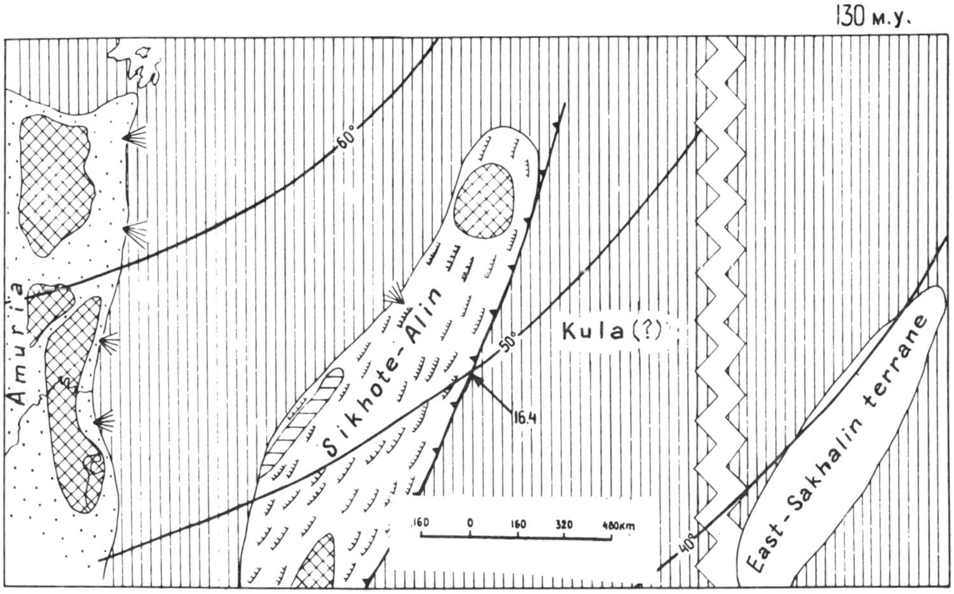

Fig. 104. Palinspastic reconstruction of the Sikhote-Alin - Sakhalin belt for 130 Ma. For explanation, see Figure 99.

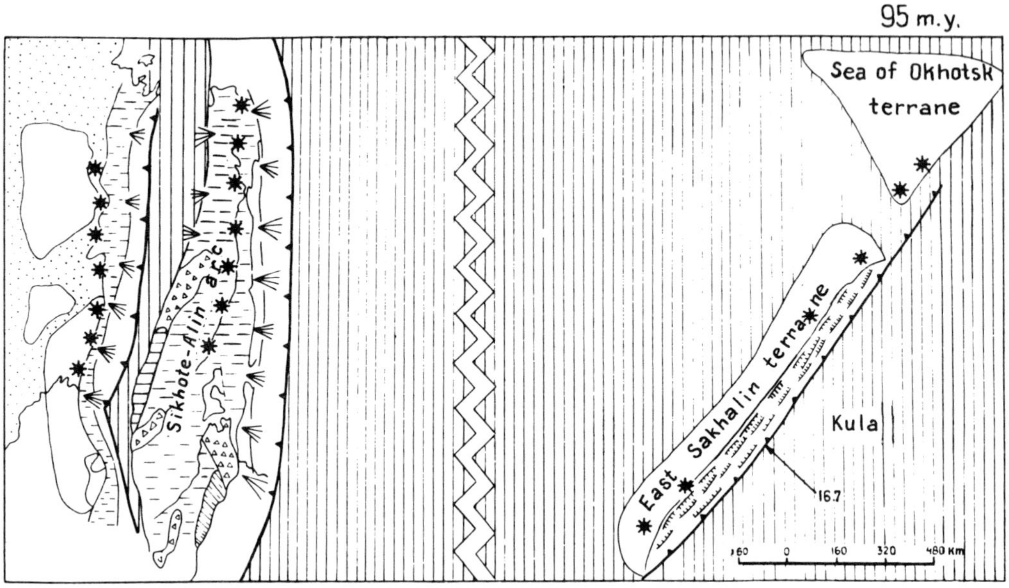

Fig. 105. Palinspastic reconstruction of the Sikhote-Alin - Sakhalin belt for 95 Ma. For explanation, see Figure 99.

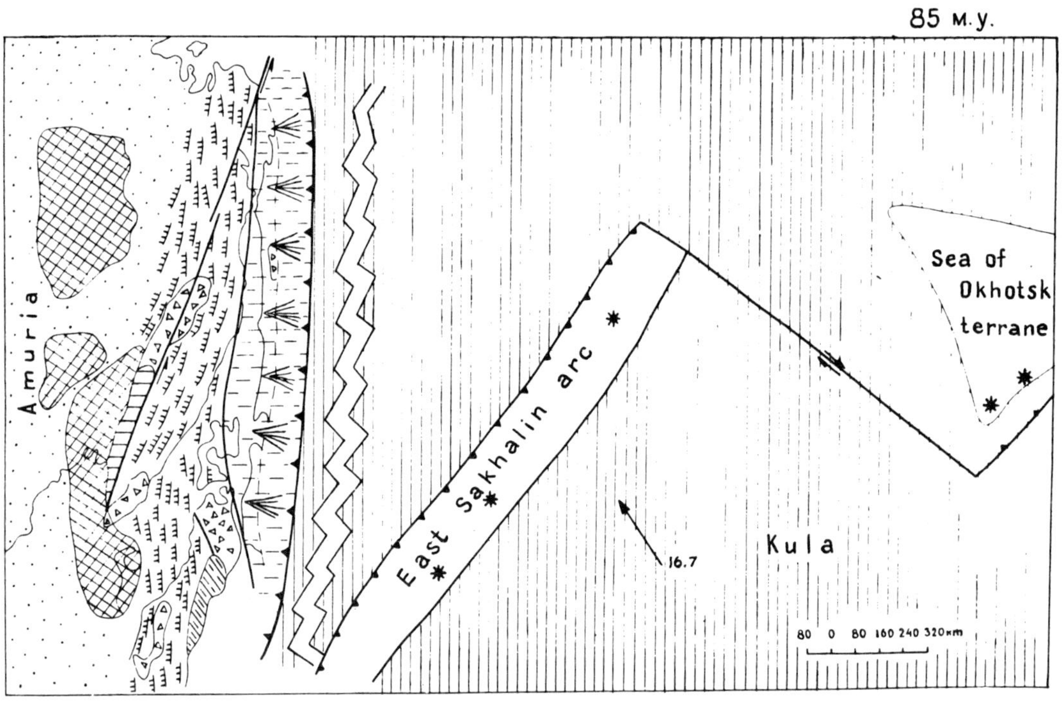

Fig. 106. Palinspastic reconstruction of the Sikhote-Alin - Sakhalin belt for 85 Ma. For explanation, see Figure 99.

We suppose that from the middle of the Late Cretaceous the East Sakhalin arc was attached to the Kula plate margin and that they moved together at the rate of 15-16 cm/yr toward the margin of Eurasia.

Latest Cretaceous, 70 Ma (Figure 107)

The continental margin affiliation of the Sikhote-Alin volcanic belt established at this time, as the arc became part of the

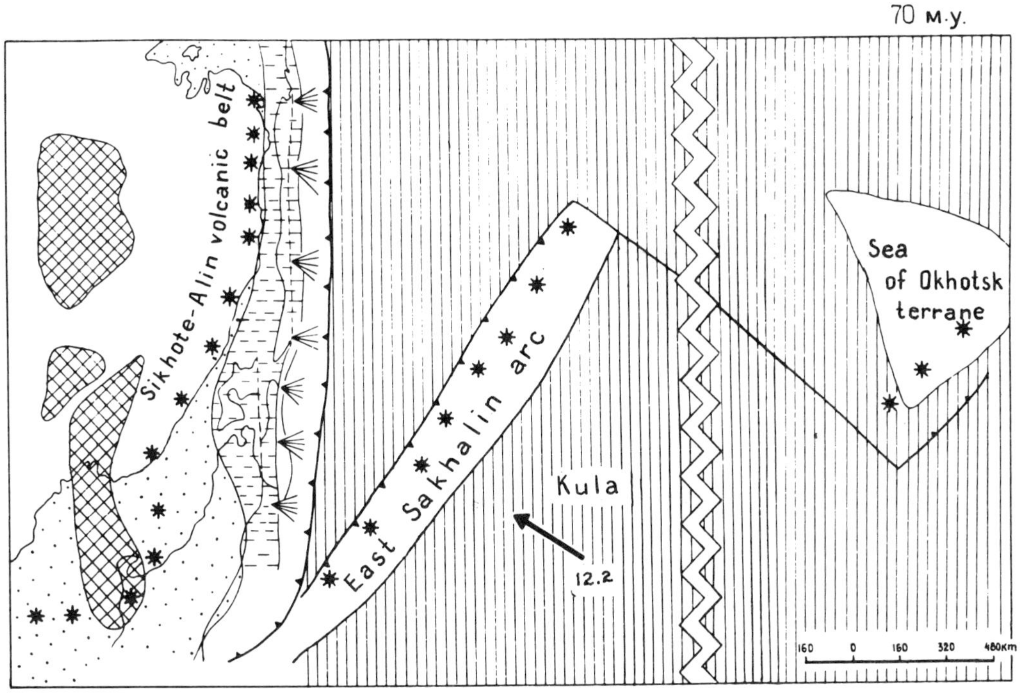

Fig. 107. Palinspastic reconstruction of the Sikhote-Alin - Sakhalin belt for 70 Ma. For explanation, see Figure 99.

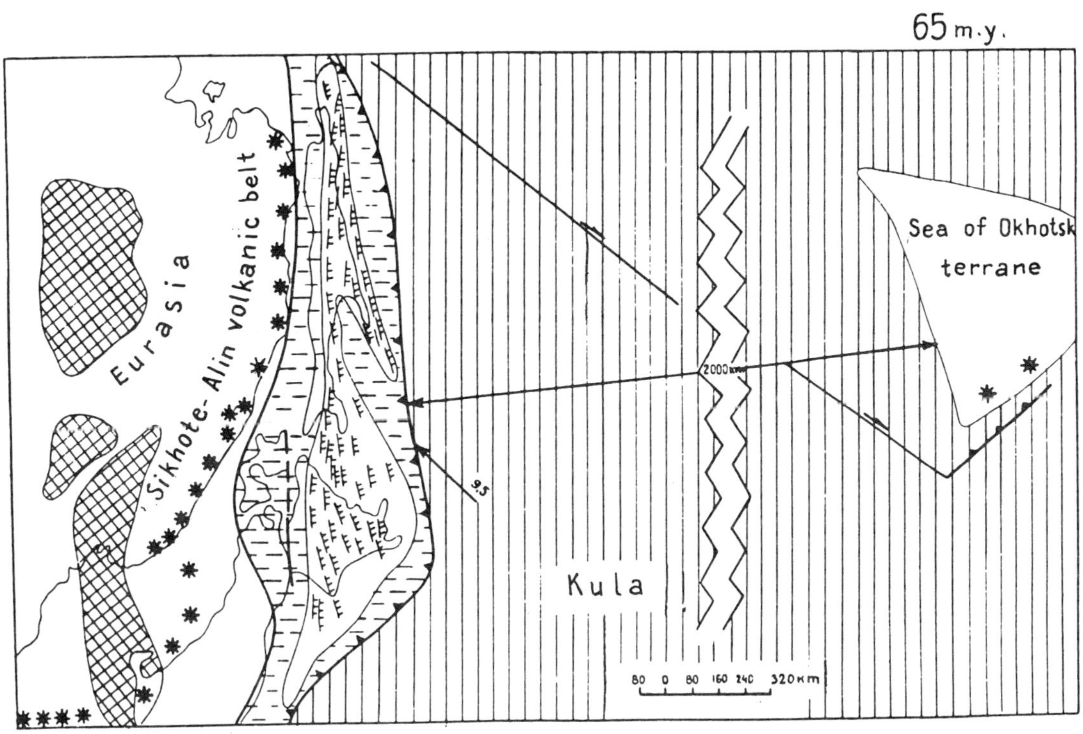

Fig. 108. Palinspastic reconstruction of the Sikhote-Alin - Sakhalin belt for 65 Ma. For explanation, see Figure 99.

newly constructed margin of Eurasia. Toward the very end of the Cretaceous, the East Sakhalin island arc collided with the subduction zone of the Sikhote-Alin active margin.

Cretaceous-Paleogene Boundary, 65 Ma (Figure 108)

The major event of this time span was collision of the East Sakhalin arc with the Eurasian active margin. Deformation occurred only within the East Sakhalin arc and did not spread into the Sikhote-Alin forearc basin. This perhaps can be explained by the brevity of collision. As soon as the East Sakhalin block approached the subduction zone and joined the near-shore trench slope, the subduction zone jumped to a new position on the eastern margin of the former East Sakhalin arc. As a consequence, the arc proper together with all accretionary complexes came to be included in the near-shore trench slope, thus accreted the Eurasia margin. Subduction proceeded further, which is proved by the continuous formation of the Sikhote-Alin volcanic belt up to the middle of the Eocene.

The subsequent (Late Cenozoic) history has been considered above. It includes formation of the Miocene Sakhalin, Kurile, and Japanese island arcs, back-arc basins of the Seas of Japan and Okhotsk, and simultaneous collision of the Sea of Okhotsk block with the Miocene East Sakhalin arc.

Chapter IX

FOLDBELTS OF THE NORTHEAST USSR, TAIMYR AND THE ARCTIC

GENERAL DESCRIPTION

The northeastern part of the USSR has been analyzed in a mobilistic approaches by Parfenov [1984], Natal'in [1984], Tilman [1973], and Natapov, et al. [1977]; these studies contain no palinspastic reconstructions, but reconstructions were presented by Fujita [1978], Fujita and Newberry [1983], Zonenshain [1984], and Natapov and Stavsky [1985]. Churkin [1972] was the first to recognize the composite pattern of the northeast USSR where he delineated the boundary between North American and Siberian domains, and he also was the first [Churkin, 1979] who postulated widespread occurrence of accreted terranes in Chukotka and Koryakya.

The northeast USSR (Figure 109) includes (i) the Verhoyansk foldbelt, which originated on the former passive margin of Siberia; (ii) the Kolymian structural loop including a number of exotic blocks overlain by Late Jurassic volcanic arc belts (Uyandina-Yasachnaya and Alazei-Oloy) and intruded by a chain of Cretaceous granitoid batholiths (Kolymian batholithic belt); (iii) the Omolon and Okhotsk massifs, microcontinents with exposures of Precambrian basement; (iv) the South Anyui foldbelt, which is an Early Cretaceous collisional suture between continental blocks; (v) the Chukotka foldbelt (together with the Arctic Islands), consisting of Paleozoic and Early Mesozoic complexes of the passive margin of the Arctida, or Hyperborean, continent; (vi) the Koni-Murgal and Okhotsk-Chukotka volcanic belts; (vii) the Wrangel-Herald-Brooks suture zone, a deformational belt of Late Cretaceous-Early Tertiary age; and (viii) the Paleozoic and Mesozoic complexes of Taimyr and Severnaya Zemlya (north part of Figure 1, near 100° longitude).

The arcuate shape of the major fold structures in the Kolymian structural loop was formerly thought to signify a rigid mass in the center of the loop--the Kolymian Median massif. We shall see that the Kolymian loop resulted from horizontal squeezing by longitudinal compression when plates of the Pacific Basin, Siberia, and Chukotka converged.

Verkhoyansk Foldbelt

This belt coincides areally with the so-called Verkhoyansk sedimentary complex deposited during Late Paleozoic, Triassic, and Jurassic times. The Verkhoyansk complex is a monotonous clastic sequence that accumulated in a shallow-water, or rarely a continental-slope, environment. The clastics were derived from the Siberian platform only. The off-shore sediments increase in thickness and amount with distance from the Siberia margin (Figure 110). Clastic sequences of the Verkhoyansk complex are interpreted as formations of huge submarine cone which formed on the shelf and continental slope of the Siberian passive margin.

Clastics of the Verhoyansk complex overlie older Paleozoic and Upper Precambrian carbonate sequences that are also characteristic of a passive continental-margin environment. Thus, the passive margin on the edge of Siberia probably existed for a long time, not less than half a billion years. The carbonate sequences contain basalt and gabbro at two levels, suggesting rifting during the Riphean, when the passive margin began to form, and during the Late Devonian when it was broken by the Vilyuy rift system. Splitting of the Siberian margin could be associated with Late Devonian rifting and moving away of a piece of the continent.

The Verkhoyansk complex underwent intense deformation. Folding was related to a detachment fault along the base of the sedimentary cover (Figure 111). Folding took place throughout the belt in the Middle and Late Jurassic and at the very beginning of the Cretaceous in the eastern parts of the belt, and in the mid-Cretaceous in the western parts [Parfenov, 1988]. It should be noted that the Verkhoyansk complex forms an intricate loop concordant with the Kolymian loop (Figure 109).

Kolymian Structural Loop

This loop is easily seen on any map (see Fig. 109) from the distribution of pre-Upper Paleozoic rocks adjacent to the area underlain by the Verkhoyansk clastic sequences. Two structural complexes, separated by a major unconformity, are easily recognized. On the periphery of the loop (Polousnyi Ridge, Chersky Range, and Pri-Kolymian region) the unconformity is pre-Bathonian, while in the center of the loop (in the Alazeya plateau) it is pre-Middle Jurassic and in places pre-Upper Triassic. This unconformity separates more ancient exotic terranes from Mesozoic island-arc complexes and records the time when those terranes were amalgamated together to provide a basement for younger island arcs. The distribution of terranes is shown in Figure 112.

The Paleozoic and Pre-Paleozoic deposits that compose the terranes within the Kolymian loop contain numerous carbonate sequences, and have for a long time been considered as either the sedimentary cover of the Kolymian massif or expo-

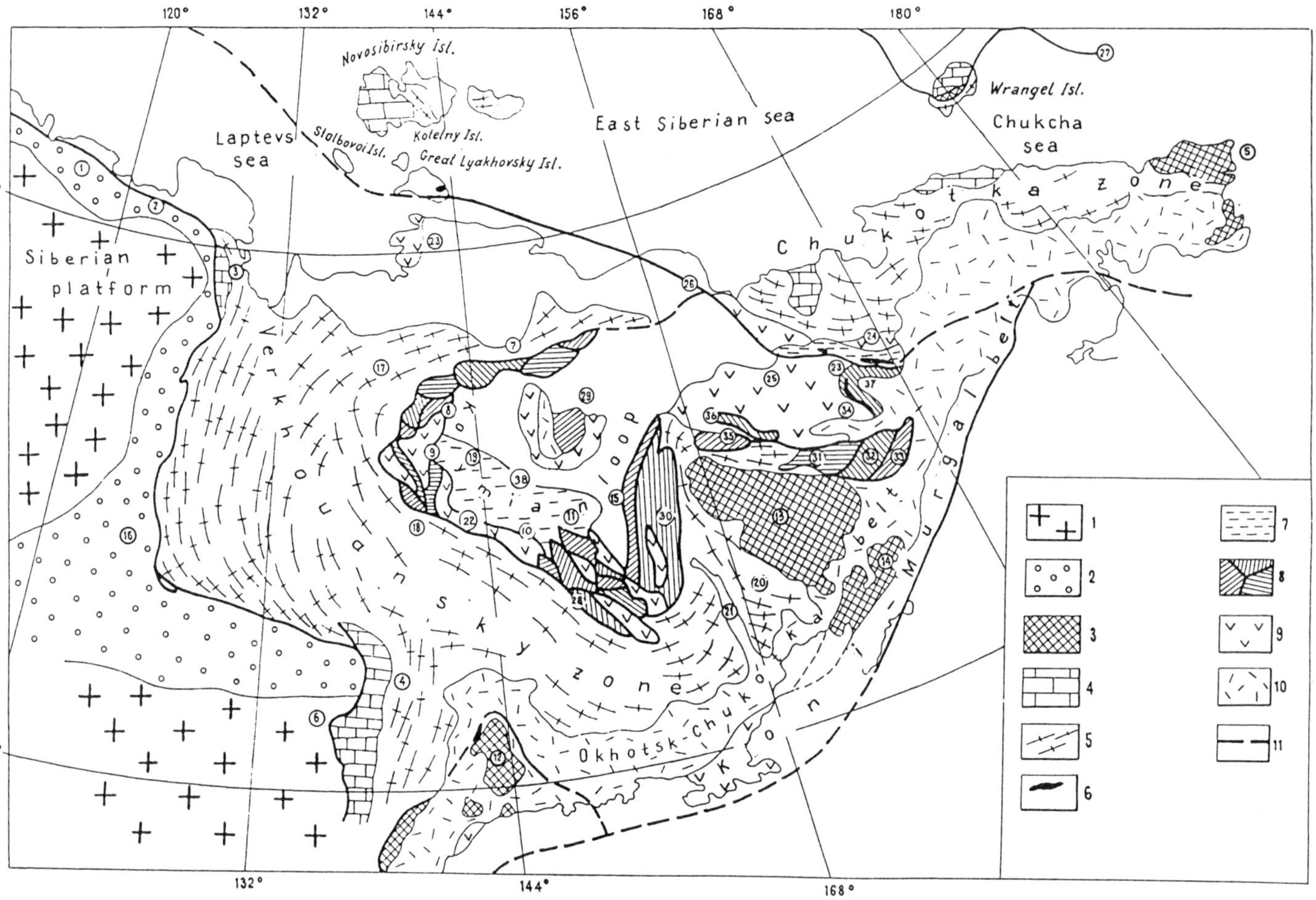

Fig. 109. Tectonics of the Verkhoyansk-Kolymian belt.
1, Siberian platform; 2, Pre-Verkhoyansk foredeep; 3, Precambrian massif; 4, Lower Paleozoic shelf carbonate complexes; 5, Upper Paleozoic and Mesozoic clastic series; 6, ophiolite; 7, South Anui suture zone (Late Jurassic-Early Cretaceous); 8, various exotic terranes; 9, Late Jurassic-Early Cretaceous subduction-related volcanics; 10, Middle Cretaceous subduction-related volcanics (Okhotsk-Chukotka volcanic belt); 11, main thrust and suture.
Numbered are: 1, Pronchishchev zone; 2, Chekanovsky zone; 3, Kharaulakh zone; 4, Sette-Daban zone; 5, East Chukotka massif; 6, Kyllakh uplift; 7, Polousny zone; 8, Selenyakh zone; 9, Tas-Khayakhtakh zone; 10, Chersky zone; 11, Omulevka zone; 12, Okhotsk massif; 13, Omolon massif; 14, Taiganos massif; 15, West Kolymian block,; 16, Pre-Verkhoyansk foredeep; 17, Oldzhoi trough; 18, In'yaly-Debinsky synclinorium; 19, Bastakh zone; 20, Sugoi zone; 21, Omsukchan graben; 22, Uyandina-Yasachanaya arc; 23, Anyui-Svyatoyannos arc; 24, Nutesin arc; 25, Alazei-Oloi arc; 26, South Anyui suture; 27, Herald-Brooks suture; 28, Kolyma-Indigirka suture; 29, Alazei terrane; 30, Pri-Kolymian terrane; 31, Ushurak-chan terrane; 32, Levo-Oloi terrane; 33, Eropol terrane; 34, Aluchin terrane; 35, Berezov terrane; 36, Siver terrane; 37, Yablonsky massif; 38, Zyryanka depression.

sures of the Siberian platform cover cropping out from beneath the Verkhoyansk complex. Recently these sequences have proved to be rather diverse and to consist not only of carbonate but also volcanic, cherty, and terrigenous sequences with ophiolites. Different facies are close together, and each block has its own composition. The pre-Upper Paleozoic complexes have a complicated inner structure, consisting of numerous slices, from kilometers to tens of kilometers across and thrust one over another. Usually, the inner structures of adjacent blocks are not alike; T-shaped junctions are typical. The general structure reminds one of a gigantic breccia. Considering the above diversity, the Kolymian loop is a deformed mosaic of tectonostratigraphic terranes.

The Paleozoic and Mesozoic sequences that occur in blocks are of the following types (Figure 112): (i) carbonate sequences (Pri-Kolymian, Omulevka sequence); (ii) Ordovician volcanics (Rassoshinsa), (iii) Devonian ophiolites (Uyandina, Uvyazka, etc.), (iv) volcanogenic-cherty sequences of Upper Paleozoic and Lower Mesozoic age (Alazeya), (v) Permian deep-water volcanogenic-cherty sequences (Shumninsky), and

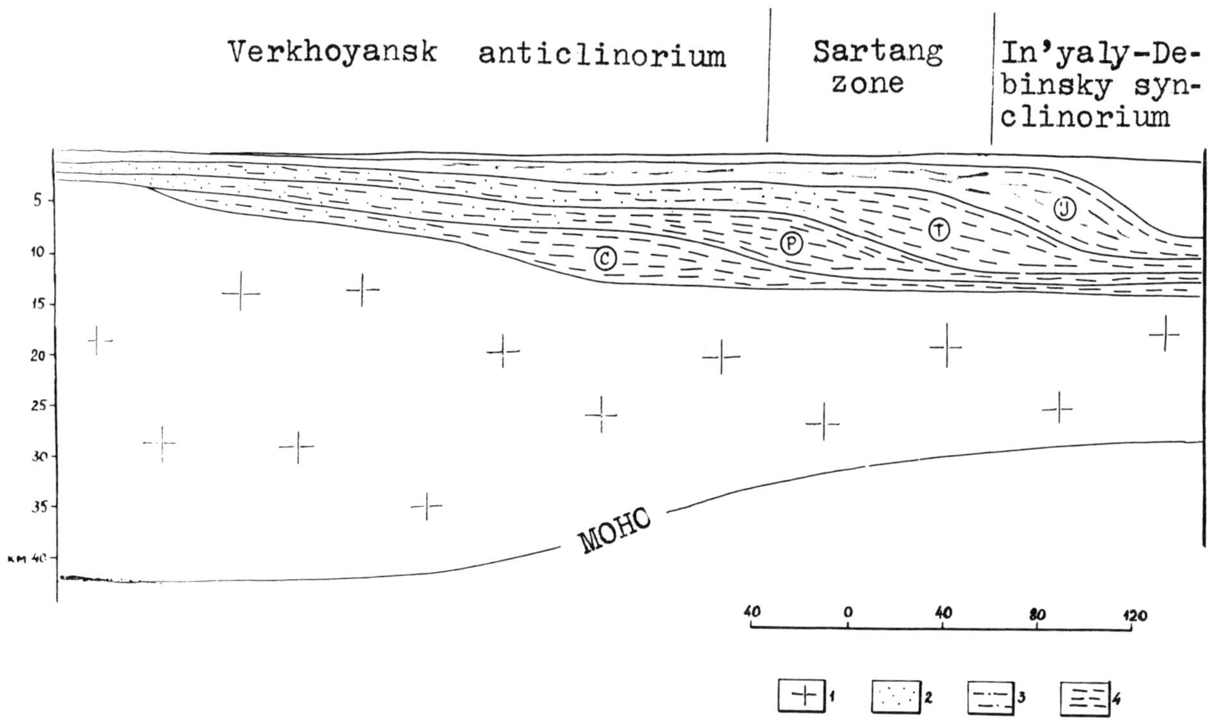

Fig. 110. Paleofacies cross-section through the Verkhoyansk-Kolymian belt.
1, Precambrian basement; 2, near-shore clastics; 3, shelf sand and silt deposits; 4, continental slope and rise deposits.

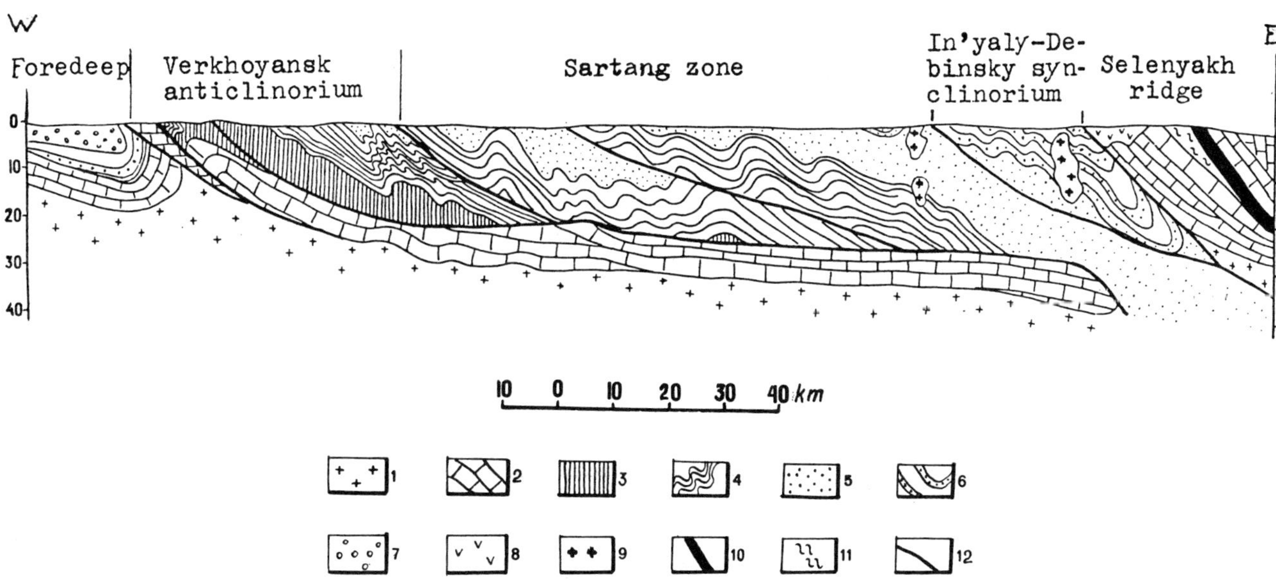

Fig. 111. Geological cross-section through the Verkhoyansk-Kolymian belt.
1, Precambrian basement of the Siberian platform; 2, Upper Precambrian-Lower Paleozoic rocks; 3, Carboniferous rocks; 4, Permian rocks; 5, Triassic rocks; 6, Jurassic rocks; 7, Cretaceous rocks; 8, Late Jurassic calc-alkaline volcanics; 9, Cretaceous granite; 10, ophiolite; 11, metamorphics; 12, thrust.

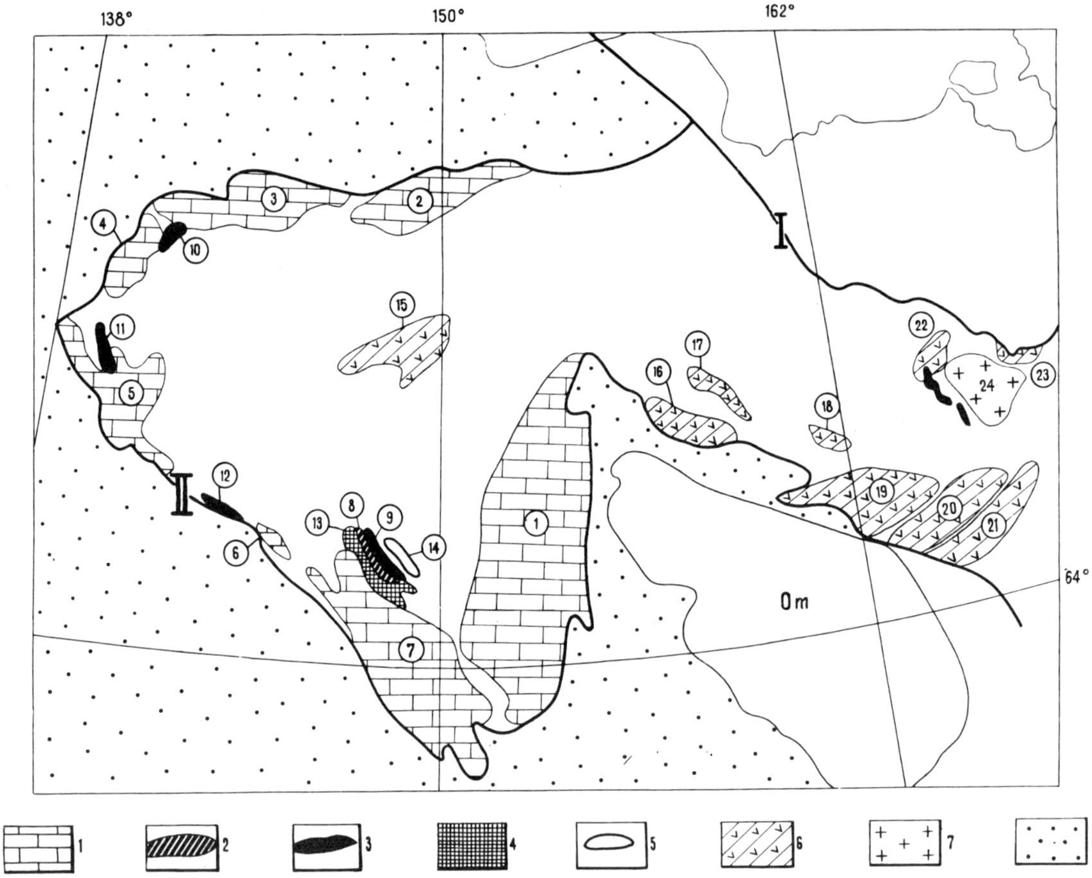

Fig. 112. Exotic terranes within the Verhoyansk-Kolymian belt.
1, Terranes of Siberian provenance; 2, Ordovician volcanics; 3, ophiolite; 4, Upper Paleozoic suspect terrane; 5, Permian deep-water chert and volcanics; 6, Upper Paleozoic and Lower Mesozoic island-arc volcanics; 7, continental fragment; 8, Verkhoyansk complex (originally a submarine fan). I - South Anui suture; II - Chersky suture, Om - Omolon microcontinent.
Numbered terranes: 1, Pri-Kolymian; 2, Ulakhan-Sys; 3, Polousny; 4, Selenyakh; 5, Tas-Khayakhtach; 6, Chemalga; 7, Omulevka; 8, Rassoshinsky; 9, Uvyazka; 10, Uyandina; 11, Munilkan; 12, Ucha; 13, Bulkut; 14, Shumninsky; 15, Alazei; 16, Berezov; 17, Siver; 18, Nadezhninsky; 19, Ush-urakchan; 20, Levo-Oloi; 21, Eropol; 22, Aluchin; 23, Yarakvaam; 24, Yablonsky.

(vi) Middle and Upper Paleozoic island-arc complexes.

The carbonate sequences which occur as blocks show similarities with the Siberian platform. Within these blocks, Famennian (latest Devonian) shelf limestones are followed by deep-water pelagic limestones and cherty deposits. There the Late Devonian was the time when these blocks were rifted from the eastern Siberian margin, where rift structures are known.

The blocks with ophiolites and volcanic-cherty sequences are apparently exotic. They appear to be remnants of former island arcs and associated fore-arc and back-arc basins.

Two Mesozoic island-arcs can be reconstructed within the Kolymian loop: the Oloi-Alazeya and Uyandina-Yasachnaya arcs. The older is the Oloi-Alazeya volcanic arc, which developed from the Middle Triassic to the Middle Jurassic. It is represented by basalts, andesites, their tuffs, tuffogenic flysch, chert, and volcanomictic rocks. The exotic blocks that form the accretional basement of the Kolymian loop were apparently carried to the subduction zone of this Oloi-Alazeya arc.

The volcanics of the Uyandina-Yasachnaya arc range from Callovian to Early Volgian in age. They are represented by a typical calc-alkaline series of differentiated rocks from basalts to rhyolites through abundant andesites (Ilin-Tas series). The southeastern segment shows a lateral magmatic zonation reflected in a successive change from tholeiitic basalts in the northeast to calc-alkaline andesite volcanics toward the southwest and subsequently to silicic lavas. Based on $K_2O:SiO_2$, Ged'ko [1988] calculated that the paleo-subducting slab dipped 45° southwestward (Figure 113). Thick Upper Jurassic flysch deposits (Bastakh suite) in a wide area northeast of the Uyandina-Yasachnaya volcanic belt are regarded as forearc sediments.

Both the volcanic (Ilin-Tas) series and terrigenous Bastakh

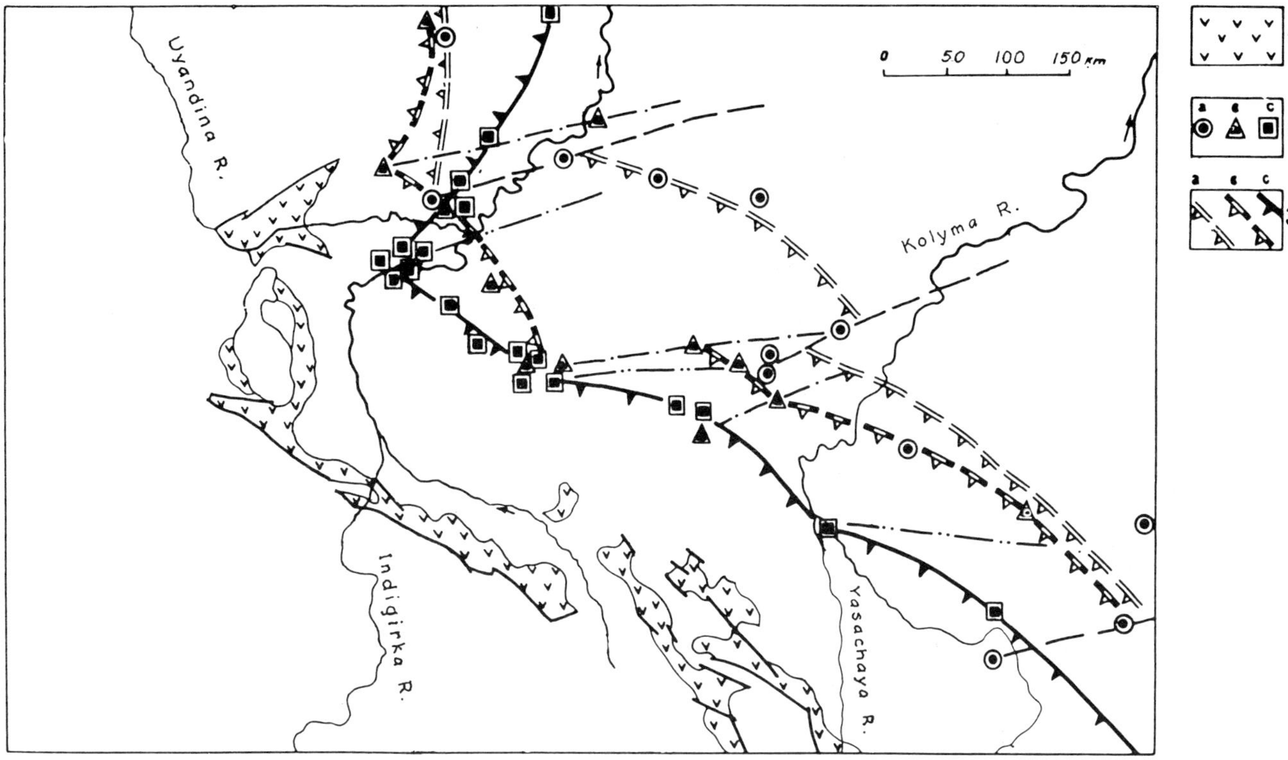

Fig. 113. Calc-alkaline volcanics of the Late Jurassic Uyandina-Yasachnaya arc, after Ged'ko [1987]. 1, Distribution of calc-alkaline volcanics; 2, points showing calculated position of the fossil subduction zone: a) early stage, b) middle stage, c) late stage; 3, inferred position of subduction zone during Kimmeridgian-Oxfordian times: a) early stage, b) middle stage, c) late stage.

suite are intensely deformed, thrust westwards, and bent into the Kolymian loop. The time of deformation is constrained by a pre-Aptian unconformity and by the intrusion of a belt of granitic batholiths. That belt is discordant to the configuration of the Kolymian structural loop in the south, but in the north it faithfully follows the bend of the loop, changing its strike from SE-NW through SW-NE to W-E (Figure 114). Hence, the bending of the loop is diachronous, older in the south and younger in the north relative to the batholith. The Kolymian batholiths are palingenetic and may be regarded as typical examples of collisional granites. They probably originated when the structures of the Kolymian loop were thrust onto the Verkhoyansk complex, causing crustal thickening and melting to form granites [Natapov and Stavksy, 1985].

Thus, the Kolymian loop is a secondary feature, a repeatedly deformed accretional mosaic of exotic terranes assembled together with several island arcs of different ages. The primary configuration of the terranes composing the loop was certainly completely different from their present shape as exemplified by the Uyandina-Yasachnaya arc, which is now convex to the west, just opposite to its original convexity to the east.

The wide distribution of Triassic and Early Jurassic calc-alcaline volcanics shows that the Oloi-Alazeya arc already existed at that time, and different terranes were carried into the accretionary prism in front of its subduction zone. In the Late Jurassic the accretionary terrane mosaic served as a basement for the Uyandina-Yasachnaya arc, which at that time was located at a significant distance from the area where the Verhoyansk complex had accumulated. In the middle Early Cretaceous, before the Aptian, the arc was broken, and the Peri-Kolymian block rotated relative to the blocks of the Chersky Range. Simultaneously, the whole arc was thrust over the back-arc basin toward the Verhoyansk complex. After that, the Kolymian granite batholiths were emplaced; then the northern segment was bent. Thus, the Kolymian structural loop was formed in several steps.

Omolon and Okhotsk Massifs

At present the Omolon and Okhotsk massifs are separated, but they have so much in common that they may be regarded as remnants of the same continental mass. However, their cover and basement are considerably different from those of the Siberian platform and the Verkhoyansk complex situated nearby. The location of these massifs in the general structure of the north-east USSR (see Figure 109) shows how important they were in forming the structural pattern of the region. The Omolon massif is indented into the Kolymian structural loop and is enveloped by narrow strips of the Verhoyansk clastics, which were rotated clockwise together with the massif. The sharp northern tip of the Okhotsk massif bifurcates the Verkhoyansk foldbelt into two parts: the South Verkhoyansk or Sette-Daban branch (see Figure 109) and the Sugoy branch.

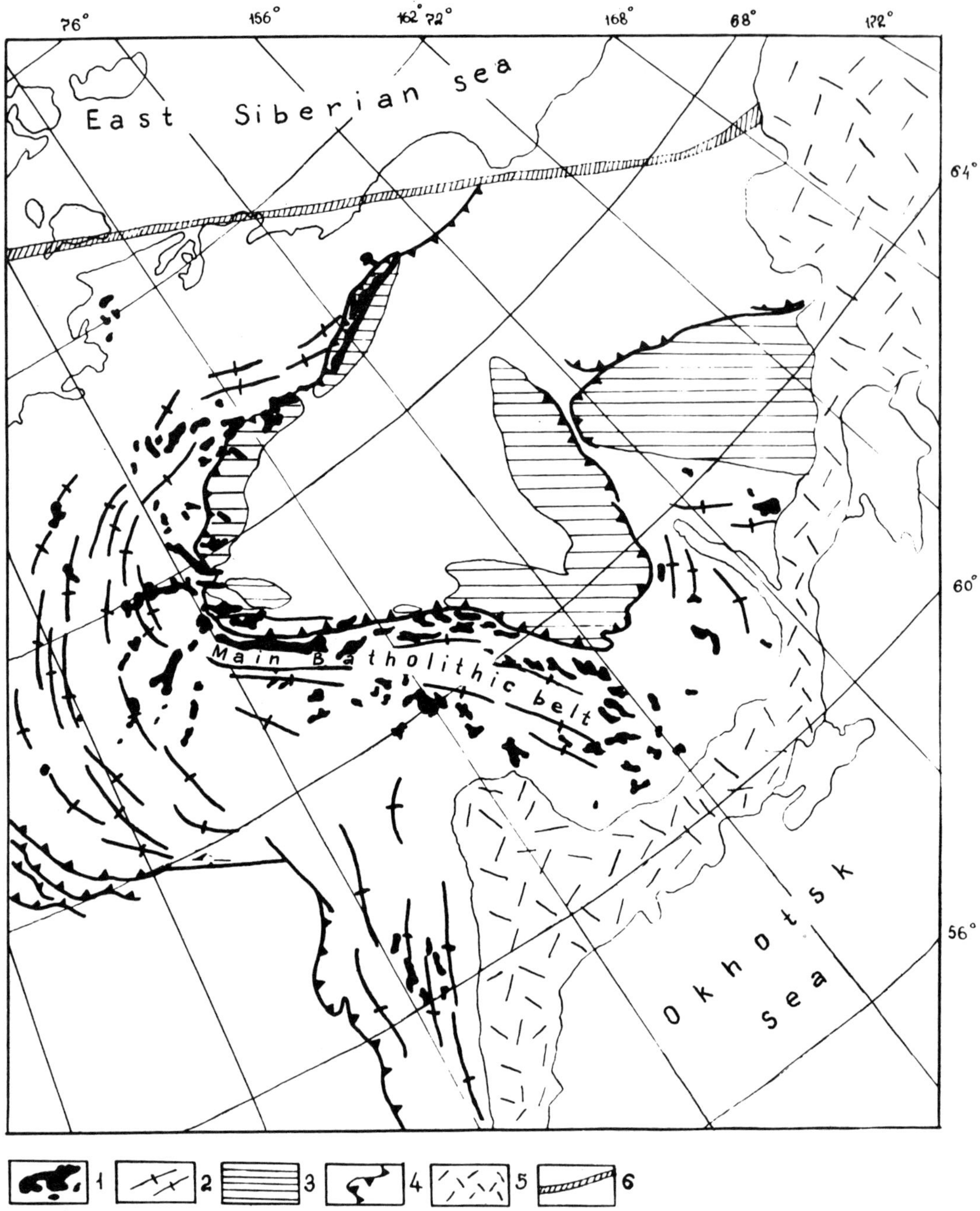

Fig. 114. Belt of the Kolymian granite batholiths and its relation to the structure of the Verkhoyansk-Kolymian belt.
1, Batholith; 2, structural trends; 3, exotic terrane (only the large ones are shown); 4, thrust front; 5, Okhotsk-Chukotka volcanic belt; 6, South Anui suture.

The metamorphic basement of the Omolon massif consists of typical Precambrian rocks of the granulite (or rarely the amphibolite facies), enderbites, granite-gneissic domes, and greenstone belts [Zhulanova, 1987]. The rocks are 3400 to 2000 Ma old based on U-Pb dating [Bibikova et al. 1978].

The sedimentary cover of the massif commences with Vendian tillites, which are followed by dominantly carbonate deposits of Cambrian age. The carbonate deposits and overlying Early and Middle Ordovician conglomerates with red beds crop out sporadically. Silurian and Lower Devonian sediments are not found. A pronounced unconformity underlies the Middle to Upper Devonian volcanic Kedon series, which overlies the massif as a thick (4 km) continuous, slightly deformed cover. The volcanic rocks belong to the calc-alkaline series and are differentiated from andesites to rhyolites; ignimbrites and silicic lavas predominate. Together with numerous subvolcanic bodies and granitoid intrusions, they make up a typical volcano-plutonic association characteristic of active margins of the Andean type.

The overlying Upper Paleozoic and Lower Mesozoic deposits are thin (a few hundreds of meters) and contain numerous erosional disconformities. The Carboniferous is typified by conglomerates and coal-bearing sequences, while the Lower Permian consists of conglomerates and siltstones. The Upper Permian is characterized by thin shallow-water carbonate-shale deposits including limestones with Kolymia fossils. The Triassic and Lower to Middle Jurassic deposits are marine sand-shale sequences with a boreal fauna. A sharp angular unconformity is observed beneath the Kimmeridgian sediments.

The Omolon massif is surrounded on all sides by shear zones (Berezov, Sugoy, Gizhiga) and can be considered as a rigid block, a typical microcontinent. As its cover differs from that of the Verkhoyansk foldbelt and the Siberian platform, the Omolon massif was probably not a part of the Siberian continent in Paleozoic time.

The Okhotsk massif is essentially similar to the Omolon massif. Upper Devonian volcanics, the analog of the Kedon series of the Omolon massif, are also widely distributed here. Thus we infer that the Omolon and Okhotsk massifs were parts of a single continent and were separated only in Post-Devonian time, possibly in the Permian. This microcontinent (Omolon-Okhotsk) was not near Siberia in the Late Precambrian, Early Paleozoic and Devonian . Its initial position is not known, for no nearby continental masses show such a big gap embracing the Late Ordovician, Silurian, and Lower Devonian or an intra-Devonian unconformity followed by Middle-Late Devonian continental-margin volcanism. The only place where Precambrian basement is overlain by Upper Devonian subaerial volcanic-plutonic sequences with little intervening section is the eastern Australian margin in Queensland where the Connors-Auburn volcanic arc of the Andean type was developed in Late Devonian time [Veevers, 1986].

South Anyui Suture Zone

The South Anyui zone [Seslavinsky, 1970] separates the Kolymian structural loop from the Chukotka foldbelt (Figure 115). According to Natal'in [1984] this zone is characterized by Late Jurassic oceanic complexes closely associated with island-arc complexes of the same age. Ophiolites are found in the basement of the Upper Jurassic-Lower Cretaceous complex. The main part of the South Anyui suture zone is occupied by the Upper Jurassic Gremuchinsk series, composed of spilites, diabases, mafic tuffs, greywackes, shales, cherts (radiolarites included), and in places glaucophane schists. It seems that the whole sequence is a chaotic complex of unknown origin. The lava horizons are composed of pillow spilites and diabases, close to MORB tholeiites in chemical composition, although some have a calc-alkaline trend. The Upper Jurassic deposits are overlain by Berriasian-Valanginian flysch, and the section is completed by Hauterivian greywackes, arkoses, and shales. Deformation of the complex took place during a short period between the Hauterivian and Aptian.

The Oloi subzone in the southern part of the South Anyui zone includes an Upper Jurassic-Lower Cretaceous volcanic complex corresponding to a volcanic island arc. North of the suture is the Nutesin subzone, also consisting of island-arc volcanics (see Fig. 109).

Thus, the South Anyui zone contains the remnants of Late Jurassic (and more ancient) oceanic basin floor that was consumed in the Late Jurassic and Early Cretaceous beneath two island arcs. The deformation occurred in a short time within the Aptian and coincided with the collision of Chukotka and greater Siberia.

Chukotka Foldbelt

The Chukotka Peninsula, situated north and northeast of the South Anyui suture zone, the Novosibirsk (New Siberian) Islands, and De Longa and Wrangel Islands, are all fragments of the former Arctida continent [Zonenshain and Natapov, 1989], which underwent strong deformation and faulting in the Late Mesozoic.

Precambrian crystalline basement is exposed in the extreme east of Chukotka where it makes up the East Chukotka uplift [Sidorenko, 1970b]. It is also found on Bol'shoy Lyakhov Island and is notably present on Wrangel Island.

Paleozoic deposits crop out in uplifts in Chukotka and are widespread on the islands. They include only sedimentary sequences and are totally devoid of volcanic rocks. They are very similar on Chukotka and in the Novosibirsk (New Siberian) Islands. The Lower Ordovician deposits are graptolitic shales; then up to the Middle Carboniferous the deposits are mainly limestones of shelf facies. Upper Carboniferous and Permian sediments are lacking.

The Mesozoic deposits of the Chukotka zone begin with thick terrigenous Triassic sequences lying on the Paleozoic formations with erosional unconformity. Lower and Upper Triassic are known on Chukotka, but Middle Triassic faunas have not yet been found. Diabase bodies of the flood-basalt type are characteristic of the Lower Triassic. The widely developed Upper Triassic rocks are clastic sequences of flysch type; they were intensely deformed in pre-Oxfordian time.

The shelf origin of the Paleozoic sedimentary cover indicates that Chukotka and the Novosibirsk Islands were undoubtedly parts of some continent. Rocks analogous to those of Chukotka and the Novosibirsk (New Siberian) Islands are also known in northern Alaska and in the northern part of the Canadian Archipelago, north of the Innuitian foldbelt. They too were parts of the Arctida continent and in the Late Paleozoic and Early Mesozoic belonged to the North American continent [Zonenshain and Natapov, 1989].

Koni-Murgal Volcanic Belt

The Koni Murgal belt includes volcanic series ranging from Permian to Lower Cretaceous (Hauterivian) and extends from the Murgal Range (Koryak Highlands) in the northeast through the Taigonos and Pyagina Peninsulas to the Koni Peninsula [Parfenov, 1984]. Nekrasov [1976] recognized four volcanic cycles--Late Permian, Triassic to Early Jurassic, Middle to early Late Jurassic, and Late Jurassic (up to the early Volgian). An independent fifth cycle is composed of Upper Volgian-Lower Cretaceous volcanogenic-terrigenous sequences of continental and near-shore facies. The section is completed by a continental coarse-clastic sequence of Barremian-Aptian and possibly Albian age. The Koni-Murgal belt includes fragments of several island arcs that formed throughout 125 million years (from 250 to 125 Ma). Most probably they were

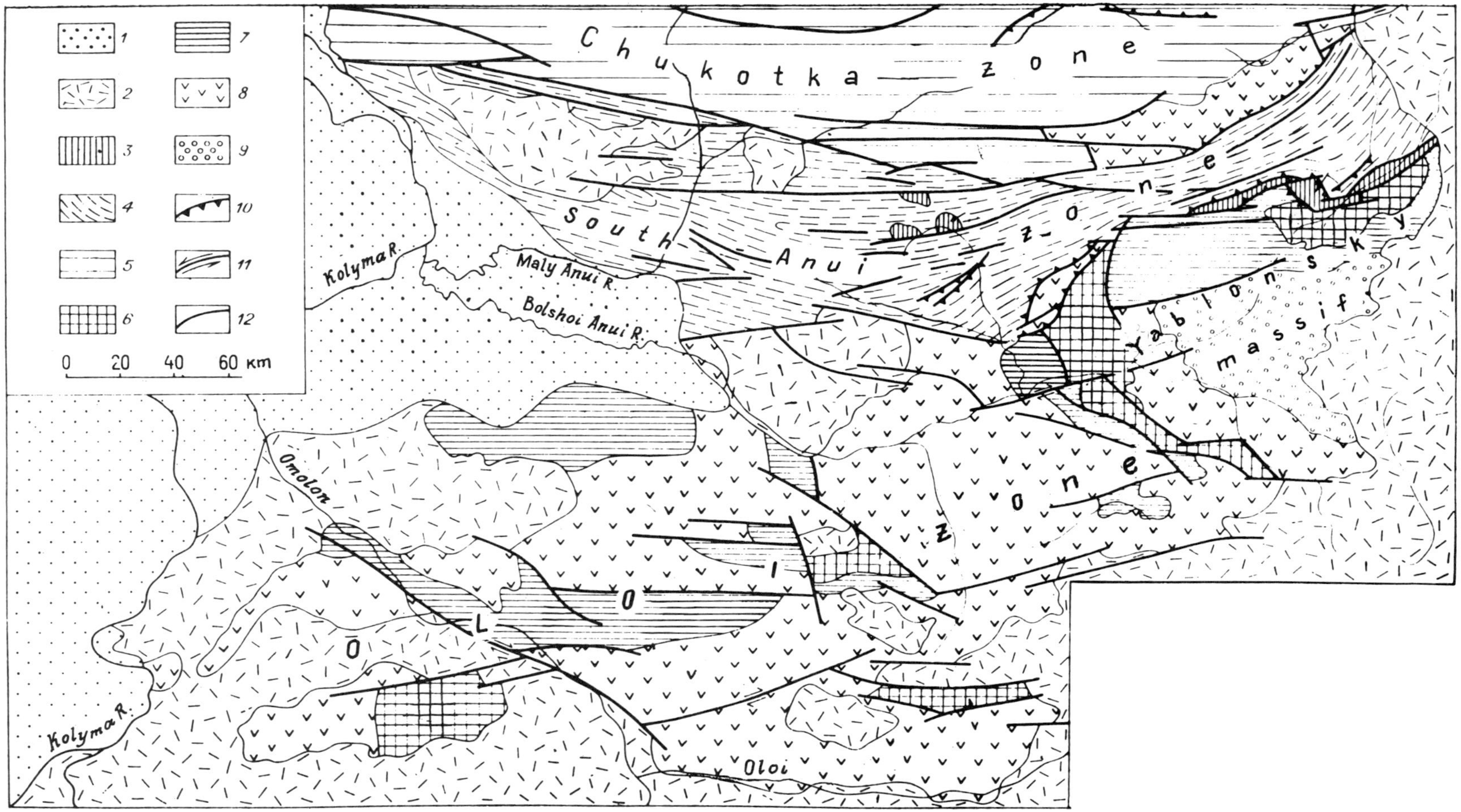

Fig. 115. Tectonics of the South Anui zone, after Natal'in [1984].
1, Cenozoic deposits; 2, Okhotsk-Chukotka volcanic belt; 3, ophiolite; 4, Jurassic and Lower Cretraceous rocks of the South Anui zone; 5, Triassic rocks of the Chukotka zone; 6, Paleozoic rocks of the Oloi zone; 8, Upper Jurassic-Neocomian calc-alkaline volcanics; 9, Barremian-Aptian rocks; 10, thrust; 11, strike-slip fault; 12 other fault.

not a single arc but were accreted to one another, for the belt has a strongly imbricated series of thrust slices.

In the northern (Murgal) segment of the belt, the volcanic complex evolved from the Upper Jurassic to the Hauterivian and consists of basaltic and andesite-basaltic tuffs [Filatova, 1979]. Filatova [1979] and Parfenov et al. [1984] calculated on the basis of $K_2O:SiO_2$ ratios that the dip of the paleo-subducting slab under the Koni-Murgal arc was 50-60°northwestward (Figure 116).

The Koni-Murgal belt is considered to be an amalgamation of arc complexes of different ages from the Permian to the Lower Cretaceous, deformed and joined to older orogenic structures of the northeast USSR in the mid-Cretaceous, evidently prior to the Late Albian, i.e., before the activity of the Okhotsk-Chukotka volcanic belt. Deformation of the Koni-Murgal belt indicates an important event when numerous arc complexes, formerly situated at an unknown distance from the newly formed Siberian margin, rapidly approached and catastrophically joined it.

Okhotsk-Chukotka Volcanic Belt

The Okhotsk-Chukotka belt shows an obvious similarity with the present-day Andean volcanic belt. It marks the Cretaceous active continental margin. The formations of the belt lie with sharp unconformity upon all older structures of the northeast USSR, including the enlarged Koni-Murgal belt. Thus, by the time the belt began to form, the new margin of the expanded Siberian continent had been completely shaped. The composition and structure of the Okhotsk-Chukotka belt are reported in a number of studies, for instance Belyi [1977], Filatova [1979], and Sidorenko [1970c].

According to Belyi [1977], the belt formed in a comparatively narrow time span embracing the Albian through the Cenomanian, i.e., through only 20 Ma. However, Filatova [1979] provides data indicating that volcanism continued up to the Senonian. Andesites, rhyolites, and in places andesite-basalt sequences, are abundant in the belt. This is a typical calc-alkaline volcanic series. A distinct east to west lateral

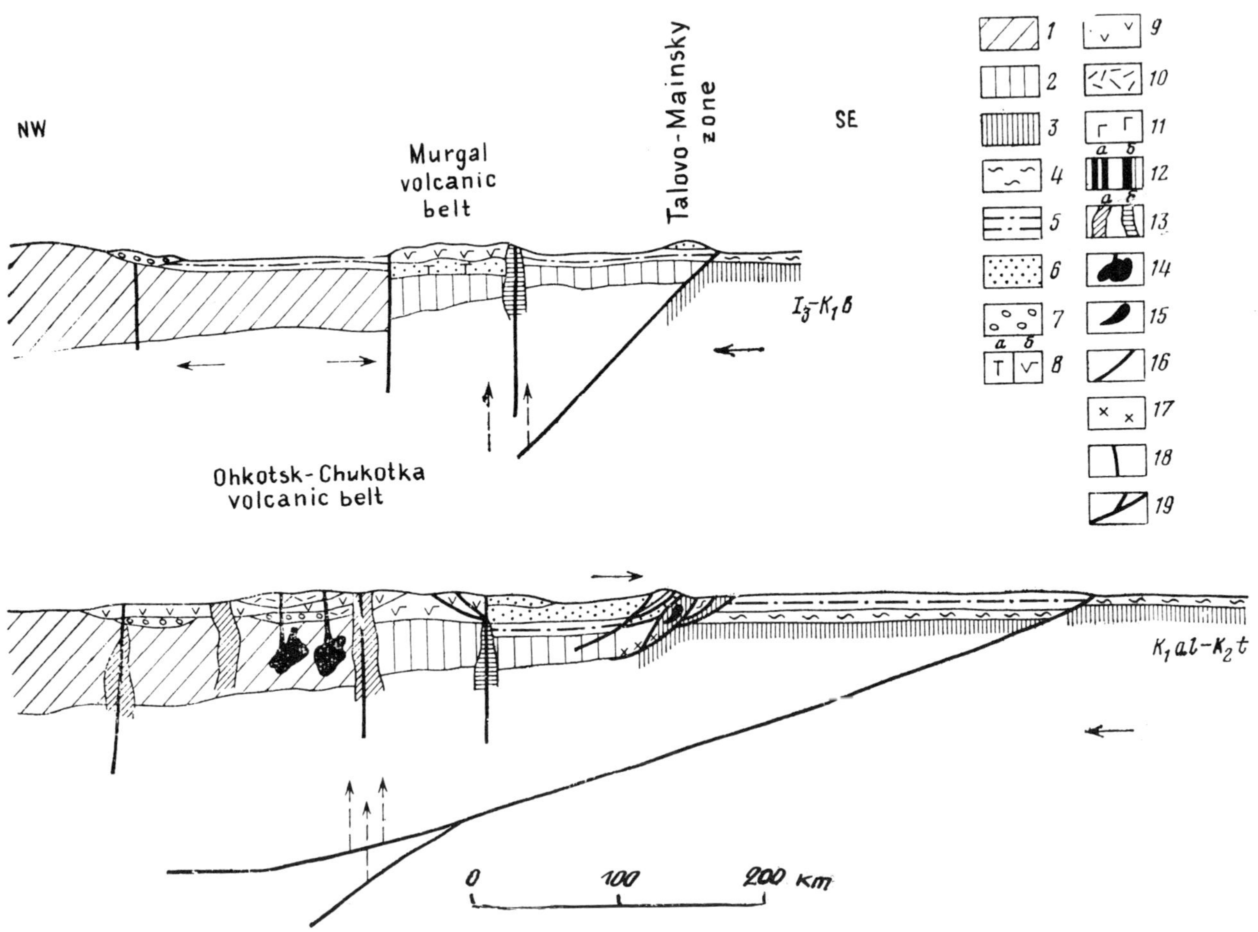

Fig. 116. Paleotectonic cross-section through the Okhotsk-Chukotka volcanic belt, after Filatova [1979]. 1, Continental crust; 2, transitional crust; 3, oceanic crust; 4, oceanic tholeiite; 5, flysch; 6, fine clastics; 7, coarse clastics; 8, Upper Jurassic-Lower Cretaceous: a) volcaniclastics, b) volcanics; 9, Albian-Cenomanian intermediate and mafic volcanics; 10, Cenomanian-Turonian felsic volcanics; 11, Senonian alkaline and mafic volcanics; 12, granite; 13, subvolcanics; 14, magma chamber; 15, ultramafics; 16, fossil subduction zone; 17, blueschist; 18, fault; 19, thrust.

zonation inward toward the continent has been recognized. The dip of the paleo-subducting slab has been calculated several times using K_2O versus SiO_2 content in the rocks [Moralev and Grigorash, 1976; Filatova, 1979; Parfenov, 1984]. All results indicate a subduction zone dipping gently (20°) northwest (Figure 116, lower section). According to these data, the trench must have been 500 km from the present edge of the belt, i.e., in the present-day Koryak Highlands and within the Sea of Okhotsk.

Wrangel-Herald-Brooks Thrust Belt

This zone is a band of deformation of Late Cretaceous-Paleogene age that runs from the western Brooks Range in Alaska across the Chukchi Sea shelf. East of Herald Island, seismic reflection reveal young deformation in the sedimentary cover, probably of thrust type [Grantz et al., 1981]. We assume that the main thrust of the Central zone on Wrangel Island is part of this foldbelt.

Taimyr and Severnaya Zemlya

The Taimyr Peninsula and the Severnaya Zemlya Archipelago north of it (Figure 117) are built of rock complexes from Archean to Jurassic in age, that were deformed during several orogenic events, from the Late Precambrian to the Middle Mesozoic [Vakar et al., 1958; Pogrebitsky, 1971; Khain, 1979a; Bezzubtsev, 1981; Zabiaka et al., 1984].

Taimyr and Severnaya Zemlya are divided into three zones: North Taimyr, Central Taimyr, and South Taimyr. They have different Precambrian and Paleozoic sections, and were joined together at the end of the Paleozoic and in the Middle Mesozoic.

The North Taimyr Zone

The North Taimyr zone includes the northern parts of the Taimyr Peninsula and the Severnaya Zemlya Islands. It is characterized by three rock complexes: (i) outliers of Early Precambrian basement, 2200-2400 Ma old [Khain, 1979a], (ii) strongly deformed Late Precambrian (and probably Cambrian) sequences, mainly flysch, and (iii) a relatively weakly deformed sedimentary cover of Ordovician, Silurian, and Devonian deposits. The major part of the zone is occupied by deformed Upper Precambrian formations. Early Precambrian rocks in the cores of granite-gneissic domes constitute the Kara massif. The Severnaya Zemlya trough in the northwest part of Severnaya Zemlya is filled with a Lower Ordovician to Upper Devonian sedimentary series [Morkovsky and Smirnova, 1982]. From the Ordovician to the Silurian, shallow-water limestones with fauna of Siberian affinities predominate. The Silurian and Devonian are represented by lagoonal and continental facies including red beds whose composition and fossils (fish) recall the Old Red Sandstone of Europe. The sediments of the Severnaya Zemlya trough were moderately deformed, evidently in the Early Carboniferous.

The North Taimyr zone is separated from the Central Taimyr zone by large overthrusts, including the Main Taimyr thrust, that were described in the early 1920s by Urvantsev and were recently studied by Bezzubtsev [1981].

It is hard to interpret the tectonic nature of the North Taimyr zone unambiguously. It is proabably a fragment of a continent that collided with some other continental margin, resulting in the flysch sequence, at the very end of the Precambrian or in the Cambrian, and then was covered by a shallow sea. In accordance with the model of Zonenshain and Natapov [1989], this block was part of Arctida, which was not far from Siberia in the Ordovician and Silurian. The early Carboniferous deformation seems to correspond to the Ellesmerian orogeny when Arctida collided with Euramerica, thus explaining why the Severnaya Zemlya deposits are similar to European facies.

The Central Taimyr Zone

The Central Taimyr zone is at present a narrow (50 km wide) strip extending through the whole Taimyr for about 300 km. Although the northern boundary of this zone with the North Taimyr zone is distinct enough and coincides with large overthrusts, its southern boundary, separating it from the South Taimyr zone, can at present be drawn only tentatively. The southern border is within the Boundary Flexure established by Pogrebitsky [1971] and is drawn approximately between two Paleozoic sedimentary facies: a shale in the Central Taimyr zone and a carbonate facies in the South Taimyr zone.

The Central Taimyr zone consists of three main complexes: (i) a Proterozoic, pre-Riphean, metamorphic complex; (ii) a Lower-Middle Riphean volcanogenic complex including ophiolites; and (iii) an Upper Riphean-Vendian-Paleozoic complex. All these complexes are essentially different from complexes of the same age of the North and South Taimyr zones.

The Proterozoic (pre-Riphean) rocks make up two massifs or outliers--the Shrenkov in the west and the Faddeev in the east (see Figure 117). They are composed of plagiogneisses with amphibolites which are not older than the Lower Proterozoic. Bezzubtsev [1981] reported the age of intruding granite-gneisses to be 1000-1500 Ma.

The Riphean volcanogenic complex constitutes the greater part of the Central Taimyr zone. The greenstone and greenschist sequences formed from spilites and diabases on the one hand, and from andesites and dacites on the other hand [Bezzubtsev, 1981]. These sequences are associated with ultramafics and some other members of ophiolite association, in particular with layered gabbros. Widely developed andesites and dacites as well as abundant felsic lavas represent an island-arc assemblage. The isotopic age of these volcanics is 1000-1200 Ma [Bezzubtsev, 1981]. The island-arc andesitic series (Oktyabrsk suite) appears at higher levels as well, intruded by 850 Ma granites. Thus, all these heterogeneous sequences were amalgamated by 850 Ma when, as a consequence of collision and crustal thickening, palingenic granites were formed.

The Proterozoic outliers and Riphean volcanogenic complexes together form the basement for the unconformably overlying younger complex, which commences with Upper Riphean rocks and terminates with Devonian or Lower Carboniferous formations. According to Bezzubtsev [1981], this complex begins with Upper Riphean coarse clastics, but the most remarkable for sediments of this age are Riphean stromatolitic and algal limestones. The Cambrian deposits, which are separated from the Precambrian by an erosional disconformity, consist of limestones and black shales. Graptolitic black shales are abundant in the Ordovician, Silurian, and Lower-Middle Devonian section. Approximately the same black shale continues up to the Famenian (uppermost Devonian) in the east and up to the Namurian (upper Lower Carboniferous) in the west. Thus, the whole rock series from

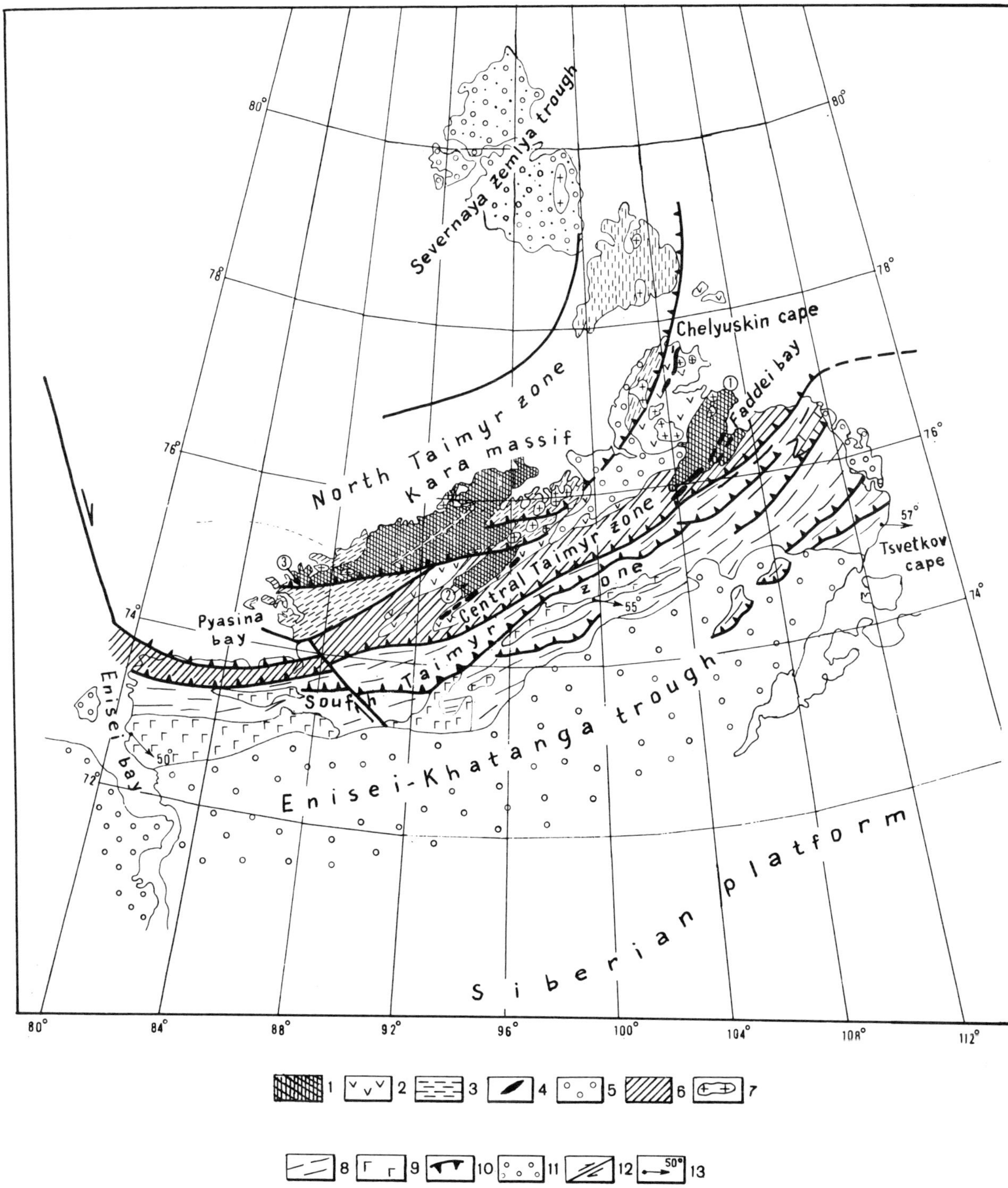

Fig. 117. Tectonics of Taimyr and Severnaya Zemlya.
1, Pre-Riphean basement; 2, Riphean calc-alkaline volcanics; 3, Riphean flysch; 4, ophiolite; 5, Lower-Middle Paleozoic shallow-water sediments and redbeds of northern Severnaya Zemlya; 6, Lower-Middle Paleozoic greywacke and shale; 7, Upper Paleozoic granite; 8, Lower-Middle Paleozoic carbonate rocks (partly overlain by the Upper Paleozoic and Mesozoic Verkhoyansk complex); 9, flood basalt; 10, thrust; 11, Cretaceous and Cenozoic rocks 12, postulated transform fault; 13, paleomagnetic vector.
Numbered are: 1, Faddeev massif; 2, Shrenkov massif; 3, Main Taimyr thrust.

the Cambrian to the Lower Carboniferous belongs to the so-called basin facies and was deposited in stagnant conditions. Normally, such basins form when continental crust extends and rifts but does not separate, and deep subsidence occurs in the thinned crust between the unaffected blocks.

No post-Namurian, Upper Paleozoic, deposits are known in the Central Taimyr zone. Porphyritic and two-mica granites and granodiorites, from 280 to 240 Ma, characterize this belt as well as the neighboring North Taimyr zone [Bezzubtsev, 1981]. Granitic massifs often lie in the cores of granite-gneissic domes with which the zonal metamorphism of the sedimentary sequences is associated. It is obvious from geological maps that the granitic massifs cut folds and thrusts. Therefore, the thrusts separating the North and Central zones, including the Main Taimyr thrust, are older than the granitic intrusions, i.e., the thrusts are Late Carboniferous or Early Permian. We assume that the granites and zonal metamorphism mark Late Paleozoic orogenic events accompanied by folding and thrusting. Consequently, the black shale section of the Central Taimyr zone was deformed in the Late Paleozoic and repeatedly disturbed in the Mesozoic. There is every reason to consider that the continental blocks of the North and Central Taimyr zones collided and joined together in the Late Paleozoic and then behaved as a single block with respect to the South Taimyr zone.

We believe that the Central Taimyr zone was part of the Precambrian basement of the Polar Urals and the Pechora lowland, i.e., Barentsia.

The South Taimyr Zone

The South Taimyr zone coincides with a band of Ordovician to Jurassic deposits. The sediments suffered strong deformation decreasing southwards from the boundary of the Central Taimyr zone. The section includes diverse rocks and lacks hiatuses. It is very similar to the Verkhoyansk complex and shows a gradual transition to platform sections south of the Enisey-Khatanga depression. Acording to Natapov [in press], in places the Paleozoic and Lower Mesozoic rocks are thrust over the Lower Cretaceous, i.e., deformation evidently lasted, as on the western slope of the Verkhoyansk Range, up to the middle of the Early Cretaceous.

The South Taimyr zone is markedly different from the Central Taimyr zone as clearly seen by comparing the carbonate and reef-type Paleozoic sequences in the South Taimyr zone, which formed in the shallow, open shelf sea of the Siberian passive margin, with the black shales of the same age in the Central Taimyr zone, which formed in a deeper, closed, and stagnant basin.

The belt of the linear folds and thrusts of South Taimyr formed by collision of the Kara block (embracing Severnaya Zemlya and Central and North Taimyr) with the passive Siberian margin. Zonenshain and Natapov [1989] suggest that this event was associated with opening of the Canadian Basin of the Arctic Ocean when the Kara block was torn off from North America and was pushed against Siberia. Taking into account the absence of Mesozoic calc-alkaline magmatism, the width of the basin that closed when the Kara block collided with Siberia could not have been greater than 150 km.

Summation

Thus, the Taimyr and Severnaya Zemlya are formed of three continental blocks. Only the southern block, the South Taimyr zone, undoubtedly belongs to the Siberian passive margin. The two other blocks show significant differences from Siberia.

In summary, the full succession of events on North Taimyr and Severnaya Zemlya is as follows: (i) formation of the basement at the end of the Precambrian and into the Cambrian, (ii) progressive platform type subsidence of the Severnaya Zemlya trough in the Ordovician and Silurian, (iii) collision with Euramerica at the end of the Devonian and beginning of the Carboniferous. In Central Taimyr: (i) activity of the Riphean volcanic arc, (ii) formation of an accretionary prism before the Vendian, (iii) separation of the block from its parental continent, most likely from East Europe (former Barentsia), (iv) formation of the Paleozoic euxinic basin. South Taimyr: part of passive margin of Siberia, with a history analogous to that of the Verhoyansk margin. At the end of the Paleozoic, Central Taimyr collided with North Taimyr and Severnaya Zemlya, i.e., with Euramerica. At the boundary between Jurassic and Cretaceous time, the Kara block separated from Laurasia and collided with the passive margin of Siberia, i.e., with the same Laurasia, but on the opposite side of the former oceanic basin (now the Arctic Ocean).

Ancient Massifs

The northeastern part of the USSR (see Figure 109) including Taimyr and Severnaya Zemlya (see Figure 117) was created by the collision of three continental blocks: the Siberian block rimmed by the Verkhoyansk fan, the Chukotka (or Arctic-Alaska) block, and the Omolon-Okhotsk block. The collision of the Chukotka massif with Siberia and the Omolon-Okhotsk massif produced the South Anyui suture zone, which continues to the northern margin of South Taimyr. The earlier collision of the Omolon-Okhotsk massif with Siberia formed the Kolymian structural loop. The Wrangel-Herald-Brooks suture zone is where the North American and Eurasian continents interacted at the end of the Cretaceous and in the Paleogene.

The available data show that Chukotka, together with the Kara massif, were part of Laurasia in the Late Paleozoic and Early Mesozoic. In the Early Paleozoic, however, they were involved in another continent--Arctida--(see Figures 118 and 119) [Zonenshain, and Natapov, 1989]. Collision of Arctida with Euramerica formed the Innuitian orogenic belt, and also the Late Devonian-Early Carboniferous deformation of Severnaya Zemlya.

The Omolon-Okhotsk massif (or microcontinent) is the only one of these large continental blocks that arrived from far away; possibly it is of Gondwanan rather than Laurasian origin.

Other possible continental fragments are present as exotic terranes within the Kolymian loop. They are evidently parts of a single greater massif that can be called the Chersky massif, which was rifted from Siberia at the end of the Devonian.

Oceanic Complexes

In the northeast USSR, oceanic complexes are restricted in distribution. The most extensive ophiolitic belt lies within the South Anyui suture zone. Late Jurassic, and also partly Paleozoic, ophiolites are fragments of the South Anyui oceanic basin which protruded inside Laurasia from the Paleo-Pacific Ocean (Panthalassa).

The recent oceanic basins of the Arctic (Figure 118), the Eurasian, Makarov, and Canadian (Amerasian) basins, are divided by the Lomonosov and Alpha-Mendeleev Ridges. The oceanic crust of the Eurasian basin was created during the last 56 Ma from the presently active Gakkel (Arctic Mid-Ocean Ridge) spreading center [Vogt et al., 1979; Karasik, 1980]. The Makarov basin probably developed from 76 to 55 Ma (anomalies 33 to 23), and the Canadian basin from 155 to 125 Ma (anomalies M25 to M12) [Vogt et al., 1984; Jackson and Johnson, 1984].

Convergence-Related Complexes

Several epochs of island-arc volcanism are distinguished in the northeast USSR: Carboniferous-Permian in the Alazean arc; Triassic-Early-Middle Jurassic in the Oloy-Alazean arc; Late Jurassic in the Uyandina-Yasachnaya arc; and Early Cretaceous in the Nutesin arc. The remnants of five volcanic island arcs (Late Permian, Triassic-Early Jurassic, Middle-Late Jurassic, Late Jurassic, and Volgian-Early Cretaceous) are incorporated in the Koni-Murgal belt. Thus, analyzing the history of the region we should consider that two and sometimes three island arcs and corresponding subduction zones separated Siberia and North America from the Panthalassa or Paleo-Pacific Ocean.

HISTORY

Kinematic Data

To understand the geological history of the region, we must know the plate motions in the Arctic and interaction of the Pacific Ocean floor with Siberia.

Available data on the Arctic basin are summarized in Figure 118. In the Canadian basin, two sets of magnetic lineations from M25 to M20 extend parallel to the Canadian shelf and Northwind Ridge from the Chukchi borderland, and a fan-shaped system from M20 to M12 branches northward from the Mackenzie delta [Vogt et al., 1979, 1984; Karasik and Sochevanova, 1981]. These anomaly systems provide evidence for two stages of basin opening: from 155 to 148 Ma and from 148 to 130 Ma.

In order to match the anomalies of the young fan-shaped lineation system, the Chukchi block together with Alaska should be rotated clockwise around a pole situated near the Mackenzie delta. The remaining gap is closed if the Chukchi borderland together with the block of the Novosibirsk (New Siberia) Islands is pushed to the edge of the Canadian shelf, parallel to the margin of the Alaska shelf (Figure 119). This reconstruction for 155 Ma serves as the base map for both younger and more ancient reconstructions. In order to obtain an Early Mesozoic reconstruction of the Arctic, two other basins, the Eurasian and the Makarov, must be closed. The Lomonosov Ridge must be moved to the edge of the Franz Josef Land shelf and Severnaya Zemlya and the Alpha-Mendeleev Ridge must be fitted to the Lomonosov Ridge along the 2000 m isobath. North Taimyr with Severnaya Zemlya, Franz Josef Land, and the Alpha-Mendeleev and Lomonosov Ridges together formed the Kara massif, which moved 600 km along a presumed NNE-trending transform fault linking the northern termination of Novaya Zemlya with the South Taimyr orogenic belt. Such a motion compensates for the relative motion between Greenland and North America (Ellesmere Island).

The tectonic history of the Arctic in the Cretaceous and Cenozoic is determined by Eurasia/North America interaction. Using plate-motion parameters by Pitman and Talwani [1972], Herron et al. [1974] concluded that, as a result of the opening of the North Atlantic, there must be plate convergence and significant crustal shortening between Chukotka and Alaska-Canada. In this connection, Herron et al. referred to the Alpha Ridge as the structural unit which rose along the convergent boundary.

We assume that no relative motion occurred between Eurasia and North America before the Early Cretaceous, and consequently all events connected with opening of the Canadian basin in the Late Jurassic and Early Cretaceous took place within the context of a single Laurasian continent. Evidently Laurasia initially fragmented in the Early Cretaceous, while the plate started to separate only in the Middle Cretaceous. Three phases of interaction of Eurasia and North America in the Arctic basin can be distinguished (Figure 120). The first phase, in the Early Cretaceous, was an initial breakup, with extension up to 200 km near the (present) North Pole. This phase is evidently associated with widespread Early Cretaceous basaltic volcanism from Greenland to the De Long Islands. In the second phase, from 110 to 55 Ma, the pole of North America/Eurasia motion was in the region of Greenland and Ellesmere Island. Consequently, south of the rotation pole toward the Atlantic, the plates diverged whereas toward the North Pole and farther toward the Bering Strait, the plates converged. In the Chukchi Sea region the amount of convergence was 900 km; near Ellesmere Island it was 350 km. It follows from the kinematic data for North America-Eurasia convergence [Herron et al., 1974] that compressive structures should have formed in the Arctic in the Late Cretaceous and Paleocene. The Wrangel-Herald-Brooks suture zone is interpreted as such a compressional belt, which may be continued further in the Mendeleev and Alpha Ridges. A convergence of 900 km also demands that, in addition to continental shortening, oceanic crust must have been consumed. Hence we suppose that the Canadian basin was 900 km wider between Alaska and the Mendeleev-Alpha Ridge. The third phase belongs to the Cenozoic, when the pole of opening of the North Atlantic and Arctic basins was within eastern Siberia. The North America/Eurasia convergence resulted in oroclinal bending of the Wrangel-Herald-Brooks suture zone (Patton and Tailleur, 1977) as well as deformation dispersed in the Kolymia Range of northeastern USSR, leading to the uplift of mountain ranges.

The interaction between plates of the Pacific basin and Eurasia was considered recently by Kononov [1984], Zonenshain et al. [1987], and Engebretsen et al. [1985]. According to Kononov [1984], from 130 to 70 Ma the Kula plate converged with Eurasia at a rate of about 16 cm/yr. According to Engebretsen et al. [1985], the Kula plate did not exist until about 85 Ma, when it separated from the Farallon plate, but from 130 to 70 Ma a succession of oceanic plates (Izanagi, Farallon, Pacific and Kula) may have converged with Eurasia at rates generally ranging from 8 to 28 cm/yr. These researchers also analyzed motion in earlier time intervals, and concluded that the oceanic plate converged with Eurasia at a much lower rate prior to 125 Ma. Thus, two constraints have been obtained from kinematic data. First, the convergence rate of the Pacific basin plates with Eurasia changed dramatically in the Early Cretaceous, possibly increasing five-fold. It should be remembered that this time span embraces

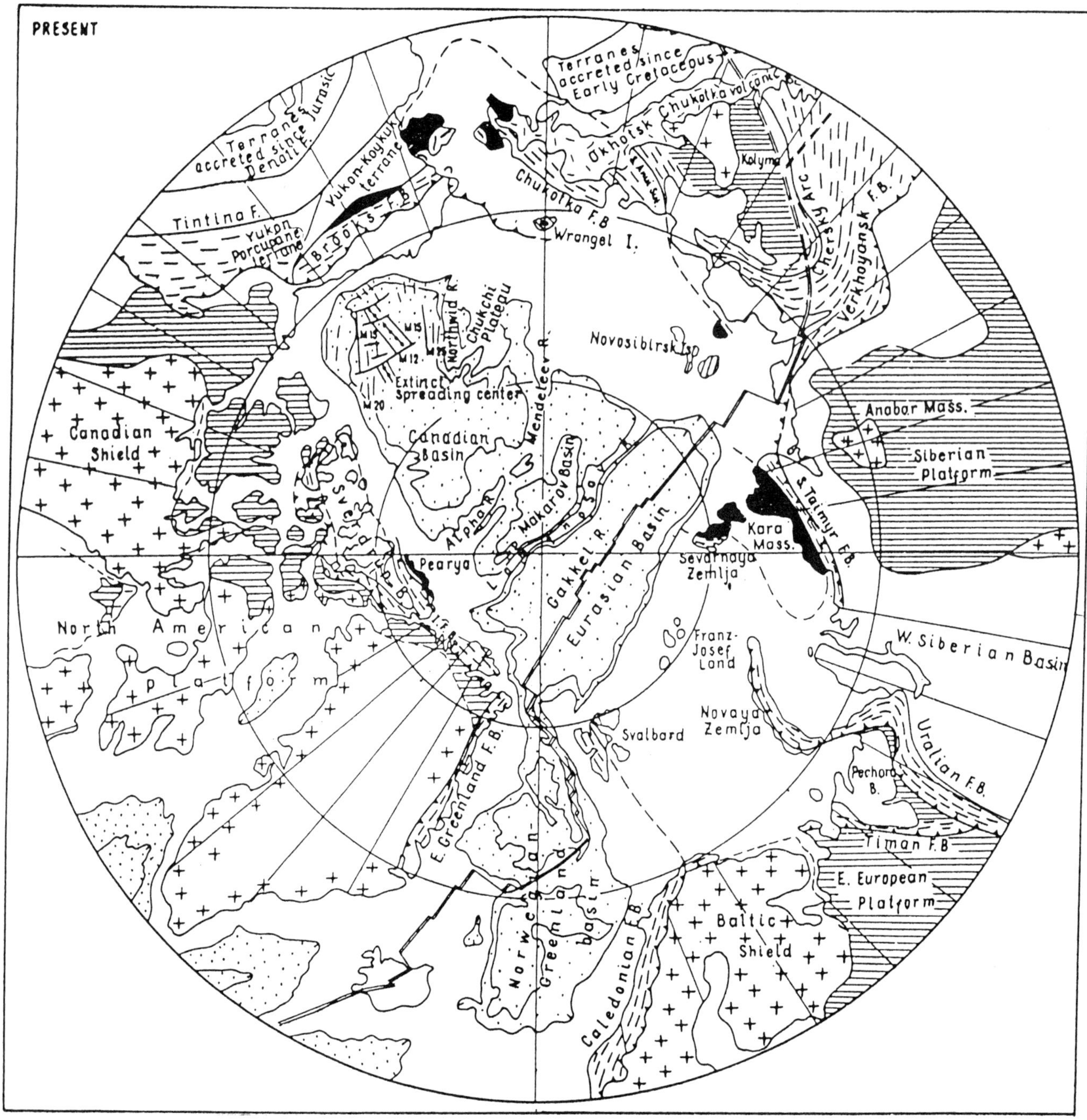

Fig. 118. Main tectonic structures of the Arctic basin, after Zonenshain and Natapov [1989]. Oceanic basins deeper than 2000 m are dotted; contours are at 2000 m and 3000 m depths. Continental shields are shown by crosses, platforms by horizontal hatching and orogenic belts by dashed lines. Ancient massifs that are remnants of the Arctida continent are in black. Tooth lines correspond to thrust fronts. Thick lines mark main sutures.

major orogenic events in the northeastern USSR, particularly folding of the Verkhoyansk belt and formation of the Kolymian loop. Second, the long-lasting convergence of the Pacific basin plates with Eurasia throughout the last 125 or 150 Ma indicates nearly continuous subduction between Eurasia and the oceanic plates. In fact, abundant calc-alkaline volcanic rocks are evidence of this continuous subduction.

Paleomagnetic Data

Apparent polar-wandering paths are reliably established

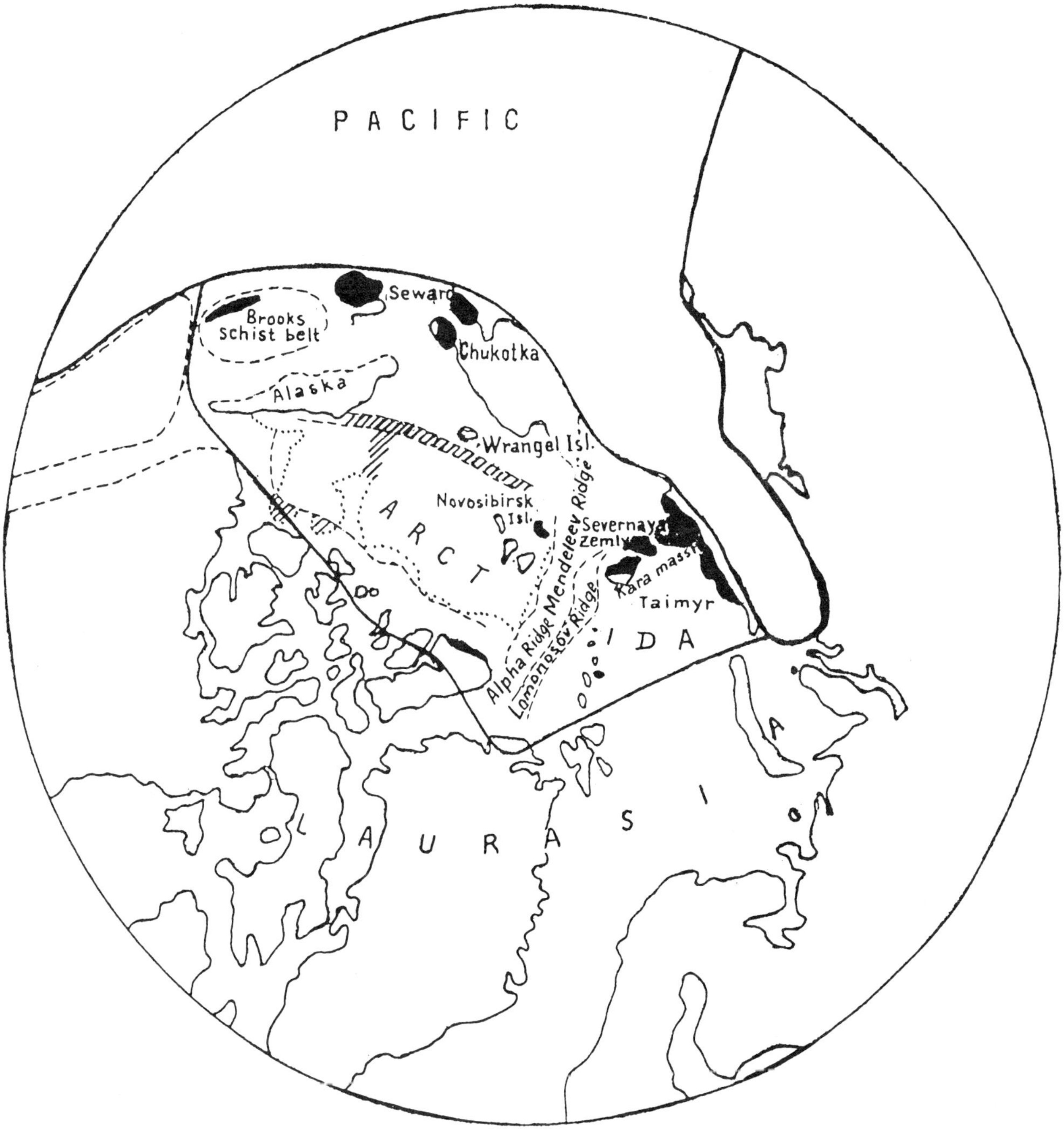

Fig. 119. Early Jurassic reconstruction showing ancient Arctic massifs (black) gathered in the Arctida continent and attached to Laurasia. Bands of diagonal lines show overlapping.

for North America [Irving, 1977, 1983] and Siberia [Khramov, 1982]. Unfortunately, no paleomagnetic data are available for the massifs representing Arctida, and only a small amount exists for the Omolon-Okhotsk and other blocks (Table 6; Figure 121).

Reliable paleomagnetic data on the Omolon massif relate only to the Late Triassic and Jurassic [Lozhkina, 1981b]. They indicate, first, that throughout the Triassic and Jurassic the massif was at a latitudinal distance of 20-30° from its present position relative to Siberia, but it was gradually converging with the latter; second, that it rotated with respect to Siberia during two stages; from the Upper Triassic to the Middle Jurassic, it rotated clockwise 189°, and then in the Late Jurassic-Early Cretaceous it rotated 77° counterclockwise. Also, two episodes of rapid displacement of the Omolon massif toward Siberia are clearly indicated: one in the Middle

Fig. 120. Trajectories of North America with respect to Eurasia within the Arctic basin, after Zonenshain and Natapov [1989]. Figures indicate ages in Ma. Positions of North America, Greenland, and Chukotka at 190 Ma are shown by dotted lines.

Jurassic, when the massif moved not less than 2000 km toward Siberia; the other in the Late Jurassic-Early Cretaceous (from 155 to 125 Ma) when the rate of motion was 6 cm/yr and the massif traveled up to 1500 km.

Scarce paleomagnetic data on the Okhotsk massif [Lozhkina, 1981a] show that in the Triassic it was at a significant distance from Siberia, as much as 6000 km, and that before it joined Siberia, it rotated 100° overall.

For the Chersky block (Yasachnaya River region), paleomagnetic data are available for Givetian and Upper Devonian deposits [Khramov, 1986]. According to these results, the Chersky block in the Middle Devonian could have been situated near the Siberian margin near 57° N, approximately 10° south of its present position, but in the Late Devonian it had shifted 1300 km southward, from its previous position. Figure 122 shows how the Omolon and Okhotsk massifs and

TABLE 6. Paleomagnetic data for the northeastern USSR

Age	Levooloy block 65°N, 162°E				Omolon massif 65°N, 159°E				Okhotsk massif 60°E, 140°E				Omulevka block 63°N, 151°E				Chukotka 64°N, 161°E				Siberia	
	Pole N	Pole E	Paleolatitude	Declination	Pole N	Pole E	Paleolatitude	Declination	Pole N	Pole E	Paleolatitude	Declination	Pole N	Pole E	Paleolatitude	Declination	Pole N	Pole E	Paleolatitude	Declination	Pole N	Pole E
	Degrees				Degrees				Degrees				Degrees				Degrees				Degrees	
D_{2-3}	23	218	35/42	112/209									44	201	57/46	100/195					20	140
													44	226	45/46	78/195						
C_{1-2}	34	232	39/42	92/209																		
	25	224	34/54	104/198																	30	150
Tr_{2-3}					61	284	42/76	32/280														
									42	252	27/83	52/309					18	162	38/75	180/84	64	127
					59	290	41/75	31/304														
J_1	67	124	73/82	293/322	80	168	72/83	6/337													70	121
J_3					64	103	65/84	298/15													69	164

Note: In denominator - paleolatitudes and declinations if the site were at the Siberian plate.

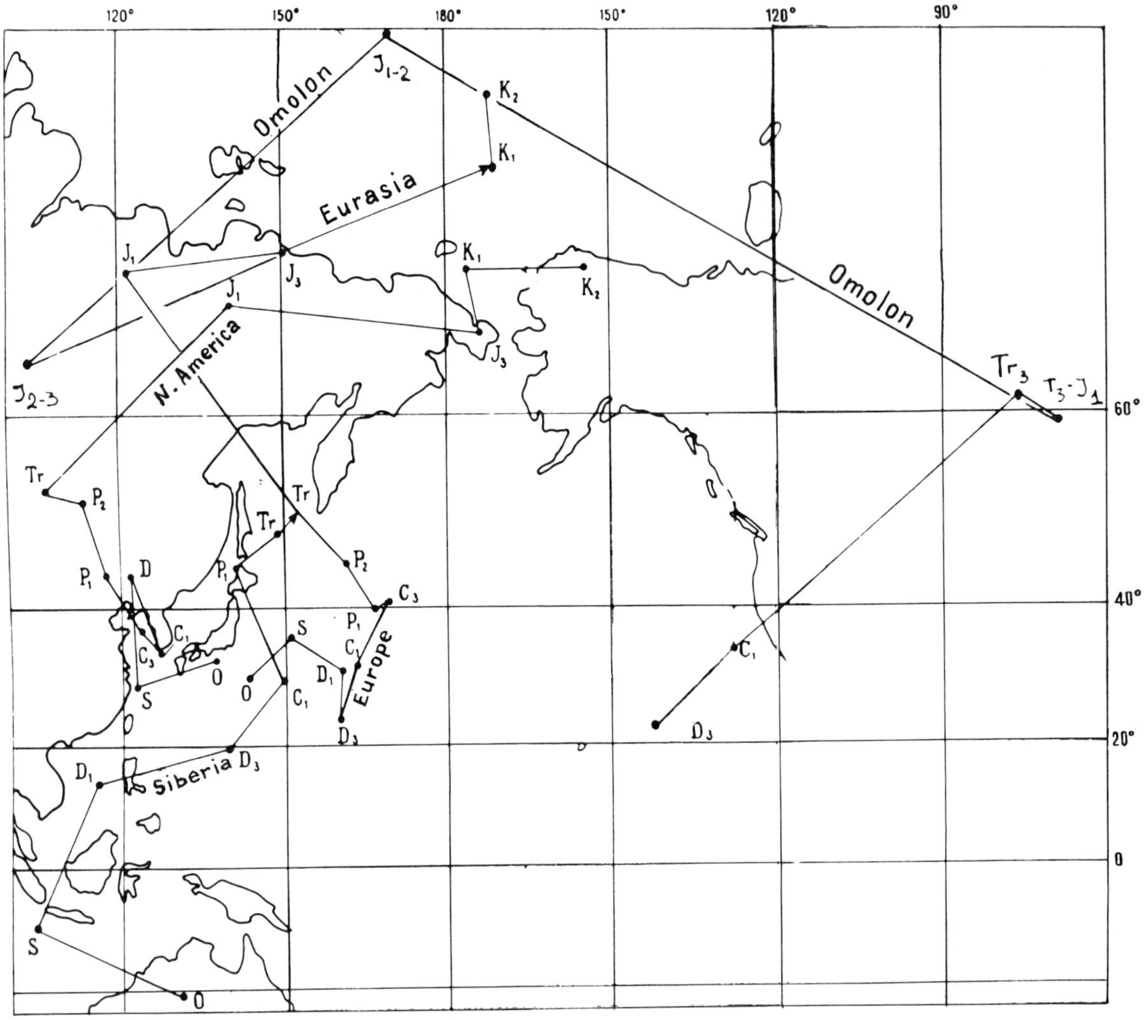

Fig. 121. Apparent polar wander paths for Europe, North America, Siberia, and Omolon, after Zonenshain and Natapov [1989].

the Chersky block drifted as determined from paleomagnetic data only.

Paleomagnetic data for Taimyr record secondary bending of the Taimyr arc in Post-Triassic time, i.e., in the period of folding approximately at the Jurassic-Cretaceous boundary. Preliminary paleomagnetic results for the Ordovician shale section of the Central Taimyr zone and carbonate section of the South Taimyr zone [Rodionov and Shemyakin, in press] indicate that the respective paleomagnetic poles were located at 31° S/126° E, and 4° S/134° E. The corresponding paleolatitude for the Central Taimyr is 17° S, and for the South Taimyr 8° N, indicating that the two regions were at least 2500 km apart.

Pre-Mesozoic History

Unfortunately, the data are too scarce to reliably define the initial positions of the numerous blocks constituting the region under consideration. In the Ordovician, eastern and northern Siberia were a passive margin that slowly subsided, accumulating sequences of limestones with reefs, which are particularly numerous in South Taimyr. The Siberian margin was quite wide and included the carbonate rocks of the future Chersky block. The changes from carbonate to shale facies in the suspect terranes of the Chersky Range indicate a transition from shelf to continental slope and rise; the Lower Paleozoic of the Novosibirsk (New Siberian) Islands shows the same change. On the Arctida margin there is a transition from shelf to deeper water facies in the basin separating these two continents. Separate from both continents was the Central Taimyr block, probably located near the east European margin; between it and the east European margin lay a closed or semi-closed basin where black shales accumulated. Presumably the Central Taimyr block split off from eastern Europe in the Early Ordovician, at the time when rifting is recorded on the western Uralian slope.

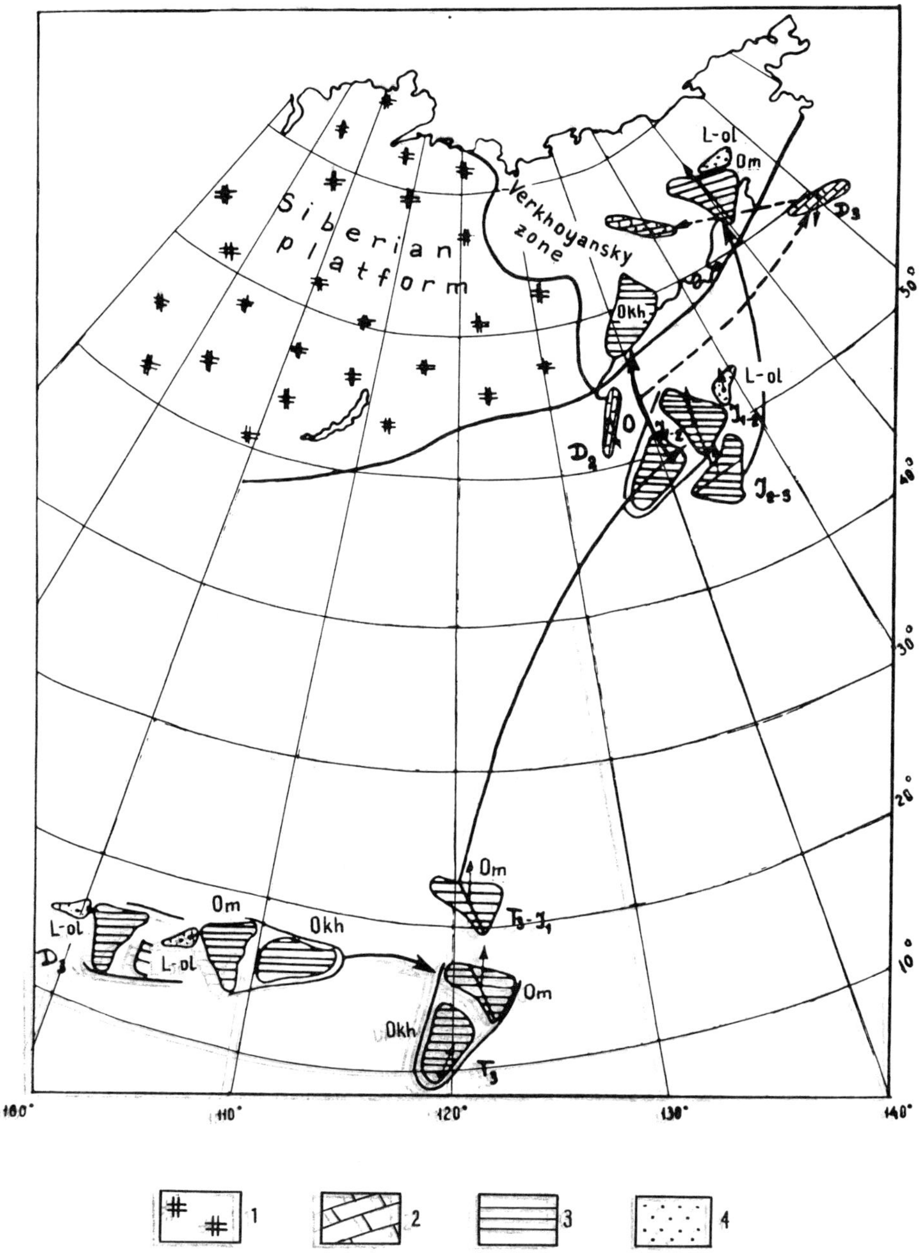

Fig. 122. Movements of the Omolon, Okhotsk and Chersky massifs with respect to Siberia, from the Middle Devonian to the Late Jurassic.
1, Siberian continent; 2, Chersky massif; 3, Omolon and Okhotsk massifs; 4, Levo-Oloi block. Om - Omolon massif, Okh Okhotsk massif, L-ol - Levo-Oloi block. Arrows show paleomagnetic vector orientations.

By the end of the Devonian or possibly at the beginning of the Carboniferous, the carbonate-covered Chersky block as well as the Omolon-Okhotsk massif had broken off from their parent continents. The Omolon-Okhotsk massif was a single microcontinent situated about 3000-3500 km from the Siberian margin.

In the Early Carboniferous, all the Arctic continents reached the Polar area. In the Late Visean, a major event began--the formation of the huge Verkhoyansk fan. In the Late Paleozoic the Chersky and Pri-Kolymian blocks, having split off Siberia, continued moving oceanward, subsiding and receiving relatively deep-water cherty sediments.

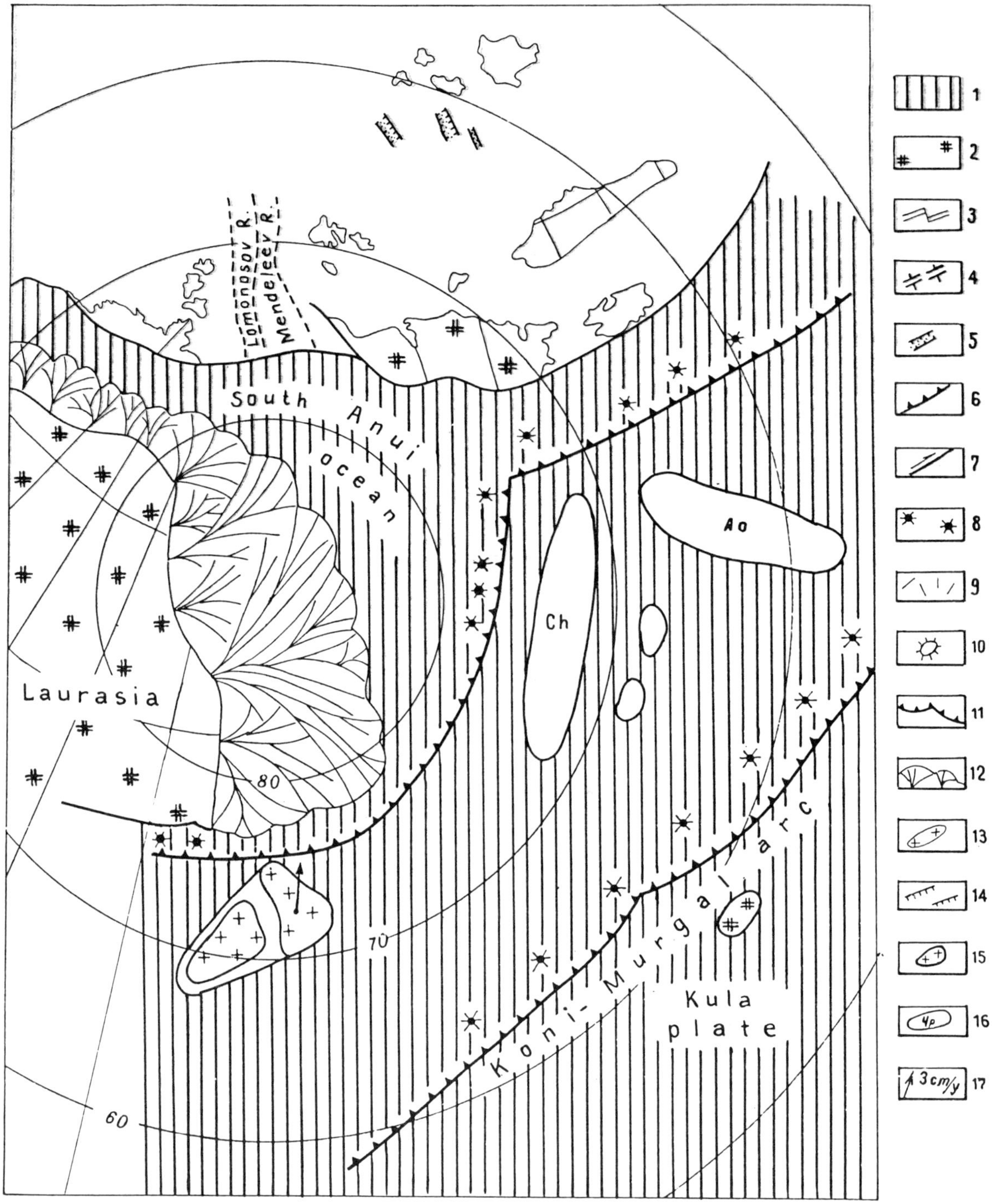

Fig. 123. Palinspastic reconstruction of the locations of NE USSR and Taimyr for Late Triassic and Lower-Middle Jurassic with respect to Siberia, which is arbitrarily left fixed. Paleomagnetic data after Khramov [1982, 1986].
1, Oceanic floor; 2, continenal crust; 3, spreading axis; 4, back-arc spreading; 5, continental rift; 6, subduction zone; 7, strike-slip fault; 8, island-arc; 9, continental-margin volcanic belt; 10, intraplate volcanics; 11, fan (deltaic deposit); 12, thrust; 13, granite batholith; 14, folding; 15, microcontinent; 16, exotic terrane; 17, plate motion vector, with rate in cm/yr. Ao - Alazei-Oloi block, Ch - Chersky block, PK - Pri-Kolymian block, Kn - Kanchalan block, Okh - Okhotsk massif, Om - Omolon massif.

We speculate that in the Carboniferous the Central Taimyr block was torn off its former position near eastern Europe and moved toward North Taimyr, i.e., to the margin of Arctida, with which it evidently collided and amalgamated in the Early Permian. Granitic intrusions and associated granite-gneissic domes mark the occurrence of this collision.

Early Mesozoic, Triassic-Middle Jurassic (Figure 123)

The Mesozoic rock assemblages of the northeastern USSR are characterized by numerous island-arc calc-alkaline magmatic complexes, from which two island arcs--Oloi-Alazeya and Koni-Murgal--can be reconstructed. They separated the Pacific Ocean plates from Laurasia and the South-Anyui basin. Their position on the reconstruction (Figure 123) is based arbitrarily on the present distribution of island-arc complexes. Accordingly, the Oloi-Alazeya arc is placed near the Siberian margin (Laurasia), whereas the Koni-Murgal arc is shown farther out in the ocean. Between the two arcs there was an oceanic area of unknown width, probably with back-arc spreading. A large number of exotic terranes lay within this ocean, the largest being the Chersky block with a carbonate

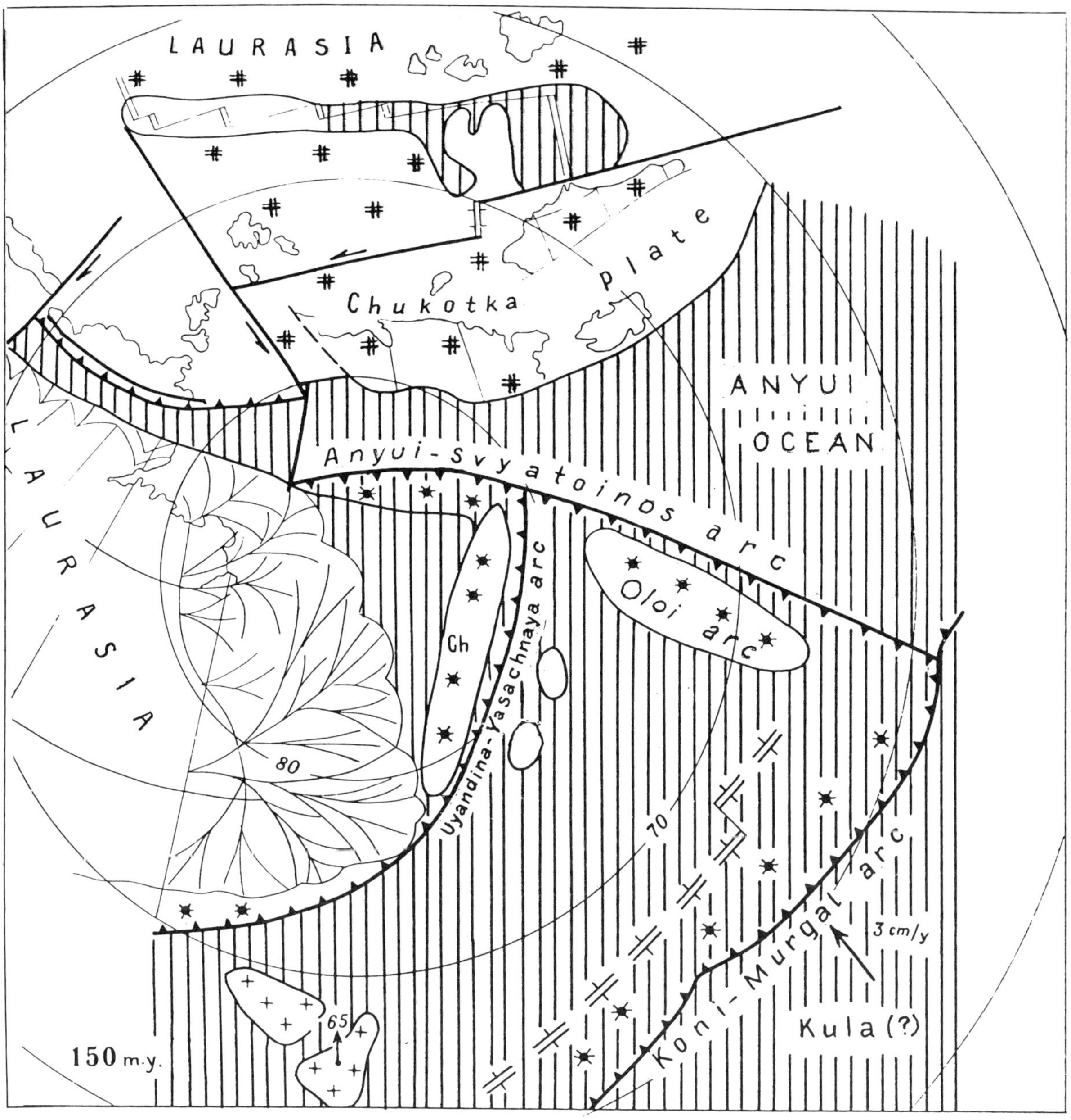

Fig. 124. Palinspastic reconstruction of the NE USSR and Taimyr for 150 Ma. See Figure 123 for explanation.

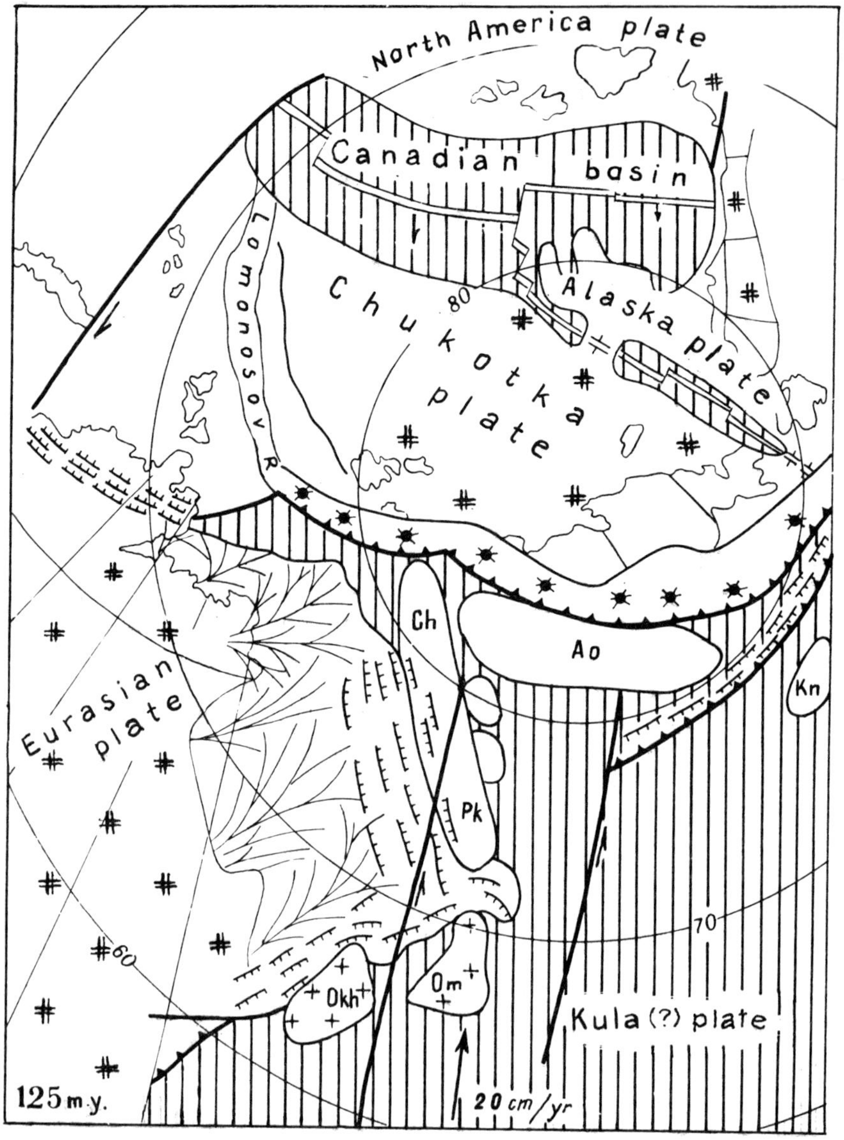

Fig. 125. Palinspastic reconstruction of the NE USSR and Taimyr for 125 Ma. See Figure 123 for explanation.

Paleozoic section, and the Alazeya block with a volcanogenic Paleozoic section. The position of the relatively large Omolon-Ohkhotsk microcontinent accords with paleomagnetic data; it was already 500 km south of the submarine Verkhoyansk cone.

Throughout the Early Mesozoic, the above exotic terranes were approaching the Oloi-Alazeya arc, and none emerged above the sea level, since they are covered by shallow-water sediments. In the Late Jurassic the Chersky and Alazeya blocks collided with the Oloi-Alazeya arc.

Late Jurassic, 150 Ma (Figure 124)

The plates were significantly reorganized in the Late Jurassic. The Chukotka and Kara blocks broke off from North America (from Laurasia) as the Canadian oceanic basin opened, forming an independent Chukotka plate. As Chukotka and Siberia converged, the South Anyui ocean started to close and an extensive subduction zone originated, above which the Oloi island arc was formed. Evidently, the Kara block was the first to approach Siberia, as its distance from Siberia was no more than 150 km.

Two island arcs existed between Siberia and the Pacific Ocean: the older Koni-Murgal arc beneath which the Pacific Ocean crust was consumed at a low rate (about 3 cm/yr), and the newly developed Uyandina-Yasachnaya arc, which formed mainly on a basement made of suspect terranes that had amalgamated in the subduction zone of the more ancient Oloi-Alazeya arc.

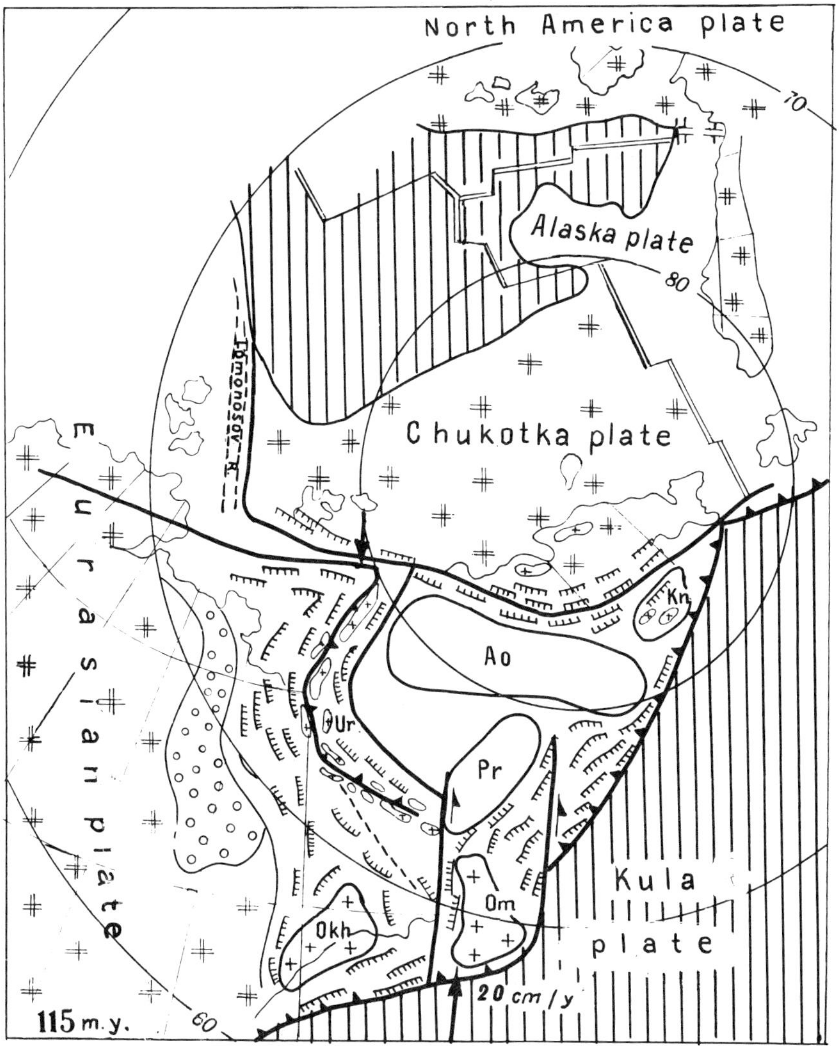

Fig. 126. Palinspastic reconstruction of the NE USSR and Taimyr for 115 Ma. See Figure 123 for explanation.

Early Cretaceous, Neocomian, 125 Ma (Figure 125)

The principal reorganization in the northeastern USSR started in the Neocomian. It was caused by the nearly simultaneous convergence of continental and different exotic terranes with each other, with Siberia, and with the Verkhoyansk fan along the Siberian margin. Hence, the Verkhoyansk fan was deformed and mountain-fold structures began to grow. These processes were associated with two events: first, the North American and Chukotka-Alaska plates continued to diverge, thereby closing the South Anyui basin entirely; and second, plate velocities within the Pacific (Kula and/or Izanagi plates) increased sharply from 3 cm/yr to 15 cm/yr. As a result, the Koni-Murgal arc was broken, and its fragments quickly approached and joined the Omolon-Okhotsk microcontient. These events are marked by thrusting, granitic intrusions, and metamorphism in the Taiganos Peninsula. For a period lasting 10-15 Ma, the Omolon-Okhotsk microcontinent was attached to the Kula plate and moved rapidly towards the Verkhoyansk fan. In the Early Cretaceous the Kara block-Siberia collision was completed and the South Taimyr foldbelt came into being.

Early Cretaceous, Aptian, 115 Ma (Figure 126)

During a period of 10 Ma truly catastrophic events took place in the northeastern USSR, obviously connected with the high (15-20 cm/yr) rate of convergence of the plates of the Pacific basin with the East Asian margin and Chukotka. At this time the Omolon-Okhotsk microcontinent, presumably together with the Izanagi or Farallon plate, moved to the north and was pushed into the Verkhoyansk fan. Moving together with the fan remnants (Sugoy and Berezov zones), it broke and split the Uyandina-Yasachnaya arc and rotated the Pri-Kolymian block counterclockwise, forming the knee-like bend in the southern part of the Kolymian loop and intensely deforming all the sequences surrounding the Omolon massif. At the same time, the Okhotsk massif was pushed into the southern part of the Verkhoyansk fan and eventually collided with the Siberian platform, producing the mountain-fold structures of the South Verkhoyansk synclinorium. Folding and crustal thickening caused melting and produced the Kolymian belt of granitoids. In the north, the collision that closed the South Anyui basin formed the South Anyui suture zone, folded the sedimentary cover of Chukotka, and led to granitoid intrusions. The only place where a deep oceanic basin with oceanic crust was preserved, was the Zyryanka depression within the Kolymian loop.

Middle Cretaceous, Albian, 100 Ma (Figure 127)

By the middle Albian, convergence and collision of all the blocks that make up the present structure of the Verkhoyansk-Kolymian area had ceased. The Omolon massif, probably attached to the Izanagi or Farallon plate, finished moving northward and rotated 60° counterclockwise. This led to the completion of the Kolymian structural loop. The series of collisions of Chukotka, Siberia, and blocks within the Kolymian loop forced the Verkhoyansk thrust front to prograde toward the Siberian platform and formed the Pre-Verkhoyansk foredeep, which filled with clastics derived from the growing Verkhoyansk mountains.

The Okhotsk-Chukotka volcanic belt formed in the Albian, nearly simultaneously throughout the whole 4000 km length of the belt. The subduction zone near the Eurasia margin originated concurrently. This new active margin indicates a marked reorganization of plate interactions. Continental collisions within the northeastern USSR ceased, for all continental massifs and exotic terranes were so closely packed against each other that relative movements between them became impossible. However, the Izanagi and Farallon plates were mov-

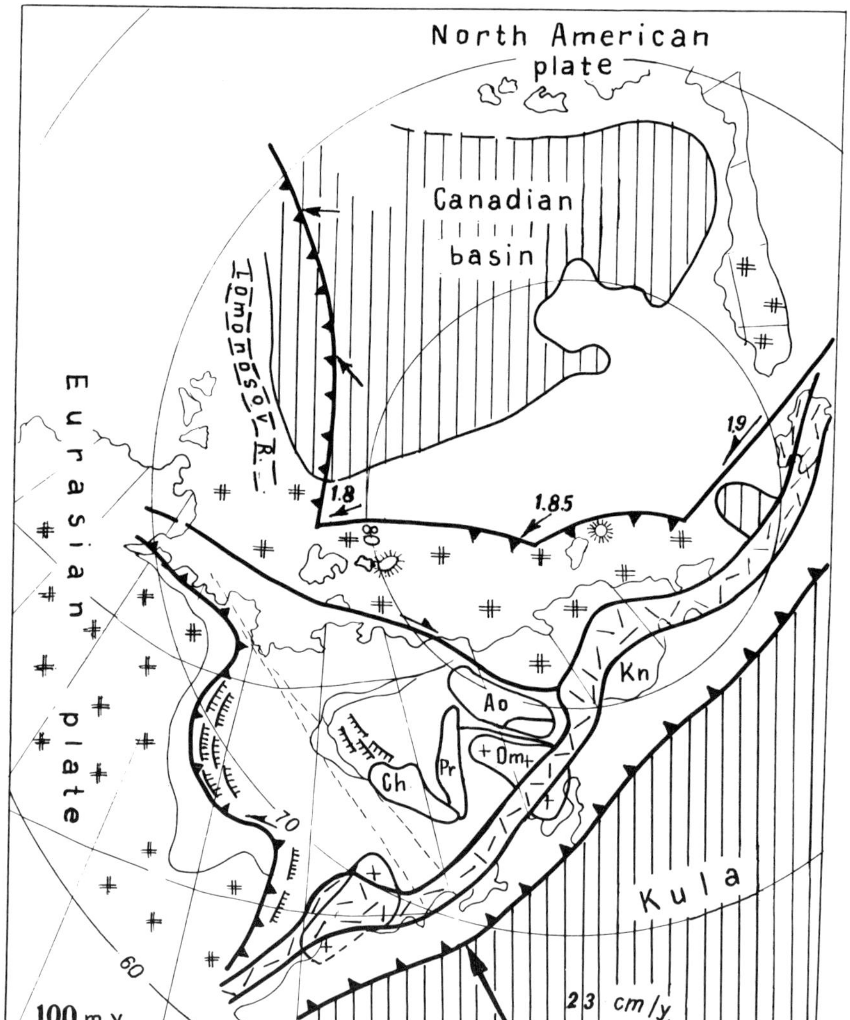

Fig. 127. Palinspastic reconstruction of the NE USSR for 100 Ma. See Figure 123 for explanation.

ing toward Eurasia at a high rate, and whichever plate was in contact with the continent had to start subducting beneath the newly shaped margin.

Late Cretaceous, Santonian, 80 Ma (Figure 128)

The Late Cretaceous-Cenozoic history of the region is closely associated with plate interaction in the Arctic region. The compression that commenced in the Early Cretaceous still continued. A convergent boundary was evidently present along the thrust front of the Brooks Range and the Herald-Wrangel zone and father west along the Chukchi and eastern Siberia shelfs, reaching to the edge of the Canada Basin. From that point it turned southward to Nares Strait between Greenland and Ellesmere Island and the region between Greenland and Svalbard.

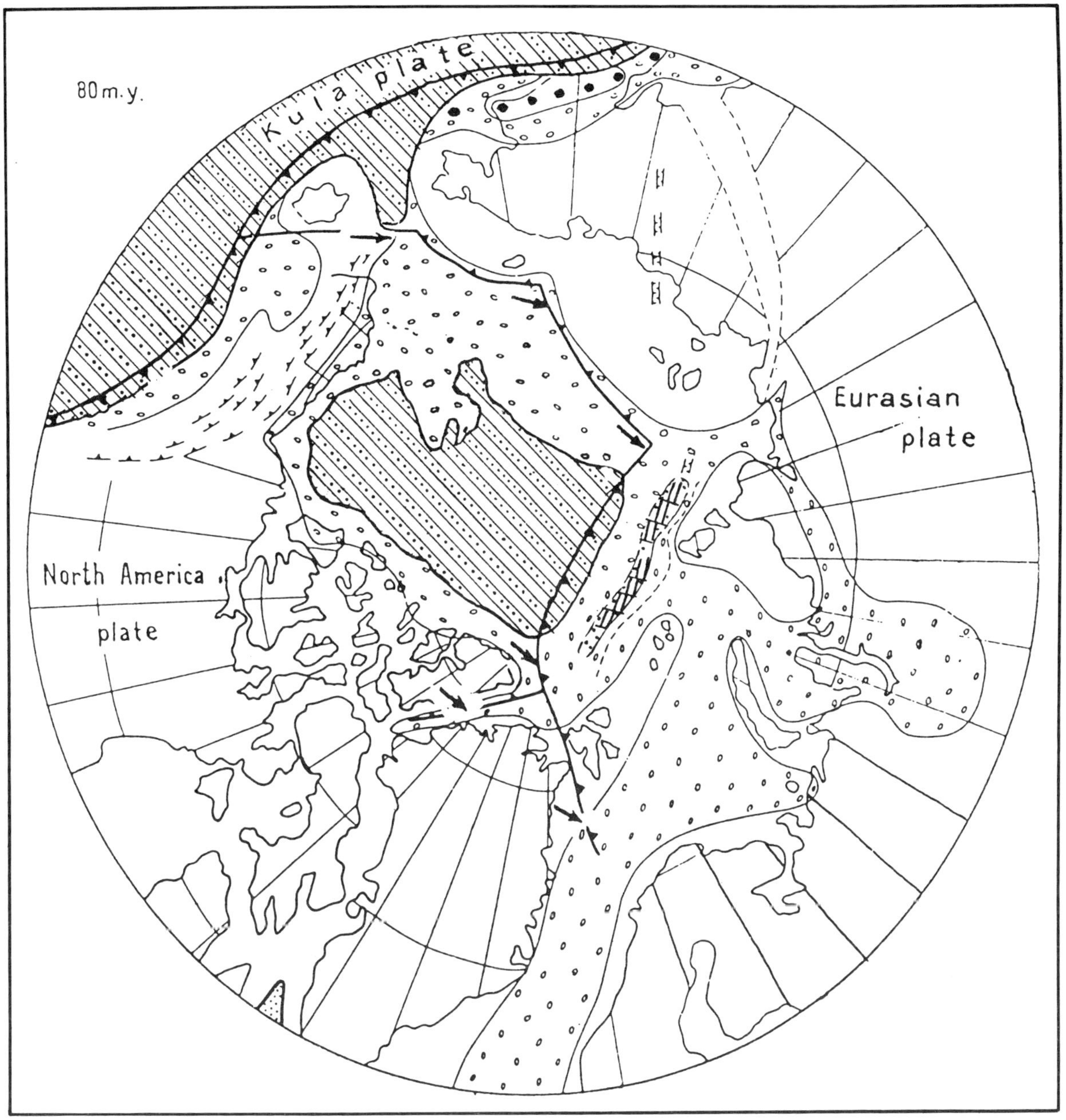

Fig. 128. Palinspastic reconstruction of the Arctic basin for 80 Ma, after Zonenshain and Natapov [1988]. Thin double lines - spreading center; thick toothed lines - subduction zone; straight line - transform fault. Arrows show relative plate motions. Black dots correspond to calc-alkaline volcanism. Short toothed lines mark folding and thrusting. Oceanic basins are diagonally hatched. Small circles - shallow sea. Dry land is left blank.

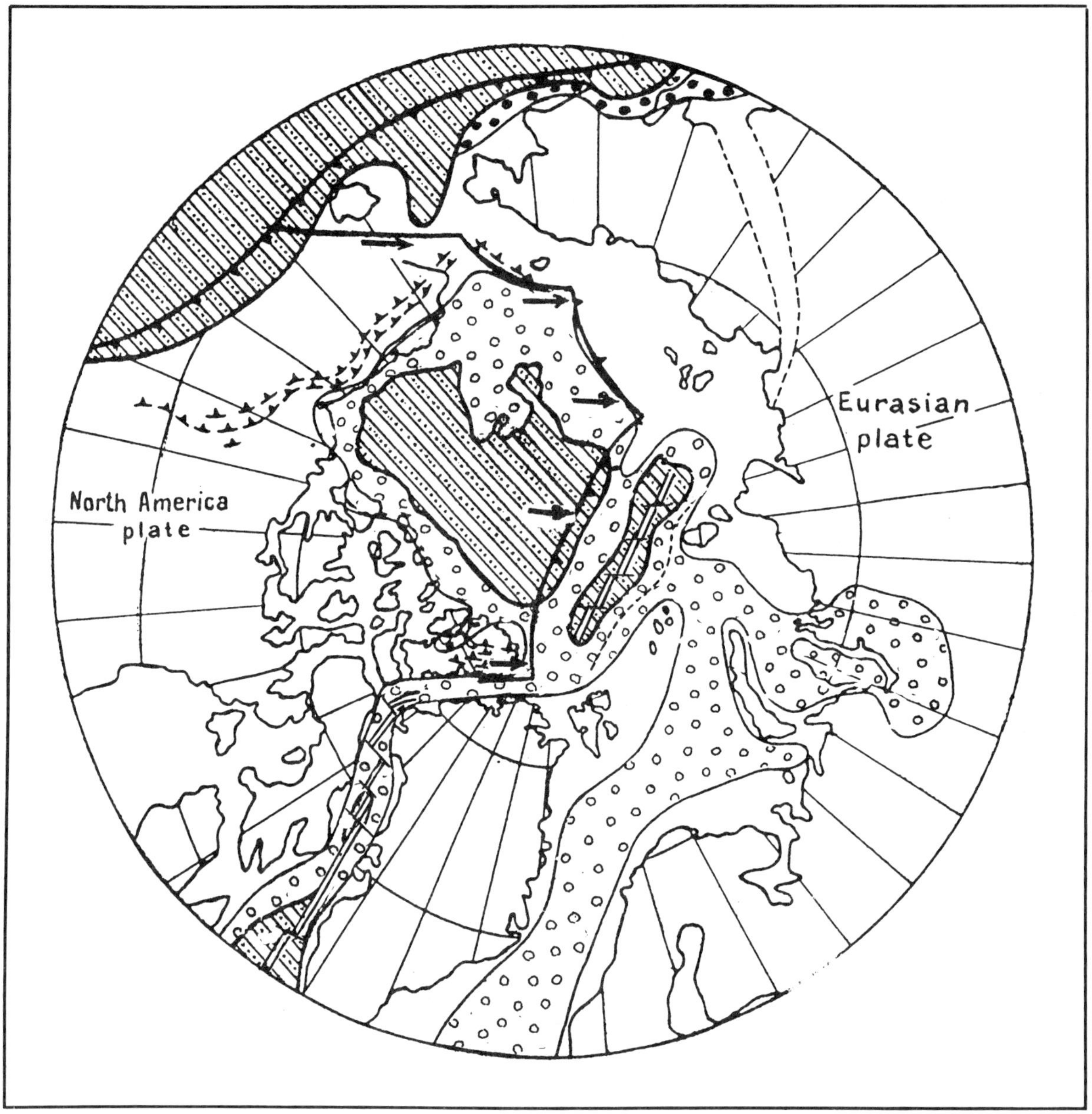

Fig. 129. Palinspastic reconstruction of the Arctic basin for 65 Ma. See Figure 128 for explanation.

Cretaceous- Paleogene Boundary, 65 Ma (Figure 129)

In the time span from 80 to 65 Ma, plate interactions were nearly unchanged. The main difference from the preceding period was that the Eurasia-North America pole of rotation shifted north to Ellesmere Island. As a result, Eurasia converged with North America somewhere between Greenland and Alaska; nearly 300 km of the Canada Basin oceanic crust was consumed during 15 Ma. South of the pole between Greenland and North America, the plates diverged forming the oceanic crust of the Labrador Sea and Baffin Bay. By the Late Cretaceous, the Labrador oceanic basin was 350 km wide.

Oligocene, 35 Ma (Figure 130)

Commencing at 55 Ma, the Eurasian basin of the Arctic Ocean opened along the Arctic Mid-Ocean Ridge spreading center, splitting the Lomonosov Ridge off from the Eurasia margin. No Paleogene structures or rock assemblages in the northeastern USSR indicate plate boundaries there. The North America-Eurasia pole of rotation was near Sakhalin; consequently, if a plate boundary had existed in the north-east, it would have been a divergent boundary. However, no grabens or other evidence of divergence are found in the northeastern USSR, except for Paleogene plateau basalts, which could be

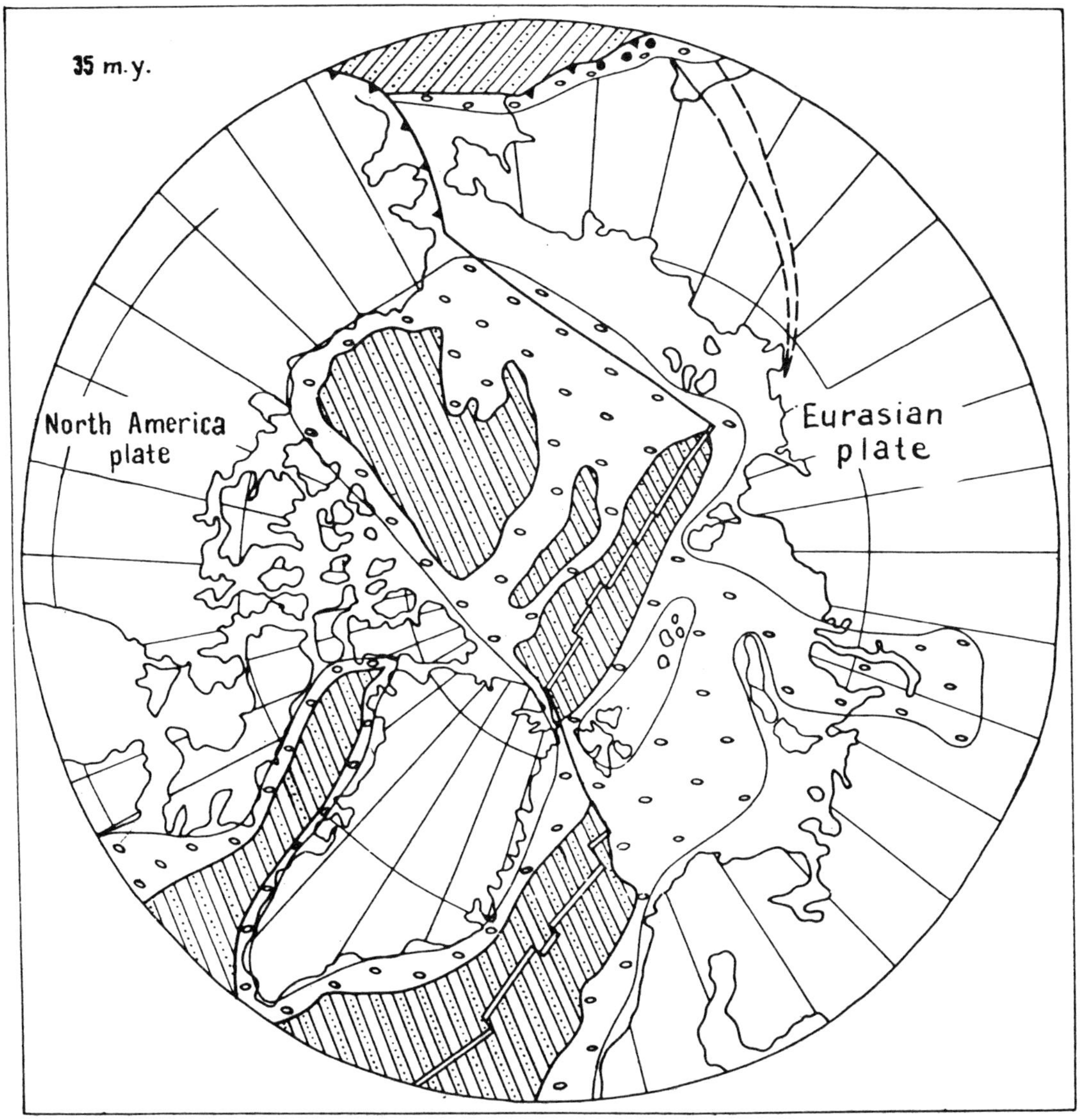

Fig. 130. Palinspastic reconstruction of the Arctic basin for 35 Ma. See Fig. 128 for explanation.

regarded as intraplate extrusives. Hence we show the plate boundary for the Paleogene along the edge of the Chukchi and East Siberia shelves as a strike-slip zone. In this case, it must have had a dextral displacement of about 250 km from 55 to 35 Ma.

During the period from 35 to 20 Ma, the North America-Eurasia pole of rotation shifted to the north, to the region of the Laptev Sea. Along the segment of plate boundary between that pole and Bering Strait, therefore, plates had to converge. We suggest that the resulting compression and shortening of about 190 km in the Bering Strait is reflected in the Alaska orocline (or Bering orocline), a sharp sigmoidal bending of structures in the extreme northwest part of Alaska [Patton and Tailleur, 1977].

At 20 Ma, the North America-Eurasia pole of rotation shifted to the south to the upper course of the Kolyma River. At this time a divergent boundary developed within the northeastern USSR and produced the Moma rift (Grachev, 1982).

Thus, the structure of the northeastern USSR was created by accretion, collision, and amalgamation of blocks of different types and sizes--blocks that were either broken-off pieces of various continents, fragments of island arcs, or remnants of the oceanic floor. They were accreted to the Eurasian margin by convergence of plates of the Pacific and Arctic in a comparatively short time interval--Late Jurassic and Early Cretaceous. Before collision, these blocks or terranes travelled different distances, probably several hundred or thousand kilometers.

Chapter X

KORYAK-KAMCHATKA FOLDBELT

GENERAL DESCRIPTION

The Koryak-Kamchatka belt (Figure 131) includes the structures lying between the Okhotsk-Chukotka volcanic belt and the Pacific Ocean. The belt was deformed during several orogenic events from Middle Cretaceous to Recent time. Recent tectonic and magmatic activity is characteristic of the belt, and is mainly connected with subduction of the Pacific plate along the Kurile-Kamchatka trench. At present this region belongs to the North American plate from which the small Sea of Okhotsk plate recently split off [Zonenshain and Savostin, 1979].

The Koryak-Kamchatka belt is a good example of accretionary structure formed by a number of terranes [Parfenov, 1984; Tilman, 1987; Stavsky et al., 1990; Zonenshain et al., 1987; Sokolov et al., 1987; Chekhov, 1982; Chekhovich and Sukhov (in preparation); Peive, 1984; Puscharovsky, 1982]. Three major types of structural-lithologic complexes are distinguished within the belt [Stavsky et al., 1990]: (i) volcanic complexes marking former continental-margin volcanic belts; (ii) flysch and associated tuffaceous-clastic complexes marking former forearc terraces, and (iii) terranes surrounded by flysch matrix and often by serpentinite melange or olistostromes, marking accretionary wedges of the near-shore slopes of trenches. All these genetically associated complexes become successively younger from west to east.

The Koryak-Kamchatka foldbelt is divided into the Koryak and Olyutorsky-Kamchatka systems. In the former, folding ceased in the Late Cretaceous, while in the latter it ceased near the Oligocene-Miocene boundary. The systems are divided by a distinct thrust front--the Vyvenka thrust--representing the collisional suture between these systems.

Koryak System

The Koryak system is bounded in the west by volcanics of the Koni-Murgal and Okhotsk-Chukotka belts. It consists of a voluminous flysch matrix and a number of terranes enveloped therein. The matrix is composed of flysch sequences making up five zones: (i) the Penzhina-Anadyr zone where mainly Early Cretaceous flysch corresponds to the forearc terrace of the Koni-Murgal arc, (ii) the Algano-Velikorechensky and (iii) the Alkatvayam zones, where pre-Maastrichtian (latest Cretaceous) flysch was deposited in the forearc basin of the Okhotsk-Chukotka belt, (iv) the Ukalayat and (v) the West Kamchatka zones, where flysch was deposited throughout the Late Cretaceous, Paleocene, and Eocene.

The flysch matrix envelopes a large number of terranes. In the Penzhin-Anadyr zone these are the Kuyul, Valizhgen, Pontonei, Vaega, Ust-Belaya, Pekulnei, and Kanchalan terranes (Figure 131).

The structure of the Kuyul, Valizhgen, and Pontonei terranes (which are often considered as a single Talovo-Mainsky terrane) as well as the Vaega and Ust-Belaya terranes is rather uniform (Figure 132). They consist of a series of nappes generally verging to the SE [Alexeev, 1981, 1982]. The underlying apparent autochthon consists of flysch ranging from Upper Volgian to Lower Albian. The Hauterivian is represented by an olistostrome composed of blocks derived from rocks which are now found in overlying tectonic slices. The youngest autochthonous members, on which the allochthonous slices are thrust belong to the Lower Albian and are overlain by a neo-autochthon of Upper Albian-Turonian age. The allochthon of exotic blocks approached the forearc terrace in the Hauterivian, but the main collisional episode occurred in the Middle Albian, i.e., just preceding the origin of the Okhotsk-Chukotka belt. The allochthonous slices have a chaotic structure. Volcanics and cherts predominate; the cherts contain Upper Jurassic-Lower Cretaceous Radiolaria. The volcanics include diverse rocks, mostly basalts with both oceanic and island-arc signatures. This sequence includes great exotic blocks; of particular importance are ophiolites of two ages: Early Paleozoic and Mesozoic. Among Mesozoic ophiolites, Late Jurassic-Early Cretaceous oceanic type rocks are recognized: gabbro and MORB tholeiite which are from 140 to 115 Ma old [Alexeev, 1987]. Glaucophane schists 330-340 Ma old are associated with the ophiolites. Numerous blocks in the serpentinitic melange include basalts, limestones, cherts and Middle Ordovician graptolitic shales, Upper Ordovician-Lower Silurian reef limestones, abundant Devonian reef limestones, Lower Carboniferous coarse clastics and continental coal-bearing deposits, and Visean-Namurian andesites. The allochthon also incorporates slabs of Upper Permian, Middle Triassic, and Norian-Rhaetian clastic deposits, the remnants of the Verhoyansk complex. Evidently rock complexes brought both from the Pacific and from the Eurasian continent were involved in the thrusting.

The island-arc volcanics indicate the existence of a Volgain-Valanginian island arc. Exotic terranes arrived in the subduction zone in front of this arc, but they had already been

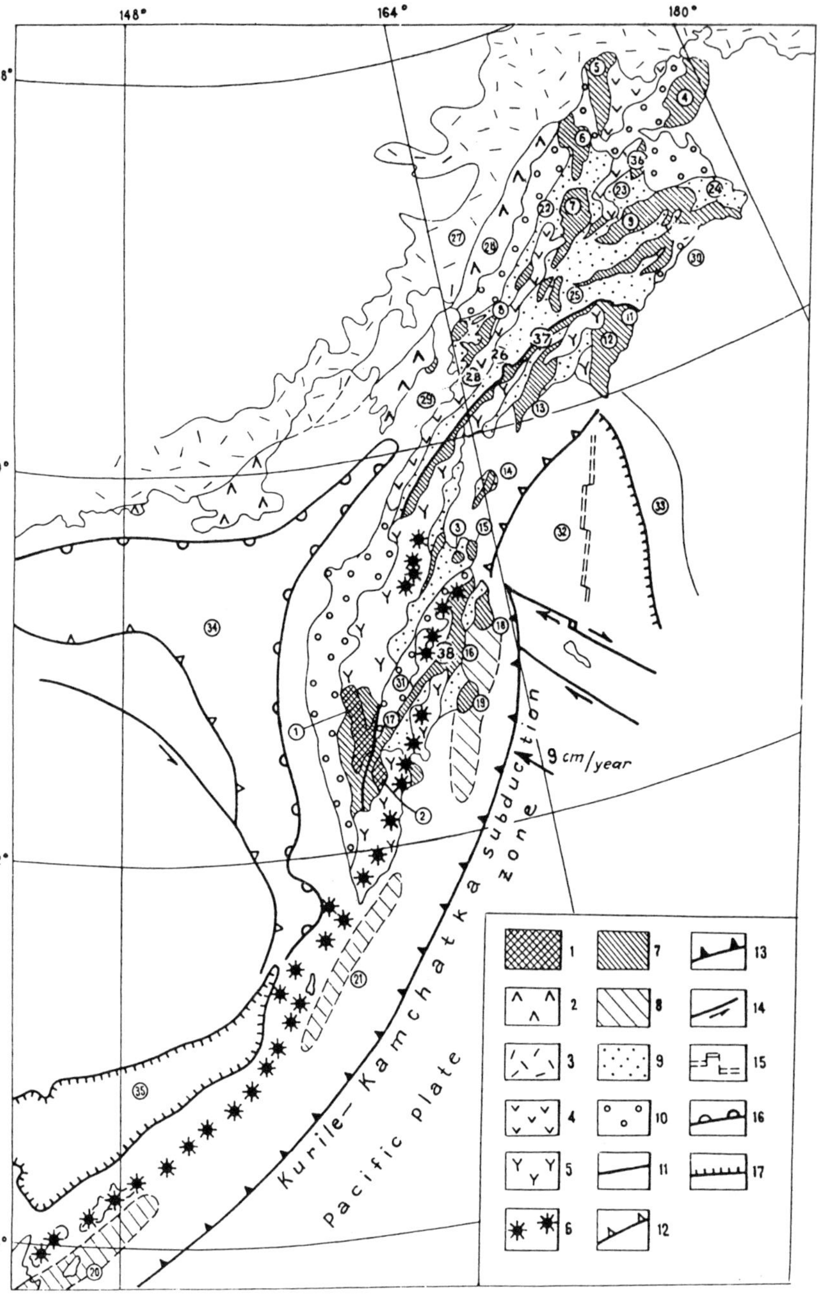

Fig. 131. Tectonics of the Koryak-Kamchatka belt.
1, Ancient blocks; 2, Jurassic-Lower Cretaceous volcanic belt; 3, Middle Cretaceous volcanic belt; 4, Late Cretaceous-Paleogene volcanic belt; 5, Neogene volcanic belt; 6, active volcano; 7, exotic terrane; 8, supposed exotic terrane; 9, flysch matrix; 10, Cenozoic depression; 11, thrust; 12, fossil subduction zone; 13, active subduction zone; 14, transform fault; 15, fossil spreading center; 16, boundary of the North Okhotsk depression; 17, boundary of oceanic crust.
Numbered are: 1, Median Range terrane; 2, Ganal terrane; 3, Khavyvensky terrane; 4, Kanchalan terrane; 5, Pekulnei terrane; 6, Ust-Belaya terrane; 7, Vaega terrane; 8, Kuyul terrane; 9, Mainitsky terrane; 10, Ekonai terrane; 11, Vatyn terrane; 12, Achayvayam terrane; 13, Goven terrane; 14, Karaginsky terrane; 15, Ozernovsky terrane; 16, Kumroch terrane; 17, Irunei terrane; 18, Kamchatsky Mys terrane; 19, Kronotsky terrane; 20, Lesser Kurile terrane; 21, Vitiaz terrane; 22, Penzhina-Anadyr flysch zone; 23, Algano-Velikorechensky flysch zone; 24, Alkatvayam flysch zone; 25, Ukelayat flysch zone; 26, West Kamchatka flysch zone; 27, Okhotsk-Chukotka volcanic belt; 28, Koni-Murgal volcanic belt; 29, West Kamchatka-Koryakia volcanic belt; 30, Lower Khatyrka depression; 31, Central Kamchatka graben; 32, Komandor basin; 33, Shirshov Rise; 34, Tinro basin; 35, South Okhotsk basin; 36, Rarytkin terrane; 37, Vyrenka thrust; 38, Valaginsky block..

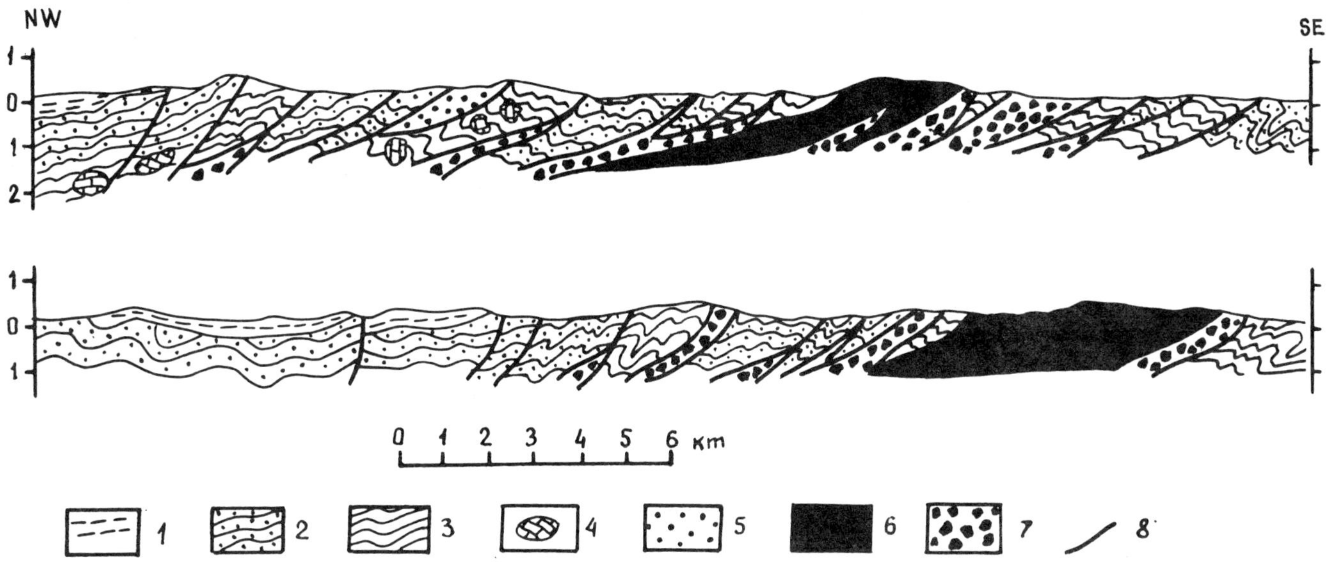

Fig. 132. Cross-section through the Kuyul block (vertical scale is exaggerated 2:1 relative to horizontal scale), after Alexeev [1981].
1, Albian and younger sediments; 2, Lower Ceretaceous rocks; 3, Upper Jurassic rocks; 4, olistolith; 5, Upper Triassic rocks; 6, ultramafics; 7, serpentinite melange; 8, thrust.

deformed before they became accreted to the arc, for they include ancient ophiolites and glaucophane schists. The terranes came from different places and contain such diverse rock types as reefs, deepwater sediments, oceanic-floor ophiolites, and shallow-water shelf sediments. Later, in the Hauterivian, the volcanic arc together with the exotic terranes was pushed against the fore-arc terrace of the Koni-Murgal belt, and an olistostrome was formed. In the last stage (Middle Albian), the main thrusting and nappe formation resulted from collision of the Koni-Murgal arc with the Eurasian margin. These movements were so great that nappes composed of the Verkhoyansk complex were thrust from the Siberian margin over the accretionary prism of the Koni-Murgal arc.

The Pekulnei terrane, unlike most of the other blocks of the Koryak system, contains no fragments with a Paleozoic fauna, but is characterized by a great body of metamorphosed gabbro with associated dunites and pyroxenites. The isotopic ages of these rocks range from 650 to 1300 Ma [Markov et al., 1982]. Another distinction of the Pekulnei block is that the matrix of melange is sheared picrite.

The core of the Kanchalan block is made up of middle-Upper Paleozoic rock complexes of varieties of arkose, felsic volcanics, and limestones. Stavsky et al. [1990] refer to this block as a microcontinent.

Flysch deposits of the Alchano-Velikorechensky and Alkatvayam zones surround the Rarytkin, Mainits and Ekonai terranes. The latter two are sometimes considered as a single Khatyrka terrane. The Rarytkin terrane, which is an allochthon lying upon Upper Jurassic-Early Cretaceous flysch, contains an Upper Jurassic volcanogenic-cherty sequence enclosing bodies of ultramafic rocks and gabbro. The Mainits and Ekonai terranes (Figures 133, 134) are included in the complex Central Koryak nappe system [Ruzhentsev et al., 1982], which, according to Palandzhyan [1986] consists of six units. The first and lowest (parautochthonous) unit is composed of tuffaceous-terrigenous deposits of Upper Jurassic-Campanian age, the second (Nakypylyakh nappe), of Carboniferous-Triassic ophiolites and cherty sequences; the third (Ekonai nappe), of Upper Paleozoic tuffaceous-terrigenous series; the fourth (Alkatvayam flysch unit), of Upper Jurassic-Cretaceous clastics with serpentinite melange at the base; the fifth (Mainits nappe), of polymictic melange with remnants of Paleozoic ophiolites as well as Upper Jurassic-Lower Cretaceous ophiolites and Lower Cretaceous greywacke sequences, and the sixth and highest (Yanranay nappe), of Upper Jurassic-Lower Cretaceous ophiolites and Lower-Upper Cretaceous greywackes. The Mainits terrane includes the fifth and sixth nappe units whereas the Ekonai terrane includes the second and third. The first and fourth units belong to the flysch matrix.

Stavsky et al. [1990] recently found that the olistostrome contains blocks of limestones and cherts with remnants of corals, Halobia, fusulinids, and Radiolaria of the Tethys biogeographic province, whereas the olistostrome matrix contains remnants of Buchia which represent the Boreal province.

Two different rock types of Late Paleozoic and Triassic age are present in the Ekonai nappes. The first is largely composed of cherts and spilites with subordinate argillites and reef limestones; the second is typified by tuffs of mafic, intermediate, and felsic composition, also of Late Paleozoic age. A fauna characteristic of the Tethys province is present in both, including fusulinids, corals, and brachiopods, [Terekhova and Epshtein, 1979; Bychkov and Dagis, 1985]. Of particular importance are paleomagnetic data on the Upper Triassic volcanogenic-sedimentary rocks, which formed at 17° paleolatitude (either northern or southern) [Lozhkina and Bychkov,

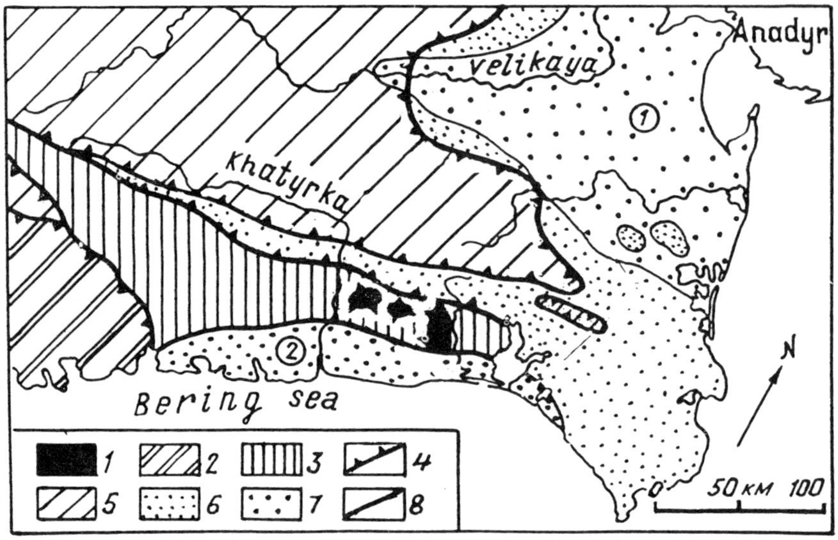

Fig. 133. Nappe units of the Koryak upland, after Ruzhentsev et al. [1982].
1, Olyutorovsky zone; 2, Yanranay zone; 3, Ekonai zone; 4, thrust; 5, Mainitsky zone; 6, Alkatvayam zone; 7, Neogene-Quaternary depressions; 8, other fault.
Numbered are depressions: 1, Anadyr, 2, Lower Khatyrka.

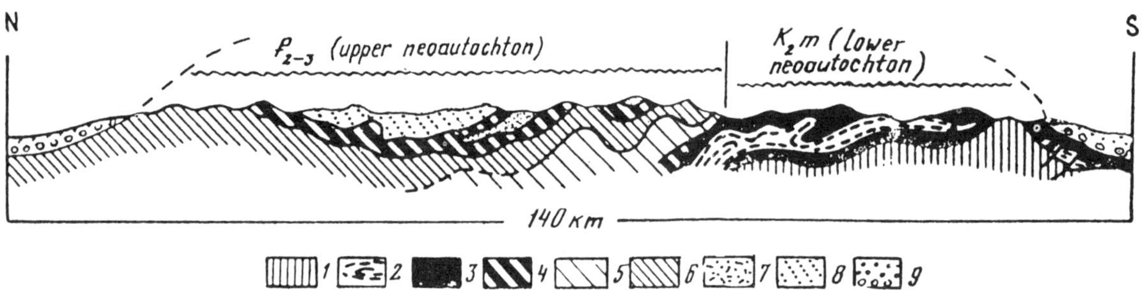

Fig. 134. Cross-section through the Koryak upland, after Ruzhentsev et al. [1982].
1, Autochthon; 2, Nakypiylyaich nappe; 3, Ekonai nappe; 4, serpentinite melange; 5, Lower Alkatvayam nappe; 6, Middle Alkatvayam nappe; 7, Lower Mainitsky nappe; 8, Upper Mainitsky nappe; 9, Neogene-Quaternary depressions.

oral communication, 1986; Stavsky et al., 1990]. A pre-Maastrichtian unconformity is well developed in the Ekonai terrane.

Olyutorsky-Kamchatka System

The Olyutorsky-Kamchatka system is arcuate, convex to the NW. The curvature is primary but it was overprinted by the recent Kurile-Kamchatka arc. In western and southern Kamchatka, the boundary between the recent and older arc complexes coincides with the base of the Oligocene, but in eastern Kamchatka and in the Olyutorsky Peninsula, it coincides with the base of the Miocene. The eastern Kamchatka peninsulas--Kamchatsky Mys and Kronotsky--as well as the Karaginsky islands are exotic terranes that joined Kamchatka in the Middle Miocene.

Three types of rock complexes have been recognized in the older arc: (i) metamorphics, (ii) volcanogenic-cherty sequences of Late Cretaceous age, and (iii) Late Cretaceous and Paleogene volcanic sequences. The Median (Sredinny) Range and Ganal massifs are composed of metamorphics. In the Sredinny Range, schists and gneisses give an isotopic age of 1300 Ma [Kuzmin and Chekhonin, 1980], but they also yield younger figures up to 120 Ma [Vinogradov et al., 1988]. According to Shapiro et al. [1986], the metamorphics are unconformably overlain by conglomerates with a warm-climate, broad-leaved flora of Senonian (Late Cretaceous) age.

Available data show that the Median Range massif is a continental fragment. Shapiro et al. [1986] found that the Median

Range massifs are overlain by allochthonous cherty and volcanic complexes of the Olyutorsky-Kamchatka system. In some places, deep-water cherts and shales prevail, intercalated with pillow basalts; Radiolaria show their age to be from Middle Campanian to Maastrichtian. In other places the cherty sequences give place to volcanics: tuffs, basalts, and andesites as well as alkaline basalts and such exotic rocks as meymechites and picrites, Maastrichtian in age. Ophiolites appear in the base of some allochthonous slices. The abundant cherts contain an extremely minor admixture of clay material, indicating that they formed on an abyssal plain [Vishnevskaya and Bernard, 1986].

In the Valaginsky and Kumroch blocks of eastern Kamchatka, the complex consists of thick flysch sequences ranging in age up to the Oligocene, or in places the Lower Miocene. The neo-autochthon begins from the Lower Miocene or often from the Upper Miocene.

The Olyutorsky and Govena Peninsulas (Figure 135) include three structural units (Vatyn, Achaivayam, and Goven), divided by strips of Oligocene-Lower Miocene flysch.

The Upper Cretaceous volcanic series of the Vatyn unit [Bogdanov et al., 1982; Shantser et al., 1985] is closely associated with ultramafics, gabbro, and other members of the ophiolite association; its structure is chaotic and consists of a number of stacked slabs. According to Bogdanov et al. [1982], the series includes (i) lower pillow lavas overlain by Albian-Coniacian cherts, (ii) a multi-colored chert sequence with numerous Coniacian-Santonian Radiolaria, and (iii) a chert-volcanic sequence of pillow and massive lavas of basalt and basalt-andesite composition associated with tuffaceous and cherty rocks of Campanian-Maastrichtian age. The lower pillow lavas are low-K (0.38% - K_2O) but high-Ti (TiO_2 - 1.98%) basalts intercalated with high-K (K_2O - 1.38%) rocks. They probably correspond to oceanic island basalts. The upper lavas are typical calc-alkaline rocks characterized by low-Ti (TiO_2 - 0.75%-1.0%) content and relatively high potassium (K_2O - 0.7%). The various compositions of the lavas indicate that the Vatyn volcanics formed in different environments, both oceanic islands and island arcs, and that the complexes of different facies were tectonically juxtaposed.

In the Achaivayam unit, typical calc-alkaline rocks predominate [Bogdanov et al., 1982; Fedorchuk et al., 1988]. The lower part is Maastrichtian-Danian, while the upper part is Paleocene-Oligocene; in places Lower Miocene is included [Bogdanov et al., 1982]. In summary, the Achaivayam unit contains remnants of latest Cretaceous and Paleogene volcanic arcs.

The Goven unit resembles the Achaivayam unit, except that the volcanic sequences begin with the Campanian and are associated with numerous ophiolitic bodies.

The Kronotsky and Kamchatsky Mys terranes are composed mainly of volcanic sequences of the Upper Cretaceous and Paleogene in a series of tectonic nappes dipping ESE at 20° [Raznitsin et al., 1985]. In the Kronotsky terrane the lower nappe unit consists of pillow lavas, tuffs, and tuff-breccias with numerous horizons of cherts, Coniacian-Campanian in age. Chemically, the lavas are island-arc tholeiites. The upper nappe unit is composed of serpentinite melange, the blocks of which include all members of the ophiolitic association. The ophiolites are overlain by a volcanic pile consisting of the same island-arc tholeiites as in the underlying nappe, with the same Upper Cretaceous age, but possibly it embraces the lowermost Paleogene as well. The Upper Cretaceous deposits are overlain with erosional disconformity by Paleocene-Eocene and Oligocene island-arc tuffs and lava flows. This complex is in turn overlain by

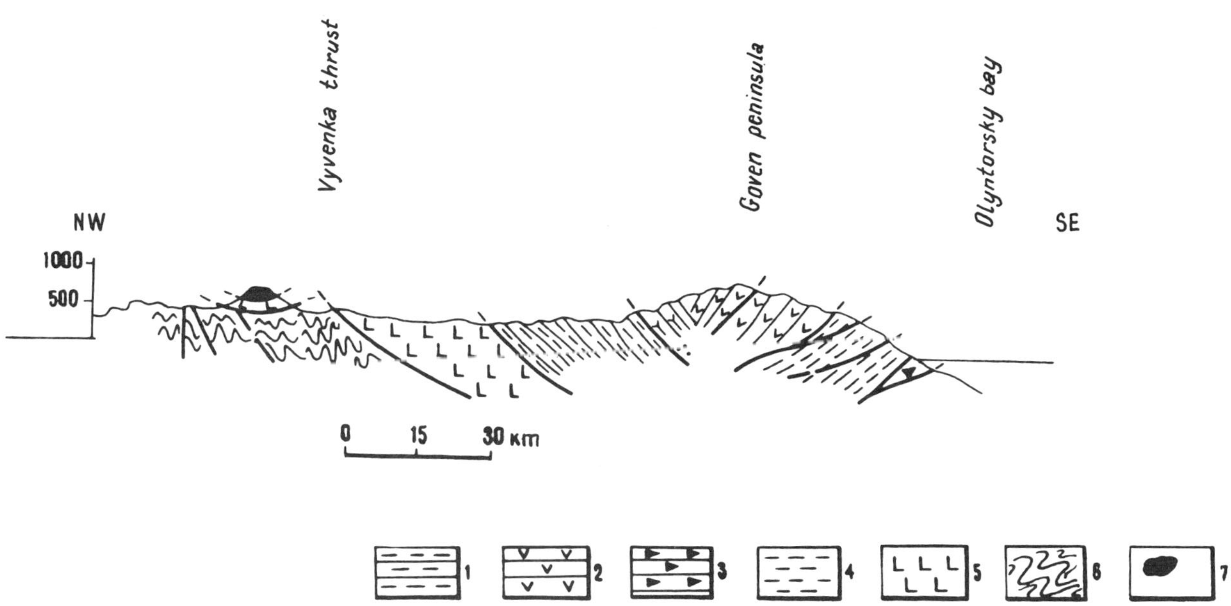

Fig. 135. Cross-section through the Olyutorsky zone, after Chekhovich (unpublished).
1, Paleogene-Lower Miocene flysch; 2, Paleogene island-arc volcanics; 3, sedimentary melange; 4, Paleogene deep-water shales; 5, Upper Cretaceous submarine volcanics and chert; 6, Cretaceous-Paleogene flysch (Ukelayat Formation); 7, ultramafics.

Lower-Middle Miocene deposits of flysch type, over which Paleogene sequences of East Kamchatka are thrust from the west along the Grechishkin thrust [Shapiro, 1976]. This thrust marks the suture between the Kronotsky terrane and Kamchatka. Upper Miocene clastic deposits are neo-autochthonous, so the Kronotsky terrane was attached by the Middle Miocene.

Thus, the Olyutorsky-Kamchatka system is mainly composed of the remnants of two island arcs of Late Cretaceous and Paleogene age. In addition, slabs of oceanic rocks, in particular rocks of oceanic island types, are present. The island-arc complexes of the Olyutorsky-Kamchatka system collided with the margin of Eurasia prior to the Oligocene, forming the west Kamchatka flysch and folding the Vatyn unit. In east Kamchatka, however, island-arc volcanicsm continued through the Oligocene and Early Miocene, i.e., the subduction zone continued to be active. The terranes of the eastern Kamchatka peninsulas were brought to this subduction zone and were accreted.

Kurile-Kamchatka Arc

Like other arcs, the Kurile-Kamchatka island arc is related to a trench situated 150-175 km from the volcanic front. A pronounced seismic belt is associated with the arc, and a distinct deep seismic zone (Benioff zone) is traceable to a depth of 650 km and is inclined on the average at 40° [Tarakanov, 1972] (Figure 136). In the back-arc area is the South Okhotsk (South Kurile) deep-water basin (Figure 137).

The Kurile-Kamchatka arc is a typical volcanic belt. Within the Great Kurile Range volcanism commenced in the Early Miocene (or possibly in the Late Oligocene, although no outcrops of rocks older than Miocene are known). The Miocene-Quaternary sequences are divided into three piles corresponding to three main stages of arc development [Frolova et al., 1985]. The lower pile embraces Lower Miocene volcanics--basalts, andesites, rhyolites, and their tuffs. Tuffs are abundant in the upper part and can be easily correlated with the Green Tuff Formation of Japan. The middle pile is Middle Miocene and consists of flysch-type sequences with conglomerates and olistostrome horizons. It is important to note that clastic material includes metamorphics, potassium granites, quartzites, cherts, jaspers, plagiogranites, and sandstones as well as volcanic rocks. No such rocks are found within the Great Kurile arc, and the material must be derived from the Sea of Okhotsk side. The middle sequence is deformed and faulted. The clastic sequences and olistostrome horizons formed in association with dissected rough topography in the back-arc area. The composition of the clastic material shows that the South Okhotsk basin did not exist yet and that clastics were carried from the Sea Okhotsk block. In our interpretation, the Miocene clastic sequences are graben facies, and their inclined position reflects block rotation along listric faults during rifting and the opening of the South Okhotsk back-arc basin. The upper pile is a volcanic sequence forming from the Late Miocene to the present, and is a typical calc-alkaline series.

Two longitudinal volcanic zones exist in the Kurile islands: the main zone associated with the axial ridge, and the western zone, which includes the western slope of the arc and seamounts of the back-arc basin. The axial zone is characterized by calc-alkalic (in places tholeiitic) island-arc volcanics, while the western zone is mainly typified by subalkaline, shoshonitic varieties. The petrochemical zonation displayed by the westward increase of potassium content is also reflected in the zonal distribution of rare elements [Popolitov and Volynets, 1981] and is correlated with the depth of the Benioff zone [Piskunov, 1975].

The Lesser Kurile Islands, the submarine Vitjaz uplift, and the east Kamchatka peninsulas (considered above) are interpreted as an outer nonvolcanic arc of the Kurile-Kamchatka system. Late Cretaceous (up to Danian) volcanics of the Lesser Kurile Islands are island-arc remnants [Streltsov, 1976]. On Shikotan island they are tectonically overlain by allochthonous layered gabbro and associated sheeted dike complex [Melankholina, 1978], which are included in a chaotic complex together with cherts and Late Cretaceous island-arc volcanics. This chaotic complex was probably formed simultaneously with the accumulation of the Maastrichtian-Danian flysch with olistostromes. These rocks were deformed in pre-Oligocene time. The Lesser Kuriles are undoubtedly an exotic terrane, like the eastern Kamchatka peninsulas, as supported by the first paleomagnetic data obtained. According to these data [Khramov, 1986], the Upper Cretaceous pole for volcanics of Shikotan island has coordinates 65° N and 247° E; thus the Lesser Kuriles were apparently at 35° N, which would be 9° S of their present position. If this block had been part of Eurasia, according to the Upper Cretaceous Eurasian pole position it would have been at 52° N. Thus, the difference in the latitudinal position amounts to 17° or about 1900 km.

The SE flank of the Kurile-Kamchatka arc slopes steeply eastward to the trench. The steep inner trench slope is 70 to 90 km wide, and its angle is 7 to 10°. The trench axis is from 7.5 to 9 km deep; its maximum is 10.5 km. The trench is distinctly V-shaped, but two flat terraces are clearly observed on traverse sections. The upper terrace, at 3000-3200 m, is a sediment-filled fore-arc basin. The lower terrace, at 5800-6300 m depth, is less distinct and is strongly dissected into small fragments, either gravitational slides or tectonic slices over the accretionary wedge. Seismic sounding [Garkalenko et al., 1977] revealed that the oceanic sediments and underlying basaltic basement of the off-shore oceanic slope continue undeformed 50 km horizontally beneath the inner trench slope. The distribution of earthquake foci shows that the outcrop of the Benioff zone is confined to a strip between the trench axis and the first terrace on the inner trench slope.

Detailed studies of trench morphology in the northern part of the trench have found sharp knee-like bends in its axis, and promontories of the inner slope project towards the ocean for some tens of kilometers. Baranov [personal communication, 1986] interprets such promontories as a result of the attachment of exotic blocks over which the subduction zone jumped. The Kurile-Kamchatka arc is broken by transverse faults into a number of segments about 100 km long which interact with the subducting plate like 'piano keys' moving either towards the arc or out towards the ocean.

The South Okhotsk back-arc depression belongs to the group of basins with a moderate heat flow (on the averager 2.3 HFU). The crust under the depression is about 10 km thick; the sedimentary layer is about 3 km. Moreover, beneath the sediments V_p crustal velocities are 6.7 km per second; thus the crust is typically oceanic. The sedimentary filling consists of two sequences: the upper layered, and the lower acoustically transparent. The transparency of the lower sequence is appar-

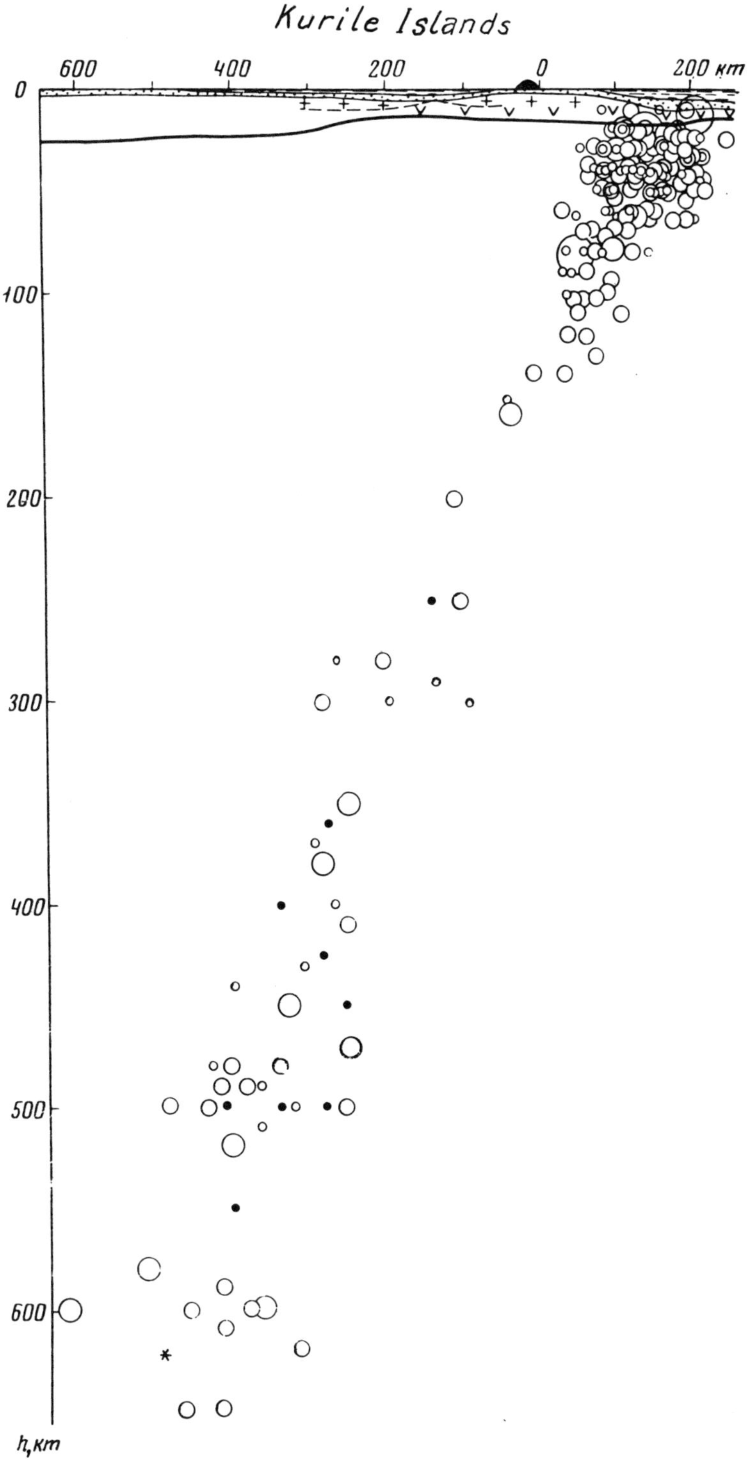

Fig. 136. Distribution of earthquake hypocenters under the middle segment of the Kurile island arc, after Tarakanov [1972]. Water shown by dashes, "granitic" layer shown by crosses, "basaltic" layer shown by v-s. Circle size corresponds to earthquake magnitude. Vertical exaggeration = 2.6.

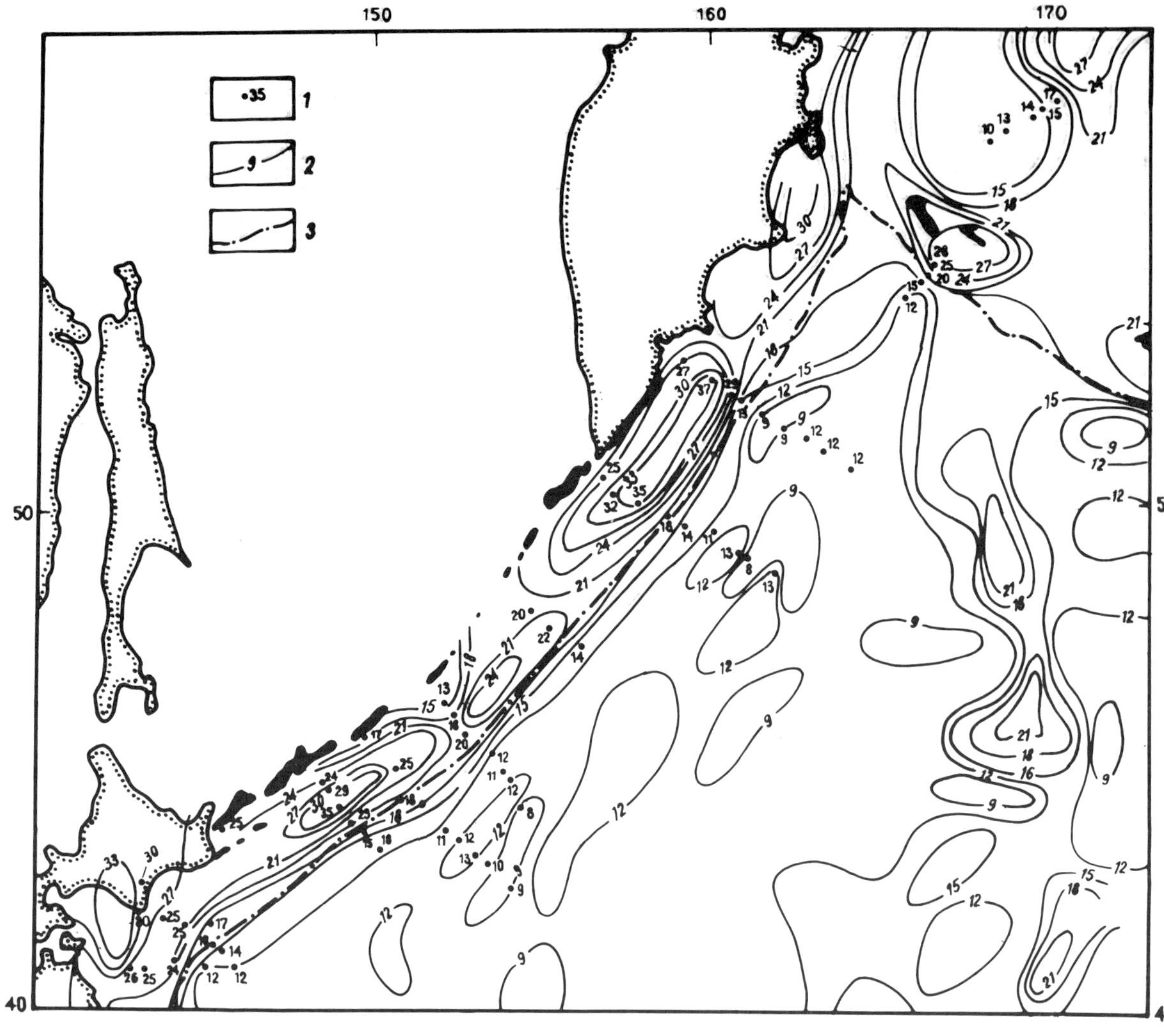

Fig. 137. Depths to the Moho discontinuity in the area southeast of the Kurile-Kamchatka arc, after Rodnikov [1983].
1, Depth to Moho according to deep seismic sounding data, in km; 2, contour of Moho depth according to gravimetric data; 3, trench axis.

ently caused by the presence of gas hydrates. The surface of the acoustic basement is markedly broken, as would be expected if it were generated at a spreading ridge. The sedimentary sequences within the basin do not seem deformed, except for a narrow zone along the Kurile Islands where anticlinal folds with limbs inclined at 20° and amplitude as much as 400 m have been found. The origin of these folds is not clear. The South Okhotsk basin cannot be ancient and probably formed 15 to 30 Ma ago, considering that the sediments are 3000 m thick and assuming a comparatively high rate of accumulation, 10-20 cm per 1000 years. If the Middle Miocene clastic sequences and olistostromes of the Kuriles are correctly interpreted as a graben facies, the South Okhotsk basin began to form about 15 Ma ago.

Komandor Islands

The Komandor Islands are the western segment of the Aleutian island arc; they too are accompanied by a trench and a seismic zone. But in contrast to the Kuriles and the eastern segment of the Aleutian arc, there is no Benioff zone here. The well known explanation is that the vector of the Pacific - North America plate motion is directed NW (320°), i.e., parallel to the arc axis in the Komandor Islands. Consequently, the boundary between the Pacific and North American plates here is of a transform fault type with dextral strike-slip. Behind the Komandor Islands is a narrow trough corresponding to another dextral strike-slip fault. Thus, the Komandor Islands are on an uplifted block bounded on both sides by strike-slip faults and obviously displaced northwestward from the Aleutian arc in the east. A short transverse SSW-NNE trough lies behind the Komandor Islands; recent volcanism and active tectonism are recorded from this trough (Baranov, personal communication, 1988), which may be interpreted as a small pull-apart basin of Late Miocene age. The Komandor Islands themselves display no recent volcanism, in agreement with the relative strike-slip motion here. The geology of the Komandor Islands seems to be simple [Schmidt, 1978; Tsvetkov, 1983]. They are largely occupied by Eocene volcanics of bimodal composition. Younger (Upper Oligocene-Lower Miocene) volcanics consist of alkalic basalts, including teschenites, that form shield volcanoes. In the last volcanic event, in the Pliocene, terrestrial andesitic basalt lavas were erupted. Only these lavas belong to the calc-alkaline series. Stavsky et al. [1990] consider that the basement of the Komandor Islands is an exotic terrane that arrived from the Pacific Ocean. Unfortunately, the origin and development of the Komandor Islands cannot be unambiguously interpreted at present.

Bering Sea

The Shirshov Ridge divides the Bering Sea into the smaller Komandor Basin in the west and the larger Aleutian Basin in the east. The latter is filled with sediments 4-5 km thick (up to 7-9 km in the Navarin trough near the Chukotka coast), which were deposited since the Late Cretaceous. Magnetic lineations in the Aleutian Basin trend N-S, and are identified as M-13 to M-1 [Cooper et al., 1976], ranging from 136 to 122 Ma in age, and becoming younger from west to east. Regardless of the origin of this oceanic crust, probably it became a part of the Kula plate and was eventually trapped behind the Aleutian arc. Before the Aleutian arc originated, there was a subduction zone near the Chukotka-Alaska continental slope.

The Komandor Basin, although the same depth as the Aleutian Basin, is filled with only 1-2 km of sediments, which are Upper Miocene-Pliocene and Upper Pliocene-Quaternary in age (Figure 138). The sediments lie on basaltic basement penetrated by DSDP drillhole N 191; the estimated age of the basement is only 9.3 ± 0.8 Ma [Harbert et al., 1987]. This estimate of a young age is supported by high heat flow values (up to 4 HFU) in the Komandor Basin. According to these data, the Komandor Basin was formed during a relatively recent spreading event.

The Shirshov Ridge has an asymmetric cross-section. Its eastern slope inclines gently towards the Aleutian Basin and is progressively overlapped by the sedimentary sequences of that basin; thus the basement of the Shirshov Ridge was welded to the basaltic floor of the Aleutian Basin before the sedimentary cover began to form in the Middle Cretaceous. The western flank of the ridge is cut by a series of steep fault scarps [Neprochnov et al., 1985], descending like a staircase down into the Komandor Basin (Figure 138). These fault scarps are interpreted as listric fault features produced as the Komandor Basin formed and extended by rifting. The crust under the Shirshov Ridge is 18 km thick, considerably thicker than the oceanic crust in the Komandor (6 km) and Aleutian (7 km) basins [Neprochnov et al., 1985]. Andesites 17 Ma old have been dredged from the Shirshov Ridge [Marlow and Scholl, 1976]; more recent dredging provided a variety of rocks, among which gabbro and amphibolites are dominant [Neprochnov et al., 1985]. Cherts are present as well, and they contain diverse Radiolaria of Middle and Upper Triassic, Upper Cretaceous, Paleocene, Oligocene, and Miocene age. Numerous island-arc type volcanics have also been dredged.

The origin of Shirshov Ridge cannot be unambiguously interpreted. Possibly it is an exotic terrane analogous to those of the Olyutorsky-Kamchatka system, which was attached to the Kula plate and moved with it to the Eurasian margin.

HISTORY

Geological Constraints

Several subduction-related volcanic belts are recognized in NE Asia (Figure 139): (i) the Koni-Murgal belt which, within the Murgal branch, is of Lower Cretaceous, Valanginian-Hauterivian age; (ii) the Okhotsk-Chukotka belt where the main volcanic events occured during Albian-Cenomanian time; (iii) the Anadyr-Bristol belt, which was mostly active in the Paleocene and Eocene; (iv) the Koryak-West Kamchatka belt, which formed during Middle Eocene and Oligocene time, and (v) the Kurile-Kamchtaka belt, which has been active since the Early Miocene. Each of these belts (except Koni-Murgal) marks the paleo-position of the Eurasian margin at a certain time and was formed during a time span of 20-30 Ma. These belts prograded towards the Pacific Ocean. The Koni-Murgal belt, as mentioned above, possibly had an intra-oceanic position. All the belts display a clear polarity of volcanics, which allows us to reconstruct successive positions of the subduction zone. Activity in each belt terminated when an exotic terrane arrived and became "wedged" in the subduction zone. A younger belt originated when the subduction zone jumped to a new position oceanward.

In addition to volcanic belts marking the Eurasian margin, island-arc volcanics are widespread within the exotic terranes of Koryakia and Kamchatka, recording intra-oceanic island arcs. These arcs were located somehwere within the Pacific Ocean, at some unknown distance from the Eurasian margin from which they were separated by various basins. The most abundant remnants of such arcs are composed of Upper Jurassic-Lower Cretaceous (Volgian-Neocomian) and the Upper Cretaceous (Campanian-Maastrichtian) volcanics. The Upper Jurassic-Lower Cretaceous volcanics allow us to reconstruct the Koryak intra-oceanic arc. The Upper Cretaceous volcanics have a different affiliation; they appear to be part of the former Olyutorsky arc, now known in the Late Cretaceous volcanic complexes of Bowers Ridge (in the Aleutian Basin), the Olyutorsky zone, east and central Kamchatka, the eastern Kamchatka peninsulas, and the Lesser Kurile Islands. The Olyutorsky arc originated somewhere in the Pacific Ocean, and the Kula plate was subducted under it. After subduction ceased in the Late Cretaceous, the arc was broken into three segments. In the Eocene one of them--Bowers Ridge--was trapped, together with a piece of the Kula Plate, in the Aleutian Basin; another segment collided with the Koryak-West Kamchatka arc in the Oligocene to form the Olyutorksy-Kamchatka system; and the third segment--east Kamchatka peninsulas and Lesser Kuriles--remained behind within the Pacific and only approached the Eurasian margin in the Miocene.

The whole Koryak-Kamchatka belt is an agglomerate of exotic terranes. Three charactristics can be used to classify these terranes: (i) the composition and time of formation of the constituent rocks, (ii) the time of their accretion in the subduction zone in front of the respective intra-oceanic island arcs, and (iii) the time of their final attachment to the Eurasian margin. With regard to the time of rock formation, there are (i) microcontinental blocks with Precambrian continental basement (e.g., the Median Range of Kamchatka), (ii) terranes that contain ophiolites, i.e., remains of former oceanic crust, (iii) terranes with oceanic island or other intraplate volcanics,

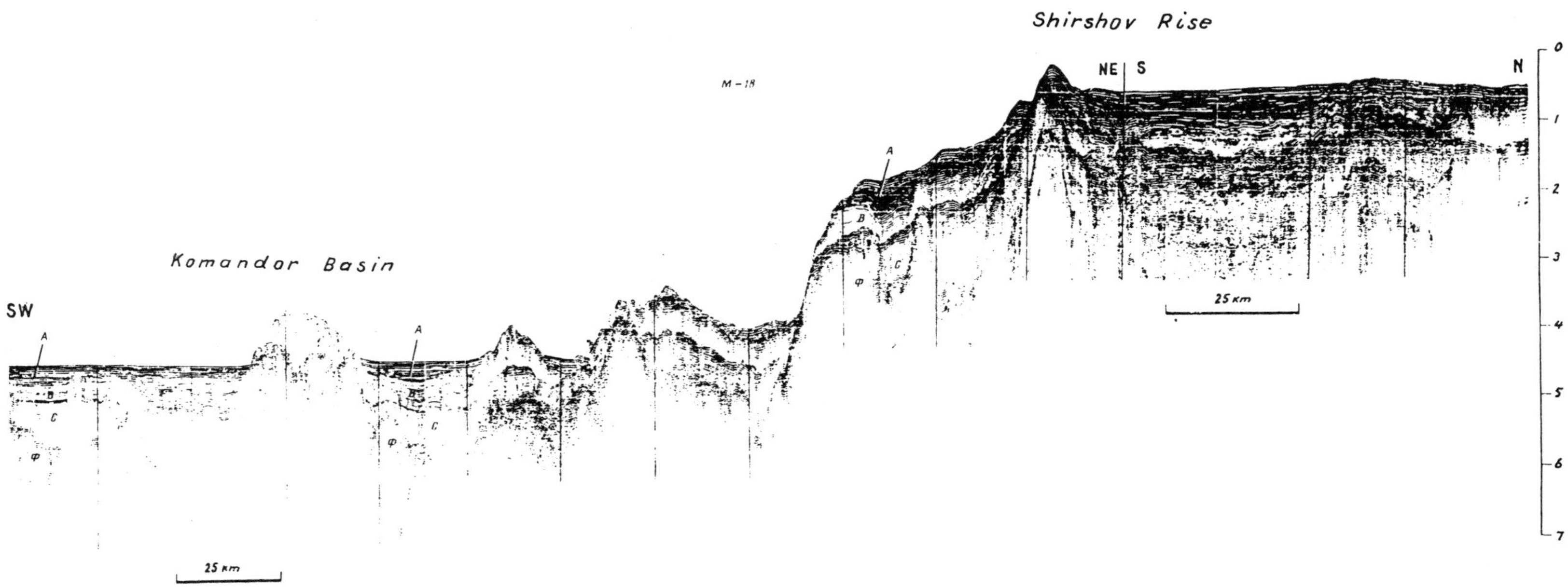

Fig. 138. Seismic reflection profile through the Shirshov Rise and Komandor Basin, after Neprochnov et al. [1985]. A, B, and C - various reflectors, ϕ - basement.

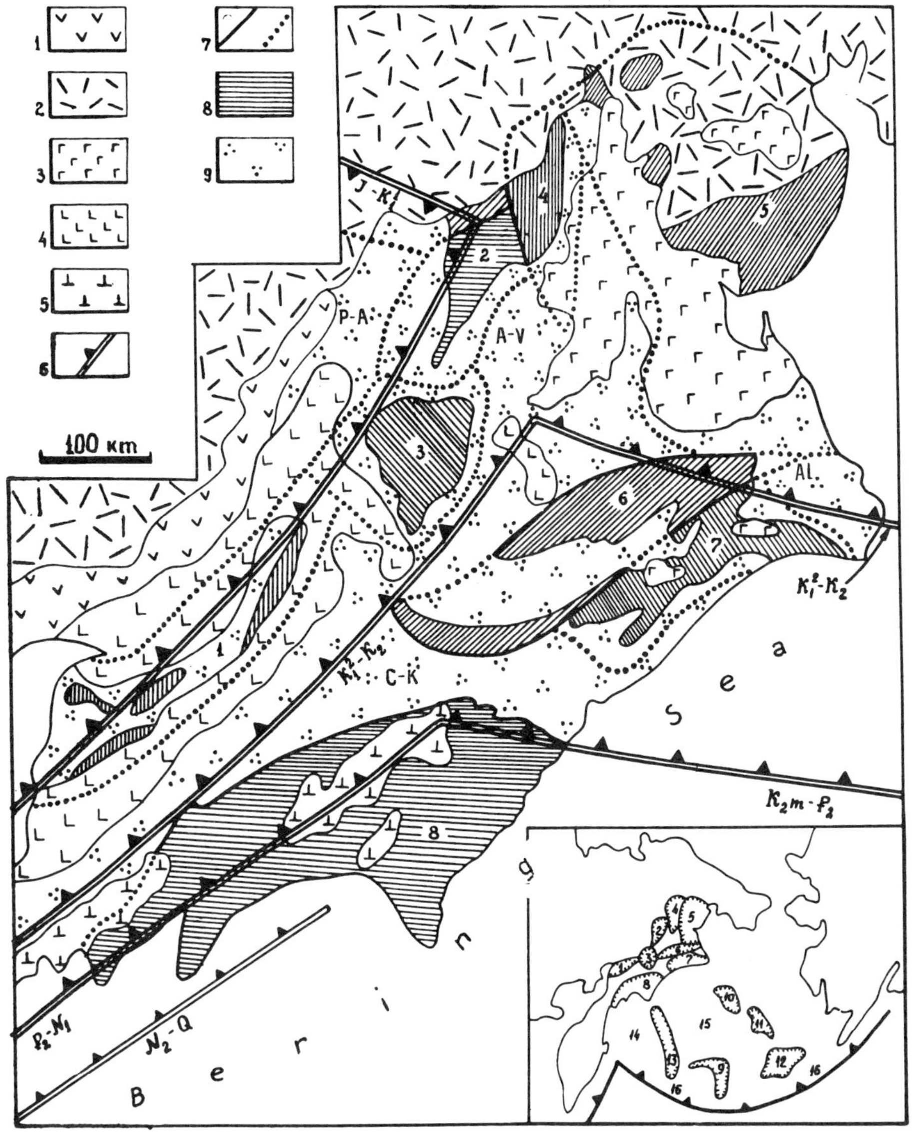

Fig. 139. Location of continental-margin volcanic belts and exotic terrains within the Anadyr-Koryak region, after Stavsky et al. [1990] Insert shows locations of exotic terranes around the Bering Sea.
1, Koni-Murgal belt; 2, Okhotsk-Chukotka belt; 3, Anadyr-Bristol belt; 4, West Kamchatka-Koryakia belt; 5, Apuka-Vyvenka belt; 6, inferred position of subduction zone; 7, exotic terrane boundary (dotted where concealed under younger deposits); 8, exotic terrane; 9, flysch and olistostrome.
Numbered are exotic terranes: 1, Penzhina; 2, Ust-Belaya; 3, Vaega; 4, Pekulnei; 5, Kanchalan; 6, Mainitsky; 7, Ekonai; 8, Olyutorsky (Vatyn+Achayvayam+Goven); 9, Bowers; 10, Navarin; 11, Pribylov; 12, Umnak; 13, Shirshov; 14, Komandor Basin; 15, Aleutian basin; 16, Aleutian island arc. Al - Alkatvayam zone, A-V - Algano-Velikorechensky zone, C-K - Central Koryak (Ukelayat) flysch zone, P-A - Penzhina-Anadyhr zone.

(iv) terranes composed of shallow-water shelf complexes, e.g., reef limestones of all periods from the Ordovician to the Triassic, mainly contained in blocks of the Koryak system.

With respect to their time of accretion in front of intra-oceanic island arcs, the exotic terranes can be divided into two groups. The first group includes terranes that amalgamated near the Koryak arc in Late Jurassic-Early Cretaceous time; these are terranes of the Koryak system. The second group consists of terranes that arrived against the Olyutorsky arc in the Late Cretaceous, and which are now represented by blocks within the Olyutorksy-Kamchatka system (Vatyn, Median Range, etc.).

According to their time of attachment to the active Eurasian margin, we distinguish (Figure 140) (i) terranes that joined the Koni-Murgal arc in the Early Cretaceous, (ii) terranes that joined the Okhotsk-Chukotka volcanic belt in

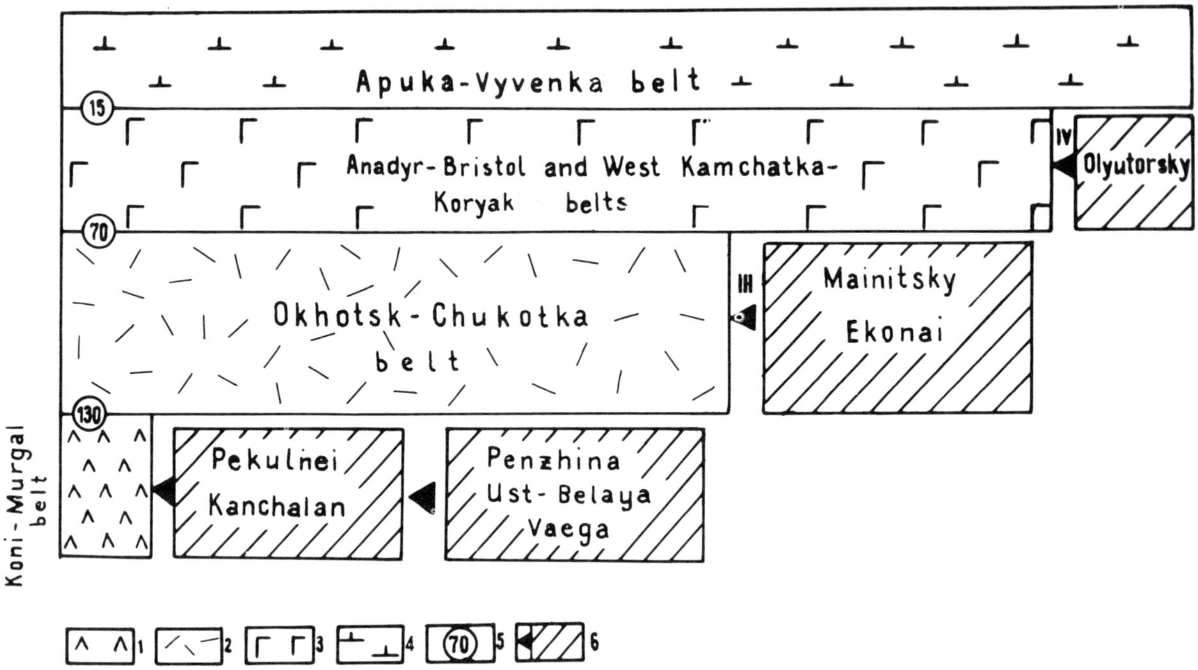

Fig. 140. Diagram showing successive stages of attachment of exotic terranes of Koryakia and Kamchatka to the Eurasia margin during Mesozoic and Cenozoic times. Terranes are cross-hatched. Encircled numbers show time of attachment (in Ma) to the Eurasian margin.

the Late Cretaceous, (iii) terranes that joined the Anadyr-Bristol and Kamchatka volcanic belts in the Late Eocene and Oligocene, and (iv) terranes that joined the Kurile-Kamchatka arc in the Middle Miocene.

Kinematic Data

By 150 Ma ago, Eurasia was interacting with successive plates in the Pacific Basin. At about 135 Ma [Engebretson et al., 1985], the rate of relative motion between the Izanagi plate and Eurasia increased drastically from 2.6 cm/yr to 20 cm/yr (Table 7), and this increase was followed by the rapid arrival of numerous exotic terranes at the subduction zone. From 135 to 70 Ma, the Izanagi plate moved almost due north relative to Koryakia at 20-16 cm/yr. From 70 to 55 Ma the Kula plate, which had succeeded the Izanagi in interaction with Eurasia, moved relatively NW (320°) at 15-12 cm/yr. From about 50 to 30 Ma (in the Eocene) the Kula-Pacific mid-oceanic ridge was progressively consumed, at first under the Olyutorksy arc, then under the Aleutian arc. Evidently this time span must have been one of the most dramatic in the history of the Koryak-Kamchatka belt. At that time the Koryak and Olyutorsky-Kamchatka systems collided along the Vyvenka suture. After the Kula/Pacific spreading center was consumed, Eurasia came in contact with the Pacific plate, which moved WNW (290°-300°) with respect to Eurasia. The convergence rate decreased considerably in the interval of 50 to 35 Ma, i.e., in the second half of the Eocene, falling as low as 3-4 cm/yr; then it accelerated to rates close to the recent rate (10 cm/yr) in the Oligocene-Early Miocene (35-20 Ma), and the Kurile-Kamchatka island arc probably originated as a result.

Paleomagnetic Data

Paleomagnetic data are available for Upper Cretaceous rocks of the Lesser Kurile and Late Triassic rocks of the Ekonai terrane; they indicate a southerly origin for both blocks. In the Late Cretaceous the Lesser Kurile block was 17° south of its present position relative to Eurasia. The Triassic deposits of the Ekonai block formed in the tropical belt at 27° paleolatitude, as shown also by the Tethyan tropical fauna. New paleomagnetic results were obtained just recently. A. N. Didenko and D. M. Pechersky (oral communication, 1989) reported that Late Jurassic-Early Cretaceous oceanic-type rocks of the Mainits terrane in Koryakya were formed at 29° north, no less than 5000 km south of the Eurasian margin. This coincides well with the inferred position of the Mainits terrane according to kinematic data. Kheifets [Savostin and Kheifets, 1988] established that the Late Campanian volcanics of the Vatyn unit of the Olyutorsky Peninsaula were erupted at 32° north, and the Maastrichtian-Danian volcanoclastics of the Achaivayam unit were deposited at 55° north. Meanwhile the corresponding point of the stable Eurasian margin would have been at 73° north. So, the oceanic-type Vatyn unit is postulated to have originated at the Kula-Pacific spreading center, and the island arc Achaivayam unit was initially located at least 2000 km off the Eurasian margin. D. Kovalenko (oral communication, 1989) reported that the Late Cretaceous-Paleocene island arc volcanics of Karaginsky Island also were formed far south of the Eurasian margin.

TABLE 7. Differential Motions of the Plates of the Pacific Basin (Pacific, Kula, Izanagi) with Respect to Eurasia

Age (Ma)	Pole Position Deg. N (Degrees)	Pole Position Deg. E (Degrees)	Angle Rotation (Degrees)	Motions at points with coordinates 45°N, 150°E for the Pacific and 65°N, 150°E for the Kula plate: Azimuth (Degrees)	Rate (cm/yr)
Pacific Plate (after Zonenshain et al. [1987])					
10 - 0	66.9	-70.6	9.4	286.5	9.4
20 - 10	71.1	-74.9	9.6	285	9.2
35 - 20	62.5	-47.1	17.6	278	12.4
43 - 35	77.58	-128.65	4.02	293.3	3.94
53 - 43	47.66	-143.52	5.23	331.7	4.1
66 - 53	12.4	-81.4	11.5	322	9.5
Kula Plate (after Zonenshain et al.[1987])					
70 - 50	6.81	-64.7	25.3	325	14
111 - 70	21.2	-108.4	60.4	358	15.5
130 - 110	22.6	-104.1	29.7	354	15.3
Izanagi Plate (after Engebretsen et al.[1985])					
56 - 53	8	-53	6.0	304	21.7
61 - 56	17	-71	6.1	318	13.6
66 - 61	17	-63	7.1	311	15.8
74 - 66	22	-56	12.8	302	17.8
85 - 74	38	-44	20.9	238	20.4
95 - 85	22	-117	21.5	356	22.1
100 - 95	9	-118	10.5	2.6	23
115 - 100	0	-139	27.6	28	20
119 - 115	4	-133	7.4	19	20
127 - 119	14	-131	15.2	12	19.7
135 - 127	14	-128	21.6	10	28.3
145 - 135	58	-155	6.1	357	2.6

Chronology of Events

We here follow the scheme by Stavsky et al. (1990), with significant changes.

Pre-Late Jurassic History

Although the Koryak-Kamchatka belt contains Precambrian, Paleozoic, and Triassic rock assemblages, it is imposible at present to reconstruct their origin. What may be said is that most are exotic terranes that arrived from low latitudes. Stavsky et al. [1990] placed the Triassic-Late Jurassic arc in the Pacific, in the southern hemisphere.

Late Jurassic (150 Ma)

This was the period of slow Izanagi/Eurasia relative motion. The Koryak volcanic arc was supposedly south of the Koni-Murgal arc within the Pacific Ocean; on Figure 141 it is placed 4000 km from the Siberia margin and 2500 km from the Koni-Murgal arc. By 150 Ma, the convergent zone in front of the Koryak arc had received nearly all exotic terranes of the

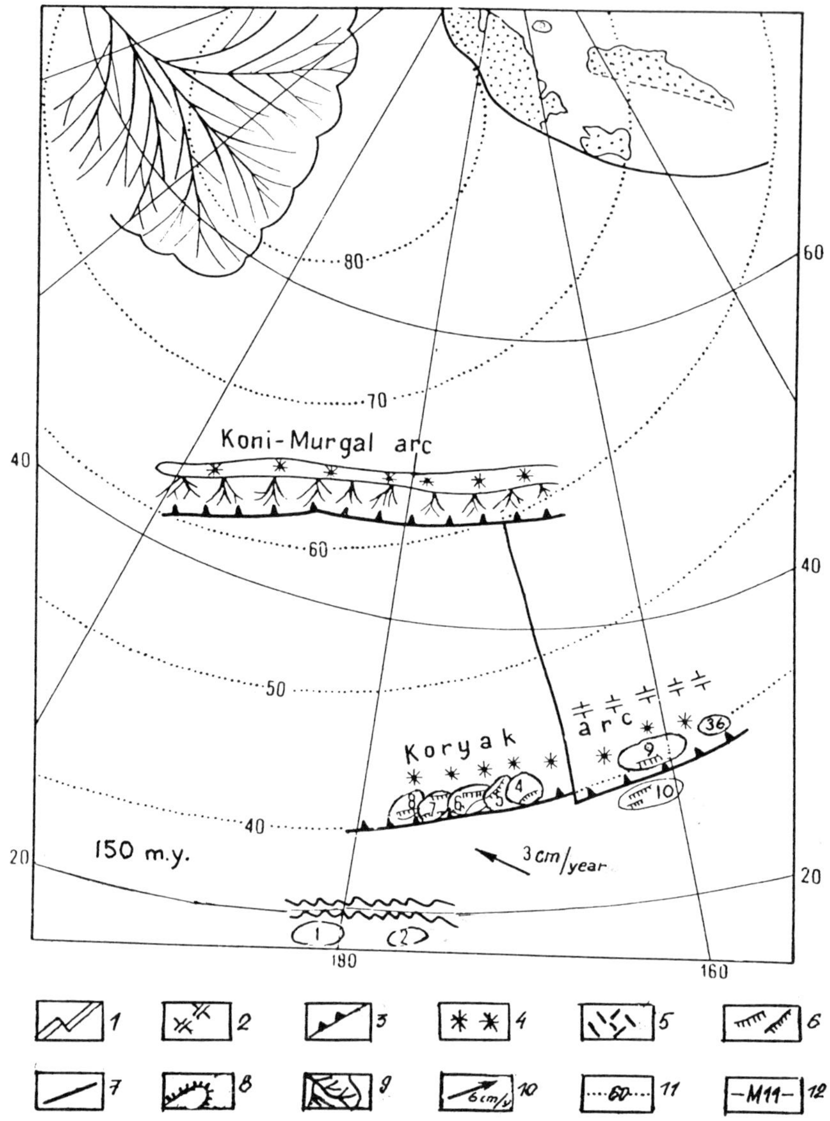

Fig. 141. Palinspastic reconstruction of the north-west Pacific region for 150 Ma. The reconstruction is made with respect to Siberia, which is arbitrarily left fixed.
1, Spreading axis; 2, back-arc spreading; 3, subduction zone; 4, island arc; 5, Okhotsk-Chukotka continental-margin volcanic belt; 6, folding and metamorphism; 7, transform fault; 8, intraplate volcanics; 9, fan; 10, vector of the Pacific/Eurasia motion, with velocities in cm/yr; 11, paleolatitude; 12, Mesozoic magnetic anomlay and its number. Continents within the present coast line are dotted. Exotic terranes are numbered as in Figure 131.

Koryak system except for the Ekonai terrane, which was still within the Pacific. The Koryak arc was situated in moderate latitudes of the northern hemisphere (approximately 40°), and already was within the area of the boreal Buchia fauna.

Early Cretaceous, Neocomian (130 Ma) (Figure 142)

At 135 Ma, when the velocity of the Izanagi: Eurasia relative plate motions increased greatly to 16-20 cm/yr, various exotic terranes immediately began to converge rapidly with Eurasia. The western part of the Koryak arc, under which subduction ceased, became attached to the Izanagi plate. Together they moved toward the Koni-Murgal arc, with which the west Koryak arc eventually collided in the Hauterivian Stage of the Early Cretaceous. The Ekonai, Mainits, and Rarytkin blocks, now amalgamated into the Khatyrka terrane, remained in the eastern part of the Koryak arc under which the Izanagi plate continued to subduct. Hence a transform fault connecting the Koni-Murgal and eastern Koryak arcs can be inferred.

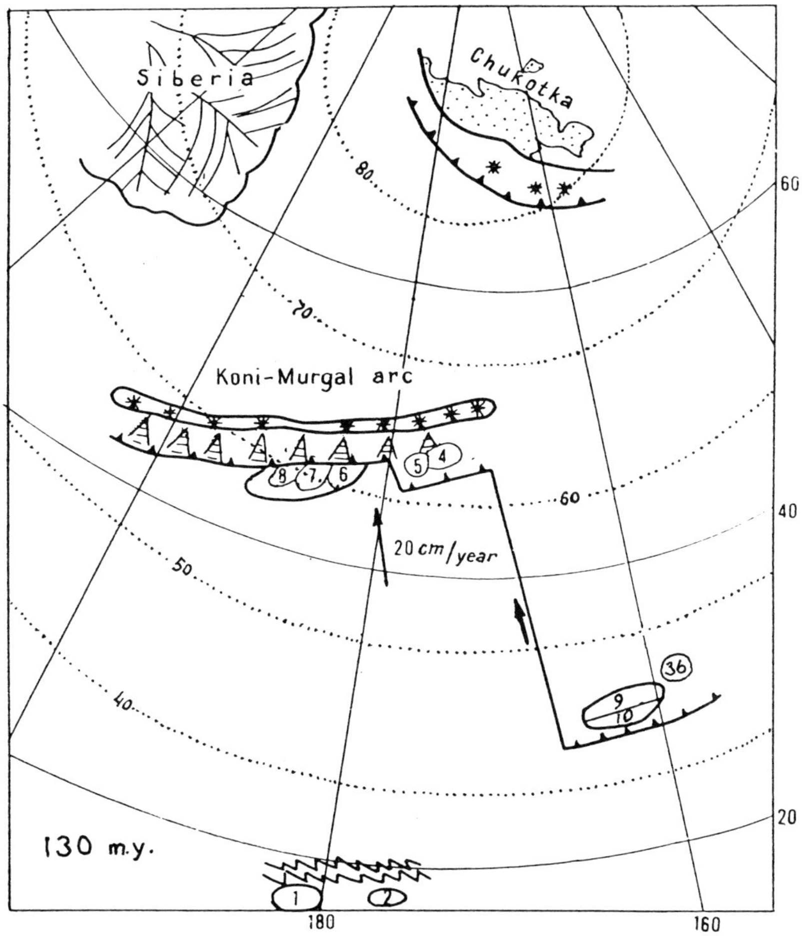

Fig. 142. Palinspastic reconstruction of the northwest Pacific region for 130 Ma. See Figure 141 for explanation.

Mid-Cretaceous, Late Albian (100 Ma)

This time was also characterized by rapid movement of the Izanagi plate towards Eurasia (Figure143). From, 130 to 100 Ma, dramatic tectonic events took place. The Koni-Murgal and Koryak arcs were broken up by the rapid convergence, and the Izanagi plate "bulldozed" the pieces towards Eurasia. In the middle Albian, the Koni-Murgal arc together with incorporated exotic blocks approached and became attached to Siberia. The western Koryak system also collided with Siberia, causing thrusting and the development of nappes. As a result, nappes of Verkhoyansk clastics were thrust from Siberia onto the exotic terranes. Meanwhile, the eastern part of the Koryak arc containing the large Khatyrka terrane, still remained in the ocean. We suppose that the intraplate volcanic massif corresponding to the present Vatyn terrane was forming near the Izanagi/Pacific spreading center. It should be noted in this connection that at this very time the Shatsky Rise, which is now on the Pacific plate, was forming on the same spreading center. Thus the intraplate oceanic-island volcanic complexes of the Vatyn zone may well represent part of the volcanic massif which was on the Izanagi plate and which was split off the Shatsky Rise by the spreading.

Late Cretaceous, Maastrichtian (70 Ma)

After 80 Ma, the Kula plate converged with Eurasia at a high rate of 16 cm/yr towards the NW (see Table 7). The subduction zone along the Eurasian margin had been continuosly active up to the middle Late Cretaceous, producing the Okhotsk-Chukotka volcanic belt, but approximately 80 Ma ago, maybe earlier, a new intra-oceanic subduction zone started to operate, resulting in the Olyutorsky island arc (Figure 144).

The initial position of the Olyutorsky island arc is reconstructed using the following data. The time span between the end of volcanism in the Early Eocene (50-53 Ma) and the collision of the arc with the Koryak-West Kamchatka continental-margin arc in the Late Eocene (38-40 Ma ago) was about 15 Ma. During that period, the island arc was included in the Kula plate and was carried by that plate to the Eurasian margin. As the relative velocity of the Kula and Eurasia plates was 16 cm/yr, the distance between the arc and the continent was not

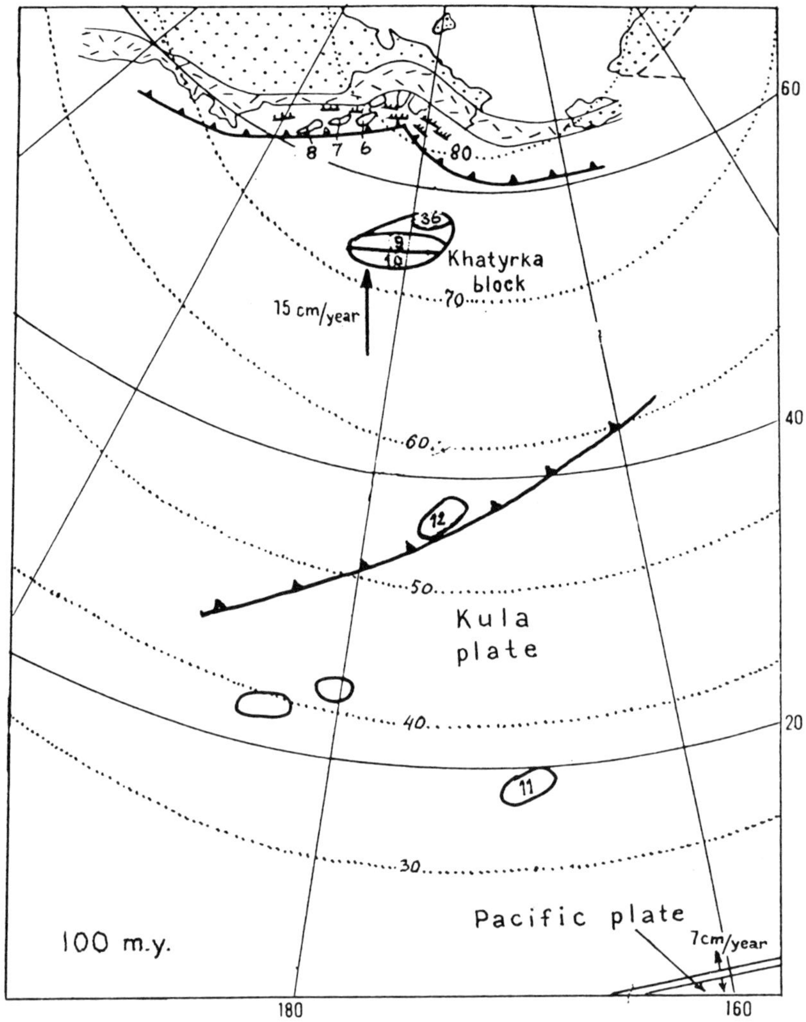

Fig. 143. Palinspastic reconstruction of the northwest Pacific region for 100 Ma. See Figure 141 for explanation.

less than 1400 km; the only available paleomagnetic results from the Lesser Kuriles show that in the Upper Cretaceous they were located 17° (i.e., 1900 km) south of Eurasia.

By the end of the Late Cretaceous, the Median Range (Sredinny block) of Kamchatka as well as the Vatyn volcanic massif split off the Shatsky Rise, approached the Olyutorsky arc, and collided with it. Behind the Olyutorsky arc there was a deep-water marginal sea basin, the remnants of which are preserved as the deep basin behind Bowers Ridge.

Early Eocene (55 Ma)

The Kula plate continued to subduct beneath Eurasia at a high velocity (15 cm/yr). In the Early Paleocene the new Anadyr-Bristol subduction zone and the corresponding arc came into existence along the Eurasian margin, taking up the Kula/Eurasia motion (Figure 145). At this time, the eastern part of the Olyutorsky arc, which included Bowers Ridge, became attached to the Kula plate and started to converge with the Anadyr-Bristol arc. As a consequence, a transform fault appeared, connecting the western end of the Anadyr-Bristol arc with the eastern end of the Olyutorsky arc.

The plate interactions in the NW Pacific Ocean began to reorganize at this time, for the Kula/Pacific spreading center was gradually consumed under the Olyutorsky arc. As a result, instead of the Kula plate, Eurasia started interacting with the Pacific plate and the velocity of convergence dropped. The subduction rate in the western part of the Olyutorsky arc was 7.0 cm/yr, and eventually decreased to 3-4 cm/yr. This part, now the Lesser Kuriles and the East Kamchatka peninsulas, moved together with the Pacific plate toward Eurasia. In the meantime, the eastern part of the Olyhutorsky arc continued to be part of the Kula plate.

Middle Eocene (45 Ma)

The Koryak-West Kamchatka and Aleutian arcs originated at this time (Figure 146).. They were evidently connected by a

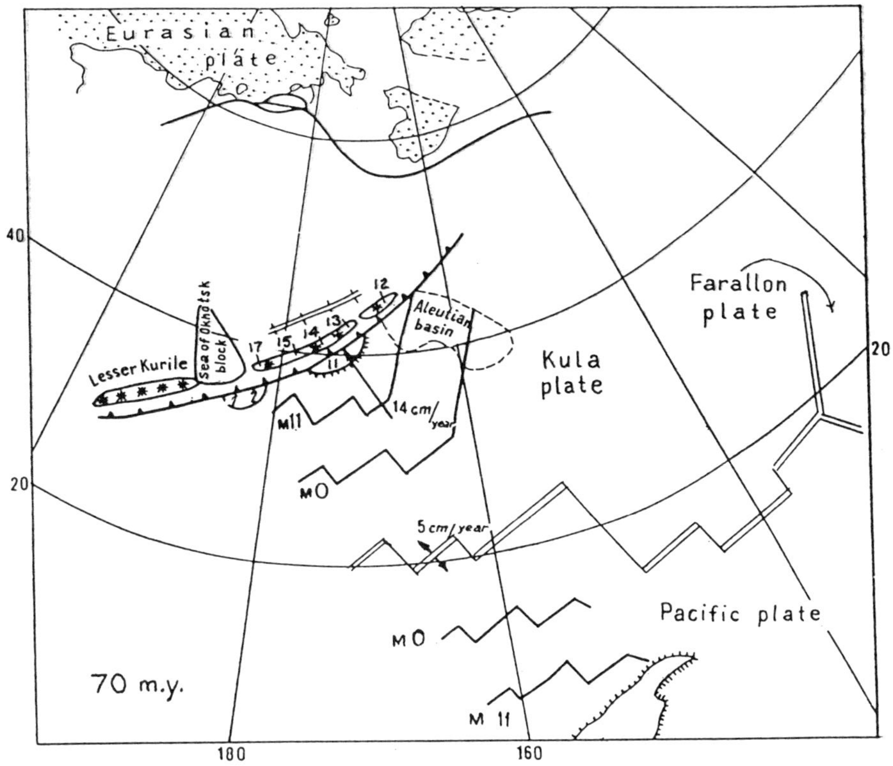

Fig. 144. Palinspastic reconstruction of the northwest Pacific region for 70 Ma. See Figure 141 for explanation.

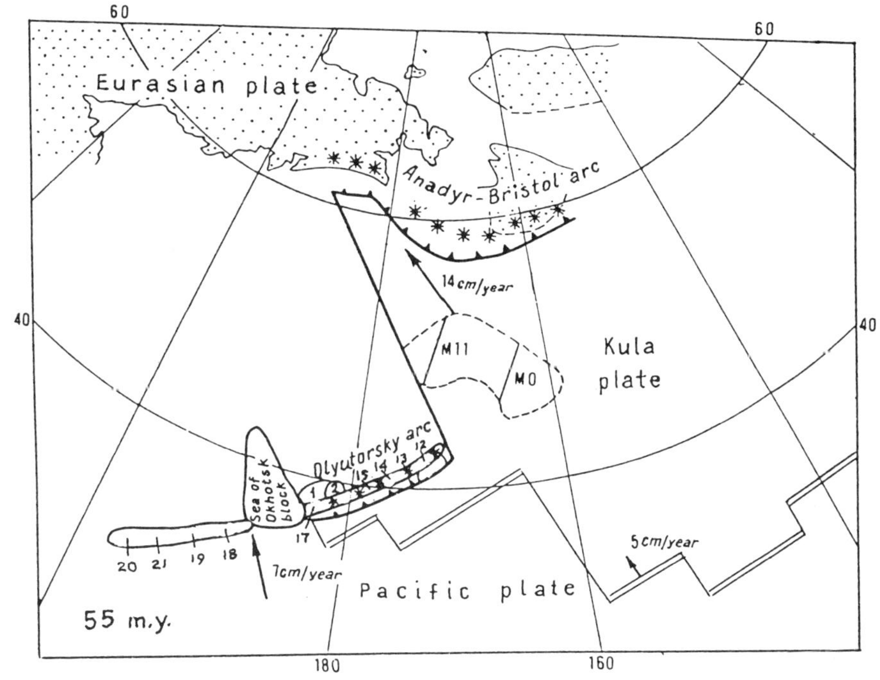

Fig. 145. Palinspastic reconstruction of the northwest Pacific region for 55 Ma. See Figure 141 for explanation.

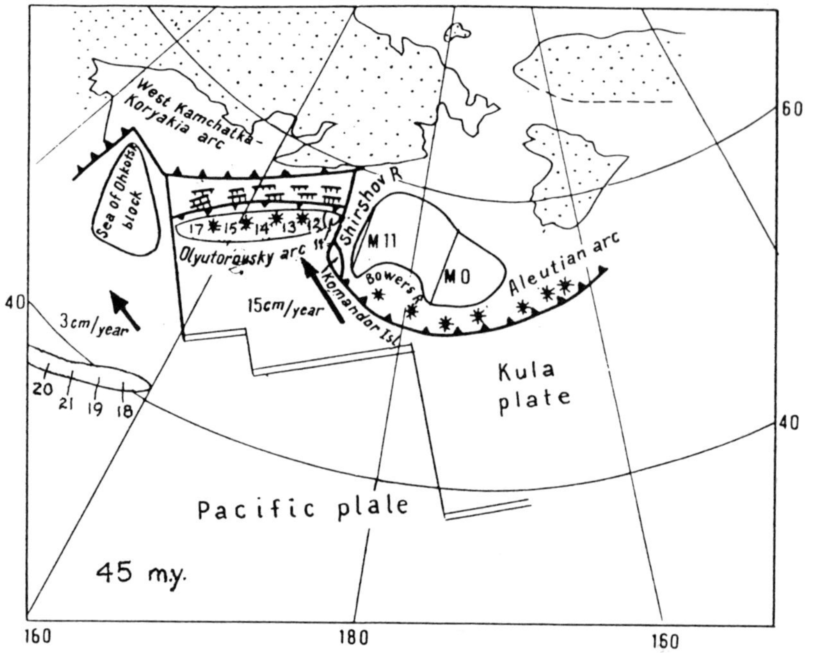

Fig. 146. Palinspastic reconstruction of the northwest Pacific region for 45 Ma. See Figure 141 for explanation.

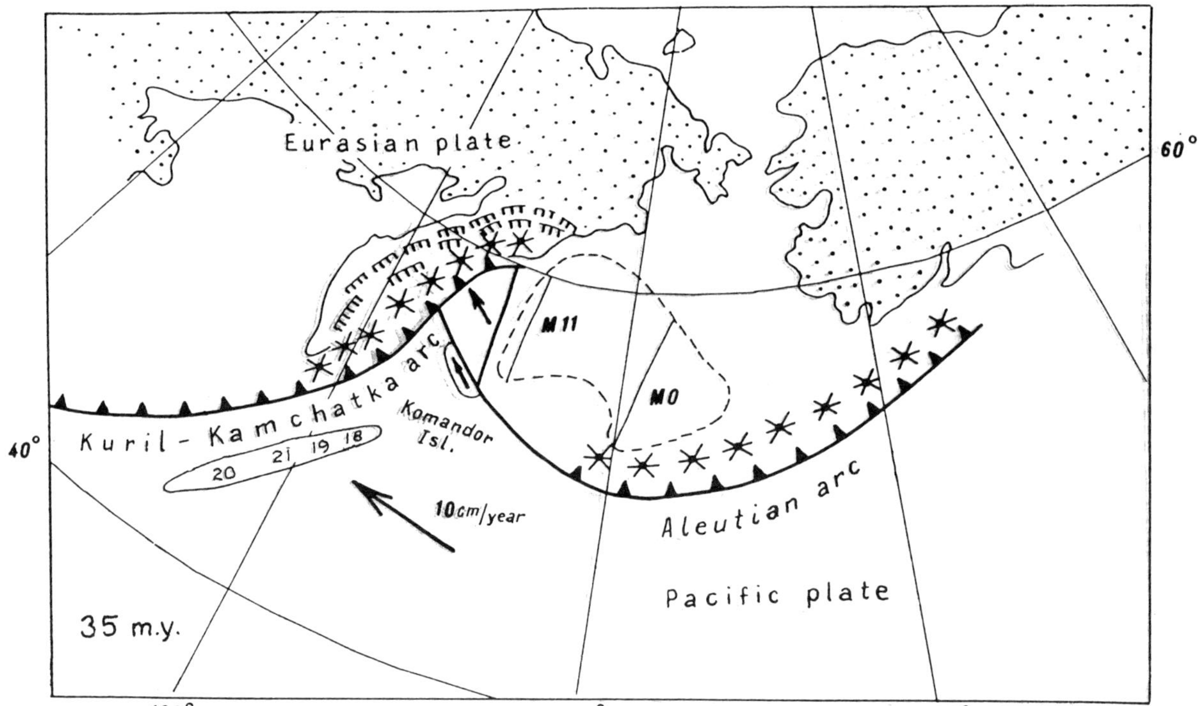

Fig. 147. Palinspastic reconstruction of the northwestPacific region for 35 Ma. See Figure 141 for explanation.

Moreover, the vector of the Pacific/North America relative motion was so oriented that there was an extensional component. Probably because of this, the Komandor Islands block was torn from the Aleutians and displaced westward as it moved with the Pacific plate.

Early Miocene (20 Ma)

This time was remarkable for two major events in the history of the region. First, the terranes of the Lesser Kuriles and East Kamchatka peninsulas collided with the Kurile-Kamchatka island arc. Second, the South Okhotsk back-arc basin opened behind the Kurile island arc (Figure 148). The Komandor Basin opened in the Late(?) Miocene.

Thus, the Koryak-Kamchatka belt was continuously an accretionary one because it was always near the active continental margin facing the Pacific Ocean. Throughout its whole history, at least since 150 Ma, Pacific Ocean plates constantly converged with this margin and, while descending into the mantle, left in the subduction zones "buoyant" accreted material which had been carried for many thousand kilometers. transform fault approximately coinciding with the Shirshov Ridge. The Komandor Islands were part of the Aleutian arc, for in both regions volcanism commenced simultaneously and similar sequences accumulated in both regions. At the same time, the intra-oceanic Olyutorsky arc was still active within the ocean and it converged with the Koryak-West Kamchatka arc at a velocity of 15 cm/yr. They collided in the Middle Eocene or slightly later. At this time the Aleutian arc had already isolated the Aleutian basin floor (a relic of the Kula plate), together with Bowers Ridge. Thus, the former Olyutorsky arc was divided at least into three parts: one part joined Eurasia as Bowers Ridge; the second part converged with Eurasia as the active Olyutorsky arc,; and the third (East Kamchatka peninsulas and Lesser Kuriles) was still part of the Pacific plate.

Oligocene (35 Ma)

By this time the Kula plate was entirely consumed in subduction zones, and all the events in the area under study were caused by interaction of the Pacific plate with Eurasia (Figure 147). The Pacific plate was moving relatively WNW at a rate of 7-10 cm/yr. The collision of the Olyutorsky arc with the Koryak-West Kamchatka arc was completed, and the subduction zone shifted to the Pacific Ocean, where the Kurile-Kamchatka arc originated above it. On the Kuriles, however, Oligocene volcanism is not reliably recorded, probably because a certain time is needed for the oceanic plate to subduct to depths where calc-alkaline magma can begin to form. Volcanic activity continued in the Aleutian arc, but was absent in its western part (Komandor Islands), where the relative plate boundary motion was approaching a transform type.

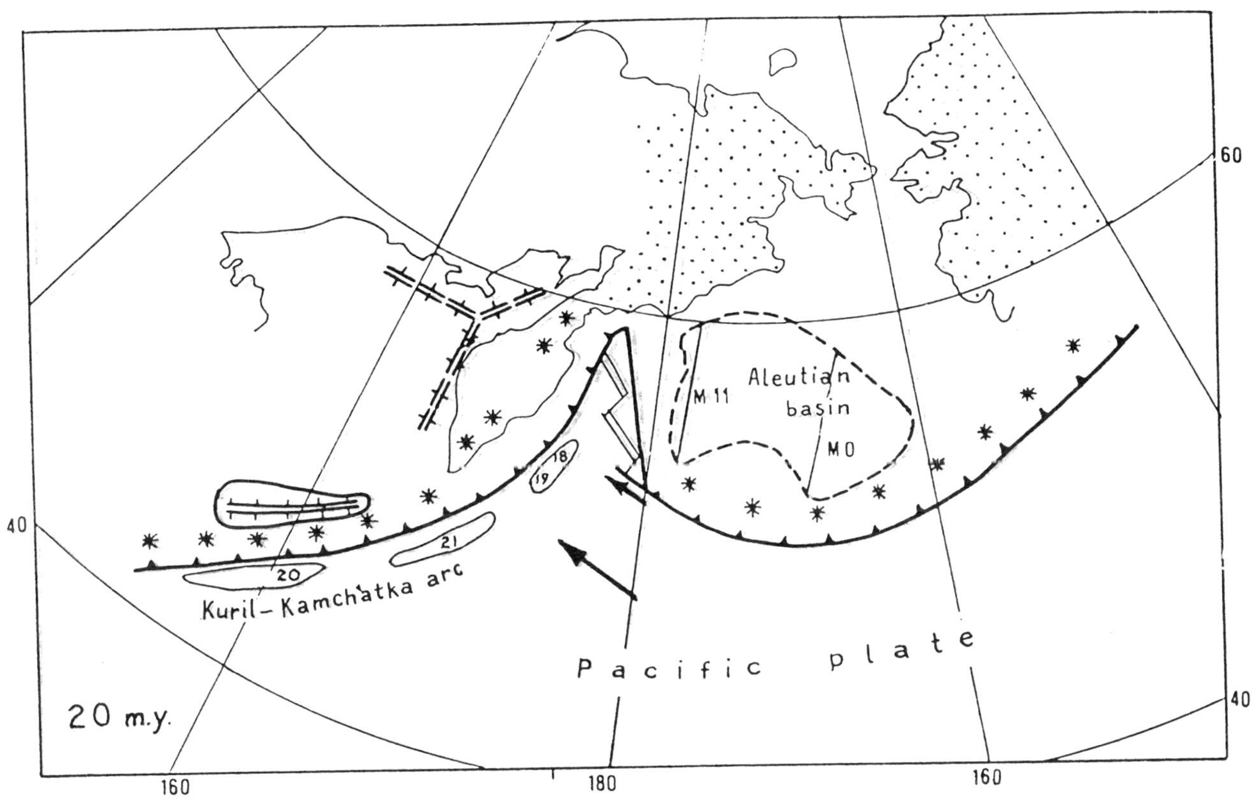

Fig. 148. Palinspastic reconstruction of the northwest Pacific region for 20 Ma. See Figure 141 for explanation.

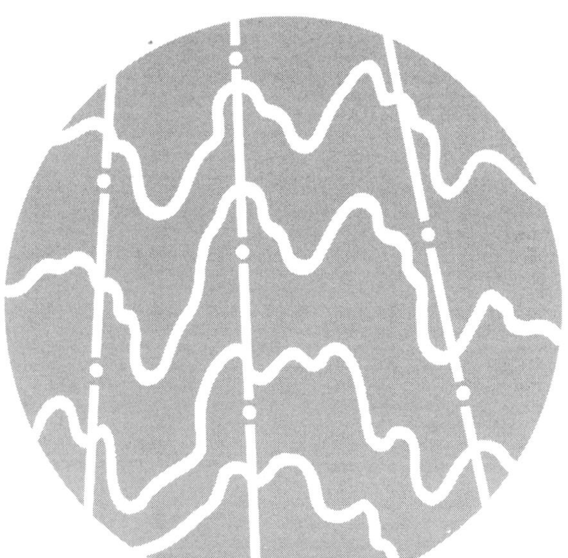

Chapter XI

ALPINE-HIMALAYAN FOLDBELT WITHIN THE USSR

GENERAL DESCRIPTION

The mountain ranges of the Alpine-Himalayan belt were created by the continental collision of Africa, Arabia, and India (i.e., Gondwanaland) with Eurasia. The USSR contains the northern part of the middle segment of the foldbelt, i.e., the East Carpathians, Dobrogea, Crimea, Caucasus, Kopeh-Dagh, and Pamirs. Parts of these fold chains surround the Black Sea and South Caspian basins, which are underlain by oceanic crust.

The tectonics and development of the Alpine-Himalayan belt within the USSR have been reinterpreted using mobilistic and particularly plate-tectonic concepts by Byzova and Beer [1974}, and Balla [1984, 1986] for the Carpathians; Khain [1975], Knipper [1975], Belov [1981, 1984, 1986], Adamia et al. [1974, 1982], Lordkipanidze [1980], Gamkrelidze [1986], Lomize [1987] and Baranov et al. [1976] for the Caucasus; and Shvolman [1977, 1980], Ruzhentsev [1986], Zakharov [1980], Burtman [1982], Legler and Przhiyalgovsky [1979] for the Pamirs. A plate-tectonic analysis of the evolution of the foldbelt was made recently by a Soviet-French team during a joint "Tethys" Project [Aubouin et al., 1986; Dercourt et al., 1986].

The main structural units of the middle segment of the Alpine-Himalayan belt are shown in Figure 149. The southern boundary, which is very distinct, lies beyond the USSR border and coincides with the thrust front stretching along the southern Zagros and the Himalayas. The northern boundary passing through the USSR territory is more diffuse, as the belt is rimmed here by Paleozoic structures separating it from the cratonic areas of Eurasia. These Paleozoic structures are largely hidden under the Late Paleozoic and Meso-Cenozoic sedimentary cover of the Scythian and Turan platforms. Early Mesozoic (Cimmerian) foldzones, in deformed Triassic and Late Paleozoic sequences, are typical for the northern rim of the belt.

Pre-Mesozoic rocks are exposed in many parts of the Alpine-Himalayan belt. They include the Precambrian and Paleozoic structures of the Peredovoy and Main Ranges of the Great Caucasus, the Dzirula massif of Georgia, the Daralagez (Nakhichevan) block of the Lesser Caucasus, Paleozoic structures of the North Pamirs and the Hindu-Kush, and the crystalline massif of the SW Pamirs. These Pre-Mesozoic outcrop areas are divided into two groups: (i) those with abundant plutonic and volcanic rocks, and with rocks which were deformed at the end of the Paleozoic during orogenics that are sometimes loosely called "Hercynian" and (ii) those with sedimentary, mainly carbonate Paleozoic sequences (e.g., the Daralagez block and South Pamirs), very similar to those of the Gondwanan passive margin.

Mesozoic and Cenozoic formations can be assigned either a Gondwanan or a Eurasian provenance, depending upon whether they formed on the southern or northern margin of the Tethys Ocean. Like the Paleozoic rocks, those from the southern margin are represented by purely sedimentary, largely carbonate sequences whereas volcanic rocks are typical for the northern margin.

Ophiolitic belts are a characteristic feature of the Alpine-Himalayan belt. They mark collisional sutures of different ages, e.g., Late Paleozoic (Peredovoy Range of the Great Caucasus), Triassic-Jurassic (Dobrogea, Crimea, North Caucasus, North Pamirs), Cretaceous (Central Pamirs, Lesser Caucasus), and Tertiary (Carpathians).

Carpathians

The Soviet Union includes a complete section of the Eastern Carpathians, from the foredeep in the east to the Pannonian depression in the west. The Carpathians consist of a series of nappes thrust one over another and displaced to the northeast toward the East European margin (Figure 150) [Byzova and Beer, 1974; Balla, 1984]. The nappes are subdivided into external, central, and internal. The external nappes (Skibovaya, Silesian Chernogorsky zones) are composed of flysch and molasse sequences ranging from Lower Cretaceous to Holocene in age, representing deposition throughout 100 Ma. Evidently this thick clastic sequence is a continental slope and rise sedimentary prism that accumulated near the East European passive margin. Deformation in the external zone started only in the Miocene and continues up to the present. In the central nappes (Burkut, Rakhov, and Smolyansky zones) Upper Jurassic oceanic-type rocks appear sporadically. The internal nappes (in the Pienin, Marmarosh, and Dragovsky zones) display a chaotic arrangement of various rock complexes. This belt exhibits so-called "klippen", blocks of Jurassic and Upper Triassic limestone and shale, Jurassic radiolarite, ultramafics, and other rocks incorporated in a flysch matrix. The internal nappes differ

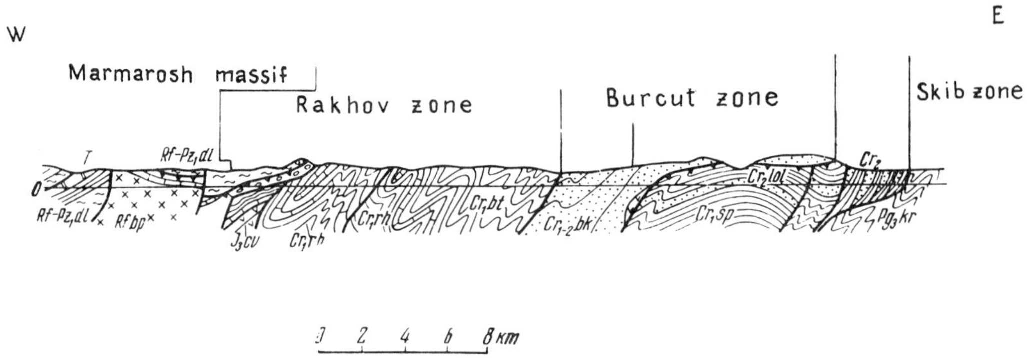

Fig. 150. Geological cross-section through the Carpathians, after Byzova and Beyer [1974].

from the external ones in that they contain older structures, formed between the Neocomian and Albian. The whole series of internal nappes was repeatedly deformed in the Late Tertiary, after the Oligocene. Farther west, behind the Carpathian chain, a zone of calc-alkaline Pliocene volcanics within the Trans-Carpathian (Pannonian) depression extends parallel to the Carpathians.

The paleomagnetic data on the Carpathian Cretaceous rocks [Burtman, 1984] and the plate-tectonic analysis of the Cenozoic development of this area [Balla, 1986] showed that the Carpathian loop did not exist,in pre-Miocene time (Figure 151). The loop is undoubtedly a secondary feature that originated after the Late Cenozoic collision of Africa with Europe, and it formed largely after nappe formation.

Dobrogea

In Dobrogea the basement crops out mainly south of the Danube River, beyond the USSR border. South Dobrogea is underlain mainly by Late Precambrian greenschists, which form the basement of the Moesia platform. Between the Late Precambrian metamorphic rocks of South Dobrogea and the East European platform is a band of deformed Paleozoic and Early Mesozoic rocks overlain by undeformed Cretaceous and Cenozoic deposits [Sljusar, 1984]. Widespread Permian and Triassic shale-slate successions are common in North Dobrogea. They include thick greenstones (spilites and diabases) having an oceanic geochemical signature. Paleozoic and Mesozoic sequences of North Dobrogea were thrust as a series of flat-lying nappes [Sljusar, 1984] northward onto the East European margin in the Late Jurassic. Folded structures of Dobrogea originated when the Moesia microcontinent collided with the East European continent. Available data show that the collision terminated by the Early Cretaceous.

Crimea

The basement of the Crimea is composed of flysch of the Taurian suite of Triassic-Lower Jurassic age. The flysch sequence is of continental-rise provenance, but the source of clastic material (south or north) is unknown. The flysch is overlain by a Lower Jurassic olistostrome (Eskiorda suite) with olistoliths of Permian limestone containing a warm-water Tethian fauna. Middle Jurassic calc-alkaline lavas are widespread in the Crimea. An important unconformity appears in West Crimea beneath thick Upper Jurassic conglomerate. Significantly, igneous and sedimentary rock fragments in the conglomerate are derived from the south, from the Pontian region in Turkey [Adamia et al., 1974], implying that the deep Black Sea basin did not exist at that time. The pre-Upper Jurassic folds beneath the unconformity mark the collision of some southern continent with the Crimean volcanic arc and with Eurasia. In the eastern part of Crimea the pre-Upper

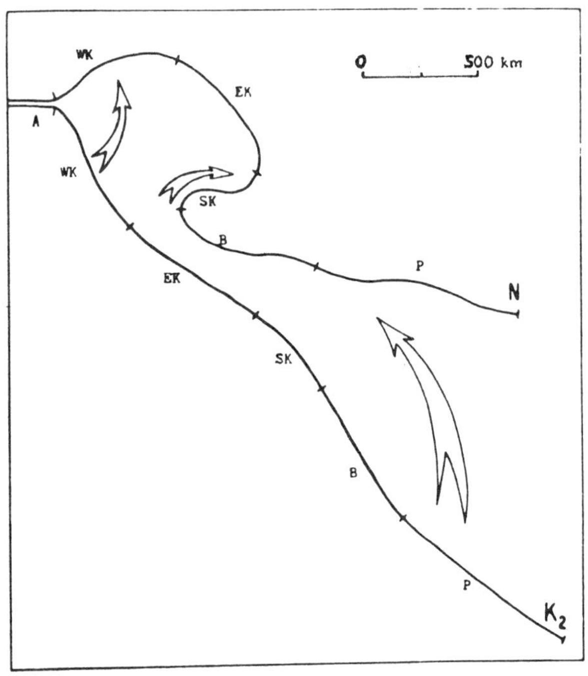

Fig. 151. Reconstruction of the Carpathian loop, after Burtman [1984]. Arrows show rotation directions. K_2 - position of the loop in the Late Cretaceous and N - in the Neogene; A - Eastern Alps, B - Balkans, EK - East Carpathians; P - Pontides, SK - South Carpathians, WK - West Carpathians.

Jurassic unconformity disppears, indicating that no continental collision occurred there.

Scythian Platform

The basement of the Scythian platform, which lies west of the Caspian Sea, comprises metamorphic sequences mainly composed of greenschists 410-470 Ma old [Belov, 1981]. The widespread Devonian and Lower Carboniferous deposits were deformed and intruded by granites between the Early and Middle Carboniferous. Upper Paleozoic deposits consist of dark-colored molasse as well as terrestrial andesite and rhyolite. Marine Permian shale and chert with interbedded limestone are also common. The Permian rocks are unconformably overlain by marine Triassic deposits, in places associated with calc-alkaline volcanics. The platform cover.sequence begins with Jurassic rocks.

Turan Platform and Mangyshlak

The Turan platform is situated east of the Caspian Sea and north of the Alpine-Himalayan belt, and is covered by undeformed Jurassic, Cretaceous, and Cenozoic deposits. The basement crops out in several isolated mountainous massifs just east of the Caspian Sea, in the Great and Lesser Balkhan, Tuarkyr, and Mangyshlak Mountains. Drilling data show that the greater part of the platform is underlain by lower-Middle Paleozoic foldbelts. They crop out near Kopeh-Dagh in the Tuarkyr Mountains and exhibit typical Eurasian Variscan-type volcanogenic-sedimentary series intruded by granites. This Variscan-type basement is unconformably overlain by Permian-Triassic molasse and intermediate and felsic volcanics, which are intruded by alaskite granites. The age of the volcanics ranges from 205 to 275 Ma, that of the alaskite granites from 180 to 207 Ma [Belov, 1981].

In Mangyshlak (northwest part of the Turan platform), a strongly folded Triassic and Permian flysch-like series crops out. The flysch series is unconformably overlain by Lower Jurassic rocks. These sequences evidently formed in a continental-rise environment along the margins of a basin that separated the Karabugaz massif (microcontinent) of the Turan platform from the Ustyurt massif of the East European platform.

Caucasus

Cross-sections through the Caucasus display the evolution of the Mesozoic active margin of the Tethys Ocean.

The Great Caucasus is built of a number of nappes thrust from north to south (Figure 152). Particularly remarkable is the Great Caucasus south slope thrust along which the Cretaceous rocks were thrust over Pliocene deposits. Continuing nappe formation and tectonic activity are reflected in present-day seismicity. The zone of earthquake hypocenters running along the southern slope of the Great Caucasus dips northward down to a depth of 100 km. Fault plane solutions indicate compression across the Caucasus [Vardapetyan, 1979]. The highest peaks of the Caucasus--Elbrus and Kazbek--are Pliocene-Pleistocene volcanoes erupting lavas of calc-alkaline composition. All these data show quite recent subduction from the south underneath the Great Caucasus.

The core of the Great Caucasus consists of the Precambrian and Paleozoic sequences of the Main and Peredovoy Ranges. The Main Range displays highly metamorphosed rocks (Makersky series), the Rb-Sr age of metamorphism being 790 Ma [Belov, 1981]. The basement is intruded by plagiogranites 360-370 Ma old and is overlain by Late Paleozoic marine deposits. These are unconformably succeeded by the Lower Jurassic rocks. The Peredovoy Range, north of the Main Range, contains Paleozoic ophiolites and island-arc complexes, which occur in a series of nappes [Khain, 1979]. A major unconformity underlies Upper Visean rocks. The formations above the unconformity include continental coal-bearing Carboniferous deposits and Permian red beds. Calc-alkaline volcanic rocks of Upper Carboniferous and Permian age are locally present. Thus, the Upper Paleozoic section has a typically European aspect.

The metamorphic complexes of the Main Range may be considered as the remnants of the basement of a microcontinent [called Makersky by Baranov et al., 1976]. The ophiolites and island-arc complexes of the Peredovoy Range with their tightly folded structure undoubtedly represent a suture zone. A magnetic high coincides with this suture zone; it can be traced farther east to the Caspian Sea and toward the northern front of the Kopeh-Dagh near Krasnovodsk. The suture is apparently a continental collision zone, possibly between the Makersky microcontinent and the continental basement of the Scythian platform.

Pre-Jurassic rocks are also known in the southern slope of the Great Caucasus in Svanetia where the strongly deformed shale-slate Diz series crops out below the Jurassic [Somin, 1971; Belov, 1981]. The Diz series contains Devonian-Triassic fossils. From its composition and structural features Kazmin (oral communication, 1986) interpreted it to be a former accretionary prism that originated near the continent in the north. Another alternative is that the Diz series can be considered to be a suspect (possibly exotic) terrane.

The major part of the Great Caucasus is underlain by Jurassic and Cretaceous rocks. The Lower and Middle Jurassic sequences have two particular features: First, they consist largely of shales and slates. Second, the shale sequences contain voluminous lavas ca. 3 km thick. Volcanic rocks occur at several stratigraphic levels. The oldest are calc-alkaline in composition and are Sinemurian-Pliensbachian (Early Jurassic) in age. According to Lordkipanidze [1980], they are remnants of a Great Caucasian island arc. In Toarcian, Aalenian, and Bajocian times (Early and Middle Jurassic), basalts of the Goitkh suite were abundantly erupted. These occur in the central Great Caucasus. The basalts are tholeiites and some are MORB-type [Lordkipanidze, 1980]. They are associated with rhyolite, and this basalt-rhyolite series is interpreted to indicate extension related to the formation of the Great Caucasian sedimentary basin [Adamia et al., 1977, 1982; Gamkrelidze,.1982]. The post-Middle Jurassic history is different for the southern and northern parts of the Great Caucasus. In a narrow band along the northern slope, there is a zone of pre-Late Jurassic deformation marked by an angular unconformity where Upper Jurassic limestones lie on imbricated Lower and Middle Jurassic shales. On the southern, much larger portion of the Great Caucasus, sedimentation was continuous throughout the Jurassic, Cretaceous, and Paleogene. Detailed investigations of facies distribution (Figure 153) [Kopp and Shcherba, 1985; Dotduev, 1986] show convincingly that before the end of the Paleogene no topograhic uplift existed and no clastics were supplied from

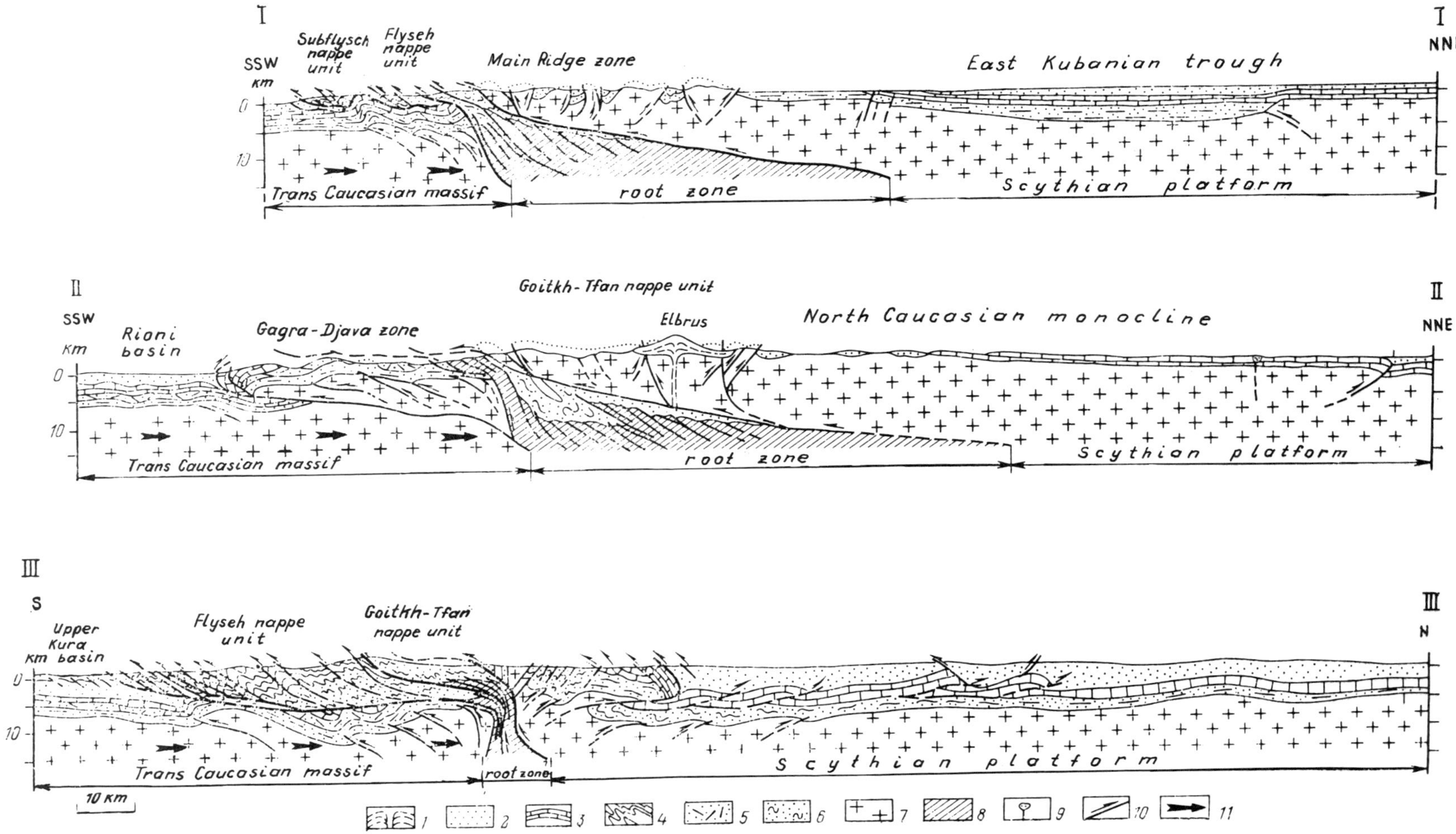

Fig. 152. Cross-section through the Great Caucasus, after Dotduyev [1986]. I-I - Sochi transect, II-II - Elbrus transect, III-III - Terek section.
1, Recent lava of Elbrus; 2, molasse; 3, Eocene shallow-marine sediments; 4, flysch; 5, Bajocian and Bathonian deposits of the Trans-Caucasian massif; 6, Lower and Middle Jurassic rocks; 7, pre-Jurassic basement; 8, nappe root zone; 9, Alpine pluton; 10, thrust; 11, direction of tectonic movements.

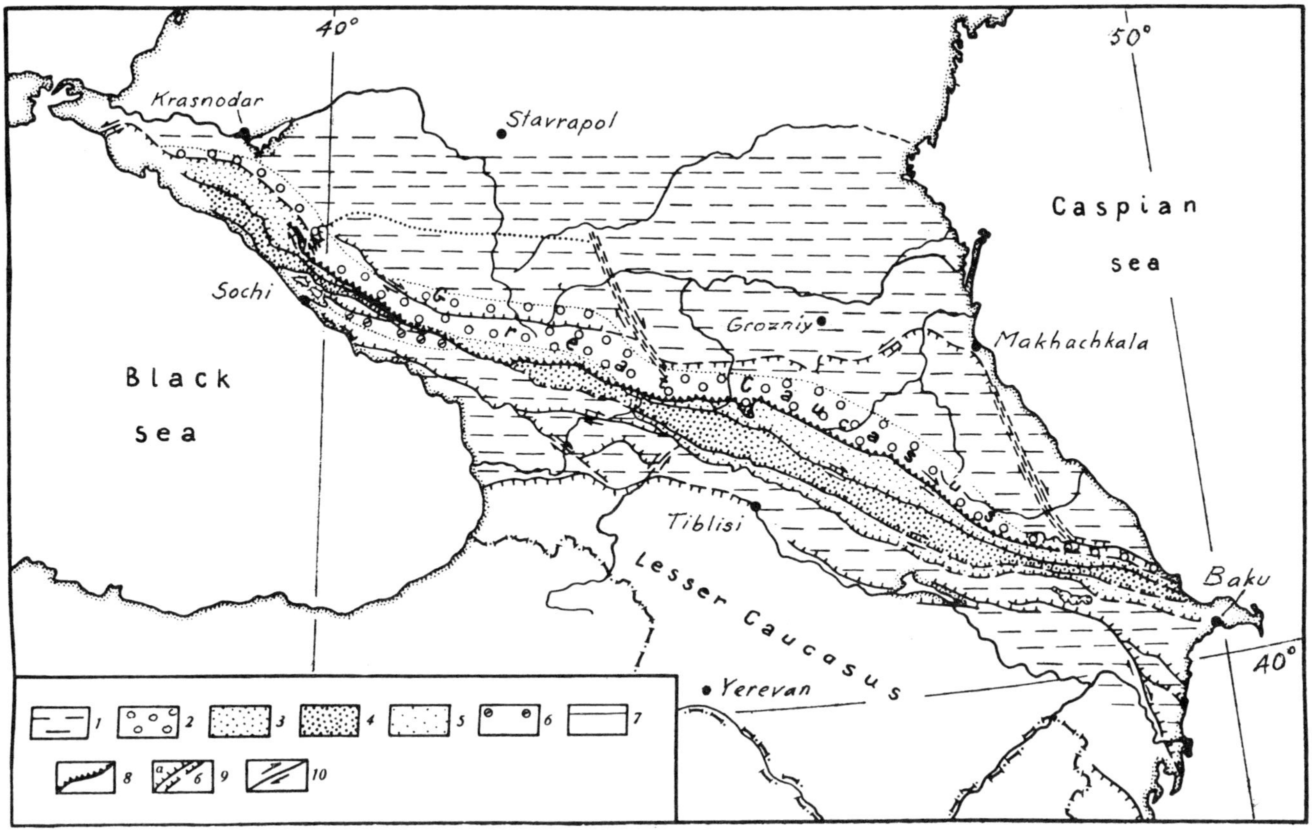

Fig. 153. Facies distribution for the Late Jurassic to Eocene in the Great Caucasus, after Doitduyev [1986]. 1, North Caucasian lagoon (shallow marine); 2, Shakhdag Fikhtin barrier reef; 3, north subflysch zone; 4, Novorossisk flysch zone; 5, south subflysch zone; 6, Akhtsu-Dzykhra barrier reef; 7, Trans-Caucasian lagoon (shallow marine); 8 Main Great Caucasian thrust; 9, thrust; 10, strike-slip fault.

the Great Caucasus (Figure 154). The uplift of the Great Caucasus began only in the middle-Late Miocene (Sarmatian) and it has continued to the present.

Thus, the Great Caucasus formed on the site of a wide marine basin which had resulted from extension in the Early and Middle Jurassic and which was filled with thick clastic deposits up to the Middle Miocene. This basin came into existence as a marginal sea in the rear of the Lesser Caucasian island arc.

The next structural unit of the Caucasus to the south is the Dzirula (or Trans-Caucasian) massif. Its basement consists of rocks metamorphosed to the amphibolite and greenshist facies. Schists are accompanied by marbles with an Early Cambrian fauna, as well as by conglomerates with plagiogranite pebbles 517 Ma old [Gorokhov et al., 1978] and by serpentinite bodies. As the different rock types are separated by shear zones and mylonites, their original relations are unknown. Most likely they are part of an accretionary subduction melange. Rb-Sr data show that the metamorphic age of schists is 430-410 Ma [Gorokhov et al., 1978]. The metamorphic complex is unconformably overlain by Carboniferous clastic and coal-bearing deposits and is intruded by Late Paleozoic alaskite granites. The Dzirula massif is very similar to the Hercynian structures of Europe and is essentially different from microcontinents of Gondwanan origin.

One of the most significant structural units of the Trans-Caucasian area is the Lesser Caucasian (Somkheto-Agdam) volcanic arc. The arc formed in the time span from the Jurassic to the Early Cretaceous, just before it collided with the Daralagez block. It is characterized by a basalt-andesite-dacite-rhyolite series. In the south, the arc lavas are associated with relatively deep-water deposits, whereas in the north they are supplanted by shallower-water volcanogenic clastic series of marginal-sea type. The Saatla deep drillhole situated NE of the Lesser Caucasian arc penetrated almost 4 km of calc-alkaline volcanics 146 Ma old, hidden under Lower Cretaceous and Tertiray sedimentary sequences [Shakhilibeily et al., 1988]. This means that the Lesser Caucasian arc continues eastward to the South Caspian basin.

The main structural boundary of Mesozoic age in the Caucasus is the Sevan-Akera ophiolite zone [Knipper, 1975; Satiani, 1984]. Knipper [1975] showed that the Lesser Caucasus ophiolites belong to a nappe unit that was broken by subsequent deformation into a number of slabs. The main deformation was associated with continental collision in Coniacian (Late Cretaceous) time. Deep-water cherty sedi-

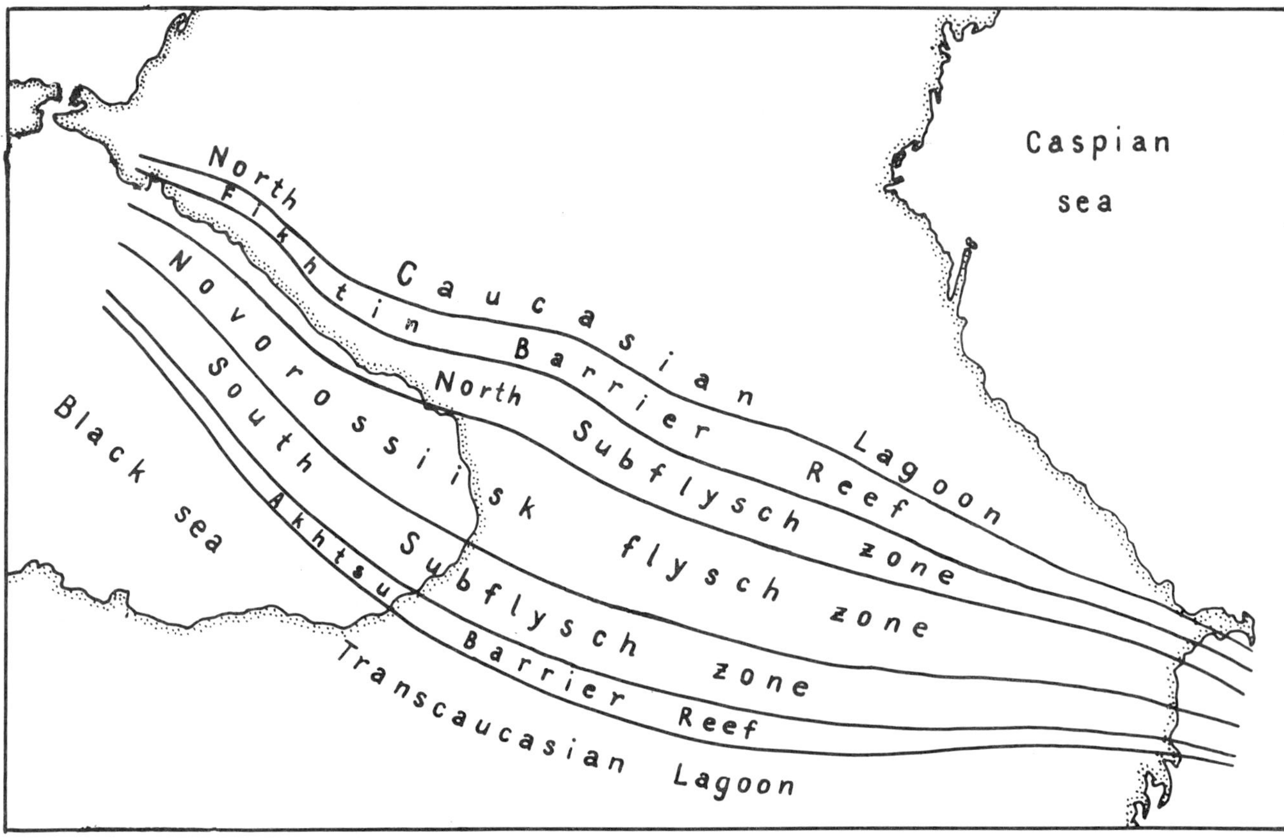

Fig. 154. Reconstruction of the initial position of the Late Jurassic-Eocene facies zones in the Great Caucasus, after Dotduyev [1986]. The northern slope of the Caucasus is arbitrarily taken as fixed. For explanation see Figure 153.

ments and turbidites of the Late Jurassic and Early Cretaceous are included in the ophiolite slabs.

The Daralagez block, situated south of the suture has a Paleozoic core of sedimentary rocks, mostly limestones. A Middle to Late Carboniferous hiatus is marked by an unconformity between overlying Permian and Lower Triassic deposits and underlying Devonian and Lower Carboniferous rocks. The Permian is represented by algal and foraminiferal limestones typical of the Tethys and characteristic of the tropical zone. The entire succession up to the Lower Triassic is significantly different from the Paleozoic and Lower Mesozoic of more northerly regions of the Caucasus, but is very similar to rocks of the southern Tethyan margin.

After collision of the Daralagez block with the Lesser Caucasian arc, the whole area of the Lesser Caucasus was occupied by the newly formed Adzharo-Trialetan arc. The arc originated in the Campanian, but the main episode of volcanism occurred in the Eocene. Adamia [1974, 1982] and Lordkipanidze [1980] found that the Adzharo-Trialetan arc has a pronounced polarity, the alkalinity increasing from south to north. They also showed that a back-arc rift can be reconstructed from a bimodal basalt-rhyolite series that formed behind the Adzharo-Trialetan arc in latest Cretaceous and Eocene time.

A new stage of volcanic activity began in the Pliocene when the Armenian highland was flooded by calc-alkaline basalts and andesites, implying that a new convergence-related event had occurred.

Kopeh-Dagh

The Kopeh-Dagh Range drops off in a high tectonic scarp which slopes steeply northward toward the Turan platform. The seismic belt which runs east from the Great Caucasus coincides with the scarp. Focal-mechanism solutions [Sborschikov et al., 1981] indicate significant dextral motion of the Kopeh-Dagh relative to the Turan platform. This fault scarp follows an early suture, because it separates two significantly different Meso-Cenozoic successions: (i) a northern one, largely concealed under the Turanian platform cover, represented by sandstone-slate sequences deformed prior to the Cretaceous, and (ii) a southern one,--the Kopeh-Dagh succession in which no folding occurred from the Jurassic to the Early Tertiary. This second succession consists of carbonate and terrigenous Jurassic to Miocene deposits that were folded only in the Late Cenozoic. Evidence suggests [Prozorovsky, 1985] that the Kopeh-Dagh structure consists either of a single great nappe or a series of smaller nappes thrust northward over the cover and basement of the Turan platform. On the whole the Kopeh-Dagh may be regarded as a fold chain that originated on the site of a former Mesozoic and Early Cenozoic passive margin, due to Late Cenozoic motion of the Iranian block relative to Eurasia.

Black Sea and South Caspian Sea Basins

Oceanic-type crust with V_p = 6.8-7 km/s underlies sedimentary sequences 15 km thick in the Black Sea and 20 km thick in the South Caspian basin [Gegelyants et al., 1958; Neprochnov, 1966].

Three sedimentary troughs open from dry land toward the Black Sea: the Karkinitsky, Nizhnekamchiisky, and Rionsky troughs (Figures 155, 156). The Karkinitsky trough runs westward from the Crimea onto the northern Black Sea shelf. It is filled by Cretaceous (possibly including Upper Jurassic) and Cenozoic deposits lying unconformably on the Variscan-type basement of the Scythian platform and the Cimmerian basement of the Crimea. Multi-channel seismic reflection data [Tugolesov et al., 1985] indicate that the basement of the Karkinitsky trough is broken by a number of listric faults into blocks that are back-tilted northward, thus displaying a typical pattern of extension and thinning of the continental crust. The Nizhnekamchiisky trough opens into the Black Sea from the west. It contains Late Cretaceous, Paleogene and Neogene deposits [Bomarsky, 1979]. The Rionsky depression opens into the eastern Black Sea. The Ochamchiri borehole was drilled on the Black Sea coast through 4 km of sedimentary rocks. It did not reach the basement, but stopped in Middle Jurassic shale. The section resembles sedimentary sequences of the Dzirula massif. The Chaldadidi hole drilled farther south penetrated the Cenozoic and Upper Cretaceous sediments, Albian volcanics, and Lower Cretaceous limestones Drilling was stopped in Upper Jurassic basalt at 4 km depth. Studies by Lordkipanidze [1980] indicate that the lavas belong to the alkaline basaltic series and indicate an extensional environment.

Detailed multi-channel reflection mapping of the Black Sea [Tugolesov et al., 1985] has revealed two deep basins in the central part of the sea: the Western and Eastern Black Sea basins (Figure 156). The basement subsided to a depth of 16 km in the Western basin and to 13 km in the Eastern basin.

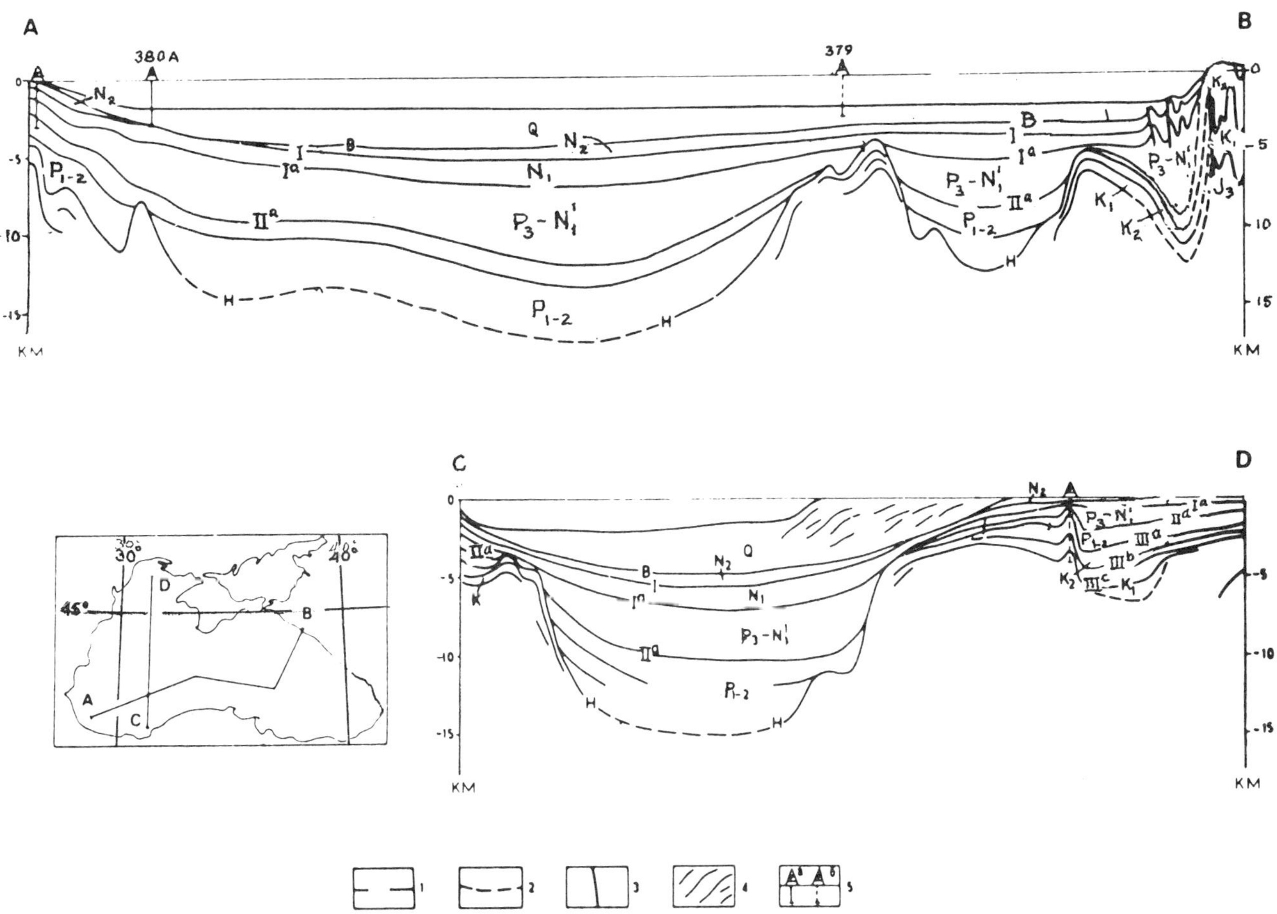

Fig. 155. Interpretive geological cross-section through the Black Sea basin inferred from multi-channel seismic and drilling data, after Tugolesov et al [1985]. See insert for location of cross-sections.
1, Reflector; 2, uncertain reflector; 3, fault; 4, cross-bedding within the Danube delta deposit; 5, drill-site.

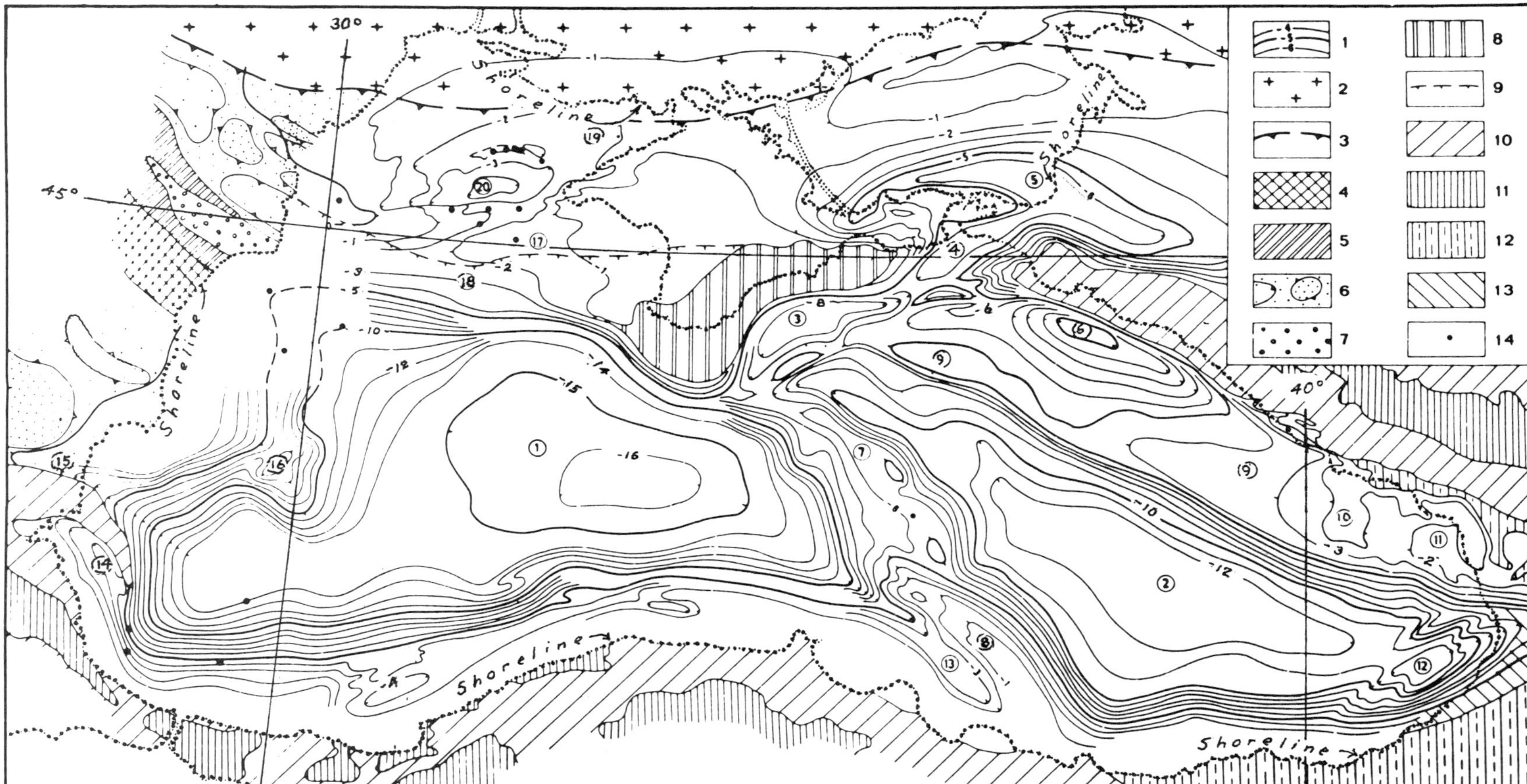

Fig. 156. Structure of the Black Sea, after Tugolesov et al. [1985].
1, Isobaths (in km) of Cretaceous-Tertiary boundary; 2, East European platform; 3, limit of East European platform; 4, Riphean rocks of Dobrogea; 5, Paleozoic rocks of Dobrogea; 6, Upper Paleozoic rocks of Dobrogea; 7, Maindanian graben; 8, Crimea; 9, limit of the Crimea-Dobrogea Triassic trough; 10, Alpides, 11, Paleozoic of Peredovoi Ridge in the Greart Caucasus; 12, Dzirula massif; 13, Adjaro-Trialetian zone; 14, boreholes.
Figures in circles: 1, West Black Sea basin; 2, East Black Sea basin; 3, Sorokin trough; 4, Kerch-Tamanian trough; 5, Indolo-Kubanian trough; 6, Tuapse trough; 7, Andrusov Swell; 8, Arkhangelsky Swell; 9, Shatsky Swell; 10, Gudauta Swell; 11, Ochamchiri Swell; 12, Gurian trough; 13, Sinopan trough; 14, Burgas trough; 15, Lower Kemchian trough; 16, Polshkov Rise; 17, Kalamitsky Swell; 18, South Kalamitsky scrap; 19, Karkinitsky trough; 20, Mikhailovsky depression.

The two basins are separated by the Central Uplift consisting of the Arkhangelsky swell in the south and the Andrusov swell in the north. Between the Eastern basin and the coast there is another uplift of the basement--the Shatsky swell or Gudauta uplift. Despite a reduced thickness of sediments, more reflectors appear on the uplifts than in the deep basins. The deeper reflectors can be fairly well correlated with stratigraphic units which were penetrated by drill holes on the Black Sea coast and which include the Lower and Upper Cretaceous boundary and extend down to the top of Upper Jurassic basalt. Thus the uplifts contain older rocks than the deep basins. The Gudauta uplift and Shatsky swell represent an underwater continuation of the Dzirula massif. A great magnetic high mapped over these uplifts evidently reflects the presence of volcanic rocks [Ross et al., 1974]. The uplifts may be regarded as a link between the contemporaneous Middle Jurassic volcanic sequences of the Dzirula massif and Crimea. Seismic reflection has revealed zones of strong deformation with chaotic piling of blocks along the Crimean and Caucasian coasts. These chaotic structures may be interpreted as an accretionary prism associated with subduction of the Black Sea crust underneath the Crimea and Caucasus [Ushakov et al., 1977]. Continued convergence of the Great Cacasus with blocks from the south is confirmed by deep seismicity beneath the Great Caucasus. This seismicity becomes deeper from south to north, and reaches depths of 100 km [Khalilov et al., 1987].

The floor of the Black Sea deep basins formed during three extensional events. During the first (Early Jurassic) event, the Great Caucasian basin was formed. Its sedimentary sequences descend under the Black Sea in the Tuapse trough. The second event occurred in the Late Jurassic and Early Cretaceous when alkaline basalts were erupted on the basement of the Rionsky depression. Extension and formation of the Karkinitsky trough was also related to this event. Evidently the Karkinitsky trough and Andrusov and Shatsky swells have continental crust that was thinned during rifting and subsequently subsided below sea level. The third, evidently main episode of extension took place in the Late Cretaceous and perhaps in the early Paleogene, when two Black Sea deep basins formed by back-arc spreading behind the Adzharo-Trialetan volcanic arc.

According to deep seismic sounding, the South Caspian basin crust consists of (i) an upper layer 20 km thick with a seismic P-wave velocity of 3.5-4 km/s, and (ii) a lower layer with a P-wave velocity of 6.6 km/s [Gegelyants et al., 1958]. Under the Turkmenian east Caspian coast the thickness of the lower layer is 6-7 km [Kurbanov and Rzhanitsin, 1982]. The upper layer is sedimentary and the lower one is oceanic basement. In the North Caspian Sea the crustal thickness increases to 40 km due to the presence of an intermediate layer with a seismic P-wave velocity of 6 km/s.

The South Caspian basin is characterized by large negative free air gravity anomalies (from 150 to 250 mgal) that indicate the lack of isostatic compensation in the basin [Artemyhev, 1975] and imply continuing subsidence. Drilling in Azerbaidzhan and Turkmenia [Alikhanov, 1978] did not reach the basement but revealed a 2500-3500 m sequence of shallow-water sediments ranging from Late Jurassic to Early Pliocene in age. A sharp change in sedimentation occurred in Middle Pliocene time: a red bed sequence (1500-3500 m thick) that is now oil-producing began to accumulate. It is overlain by Upper Pliocene and Quaternary deposits. The sediments, including those of Pleistocene age, arc intensely folded and involved in nappes. Seismic reflection allows the tracing of upper sedimentary horizons (above the oil-producing suite) from the shore to deep parts of the basin. The thickness of sediments above the oil-producing suite increases to 6000 m in the center of the basin. There is no evidence that the Jurassic and Cretaceous deposits revealed by drill holes in the Kura depression and in Turkmenia become thinner toward the sea; hence it is asssumed that the oceanic crust in the South Caspian basin formed in the Late Jurassic as a result of back-arc spreading behind the Lesser Caucasian island arc. Seismic reflection shows that the youngest sedimentary layers of the South Caspian basin are deformed into a number of folds trending NNW-SSE. Apparently deformation continues up to the present, because this fold zone coincides with a seismic belt. Focal mechanism solutions indicate compression perpendicular to the fold trend [Vardapetyan, 1979].

Pamirs

The structure of the Pamirs resulted from the India/Eurasia collision, so in this respect the Pamirs are similar to the Himalayas and South Tibet. Fold structures developed on the site of the widest part of the Tethys Ocean (6000 km), yet at present, the belt is only 400 km wide (Figure 157). Consequently, ca. 5000-5500 km of Tethys oceanic crust must have been consumed.

At present in the Pamirs area the Indian plate moves 4.4 cm/yr relative to Eurasia, at an azimuth of 345°. Zonenshain and Savostin [1979] believe that some part (approximately half) of this movement is taken up by frontal southward overthrusting in the Punjab Himalayas while the other half creates convergence in the Pamirs and Tien Shan. This interpretation is supported by repeated geodetic measurements in the Garm polygon area near the Pamir/Tien Shan boundary. The measurements show that the Pamirs and Tien Shan converge at 2 cm/yr [Pevnev et al., 1978], so we may suppose that 25 Ma ago the distance between the Pamirs and Tien Shan was no less than 500 km (Figure 158).

The structure of the Pamirs is arcuate, convex to the north. According to paleomagnetic data [Bazhenov and Burtman, 1982] the arc is a secondary feature (Figure 159), and in the Late Cretaceous it was convex to the south rather than to the north. The recent Pamir arc originated only in Neogene time.

The Pamirs consist of a number of nappes thrust one over another from south to north [Ruzhentsev et al., 1977; Belov, 1981; Shvolman, 1977, 1980; Pashkov and Shvolman, 1979]; the main structural zones are shown in Figure 157. The Pamirs may be divided into Northern, Central, and Southern segments. The northern segment underwent Late Paleozoic deformation (sometimes called "Variscan"), which is not recorded in the Central and Southern Pamirs. A very significant suture lies between the Northern and Central Pamirs along the Tanymas fault zone, which continues to the west into the Hindu Kush fault zone and to the east into the Dzhang-Tang zone of Central Tibet. North of this suture the nappes contain geological complexes which, in the Late Paleozoic and Mesozoic, were situtated near the northern margin of the Tethys Ocean. The nappes south of the Tanymas suture are composed of complexes of Gondwanan type.

Lower Carboniferous volcanics are characteristic of the Northern Pamirs. They contain both tholeiites close to MORB in composition, and calc-alkaline basalt-andesite-dacitic series. Ultramafics are locally associated with the

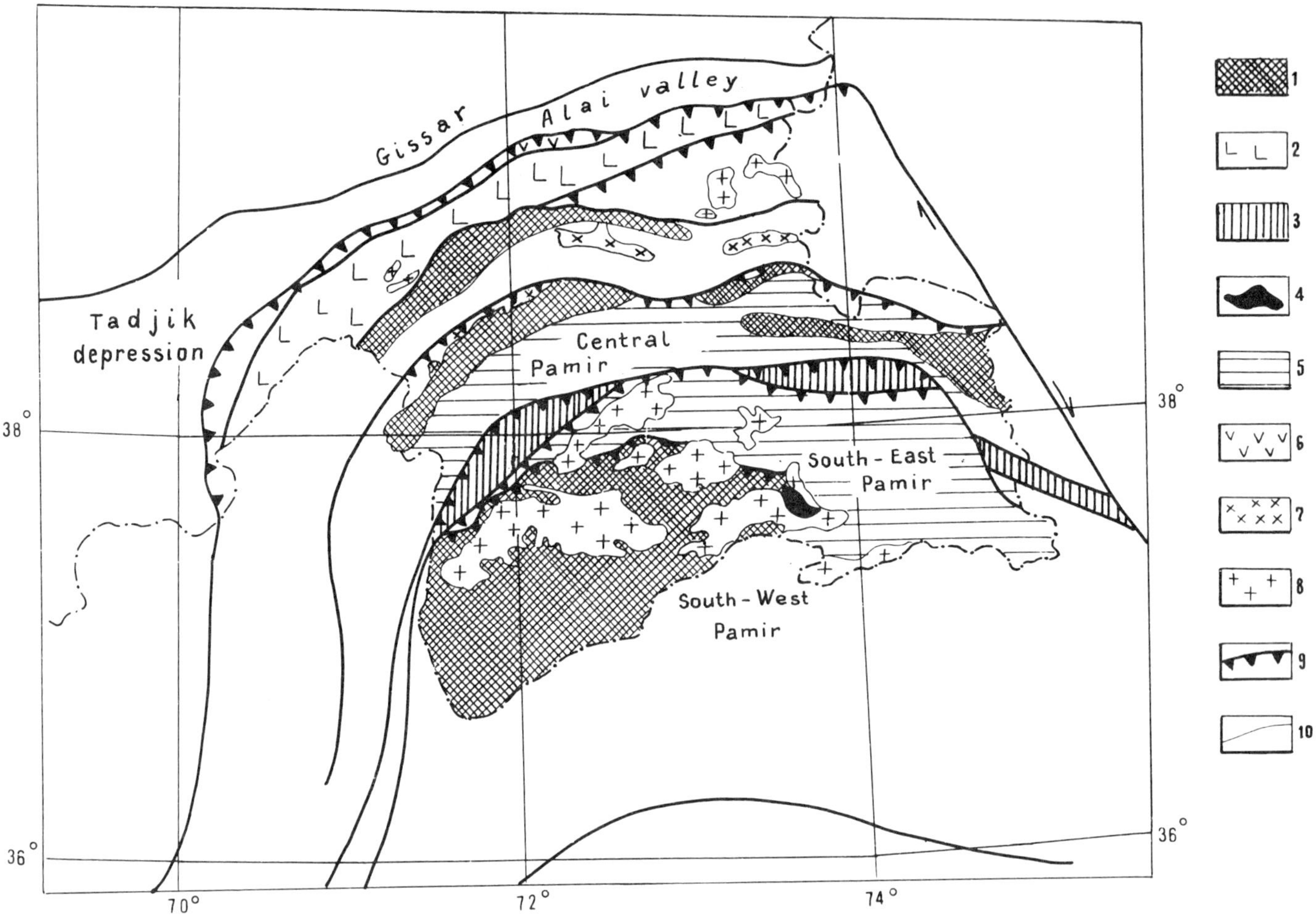

Fig. 157. Structure of the Pamirs.
1, Precambrian basement; 2, Paleozoic oceanic and island-arc complexes; 3, Mesozoic Rushan-Pshart zone; 4, Mesozoic ophiolite of the Bashbumbez window; 5, rock complexes of Gondwanan provenance; 6, Triassic calc-alkaline volcanics; 7, Lower Jurassic Karakul granite; 8, Cretaceous granite of the South Pamirs; 9, main suture; 10, continuation of suture beyond the USSR boundary.

volcanics. Evidently both oceanic and island-arc complexes are present in the Carboniferous section. In these respects the rocks of the Northern Pamirs resemble those of the Gissar area in the South Tien Shan Range; in both areas, remnants of the Paleo-Tethys Ocean floor are preserved. The volcanics are unconformably overlain by flysch-greywacke sequences of Middle to Upper Carboniferous and Permian age, and these sequences include olistostromes. All of the foregoing were deformed in latest Paleozoic time.

In the Northern Pamirs the Kurgovat subzone is of particular interest. In it, Precambrian crystalline schist, paragneiss, and marble are exposed. These rocks are unconformably overlain by Middle to Upper Carboniferous limestone, Permian flysch, and Triassic olistostromes that contain marble blocks with Upper Permian fusulinids. This subzone was evidently a microcontinent with its own sedimentary cover, and its composition is noticeably different from that of adjacent units. It may be regarded as an exotic terrane.

Zaalay and Darvazsky Ridges of the Northern Pamirs also contain a narrow band of Middle to Upper Triassic basaltic andesite. The andesite lies unconformably on Paleozoic rocks and consists of a typical calc-alkaline series. The Triassic volcanics continue into North Afghanistan, the Hindu Kush, and the Parapamisos. If the Carboniferous sequences are correctly interpreted as a mix of oceanic and island-arc complexes of the paleo-Tethys, the Triassic volcanics represent an active margin of Eurasia under which the Paleo-Tethys Ocean floor was being subducted. It seems that the Kurgovat microcontinent was the first exotic terrane to arrive in this subduction zone.

The Central Pamirs occupies the 50 km -wide zone between the Tanymas fault zone and the Rushan-Pshart suture zone. It consists of two different units. The first of these is the Sarez unit, comprising a thick terrigenous sequence of Early Paleozoic to Early Mesozoic age. Its most distinctive feature is a flysch-like Triassic sequence 2 km thick. The second unit, which ranges in age from Vendian to Middle Triassic, is predominantly a carbonate sequence, but it includes such charac-

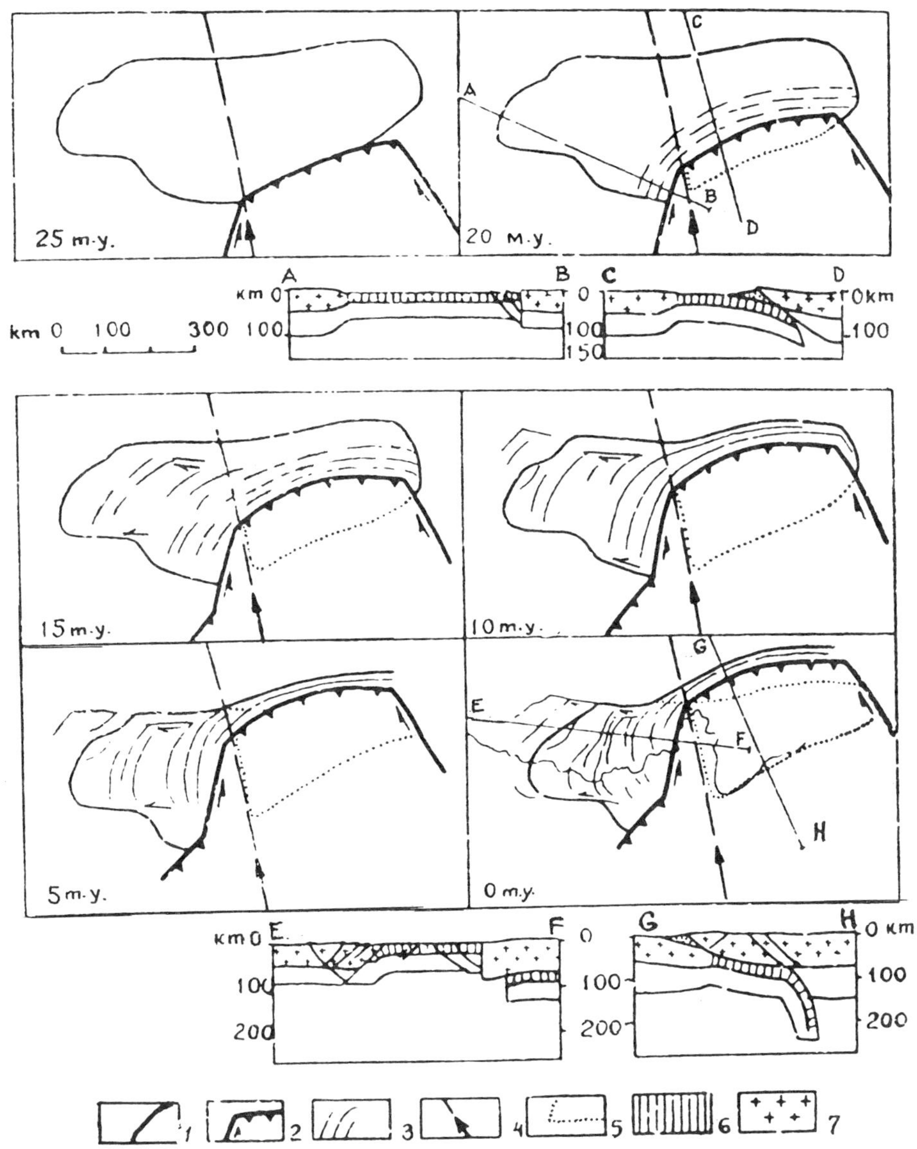

Fig. 158. Paleotectonic maps and cross-sections showing development of the Tadjik depression in Neogene-Quaternary times, after Legler and Przhiyalgovskaya [1979].
1, Depression boundary; 2, boundary of Pamir block; 3, structural trend; 4, direction of motion of the Pamir block; 5, outlines of suboceanic plate overriden by the Pamir block; 6, suboceanic crust (on cross-section); 7, continental crust (on cross-sections).

teristic formations as Lower Cambrian quarzite, analogous to the Lalun Formation in Iran. This mainly carbonate succession reminds one of the analogous Gondwanan sections of Afghanistan, Iran, and the Arabian Peninsula. We conclude that the Central Pamirs (or at least part of it) is a fragment of Gondwanaland. The Sarez unit is interpreted as a continental rise, and the second unit as part of the Gondwanan shelf. This block was attached to the Northern Pamirs along the Tanymas suture in Triassic-Early Jurassic time.

The Rushan-Pshart suture zone [Pashkov and Shvolman, 1979] consists of at least four nappes. Although the average width of this zone is only 10 km (maximum 20 km), each nappe has an individual geological section representing different initial depositional conditions, even though the age range in all four is approximately the same, from Carboniferous to Jurassic. The complex inner structure of one of the nappe units is shown in Figure 160. Two types of the successions are most interesting. In the West Pshart section lying close to the north front of the zone, fusulinid- and crinoid-bearing Permian-Carboniferous limestones are associated with basalt flows (picrites included). Pashkov and Shvolman [1979] consider this section to be a rift complex associated with the

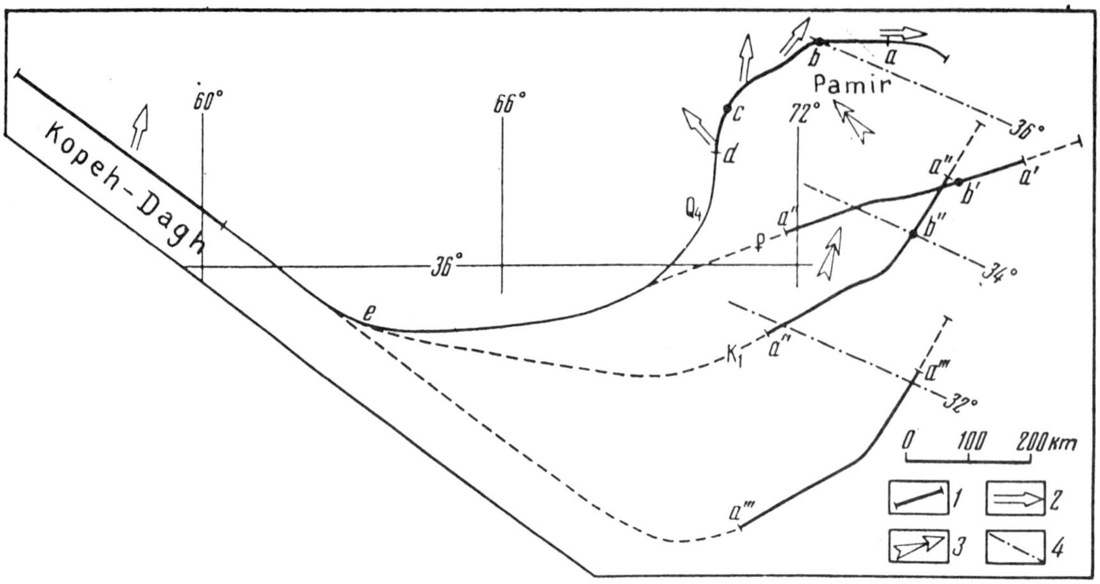

Fig. 159. Reconstruction of the position of the North Pamirs in the Early Cretaceous and Paleogene, after Bazhenov and Burtman [1982].
1, Studied segments of the Pamirs and Kopegh-Dagh; 2, paleomagnetic declination in Early Cretaceous rocks; 3, movement direction; 4, Early Cretaceous paleolatitude.

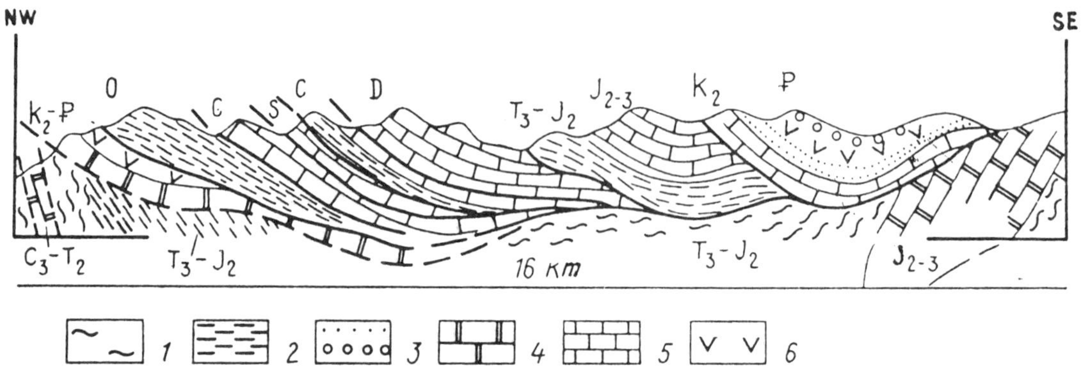

Fig. 160. Geological cross-section through the Rushan-Pshart zone in the Central Pamirs.
1, Schist; 2, phyllite, shale; 3, conglomerate; 4, marble; 5, limestone; 6, andesite, basalt.

breakup and divergence of fragments of the Gondwanan continent. The East Pshart complex contains deep-water Radiolarites and cherts of Late Permian, Early and Middle Triassic, and Early Jurassic ages. This section represents bathyal conditions most likely existing on a continental rise but perhaps on an abyssal plain. The major folding, faulting, and nappe formation in the Rushan-Pshart zone occurred near the boundary between Jurassic and Cretaceous time or in the Early Cretaceous, but the deformation was rejuvenated and intensified later. This zone is the suture along which the Southern Pamirs joined the Central and Northern Pamirs.

The Southern Pamirs are divided into SW and SE zones. The SW Pamirs is a block of Precambrian metamorphic rocks comprising two complexes that respectively yield ages of 2700 Ma [Khoreva and Bljuman, 1974] and 1000 Ma [Moskovchenko, 1978]. The SE Pamirs consist of Carboniferous-Permian and Triassic-Jurassic deposits of Gondwana type, including Upper Carboniferous glacial boulder deposits. Evidently the SE Pamirs are a fragment of the passive Gondwanan margin.

All the Pamir zones discussed above were probably stacked one onto another in the first half of Cretaceous time, after which a new stage commenced [Shvolman, 1977]. After the Early Cretaceous, red-colored and variegated coarse clastic deposits and subaerial volcanics of felsic and intermediate composition (Iryakyak suite) accumulated in the Pamirs. These rocks are intruded by large porphyritic granitic and granodioritic batholiths, whose components yield isotopic ages ranging from 100 to 130 Ma (Early Cretaceous). Upper Cretaceous volcanics are possibly present, as well as the pre-batholithic series. In the Cretaceous and Early Paleogene, the Pamirs were probably part of the Kohistan-Ladakh magmatic arc in the Himalayas under which the Tethys oceanic crust was subducted. Deformation in the Pamirs associated with the India/Eurasia continental collision began in the Oligocene [Shvolman, 1977]. In late Cenozoic time, the tectonic nappes acquired their present shape as the Pamirs arc became convex to the north.

Ancient Massifs

The above review indicates that the commonly accepted division of ancient massifs in this fold belt into two groups, of Gondwanan and Eurasian origin, is insufficient. Some blocks definitely belonged to Eurasia, e.g., the Skythian platform basement, while others belonged to Gondwanaland, e.g., Daralagez block, Iran, South Pamirs. But at the same time, some of the blocks occupied an intermediate position, e.g., Moesia platform, Dzirula massif, Kurgovat microcontinent in the Pamirs, and some others. Several blocks were first attached to Eurasia in the Late Triassic and Jurassic. Their collision with Eurasia caused the Cimmerian orogeny in Dobrogea and North Pamirs.

Oceanic Complexes

Within the USSR, only dismembered ophiolites are known in the Alpine-Himalayan belt. Two different ages of ophiolites have been established, Upper Paleozoic and Jurassic-Early Cretaceous.

Paleozoic ophiolites crop out in the Peredovoy Ridge of the Great Caucasus (Figure 161), and they form part of an ophiolite belt that can be traced eastward through the Kopeh-Dagh into north Iran (Binalud). But the best ophiolite outcrops appear from north Afghanistan (Hindu-Kush), to the North Pamirs. As a first approximation, the Paleozoic ophiolite belt may be regarded as the southern boundary of Eurasia in Carboniferous and Permian time.

The Mesozoic ophiolites are of Late Jurassic and Early Cretaceous age. Within the USSR they are present only in the Sevan-Akera zone of the Lesser Caucasus and, strongly dismembered, in the central nappes of the Carpathians and within the Rushan-Pshart zone in the Pamirs.

The ophiolites in the Sevan-Akera were studied in detail by Knipper [1975], who recognized three stages of their formation. The first stage was the formation of mantle harzburgites that were intruded by 168 Ma plagiogranitic dikes. The second stage was the accumulation of ophicalcite breccias with serpentinite fragments and extrusion of Oxfordian-Tithonian alkaline and tholeiitic basalts. The third stage is remarkable because a basalt-rhyolite volcanic series of Albian, Senomanian, and Turonian age accumulated on a more ancient oceanic basement that had experienced pre-Albian deformation. Knipper postulates that the deformed ophiolites were exposed on the oceanic bottom in a transform-fault zone.

Magmatic Arcs Implying Subduction

Mesozoic and Cenozoic calc-alkaline magmatism was of major importance in the Alpine-Himalayan belt. Five belts of calc-alkaline igneous rocks (Triassic, Early Jurassic, Middle Jurassic to Neocomian, Aptian to Senonian, and Eocene in age) have been recognized between Crimea and Turkmenia [Kazmin et al., 1986]. In the Carpathians, only Late Cenozoic calc-alkaline volcanics are developed. In the Pamirs the record of subduction is still poorly known.

The Triassic volcanic belt described by Khain [1979b] (Figure 162), stretches from North Dobrogea through the Crimea and Peri-Caucasus to the Hindu Kush and North Pamirs. Two arcs evidently existed in Early Jurassic time: the Lesser Caucasian arc, which stretched from the Pontides in Turkey to the Dzirula massif, and the Great Caucasian arc, which is typified by Sinemurian and Pliensbachian rhyolites and dacites (Figure 163). An Aptian-Senonian volcanic belt was added to the Lesser Caucasian arc (Figure 164). It formed at the time that the Daralagez block collided with Eurasia and the Sevan-Akera suture zone formed. The Late Cretaceous-Eocene Adzharo-Trialetan arc is part of an extensive magmatic belt stretching from Bulgaria (Burgas synclinorium) through Turkey (Pontides) to the Lesser Caucasus, Iran, Afghanistan, and eventually the Himalayas. As seen from Figure 165, in Iran this belt is divided into two branches--Central Iran and Elburz; the latter extends into USSR territory in the Badkhyz region (East Turkmenia). Arcs of Miocene to Quaternary age occur in the Caucasus and Carpathians.

HISTORY

Kinematic Data

To understand how the Alpine-Himalayan belt developed we need to know how the African and Indian plates interacted with the Eurasian plate and, for the last 10 Ma, how the Arabian plate moved when it was detached from Africa as the Red Sea opened. The interactions of these lithospheric plates can be reliably reconstructed from the kinematic data for the Atlantic Ocean [Olivet et al., 1984] and Indian Ocean [Patriat et al., 1982].

The results were summarized in the Soviet-French "Tethys" Project [Aubouin et al., 1986]. Motions of three selected points on the African plate, with respect to Europe are shown in Figure 166. As was established long ago, relative plate motions in the Tethys realm fall into two periods separated by a principal reorganization about 80 Ma. During the first period from 180 to 80 Ma, Gondwanaland separated from Laurasia and Africa moved sinistrally eastward relative to Laurasia (which included Eurasia) for 2200 km. The second period began about 80 Ma ago when Eurasia separated from North America and the northern edge of Africa started to converge with Eurasia, leading eventually to the collision of these continents. Table 8 lists parameters of the rotation of Africa (and Arabia for the last 10 Ma) with respect to Eurasia during various time spans of the Mesozoic and Cenozoic. Table 9 gives vectors of plate motions at four selected points.

It follows from the kinematic data that the Tethys belt must be divided into western and eastern segments for purposes of tectonic interpretation.

The western segment, west of the Black Sea, is characterized by marked variations in the direction of movement of the African plate relative to Eurasia. The velocity remained more or less uniform at about 1 cm/yr. In the western segment there were seven significant changes in the motion of Africa with respect to Eurasia. During the first epoch, from the Early

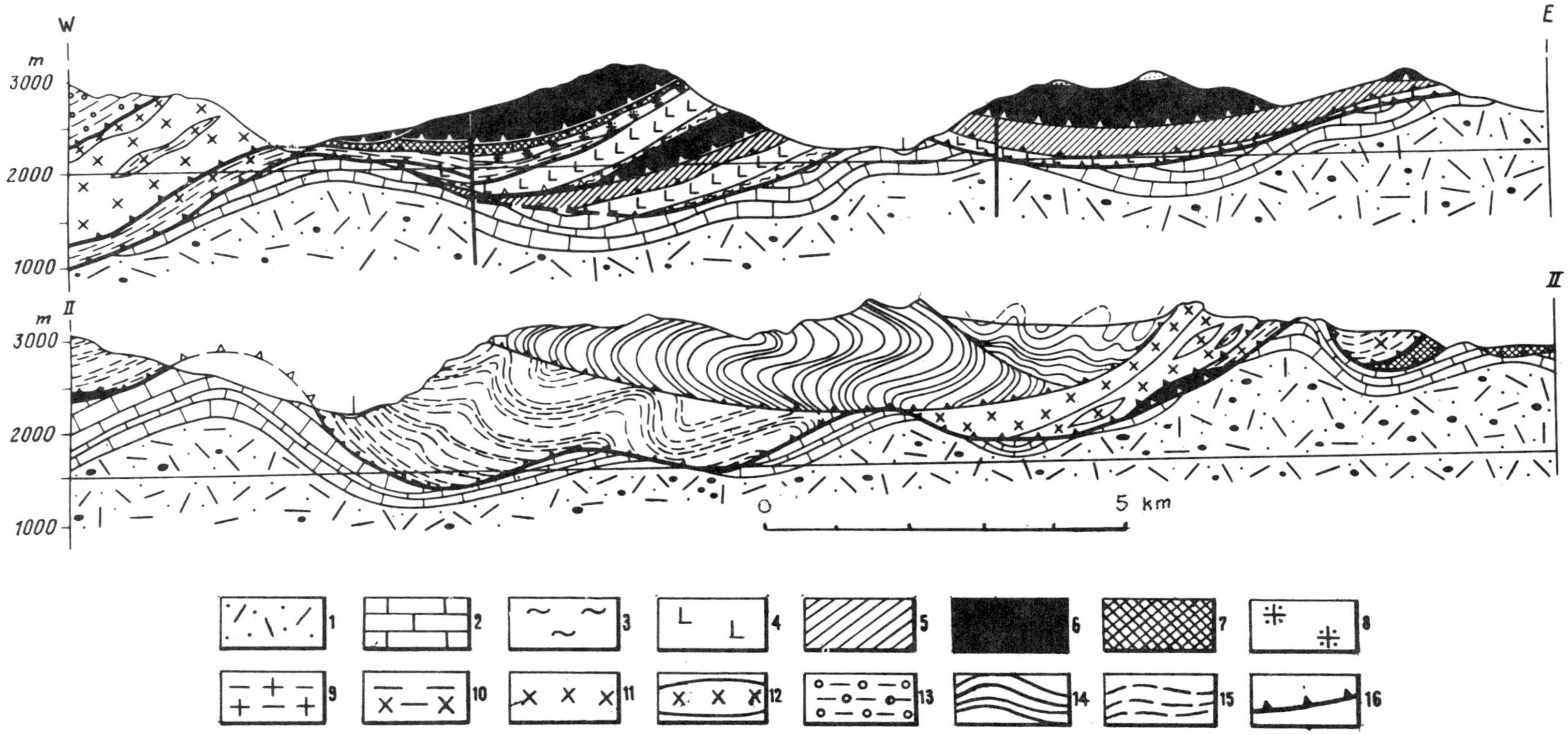

Fig. 161. Geological cross-sections through the Peredovoi Range of the Great Caucasus, after Khain [1979]. 1, Middle-Upper Devonian tuffs; 2, Famennian-Tournaisian limestone and greenschist; 3, Silurian-Devonian greenschist; 4, Silurian-Devonian spilite and diabase; 5, gabbro; 6, ultramafics; 7, amphibolite; 8, syenite; 9, granite; 10, two-mica granite; 11, grabbro-diorite; 12, tonalite; 13, schist; 14, gneiss; 15, biotite gneiss; 16, thrust.

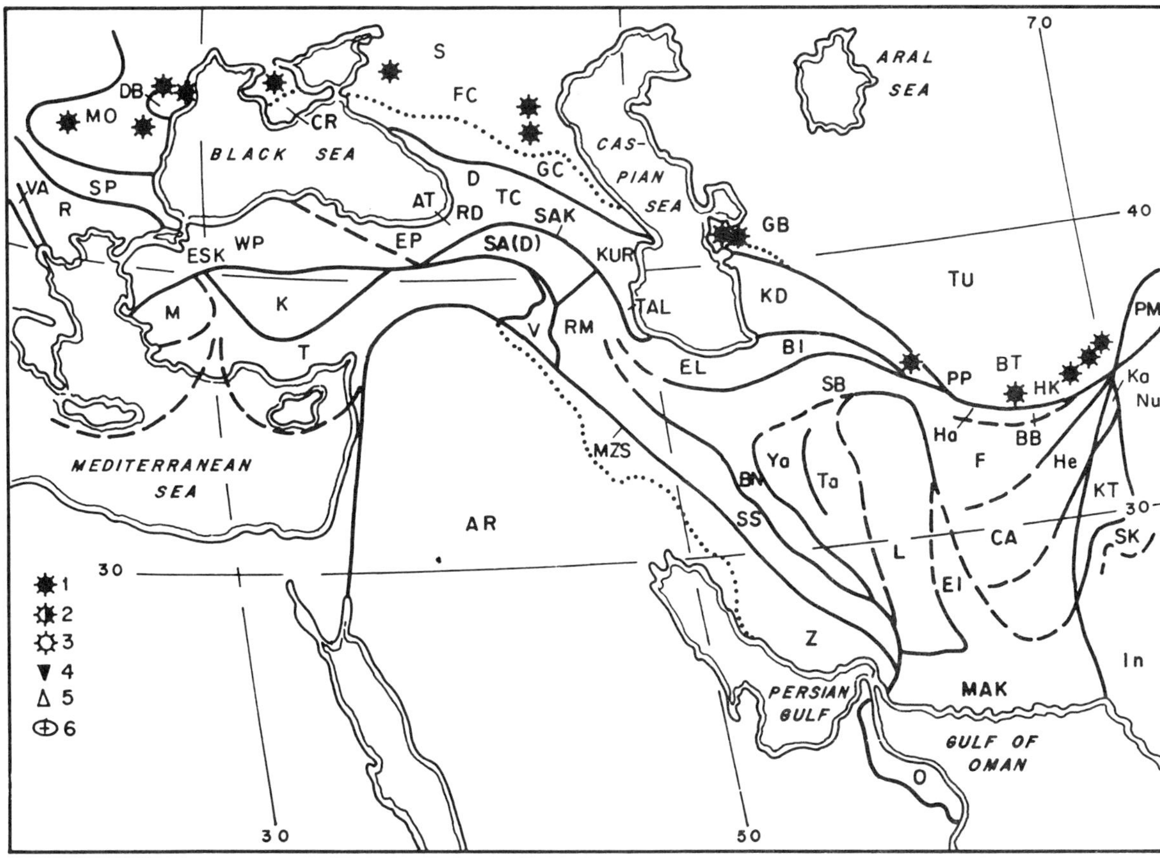

Fig. 162. Triassic volcanic belts within the Tethys realm, after Kazmin et al. [1986].

Explanation refers to Figures 162-165. 1, Calc-alkaline series; 2, shoshonite series; 3, tholeiitic series of incipient island arc and boninites; 4, tholeiitic low-potassium--high-titanium basalts (basalts of mid-oceanic ridges); 5, alkaline high-titanium basalts; 6, granite-granodiorite-diorite intrusives (numbers show absolute age in Ma).

AR - Arabia, Al - Alburz, AT - Adjaro-Trialetian zone, BB - Band-Bayan zone, BI - Binalud, BJ - Bajan-Dur-Kan zone, BN - Baft-Nain zone, BT - Bande e Turkestan, BJ - Bajan-Dur-Kan zone, BU - Burgas synclinorium, CA - Central Afghanistan, CH - Chagnai Ridge, CR - Crimea, D - Dzirula masssif, DB - Dobrogea, EI - East Iranian zone, EP - Eastern Pontides, ESK - Eskishehir zone, F - Farahrud zone, FC - Fore Caucasus, GB - Great Balkan, GC - Great Caucasus, Ha - Harirud suture, He - Helmend block; HK - Hindu-Kush, In - India, IM - Inner Makrane zone, K - Kirshehir massif, Ka - Kabul block, KD - Kopeh Dagh, KT - Katawaz, KUR - Kura depression, L - Lut, M - Menderes, MA - Maden volcanic belt, MAK - Makran, MO - Moesia, MV - Mineralnye Vody, MZS -Main Zagros suture, Nu - Nuristan, O - Oman, PM - Pamirs, PP - Parapamisos, R - Rhodope massif, RD - Rioni depression, RM - Rezaye-Makhabad area, S - Scythian platform, SA - South Armenian block (Daralagez), SAK - Sevano-Akera zone, SB - Sabzevar zone, SK - Suleiman-Kirtar, SP - Stara Planina, SS - Sanandaje, T -Taurus, Ta - Tabas block, TAL - Talesh zone, TC - Trans-Caucasian massif, TU - Turan platform, V - Van Lake, VA - Vardar zone, WP - West Pontides, Ya - Yazd block, Z - Zagros foldbelt.

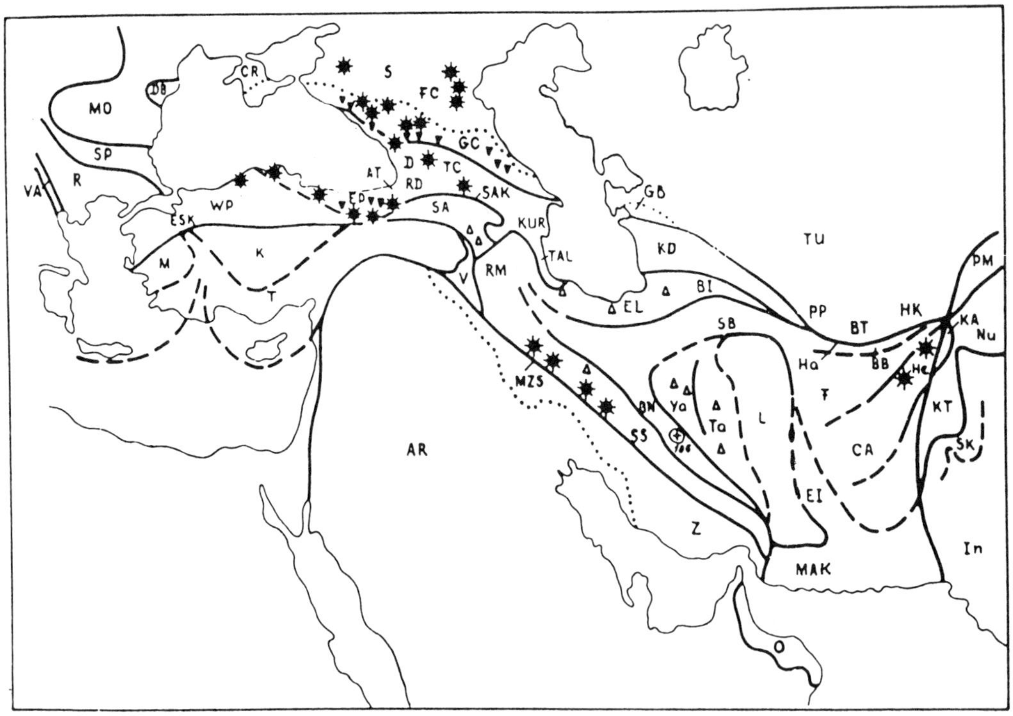

Fig. 163. Early Jurassic volcanic belts within the Tethys realm, after Kazmin et al. [1986]. See Figure 162 for explanation.

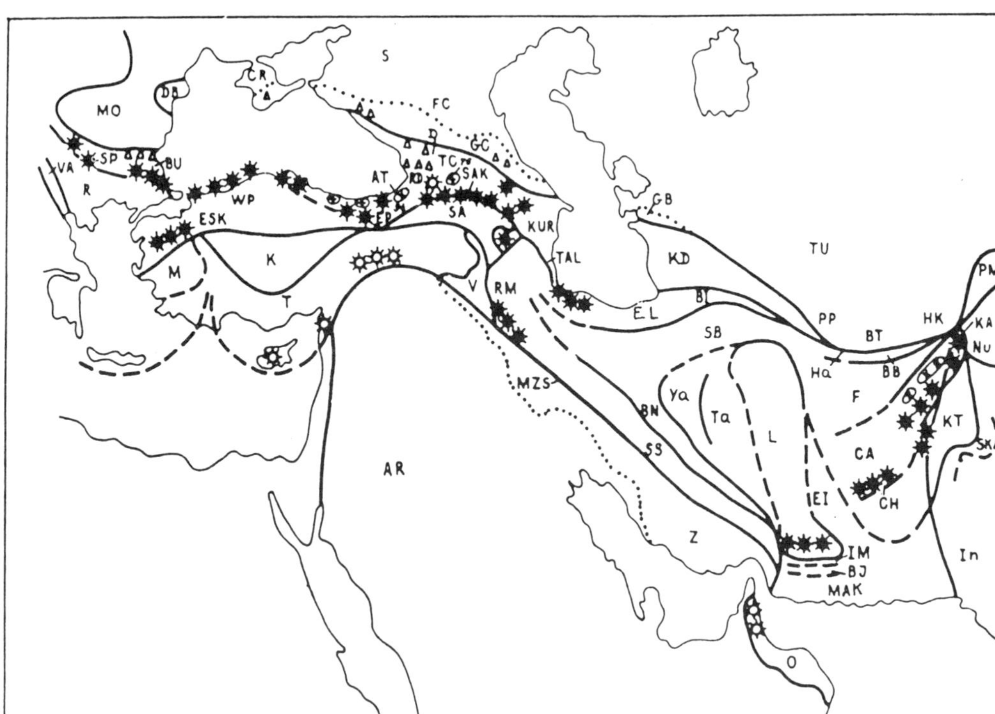

Fig. 164. Aptain-Senonian volcanic belts within the Tethys realm, after Kazmin et al. [1986]. See Figure 162 for explanation.

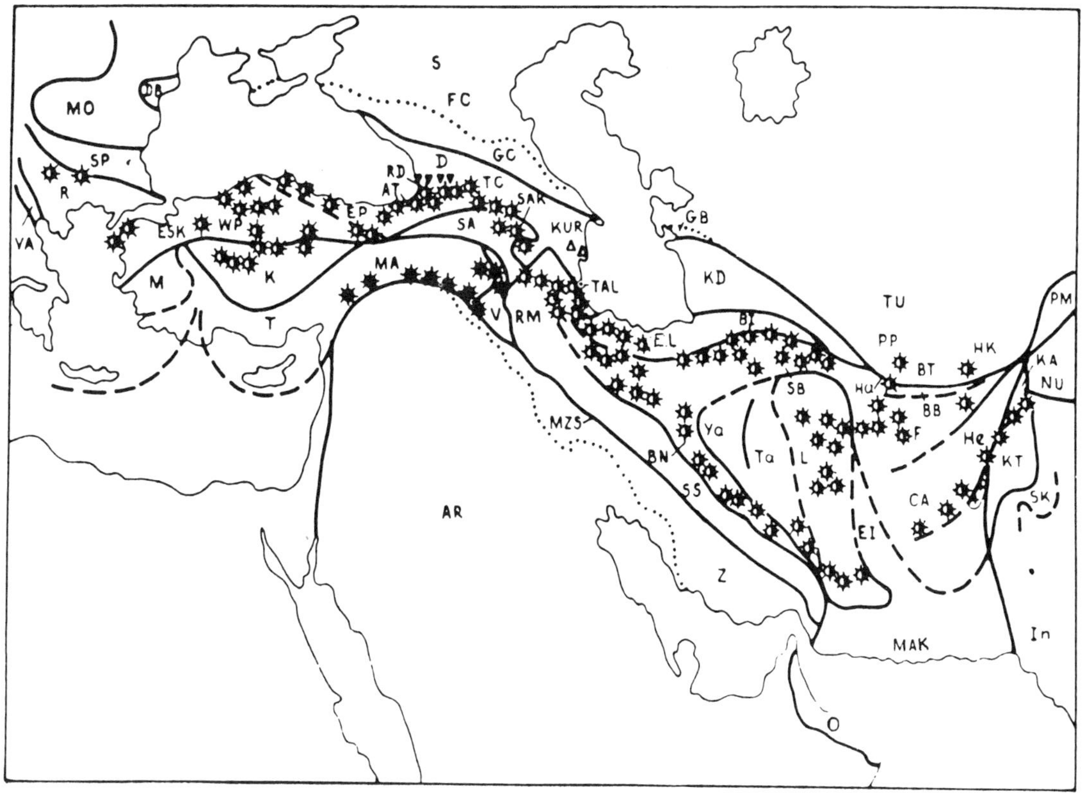

Fig. 165. Eocene volcanic belts within the Tethys realm, after Kazmin et al. [1986]. See Figure 162 for explanation.

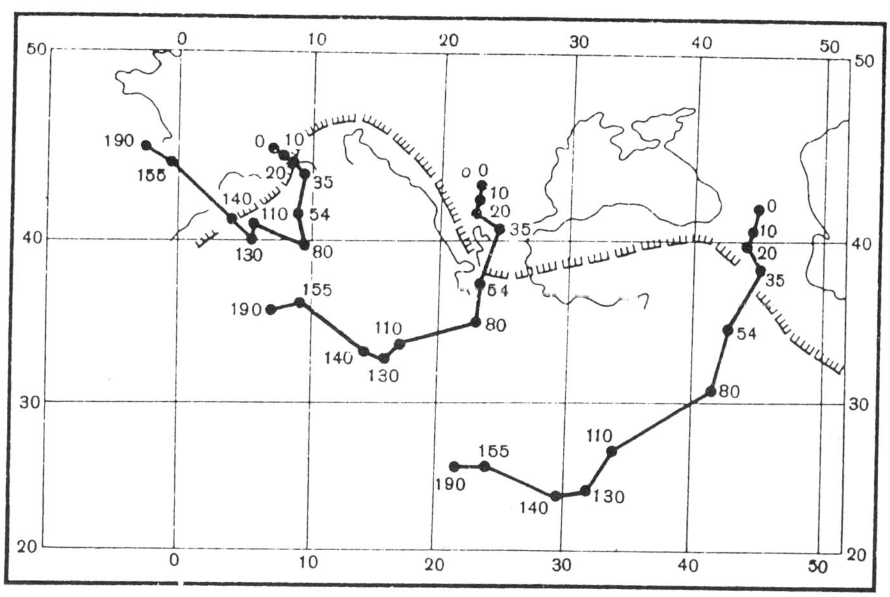

Fig. 166. Relative motion of those points on Africa with respect to Eurasia, from 190 Ma to present, according to kinematic data by Savostin et al. [1986] Figures--age in Ma. Broken toothed line marks boundary of the Eurasian continent in Mesozoic times.

TABLE 8. Parameters of Relative Rotations of the Gondwanan Continents with Respect to Eurasia

Finite Rotation

	Africa/Eurasia (after Savostin et al [1986])			India/Eurasia (after Patriat et al. [1982])		
	Pole Position		Angle of	Pole position		Angle of
	Deg. N	Deg. E	Rot'n Deg.	N	E	Rot'n.(deg.)
10	30.9	-20.4	-1.7	18.5	35.6	- 7.6
20	37.8	-19.0	-3.1	21.7	33.0	-13.0
35	20.8	-16.4	-4.1	12.3	42.2	-23.4
54	33.5	-18.0	-9.9	18.7	26.0	-37.0
65	33.5	-16.6	-12.5	17.3	19.9	-49.9
80	33.8	-14.1	-16.0	18.4	15.6	-67.7
110	45.1	-5.8	-30.0	27.0	18.6	-84.6
130	44.9	-3.5	-36.5	28.9	21.2	-90.2
141	46.2	-2.5	-39.5	30.3	22.7	-92.8
155	51.8	2.5	-46.8			
190	52.2	3.0	-51.4			
220*	52.26	2.98	-56.87			
250*	52.12	3.05	-58.34			
280*	42.86	-2.73	-68.18			

*After Zonenshain et al. [1987].

Differential Rotation

1	2	3	4	5	6	7
10 - 0	30.95	-20.42	1.71	18.5	35.6	7.6
10 - 0xx	35.7	4.43	4.79	---	---	---
20 - 10	45.89	-16.83	1.42	26.46	29.74	5.43
35 - 20	-18.32	-13.19	1.44	-0.82	49.71	11.05
54 - 35	42.3	-18.42	5.96	32.20	4.23	16.59
65 - 54	32.78	-11.04	2.61	19.61	1.63	12.65
80 - 65	33.24	-5.45	3.56	26.64	7.18	17.65
110 - 80	53.42	11.58	14.89	37.63	51.92	19.35
130 - 110	41.1	5.16	6.55	19.92	58.83	6.82
141 - 130	55.6	21.3	3.2	20.74	72.80	3.74
155 - 141	57.66	56.1	8.66	68.86	57.82	8.95
190 - 155	54.22	10.56	4.62	-76.64	112.20	6.62
220 - 190	52.79	3.24	5.47	---	---	---
250 - 220	46.81	1.38	1.47	---	---	---
280 - 250	12.30	-37.35	14.09	---	---	

*Arabia with respect to Eurasia.

Jurassic to the Early Cretaceous (to 130 Ma), Africa moved ESE, producing extension everywhere on the margins of Eurasia and Africa and causing the opening of oceanic basins in the Mediterranean realm, including the Carpathian region. In the second epoch, from 130 to 110 Ma, the African plate in the Carpathian region began to move northward. The Middle Cretaceous folding recorded in the internal nappes of the Carpathians was presumably caused by this change in plate movement. In the third interval, from 110 to 80 Ma, the African plate again moved eastward. During the fourth epoch,

TABLE 9. Azimuth (Degrees) and Rates (cm/yr) of the Motion of Africa and India Plates with Respect to Eurasia at Four Selected Points

Age (Ma)	Uzhgorod* 48.7°N 23.7°E		Tbilisi* 41.6°N 45°E		Ashkhabad* 37.9°N 58.2°E		Alai**Valley 39.5° 72°E	
	Azimuth	Rate	Azimuth	Rate	Azimuth	Rate	Azimuth	Rate
10 - 0	348	1.16	2.9	2.8	19.8	1.7	337	5.2
20 - 10	9.6	0.7	27	1.1	34.1	1.3	353	3.7
35 - 20	306	1.0	325	1.0	224	1.1	302	5.8
54 - 35	3.7	1.7	23	2.5	30.6	2.9	14.8	7.9
65 - 54	341	1.3	8	1.8	17.2	2.2	3.7	11.4
80 - 65	335	1.2	5	1.7	15.26	2.0	7.3	10.7
110 - 80	36.4	0.86	40	2.3	41.4	3.2	180	2
130 - 110	336	0.95	12	1.8	21.8	2.4	---	---
141 - 130	79	0.4	50	1.1	47.8	1.6	---	---
155 - 141	142.7	2.5	108	2.2	86.7	2.3	---	---
190 - 155	39	9.3	41	0.6	42.7	1.0	---	---
220 - 190	25	0.47	36.3	1.01	40.0	1.3	---	---
250 - 220	1.22	0.14	24.6	0.28	31.0	0.36	---	---
280 - 250	346	4.6	4.3	5.1	13.1	5.2	---	---

*Africa with respect to Eurasia.
**India with respect to Eurasia.

from 80 to 35 Ma (Late Cretaceous to Late Eocene), the African plate moved to the northwest and north toward the Eurasian margin, resulting in a continental collision.The main events of this time were thrusting, nappe formation, and metamorphism in the Alps and Carpathians, all consequences of the Africa/Eurasia collision. During the fifth interval, corresponding to the Oligocene and Early Miocene (35-20 Ma), the African plate moved WNW (290-300°) with respect to Eurasia, producing calc-alkaline volcanism along the North Carpathians [Balla, 1986]. In the sixth episode, during Miocene time (20-10 Ma), the African plate again abruptly changed direction almost 90°, from WNW to NNE. Subduction and volcanism in the Carpathian arc may be associated with this episode [Balla, 1986]. In the seventh, recent interval, Africa and Eurasia again began colliding in a NNW-SSE direction. The Adriatic promontary indented Eurasia, pushing a crustal fragment to the east into the Carpathians.

In the eastern segment, variations in the velocity of African/Eurasian convergence were of greater significance than changes of movement direction, as Africa moved more or less constantly northward. The great width of the Tethys Ocean in the eastern segment (2500 km south of the Caucasus, and over 5000 km south of the Pamirs) and the continuous convergence of the continents explain the continuous tectonic activity along the Eurasian margin, with its volcanic arcs and back-arc basins. Meanwhile, a passive continental margin existed on the African side, where the continent was distant from the convergent boundary for a long time. Plate convergence accelerated to 2.5-3 cm/yr during the intervals 155-141 Ma, 110-80 Ma, and 54-35 Ma as reflected in episodes of intensive volcanism in volcanic arcs near the Eurasian margin . Intervals of slow movement from 140 to 110 Ma and from 80 to 54 Ma, when velocities decreased to 1.2-1 cm/yr, coincided with extension, as well recorded in the Late Jurassic within the Dzirula massif, in the Kura depression, in the Kopeg-Dagh, and in the Black Sea where deep basins formed in Late Cretaceous time. The two latest intervals are particularly interesting: (i) from 35 to 10 Ma, when the velocity of convergence was as low as 1 cm/yr, and (ii) from 10 Ma up to present, when velocity again sharply increased to 3 cm/yr. However, the rate increased only north of the Arabian promontary as the Red Sea opened. In the Miocene during the time of slow convergence, volcanism ceased in the Alpine-Himalayan belt from the Balkans through the Lesser Caucasus to Iran. The recent increase in convergence rate has led to subduction of remnants of the oceanic floor, to calc-alkaline volcanism in the Caucasian segment of the belt, and to a young phase of mountain building and nappe formation in the Great Caucasus.

The parameters of India/Eurasia relative motions are given

in Table 8. Three stages of movement are distinguished. In the first stage, beginning from the breakup of Gondwana at 160 Ma up to 80 Ma ago, India together with Madagascar moved southeastward from Africa. This movement did not affect events along the northern Tethyan margin. In the second stage, beginning at 80 Ma (Late Cretaceous), the motion of India changed dramatically; Madagascar was left near Africa, and India started moving straight north with respect to Eurasia at the high rate of 12 cm/yr, resulting in rapid subduction and eventually continent-continent collision. In the third stage, beginning in the Late Eocene when India approached Eurasia and continental collision began, the velocity of India's motion decreased sharply to 5 cm/yr. Moreover, from 35 to 20 Ma, the direction of India's movement abruptly changed to northwest instead of north, according to Patriat et al. [1982] so that many blocks moved northwest and a number of dextral strike-slip faults (e.g., Karakorum and Gissar) appeared. However, for the last 20 Ma the Indian plate has again moved northward at 4-5 cm/yr, forming the strongly folded nappe structure of the Pamirs.

Paleomagnetic Data

Paleomagnetic data [Westphal et al., 1986] clearly show that there must have been large relative displacements between Eurasia and Africa. Table 10 gives Africa and Eurasia paleomagnetic poles and recalculated positions of Africa poles using Africa/Eurasia rotations. It shows that, if we use plate kinematics, the paleomagnetic poles of the two continents coincided.

Of great significance are paleomagnetic data from small blocks within the Alpine-Himalayan belt. In the Carpathians such data [Burtman, 1986] indicate that the Carpathian arc originated only during Paleogene-Early Miocene time, due to convergence of the Asia Minor-Balkan block with Europe, with 30° counterclockwise rotation of this block. According to paleomagnetic data, deformation in the Carpathians also continued after Miocene time; Middle Miocene volcanics are rotated 20° counterclockwise [Balla, 1984].

In the Caucasus, paleomagnetic data are available for the Paleozoic and Mesozoic rocks of the Main (Glavny) Range in the Great Caucasus, the Dzirula massif, and the Lesser Caucasus (Daralagez block) [Asanidze et al., 1980]. Data from the Late Paleozoic and Mesozoic rocks of the Great Caucasus indicate a constant position near Eurasia. Data from Late Paleozoic rocks in the Dzirula massif suggest an intermediate position between Eurasia and Gondwanaland. Data from the Daralagez block show a counterclockwise rotation of approximately 30° between Early and Late Cretaceous time, when this block was colliding with the Lesser Caucasian arc. Its Late Triassic and

TABLE 10. Paleomagnetic Poles for Eurasia and Africa for the Late Paleozoic, Mesozoic, and Cenozoic (After Westphal et al. [1986])

Age (Ma)	Eurasia		Africa		African poles Recalculated with respect to Eurasia	
	Deg. N	Deg. E	Deg. N	Deg. E	Deg. N	Deg. E
Permian	43	168	36	239	48	171
Lower-Middle Triassic	49	152	67	276	79	113
Lower Jurassic	69	130	69	258	75	115
Middle Jurassic	63	117	62	247	69	143
Lower Cretaceous	70	166	54	264	76	212
Upper Cretaceous	74	152	71	233	81	184
Eocene	72	137	78	178	75	134
Miocene	80	156	82	148	82	134
Pliocene	83	152	86	158	85	126

Jurassic paleomagnetic declination are close to those of Africa, although inclinations show a intermediate position between Africa and Eurasia.

In the Kopeh-Dagh [Bazhenov, 1983], data from Cretaceous-Paleogene rocks indicate that this territory was close to Eurasia and did not rotate much in contrast to the Elburz, where paleomagnetic data show 30° clockwise rotation in the Neogene.

Paleomagnetic data from the Pamirs [Bazhenov and Burtman, 1982] show that the Pamirs arc originated only in Neogene time.

Chronology of Events

Late Paleozoic (Figures 167, 168)

Several models for Late Paleozoic reconstructions have been proposed for the Caucasus [Adamia et al., 1982; Gamkrelidze, 1982]. All reconstructions show a broad (no less than 3000 km) Paleo-Tethys Ocean, bounded on the south by Gondwanaland and on the north by Euramerica and the isolated Siberian, Kazakhstan, and Chinese continents. The Tethyan oceanic lithosphere was consumed in subduction zones in the north, and Gondwanaland collided with Laurasia by the end of the Paleozoic in the west, closing that part of the ocean. A wide ocean was preserved in the east. Earlier, in the Late Carboniferous and probably in the Early Permian, Gondwanaland included the Daralagez, Iran, and South Pamirs blocks, which eventually detached from Gondwana and joined the Eurasian margin. However, the paleo-position of blocks that show an imprint of Late Paleozoic "Hercynian" orogeny is not clear. These blocks include the Moesia platform, Dzirula massif, and North Pamirs. They are commonly either included in the Eurasian margin or placed near the margin within the Tethys Ocean.

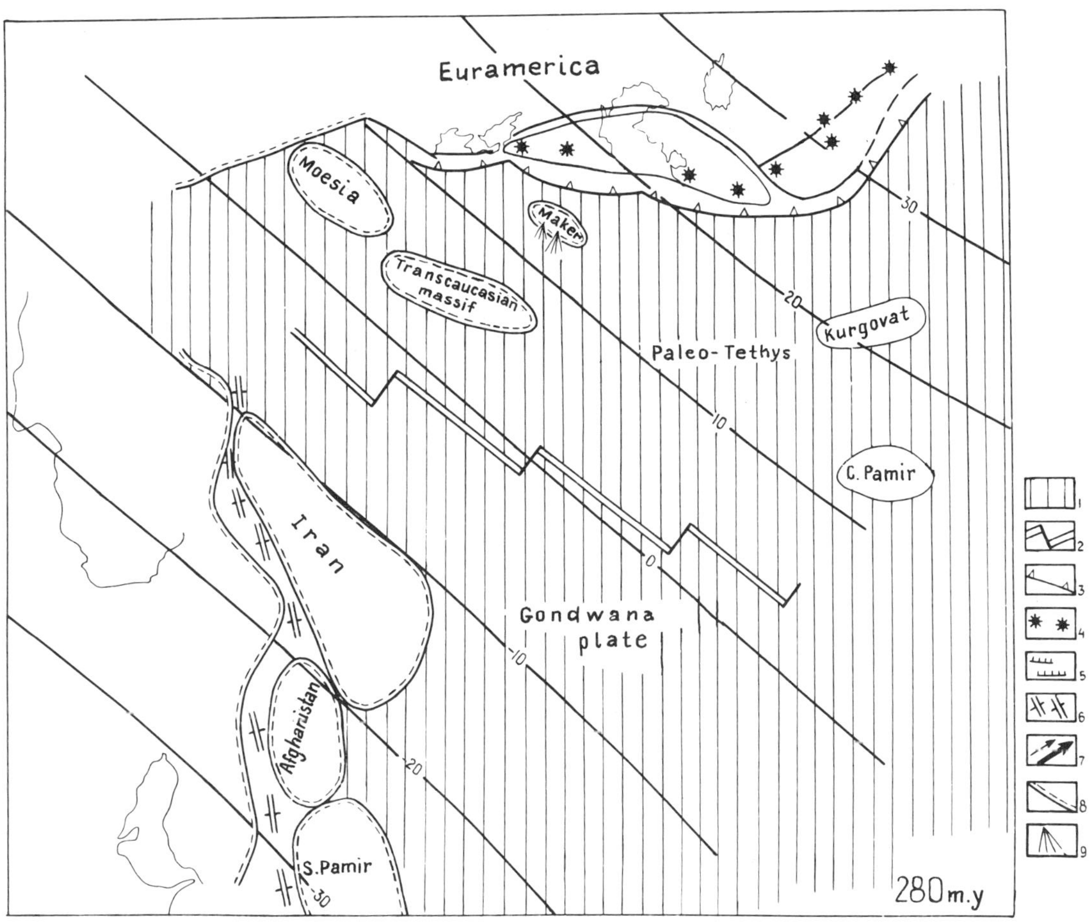

Fig. 167. Palinspastic reconstruction of the Tethys belt for 280 Ma. The reconstruction is made with respect to Eurasia, which is arbitrarily left fixed. Paleo-latitudes are shown according to data of Westphal et al., [1986]. Kinematic data of Olivet et al. [1984] and Zonenshain et al. [1987] and geological data of Dercourt et al. [1986] are used. Oblique Mercator projection with pole at 50° N and 155° W.
1, Oceanic floor; 2, spreading axis; 3, subduction zone; 4, calc-alkaline volcanics; 5, thrust front (collision zone); 6, continental rift; 7, direction of relative plate motions; 8, continental block boundary; 9, fan (deltaic deposits).

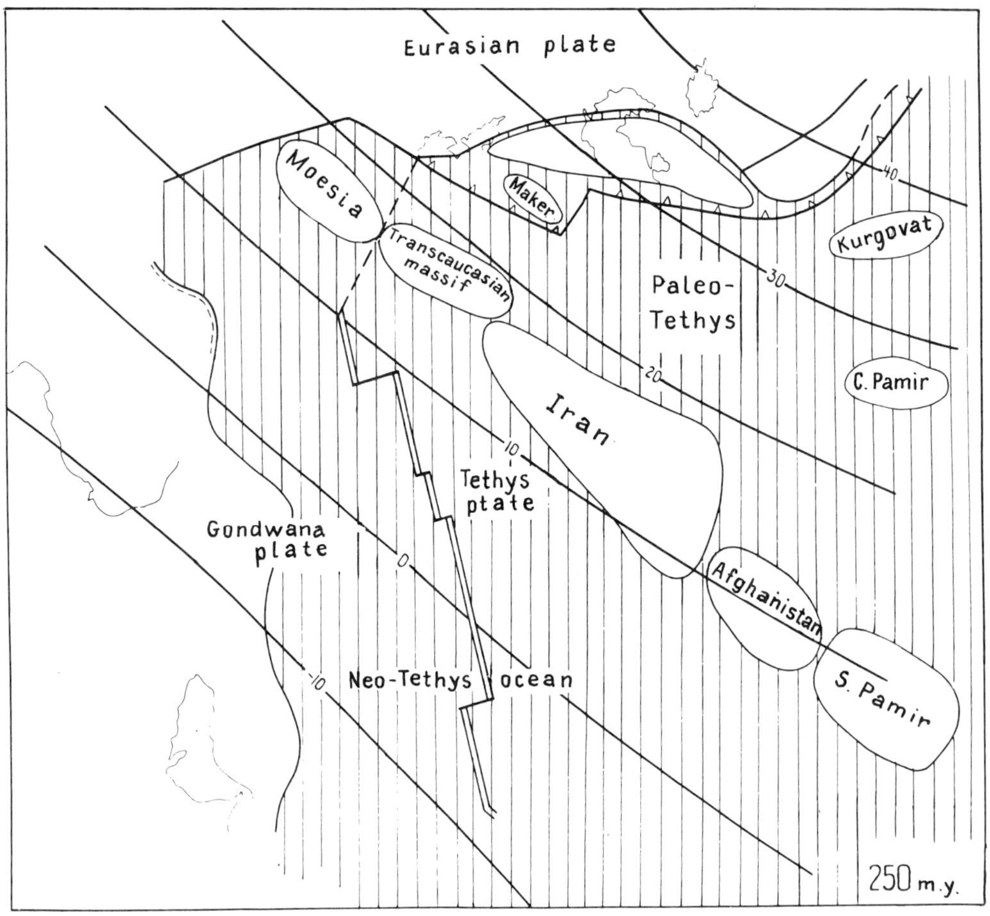

Fig. 168. Palinspastic reconstruction of the Tethys belt for 250 Ma. See Figure 167 for explanation.

As implied above, available data suggest that several continental fragments (such as Iran and South Pamirs) were rifted and separated from the Gondwanian margin. The breakup, which took place in the region of Arabia and India, is well recorded, for example in Oman [Kazmin, et al., 1986]. The northern Tethyan margin was the site of a Late Paleozoic volcanic belt [Mossakovsky, 1975], which reflects a subduction zone dipping under Eurasia.

Triassic, 220 Ma (Figure 169)

The pattern of plate positions in the Triassic was relatively simple. The Gondwanan plate, including its northern passive margin and part of the Tethys Ocean floor was situated in the south. In the center of the ocean were a number of microcontinents that had been rifted from the Gondwana margin in the Late Paleozoic; these included the Iranian, Kurgovat, South Pamirs, and other microcontinents. The spreading axis with its mid-oceanic ridge lay between these microcontinents and the Gondwana margin. The microcontinents were part of an independent Tethyan plate, which moved with them away from the Gondwanan plate northward toward Eurasia, where a long subduction zone was marked by the Cimmerian continental-margin volcanic belt along the entire Tethyan border.

Early Jurassic, 190 Ma (Figure 170)

The reconstruction in Figure 170 is in good agreement with paleomagnetic and paleoclimatic data. A coal-bearing belt extending from Dobrogea to the Turan lowland (where it is marked by the Shemshak Formation) must have formed under moderate humid climatic conditions. In the reconstruction (Figure 170) this belt lies approximately along 45° N. The northern Gondwanan margin was in an arid tropical belt, as reflected in reef limestones and evaporites.

By the end of the Triassic and beginning of the Jurassic, some microcontinents riding with the Tethyan plate approached the Eurasian margin, which was marked by the Cimmerian arc. In particular, the Iran and North Afghanistan microcontinents approached and joined Eurasia in the Kopeh-Dagh and Hindu Kush areas, and the Kurgovat and Central Pamirs massifs were accreted in the Pamirs area. As the microcontinents collided with the Eurasian active margin, the Tethys Ocean floor was entirely consumed. The collision zone is marked by the Cimmerian orogenic belt stretching from the

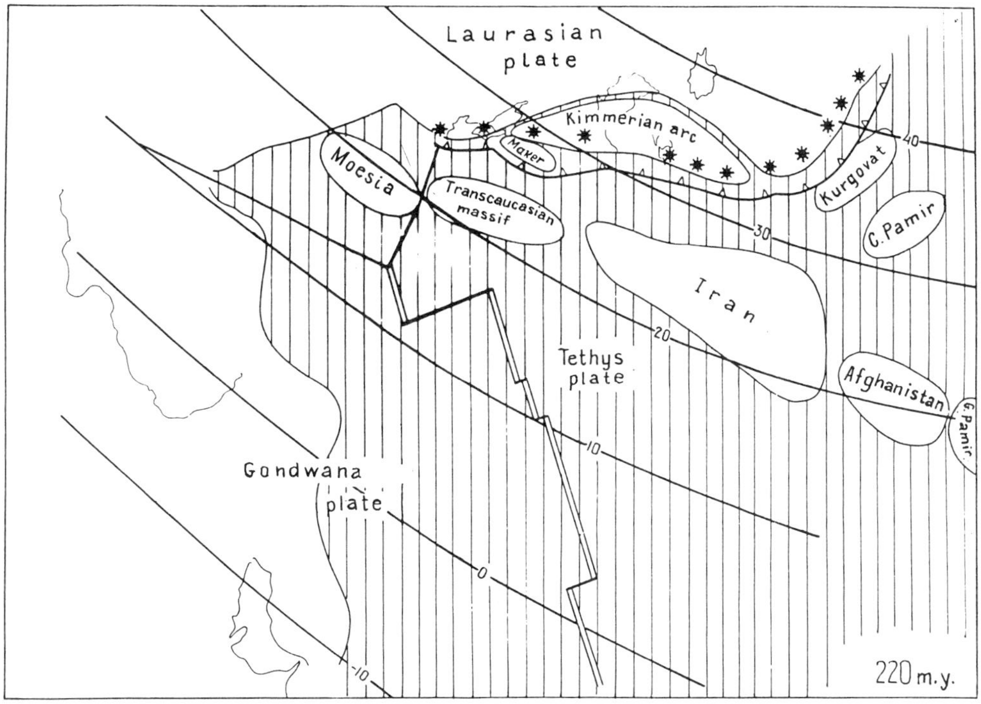

Fig. 169. Palinspastic reconstruction of the Tethys belt for 220 Ma. See Figure 167 for explanation.

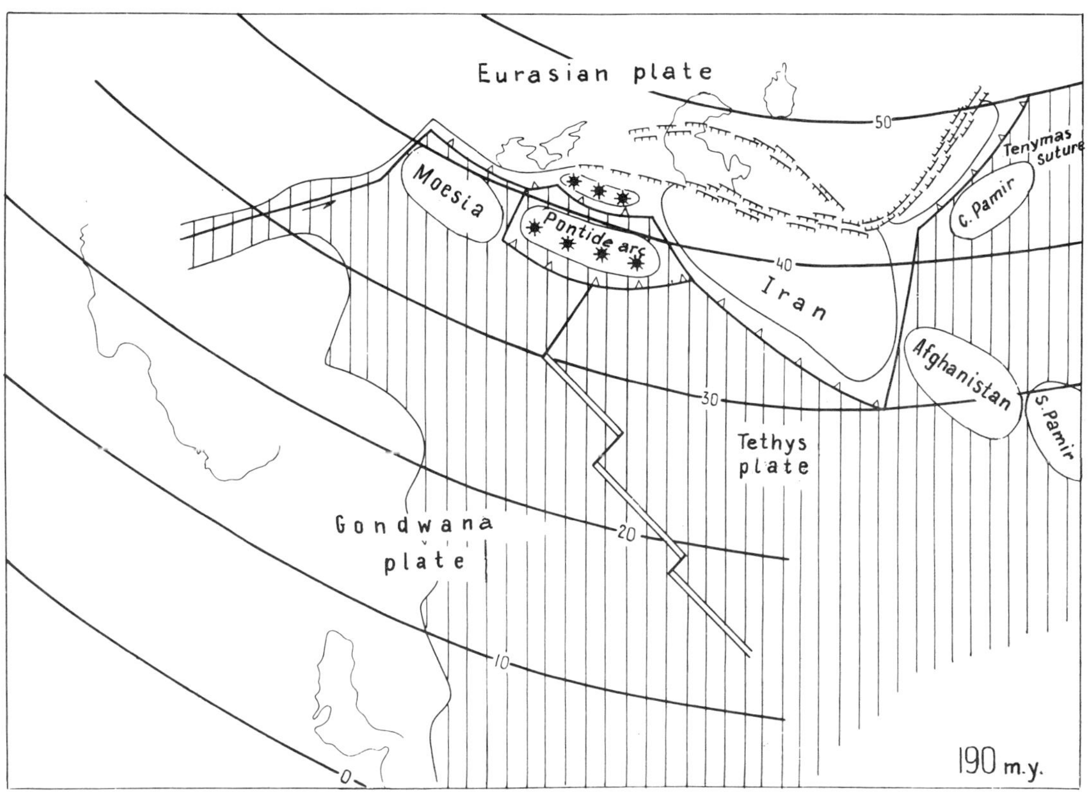

Fig. 170. Palinspastic reconstruction of the Tethys belt for 190 Ma. See Figure 167 for explanation.

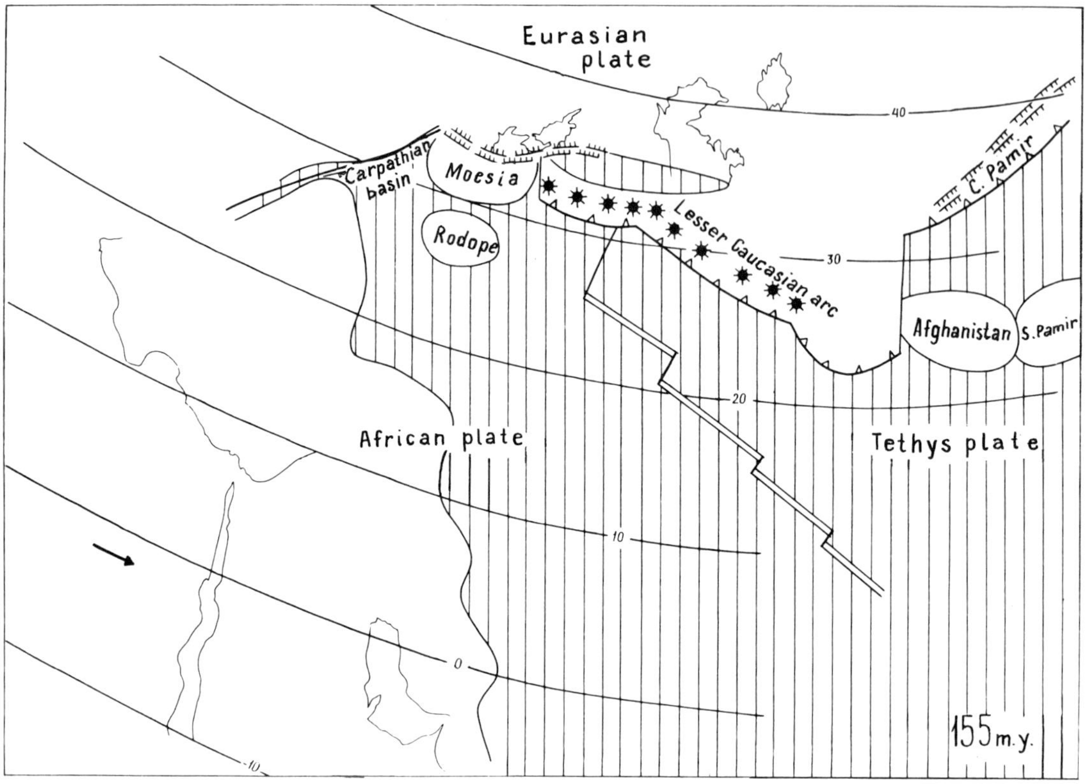

Fig. 171. Palinspastic reconstruction of the Tethys belt for 155 Ma. See Figure 167 for explanation.

North Pamirs through Afghanistan and into the Kopeh-Dagh. In the west, the Paleo-Tethys Ocean still remained, but the Moesian continent was already approaching the Eurasian margin in Dobrogea and the Crimea. In the Early Jurassic the subduction zone was reorganized into two new arcs: the Great and Lesser Caucasian (or Pontiaen).

In the Early Jurassic the Neo-Tethys Ocean was probably completely opened. If this ocean was about 4000 km wide in the east, the Gondwanan continental fragments that evidently traversed it from Gondwanaland to the Eurasian margin during 60 Ma (from 250 to 190 Ma) must have moved about 7 cm/yr. Consequently, the Neo-Tethys Ocean was formed from a comparatively fast-spreading ridge.

Late Jurassic, 155 Ma (Figure 171)

From 190 to 155 Ma, Gondwanaland moved sinistrally about 250 km southeastward with respect to Europe, creating extension in the western Tethys. However, in the eastern Tethys Gondwanaland moved northeastward relative to Europe, causing a slight narrowing of the Tethys Ocean. As in the preceding time interval, the southern edge of the Tethys was the passive margin of Gondwanaland, whereas its northern edge was an active convergent margin near Eurasia. The Lesser Caucasian arc was more extensive than before, and behind it, the Great Caucasian marginal basin opened. The Moesia microcontinent (including the Pontides) collided with the western segment of the Lesser Caucasian arc very soon after the arc was formed. The Cimmerian deformation in the segment from Dobrogea to the western Crimea is related to this collision. The large width (4000 km) of the eastern Tethys Ocean suggests that its mid-oceanic ridge continued to spread as before.

Early Cretaceous (Neocomian), 130 Ma (Figure 172)

In the Early Cretaceous, the general situation remained the same as in the Late Jurassic. Gondwanaland continued to move southeast sinistrally relative to Eurasia at 3 cm/yr in the Late Jurasic and 1-1.5 cm/yr in the Early Cretaceous, as oceanic crust was generated within the Tethys Ocean, both in its western and eastern parts. Along the Tethyan northern margin, the Lesser Caucasian island arc was still active. Extension in the back-arc area formed the pre-Black Sea basin, and presumably the oceanic floor of the South Caspian depression and a basin with thinned crust in the site of the Kopeh-Dagh were created at this time. The back-arc basin behind the Lesser Caucasian arc was about 3000 km long and 600 km wide. Available data show that oceanic crust formed in the west, in the region of the ophiolites of the Apuseni Mountains in Romania and of the central nappes and "Klippen" zones of the Carpathians. In the east, in the Pamirs, the South Pamirs continental block (earlier rifted from Gondwanaland) converged with the Eurasian margin and eventually collided, as shown by the deformation within the Rushan-Pshart suture zone. Thus the South Pamir microcontinent became attached to the pre-existing accretionary prism that had developed along this margin.

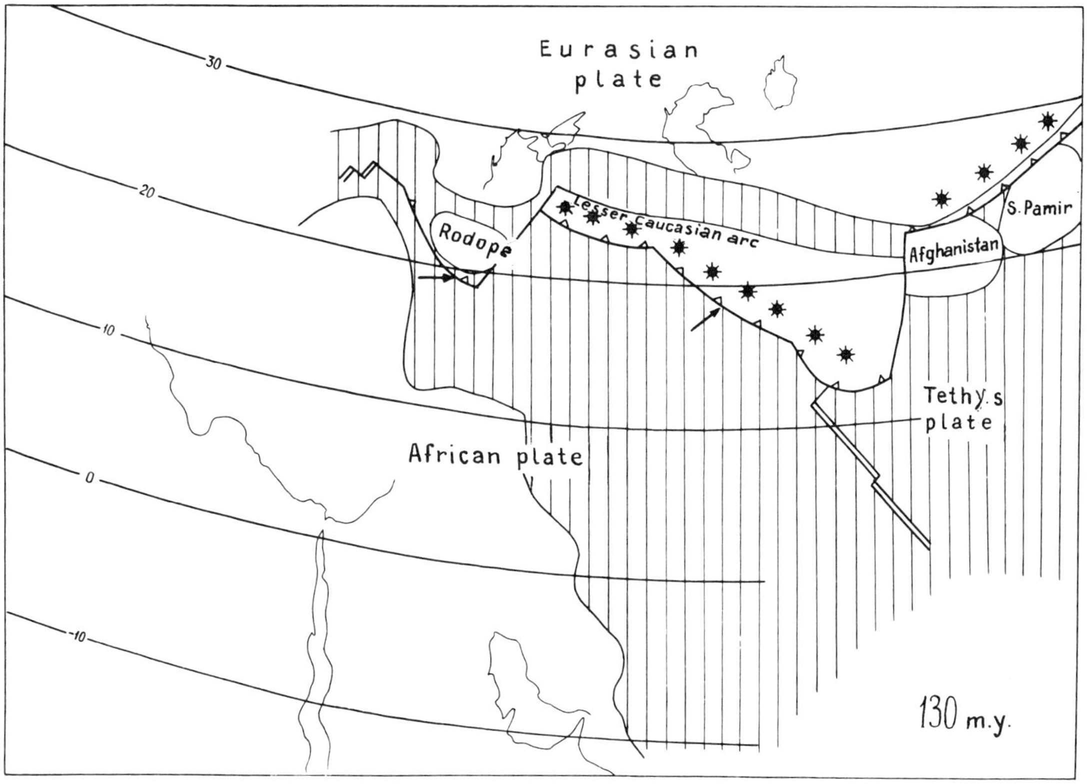

Fig. 172. Palinspastic reconstruction of the Tethys belt for 130 Ma. See Figure 167 for explanation.

End of the Early Cretaceous, 110 Ma (Figure 173)

In the Early Cretaceous, Gondwanaland continued to move sinistrally with respect to Eurasia, with concurrent extension and generation of oceanic crust in the western Tethys. However, the eastern part of the Tethys Ocean shortened by nearly 1000 km, so the subduction zone and Lesser Caucasian volcanic arc above it were still active along that part of the northern Tethys margin.

At this time the Tethyian mid-oceanic ridge apparently approached and encountered the northern subduction zone along the Lesser Caucasian arc. The Tethyian lithosphere in that sector was totally consumed under Eurasia. For a certain time, Eurasia was in contact with the Gondwanan plate, and some blocks could have been torn from the Eurasian margin and could have moved with the Gondwanan plate. For example, the Daralagez block may have been separated in this way from the Iranian plate (or microplate) and may have been displaced northwestward along the Eurasian margin, eventually colliding with the Lesser Caucasian arc. The direct contact of the Gondwanan plate with the Eurasian plate was evidently short-lived. As the Alpine-Himalayan belt contains widespread Middle Cretaceous ophiolites, particularly in the Peri-Arabian part of the belt (Oman, Troodos, Zagros [Aubouin et al., 1986]), it can be supposed that, soon after the disappearance of the former spreading ridge, a new spreading axis came into existence. It was most likely situated in the South Tethys near the Gondwanan passive margin.

Late Cretaceous, 80 Ma (Figure 174)

From 110 to 80 Ma the general movement of Africa with respect to Eurasia remained sinistral. Nevertheless, it led to convergence in and near the Tethys Ocean; for example, about 200 km of shortening in the Carpathians and up to 1000 km in the Kopeh-Dagh.

As Africa and Eurasia converged, the subduction zone on the Eurasian margin, with the Lesser Caucasian arc above it, was still active. In this time span an arc originated in the Carpathian basin, caused by movment of the Apulian microcontinent toward Europe. Collisions of microcontinents and island arcs led to deformation in many segments of the Eurasian active margin. Within the territory of the Soviet Union, the most important deformation and nappe formation was along the Sevano-Akera suture, that was formed when the Daralagez block collided with the Lesser Caucasian arc in Coniacian time, 88-85 Ma ago. By the end of the Early Cretaceous, in the Albian, the Apulian block collided with the Carpathian arc, and at this time the chaotic "Klippen" zone was formed.

After collision in the Lesser Caucasus, the subduction zone jumped southward, and the Tethyan Ocean floor was subducted beneath the accreted Daralagez block. As a result, the previous collisional zone was "released" and was in extension rather than compression. At this time, the Lesser Caucasian arc gave place to the Adjaro-Trialetian arc, which was positioned somewhat to the south relative to the Lesser Caucasian arc.

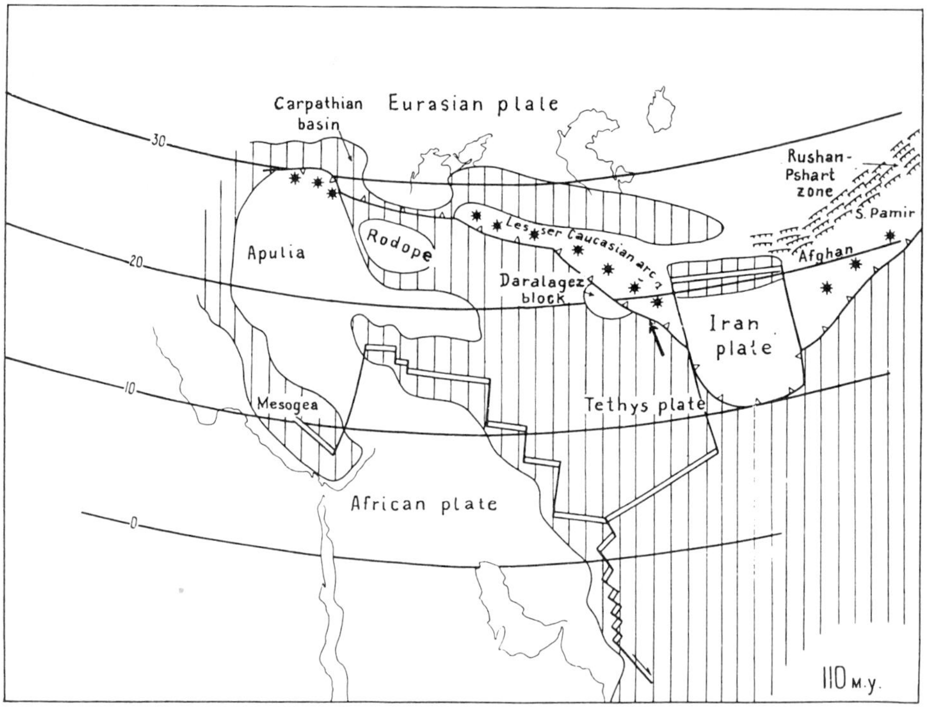

Fig. 173. Palinspastic reconstruction of the Tethys belt for 110 Ma. See Figure 167 for explanation.

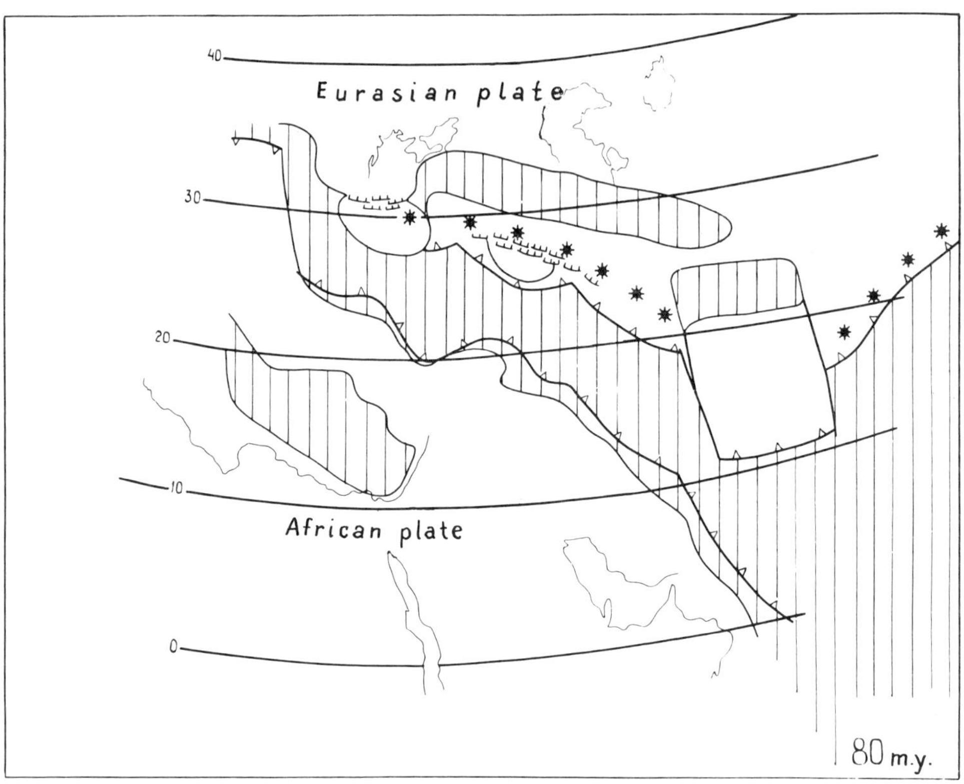

Fig. 174. Palinspastic reconstruction of the Tethys belt for 80 Ma. See Figure 167 for explanation.

Back-arc extension led to opening of the deep Black Sea basins.

A subduction zone undoubtedly existed south of the Pamirs, and produced an arc that was the western continuation of the Kohistan-Ladakh magmatic belt in the Himalayas.

Apparently, in the Late Cretaceous the mid-ocean ridge (and spreading axis) no longer existed within the Tethys Ocean, for at that time the African plate was in contact with Eurasian plate and the oceanic crust was only being consumed, - not created. In the Caucasus the ocean was 1200 km wide and in the Kopeh-Dagh it was 1600 km wide. At the same time, the ocean between the Pamirs and Indian continent was 6000 km wide.

Cretaceous-Paleogene Boundary, 65 Ma (Figure 175)

A drastic change in plate kinematics in the Tethys realm occurred in the Late Cretaceous. The pole of Africa/Eurasia rotation shifted from its former location in North Europe to the Straits of Gibraltar. Thus Africa and Arabia started moving nearly directly northward with respect to Eurasia. Consequently, the subduction zone was active along the Africa/Eurasia boundary; it is marked by the Adjaro-Trialetan volcanic arc. During 15 million years approximately 250 km of the Tethyan oceanic crust was consumed under the arc, at a rate of about 1.5-1.7 cm/yr.

The back-arc basin behind the Lesser Caucasian arc achieved its maximum dimensions--3000 km long and 900 km wide (Figure 175). This extensional depression may be called the Para-Tethys basin.

The Pamirs were evidently part of the Kohistan-Ladakh magmatic arc under which about 1600 km of Tethyan oceanic crust was consumed during the same 15 million years. During this span India, like Africa, changed its motion, which by 80 Ma became northward directly toward Eurasia at a rate as high as 11-12 cm/yr. Within the Tethys, a mid-oceanic ridge may have been active for a short time, and it may have had an eastward trend toward the Pacific Ocean. However, it was fairly rapidly brought to the northern subduction zone, and Eurasia came into contact with the Indian plate. Ophiolite obduction and the first thrusting in the Himalayas may have been caused by the encounter of the postulated mid-ocean ridge with the magmatic arc.

Eocene-Oligocene Boundary, 35 Ma (Figure 176)

During the Paleocene and Eocene, African and Indian plates rapidly converged with Eurasia; consequently, the Adjaro-Trialetan arc continuously developed along the Eurasian plate margin, and in the Caucasus the crust between Africa and Eurasia shortened by 700 km. Probably by the end of Eocene time the Tethyan oceanic crust was completely consumed, and the passive Arabian margin, with thinned continental crust, approached the subduction zone. The continental collision of the tip of the Arabian promontory started the consumption of the Para-Tethys oceanic crust under the opposite, northern side

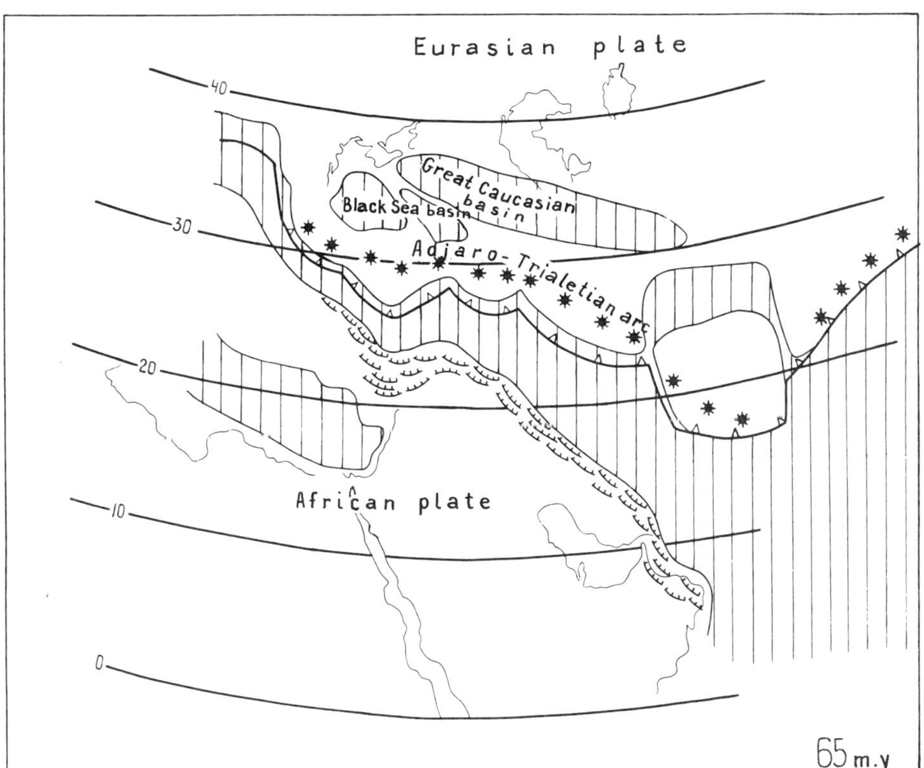

Figure 175. Palinspastic reconstruction of the Tethys belt for 65 Ma. See Figure 167 for explanation.

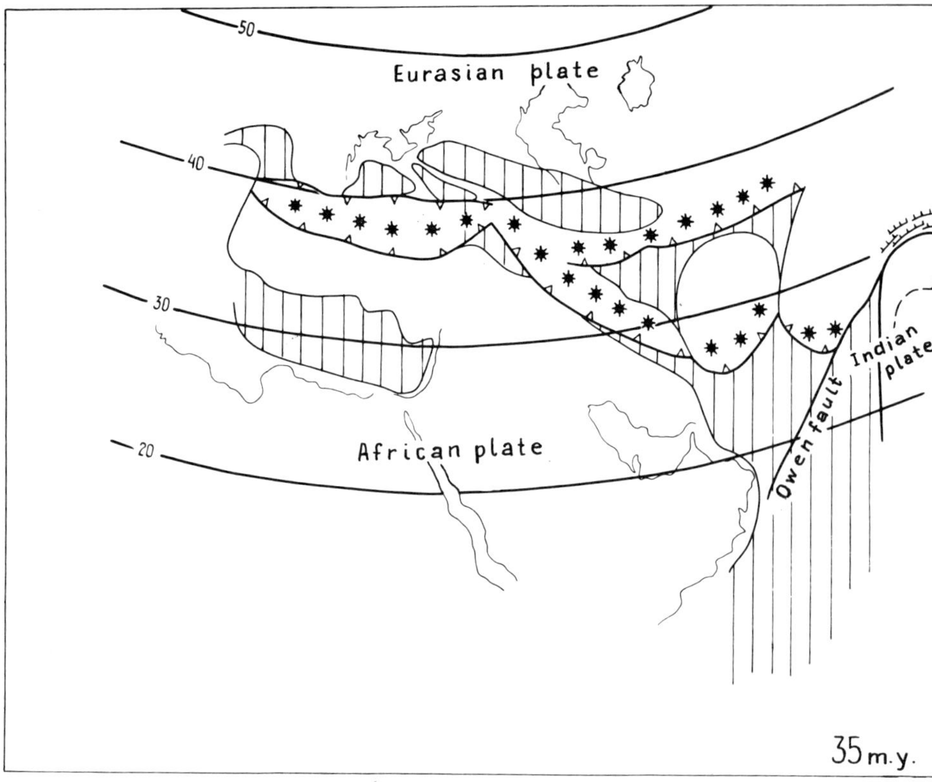

Fig. 176. Palinspastic reconstruction of the Tethys belt for 35 Ma. See Figure 167 for explanation.

of the Adjaro-Trialetan arc; this underthrusting produced the Late Eocene and Early Oligocene deformation and olistostromes of the northern slope of the Lesser Caucasus along its boundary with the Rioni depression and Dzirula massif. But a seaway from the West Tethys into the East Tethys was still open. It was interrupted only at 35 Ma (Oligocene), when the global sea level dropped 400 m. As a result of lowered sea level, an isolated basin replaced the Para-Tethys, and in it black carbonate-free shale of the Maikop suite accumulated under stagnant conditions.

The end of the Paleogene was marked by collision of India with the Pamir-Kohistan-Ladakh magmatic arc and the Eurasian margin, and the beginning of the recent stage of deformation (folding and thrusting) in the Pamirs.

Early Miocene, 20 Ma (Figure 177)

According to kinematic data (Table 8), from 35 to 20 Ma both the African (including Arabia) and Indian plates changed their motion direction relative to Eurasia. The pole of Africa/Eurasia rotation shifted from the region of the Straits of Gibraltar to Angola in the southern hemisphere, while that of India/Eurasia, formerly in NW Africa, shifted to Somalia. As a result, both Africa and India began moving northwestward with respect to Eurasia. Africa's motion, with an azimuth of 290°-320°, was relatively slow so the continent traveled only 150-160 km during 15 million years. India moved a much larger distance, 870 km, but still more slowly than in the previous period before collision. Such a movement should have caused a general displacement to the west of crustal blocks along the southern Eurasia margin, resulting in dextral motions along sublatitudinal and NW-trending strike-slip faults. Actually, in the Oligocene and Early Miocene the Afghanistan and Iranian blocks did experience such displacements, and crustal masses piled up near the tip of the Arabian promontory. These motions formed the Gissar, Gerurud, and Kopeh-Dagh strike-slip faults, and deformed and thrust the blocks westward onto the Turan platform. The Adjaro-Trialetan arc in the eastern Lesser Caucasus was sharply bent oroclinally and the Kafansky block was displaced 30 km westward during the "escape" of crustal blocks that were forced away from the collision. Westward crustal motions are also recorded west of the Arabian promontory, where the Turkish and Balkan blocks protruded towards the western Mediterranean. This is believed to have influenced the formation of Carpathian loop. The Mesozoic oceanic crust on the floor of the Carpathian basin was consumed under a newly formed volcanic arc.

Because of these movements, the Para-Tethys basin began to close, particularly near the tip of the Arabian promontory where the Great Caucasus mountain chain began to rise. The remnants of the Para-Tethys basin became entirely isolated from the world ocean, for the Tethys Ocean no longer existed as a laterally continuous body. The remnant western Tethyan basins, such as Mesogea and the Arabian basin of the Indian Ocean, were separated by juxtaposed continents.

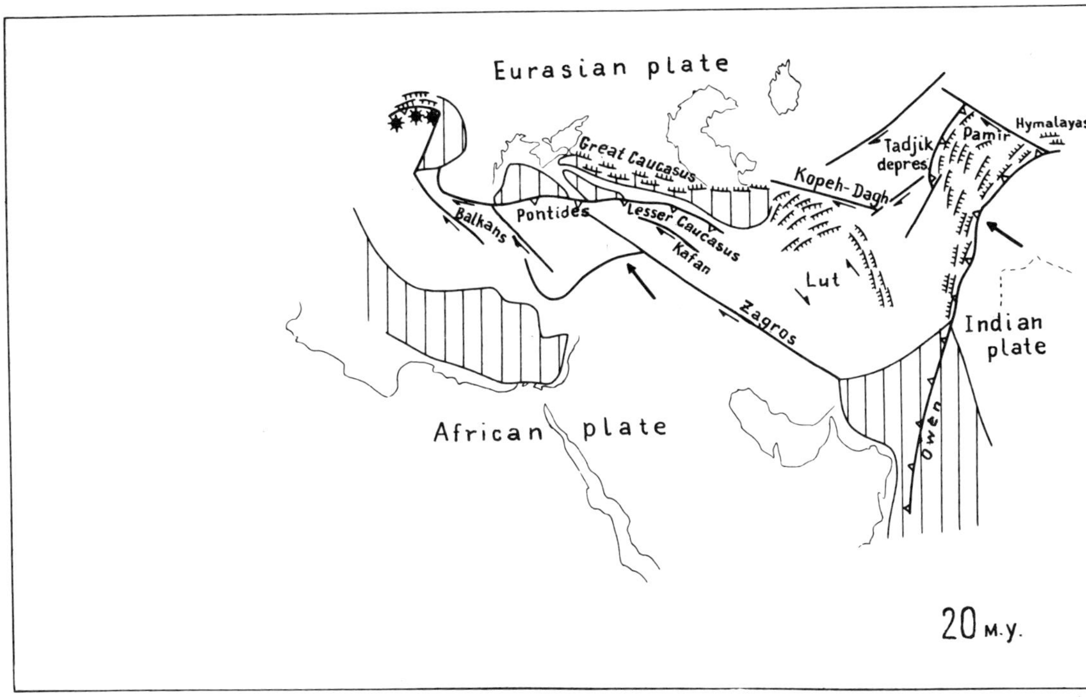

Fig. 177. Palinspastic reconstruction of the Tethys belt for 20 Ma. See Figure 167 for explanation.

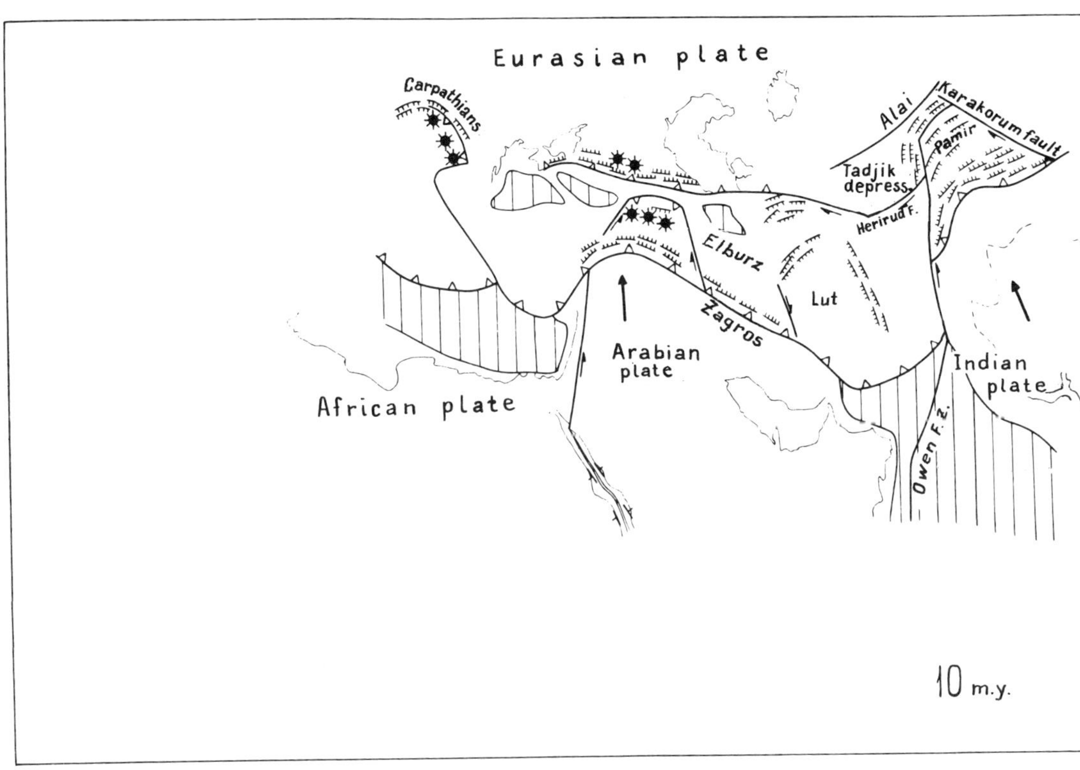

Fig. 178. Palinspastic reconstruction of the Tethys belt for 10 Ma. See Figure 167 for explanation.

Late Miocene, 10 Ma (Figure 178)

From 20 to 10 Ma both Africa and India again moved northward relative to Eurasia, but the plate motions were the lowest in the whole history of the Tethyan region, - only 0.75-1 cm/yr between Africa and Eurasia and 3.2-3.5 cm/yr between India and Eurasia. Thus, plate convergence was only 100 km between Africa and Eurasia and 300 km between India and Eurasia. In the Alpine-Himalayan belt within the USSR territory, this was a time of general compression; the major nappes were formed in the Pamirs and the nappes continued to move in the Kopeh-Dagh.

In the Late Miocene the Caucasus region was cardinally reorganized, and the Great and Lesser Caucasus ridges were strongly uplifted and compressed. Two subduction zones were formed, one dipping north under the Great Caucasus and the other dipping south under the Lesser Caucasus. They consumed the remnants of oceanic crust of the Para-Tethys basin. Volcanics were erupted above the subduction zones.

In Miocene-Pliocene time the Carpathian arc shifted to the east and the former oceanic (and thinned continental) crust of the Carpathian basin was entirely consumed.

Thus, the Alpine-Himalayan belt can be considered as a typical collisional foldbelt that originated as a consequence of convergence and eventual collision of Eurasia with Gondwanan continents.

Chapter XII

LATE PALEOZOIC AND MESOZOIC-CENOZOIC SEDIMENTARY BASINS

The territory of the USSR includes several of the greatest sedimentarty basins of the world, such as the West Siberia, Turanian, Scythian, and Pechora-Barents Sea basins. All were initiated as a result of subsidence due to extension of the continental crust within rifts and/or aulacogens. In some cases they were further developed by the generation of oceanic crust in a spreading environment. They are underlain by a heterogeneous basement incorporating both ancient platforms and Paleozoic foldbelts which had all been deeply eroded before the rifting and subsequent subsidence, and some are underlain in part by oceanic crust.

West Siberian Basin

The West Siberian basin (Figures 179 and 180) originated above Triassic grabens that are well recorded in the basement by drilling and seismic profiling. Aplonov [1987 and 1988] removed the long wave component of the magnetic field, and in the residual short wave component he found symmetric magnetic lineations (Figure 181) entirely comparable with the linear magnetic anomalies of oceans. Thus he could outline the wedge-shaped area of oceanic crust under the sedimentary cover of the West Siberian basin (so-called Ob Paleo-Ocean). Aplonov correlated the anomalies with the magnetic reversal scale and determined the age of the buried oceanic crust to be within 235-218 Ma. The amount of opening in the northern part of West Siberia was 270 km at a rate of 1.6 cm/yr. The pole of opening of the Ob Paleo-Ocean was approximately at 60° N and 80° E, and the angle of opening was 13.4°. The sedimentary cover of the West Siberian basin is up to 10 km thick in the north, where the lowest sediments are Triassic in age [Nesterov et al., 1984b; Druzhinin et al., 1988]. In the south, the section begins with Jurassic sediments and it continues to Holocene. The sequence is clastic and is characterized by an alternation of a shallow-water marine and terrestrial facies. The epochs of transgressions alternating with regressions generally coincide with the eustatic oscillations of sea level shown by the curve of Vail et al. [1977]. Some transgressions were uncompensated by deposition, as in the Late Jurassic when the sea was 700 m deep [Nesterov et al., 1984a]. At that time the famous bituminous shales of the Upper Jurassic Bazhenov suite were deposited; they are both oil-generating and oil-producing.

(Figures 179 and 180 appear on pages 200 and 201.)

The Turanian and Scythian Basins

The Turanian and Scythian basins (see Figure 149) came into existence in Jurassic time. Narrow grabens cut the base-

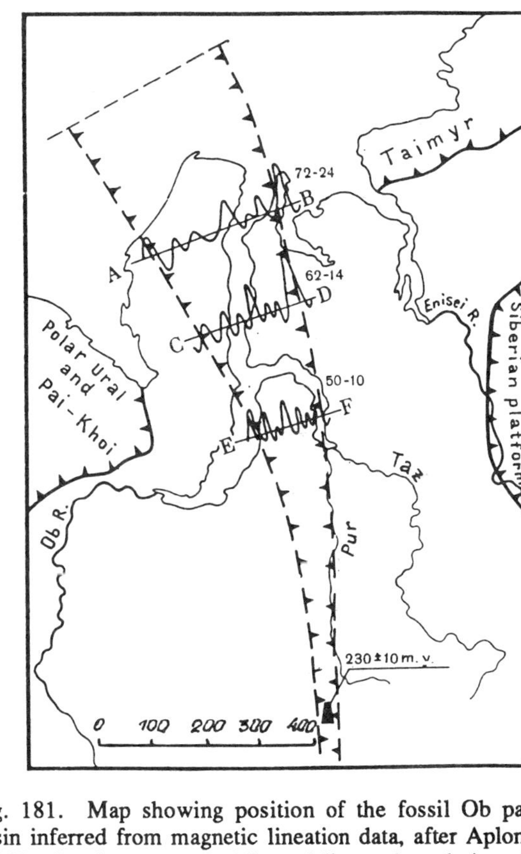

Fig. 181. Map showing position of the fossil Ob paleo-ocean basin inferred from magnetic lineation data, after Aplonov [1987, 1988]. Continuous toothed lines delineate boundaries of the West Siberian basin. Dashed toothed lines in the center show the boundaries of the Triassic oceanic floor, and three magnetic profiles displaying symmetric lineation are shown.

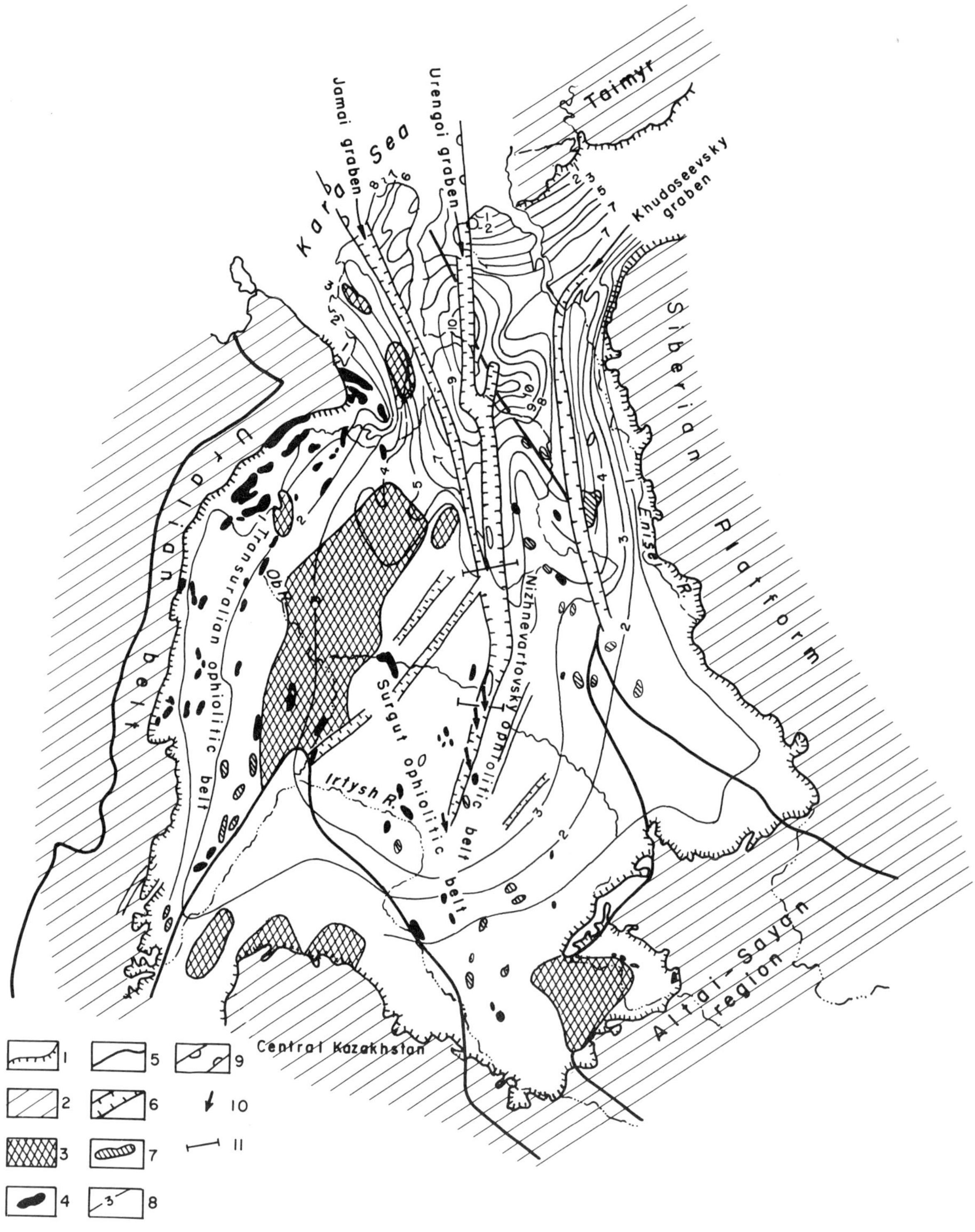

Fig. 179. Structure of the West Siberian basin.
1, Boundary of the basin; 2, Precambrian and Paleozoic basement outside the basin; 3, Precambrian massif concealed under the sedimentary cover; 4, ophiolite; 5, Paleozoic suture; 6, aulacogen and rift; 7, flood basalt; 8, isopachs of the sedimentary cover (in km); 9, boundary of the inferred Ob oceanic floor; 10, site of borehole that penetrated basalt; 11, location of cross-sections shown in Figure 180.

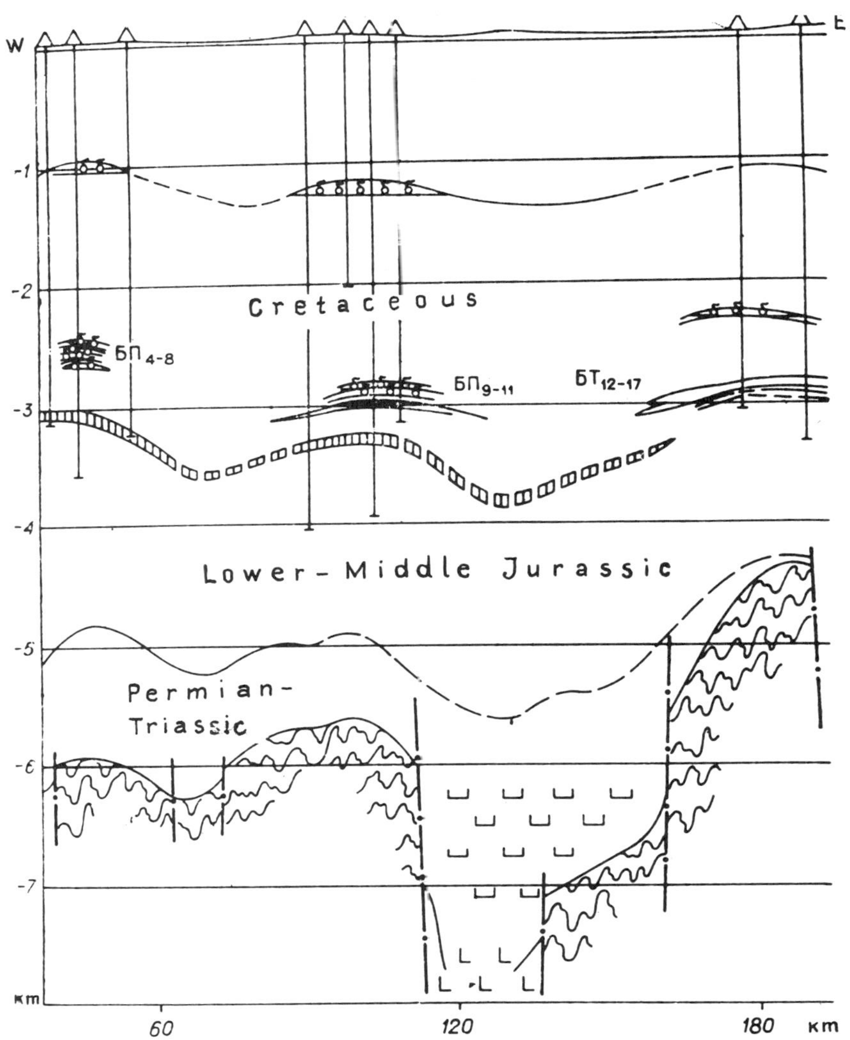

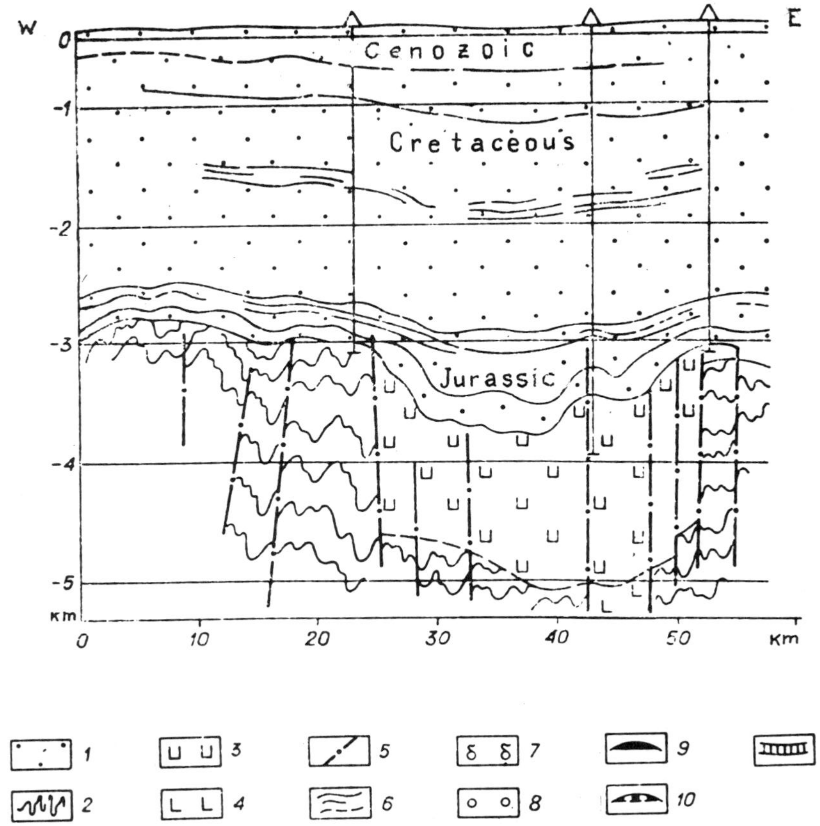

Fig. 180. Cross-sections through the central part of the West Siberia basin, after Surkov et al. [1982] See Figure 179 for location of cross-sections, along lines I-I (lower section) and II-II (upper section). Vertical exaggerations: I - 10x; II - 25x.
1, Sedimentary cover; 2, Precambrian and Paleozoic basement; 3, coarse clastics (graben facies); 4, basalt; 5, fault; 6, seismic boundaries; 7, gas deposit; 8, gas condensate; 9, oil deposit; 10, supposed oil deposit; 11, Bazhenov suite.

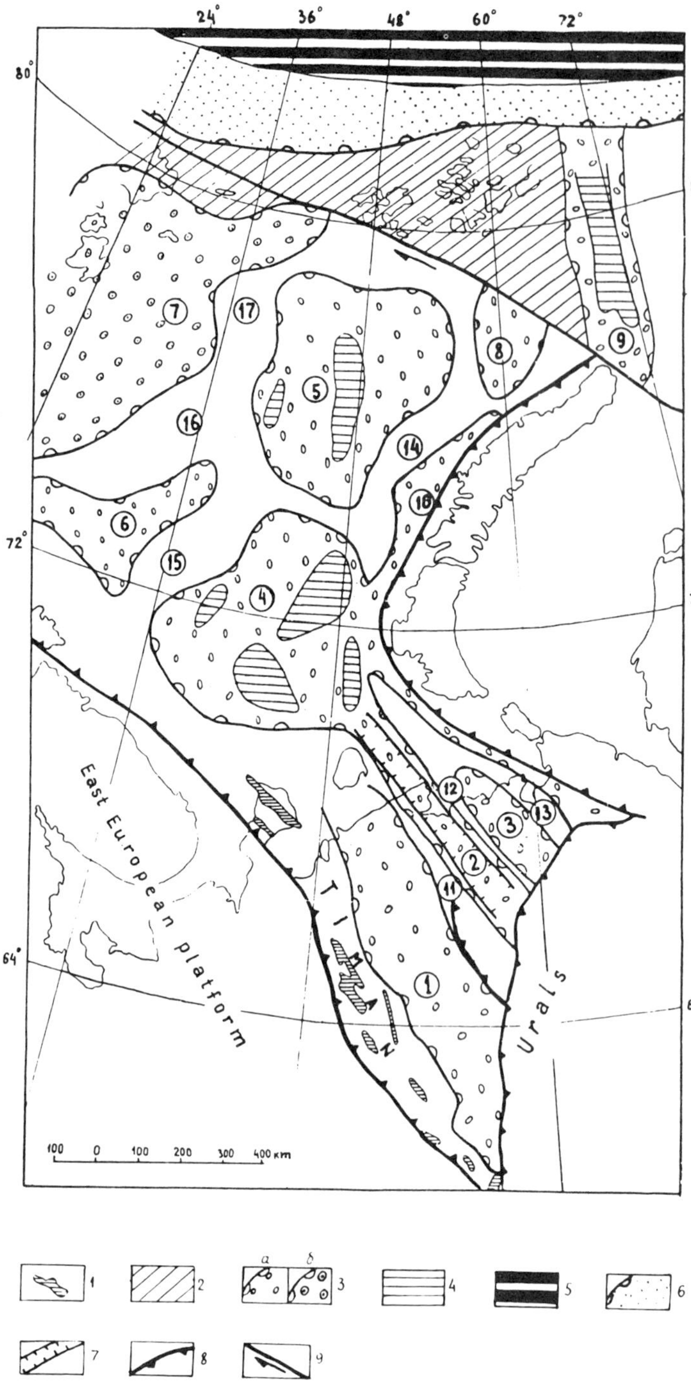

Fig. 182. Structure of the Pechora-Barents Sea basin.
1, Outcrops of the Precambrain basement within Timan; 1, uplift of Svalbard and Franz Josef Land; 3, sedimentary basin (a - evolving since the Devonian, b - evolving since the Carboniferous); 4, postulated oceanic floor; 5, oceanic floor of the Eurasian basin of the the Arctic Ocean; 6, Cenozoic passive margin; 7, graben; 8, thrust front; 9, Taimyr-Novaya Zemlya transform fault.
Depressions: 1, Izhma-Pechora; 2, Denisovsky; 3, Khoreiver; 4, South Barents; 5, North Barents; 6, North Cape; 7, Olgin; 8, North Novaya Zemlya; 9, Saint Ann; 10, West Novaya Zemlya. Swells: 11, Pechora; 12, Kolva; 13, Sorokin; 145, Admiralteistvo; 15, Fedynsky; 16, Central Barents; 17, Persei.

ment of the basins; in places listric faults have been mapped by milti-channel seismic reflection [Popkov, 1986]. For a long time, i.e., throughout the Jurassic, Cretaceous, and Paleogene, these basins occupied a vast shelf on the northern margin of the Tethys Ocean. The Jurassic to Lower Miocene sedimentary series constitutes a single complex in which continental sandstone-shale and coal-bearing sequences are replaced by marine terrigenous-carbonate deposits. The sediments attain a thickness of 8-9 km, or in places, 10-12 km. The sea remained in the Turan and Scythian basins until the Middle Miocene, when the collision of Arabia and Eurasia led to mountain building in the Kopeh-Dagh and the Caucasus.

Pechora-Barents Sea Basin

The Pechora-Barents Sea basin (Figures 182 and 183) presumably incorporates basins of several different ages [Baturin and Yunov, 1987; Kozloysky, 1984]. The earliest episode of extension in this area was as early as Early Devonian, the time of origin of the Pechora-Barents Sea rift system, when subsidence of the Pechora lowland and a major part of the Barents Sea began. The second episode occurred in the Carboniferous when the Late Devonian Ellesmerean orogeny was followed by an extensional period responsible for formation of the Sverdrup and Wandels Sea basins. In the Barents Sea this episode is recognized in the western part adjacent to Svalbard. Evidently extension was widespread in the Early Triassic, for example, as in Western Siberia. The last rifting event took place when the Eurasian basin opened in the Late Cretaceous and Early Paleogene. Geophysical data [Volk et al., 1984] indicate that in some areas of the Barents Sea the crust is no more than 7 to 9 km thick, where 'windows' of oceanic crust are preserved in the basement of the Barents Sea shelf. The windows originated as a result of the opening of Devonian rifts. The maximum thickness of sediments in the Pechora-Barents Sea basin is 16-18 km [Shipilov and Senin, 1988], the Upper Paleozoic sediments being 5-6 km and the Triassic 6-7 km thick. The structural relationship between the different stratigraphic complexes are far from clear. It is important to underline that basalts are especially abundant in the Lower Cretaceous.

Thus we see that the deepest sedimentary basins--West Siberian and Pechora-Barents Sea as well as the Peri-Caspian depression--contain remnants of oceanic crust and thus originated as "underdeveloped" oceans which failed to reach maturity.

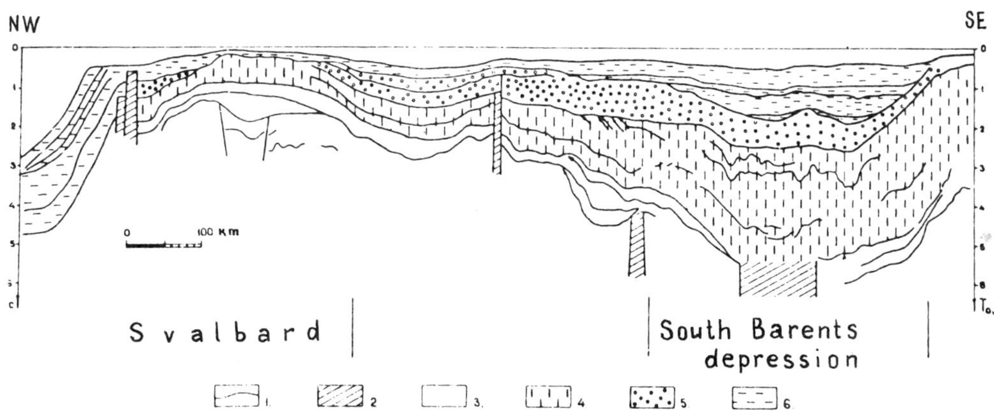

Fig. 183. Interpretive geological cross-section through the Barents Sea floor inferred from multi-channel reflection data, after Baturin and Yunov [1987].
1, Reflector; 2, fault zone; 3, Pre-Mesozoic; 4, Triassic; 5, Jurassic; 6, Cretaceous and Cenozoic. Vetical scale is two-way travel time.

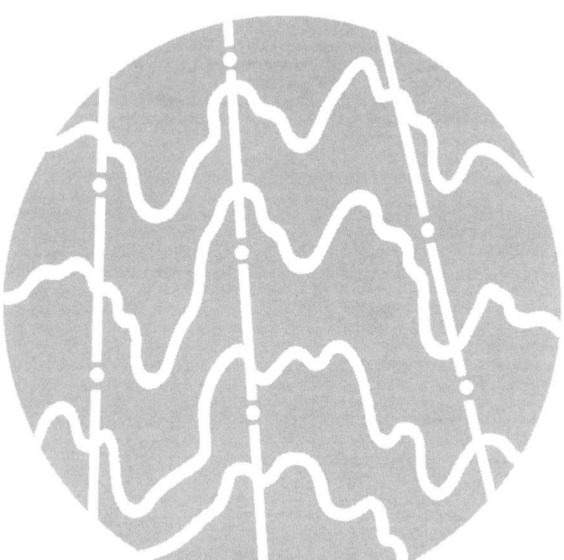

Chapter XIII

RECENT PLATE TECTONICS

At present, the largest portion of the USSR territory is in an intraplate situation, being a part of the Eurasian plate. The current plate boundaries extend along the margins of the USSR (Figure 184) and are marked by seismic belts (Figure 185). Zonenshain et al. [1983] analyzed the present plate interactions and recognized four seismic belts: Kurile-Kamchatka, Arctic, Inner Asian, and Mediterranean, each being a plate boundary.

The Kurile-Kamchatka seismic belt is confined to the island arc of the same name and separates Eurasia (actually the Sea of Okhotsk plate) from the Pacific plate which is being subducted under Eurasia at 10 cm/yr.

The Arctic seismic belt is a segment of the boundary between the Eurasian and North-American plates. It extends along the Gakkel Ridge (Arctic Mid-Ocean Ridge) in the Arctic Ocean where it coincides with the spreading axis along which the Eurasian ocean basin floor is forming. In the Lena River delta the seismic belt continues onto the continent, proceeds along the Chersky Range, and then divides: one branch goes east to join the Kurile-Kamchtaka belt, and the other goes south toward Sakhalin Island to join the Inner Asian belt. Within the continent, in contrast to the ocean, the Arctic seismic belt is dispersed in a band 100-150 km wide. The pole of rotation of the North American and Eurasian plates is situated in the region of the Verkhoyansk Range. For this reason, north of the pole, the North American and Eurasian plates diverge at 1 cm/yr but in the south, in the Chersky Range, they converge. However, these recent kinematics do not explain why the Chersky Range is cut by the Momsky Rift, a series of narrow grabens that formed by extension. This is not a contradiction if we assume that the present pattern of relative movement of the North American and Eurasian plates was established very recently, whereas the grabens of the Momsky Rift were formed in the Miocene and Pliocene when the pole of the North American and Eurasian plate motions was located south of the Chersky Range, in the Magadan region, as was calculated by Zonenshain et al. [1978].

The Inner Asian seismic belt has the mostly complicated pattern. It extends from the Sea of Okhotsk (from Udsky Bay) through Baikal and Mongolia to Tibet and the Himalayas, where it joins the Mediterrranean belt. It consists of two segments: a northeastern, or Baikal-Stanovoy, and southwestern, or Central Asian segment.

The Baikal-Stanovoy segment, particularly in the Baikal area, is characterized by concentrated seismicity. This segment separates the Amur plate from Eurasia [Zonenshain and Savostin, 1981] and includes the Baikal rift zone with its rift-like grabens (the Baikalian, Barguzin, Upper Angara, Muya, and Chara grabens, etc.) and the zone of recent structures of the Stanovoy Mountains. Zonenshain and Savostin [1981] used slip-vector orientation obtained from fault plane solutions, to calculate the Amur/Eurasia pole of rotation, which is now at 56.3° N, 118.4° E, near the eastern termination of the Baikal rift zone. Accordingly, the Amur plate is rotating counterclockwise, diverging from the Eurasian plate along the Baikal rift zone and converging along the Stanovoy Mountains.

The Inner Asian seismic zone terminates in the east, near the Sea of Okhotsk, by joining the southern part of the Arctic belt. To the east and southeast of this terminus, in the Sakhalin seismic belt, earthquakes are concentrated in the longitudinal zone stretching along Sakhalin to the western margin of Honshu Island of Japan. The aseismic area of the Sea of Okhotsk, referred to as the Sea of Okhotsk plate, is enclosed between three or four seismic belts of the eastern USSR [Zonenshain and Savostin, 1979]. Savostin et al.[1981] calculated parameters of the clockwise rotation of the Sea of Okhotsk plate with respect to the Amurian and Eurasian plates using slip-vector orientations and trends of strike-slip faults. The pole of rotation of the Sea of Okhotsk and Eurasian plate lies at a point near the east end of the Inner Asian seismic belt. For this reason, in north Sakhalin the Sea of Okhotsk plate moves dextrally, (along N-S faults) with respect to the Amur plate but in South Sakhalin it converges with the Amur plate along the NNE-trending plate boundary (Figure 184). The younger Post-Pliocene deformations of Sakhalin are caused by interaction of the Sea of Okhotsk and Amur plates.

The southwestern, Central Asian segment of the Inner Asian seismic belt begins at the southwest end of Baikal Lake. In contrast to the northeastern segment, it is characterized by diffused seismicity distributed over a vast area 1000 km wide area. Within the USSR, this band of diffused seismicity incorporates the East Sayan, Tuva, Altay, Junggar, Tien Shan, and Pamir, ranges as well as the rejuvenated mountains of Central Asia. The formation of these mountains and the seismic belt proper is undoubtedly related to the India/Eurasia collision [Molnar and Tapponier, 1975]. Zonenshain and Savostin [1981] tried to outline zones of high earthquake concentration and consider them as diffused boundaries of small plates, subplates, and crustal blocks. They distinguished in Central Asia the Mongolian, Junggar, Tarim, Tibet, Pamirs, Afghan and

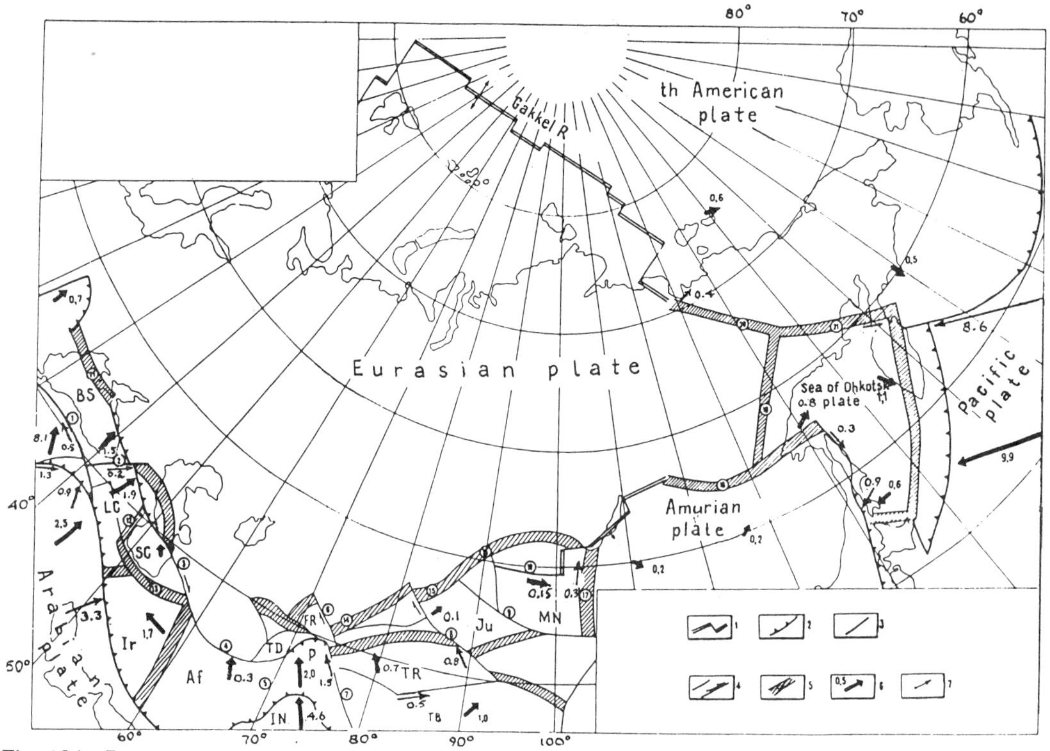

Fig. 184. Present plate boundaries within the USSR territory, after Zonenshain et al. [1983].
1, Spreading axis; 2, subduction zone; 3, transformn fault; 4, strike-slip fault; 5, diffuse plate boundary; 6, vector of plate motion with respect to Eurasia; 7, vector of relative plate motions along plate boundaries.
Plates and microplates: Af - Afghan, BS - Black Sea, FR - Fergana, IN - Indian, Ir - Iran, Ju - Junggar, LC - Lesser Caucasus, MN - Mongolian, P - Pamirs, SC - South Caspian, TB - Tibetan, TD - Tadjik, TR - Tarim.

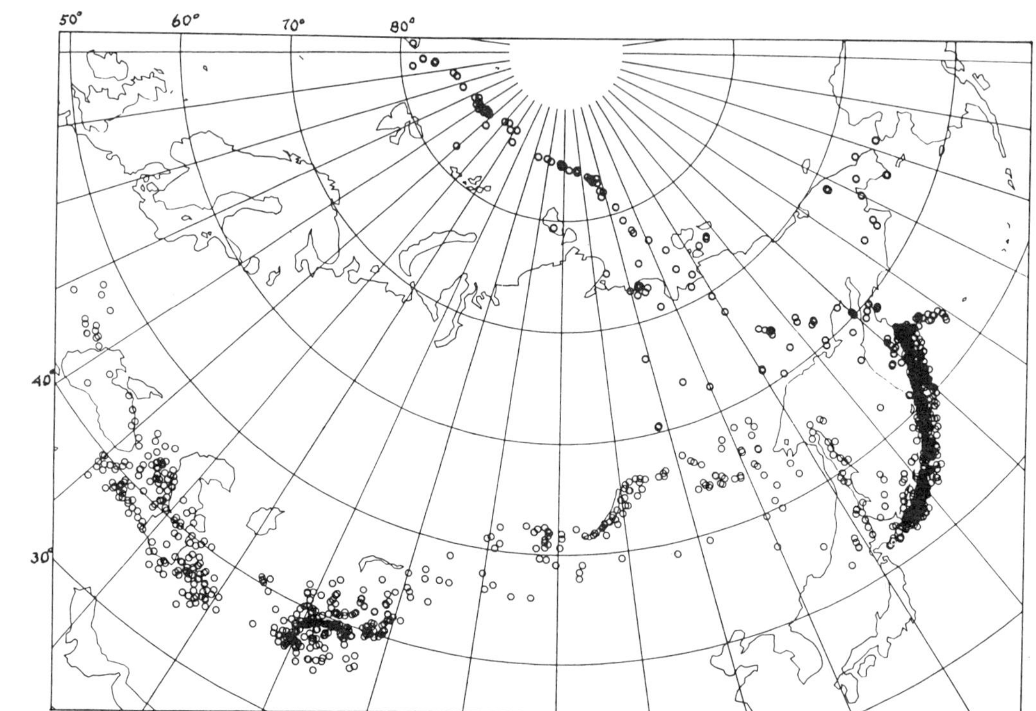

Fig. 185. Distribution of earthquake foci within the USSR territory for the period 1970 to 1985. All events of magnitude >4.5 are plotted after Lander (unpublished).

TABLE 11. Parameters of Present Plate Motions Within the USSR and Adjacent Areas

Plates and Microplates	Pole position N	Pole position E	Rate (10^{-7} degree/yr)	Source
1	2	3	4	5
North America-Eurasia	65.85	132.44	-2.31	Minster and Jordan [1978]
North America-Eurasia	59.48	140.83	-1.89	Savostin and Karasik [1981]
India-Eurasia	19.71	38.46	6.98	Minster and Jordan [1978]
Arabia-Eurasia	35.7	4.4	4.8	Patriat et al. [1982]
Arabia-Eurasia	29.82	-1.64	3.57	Minster and Jordan [1978]
Africa-Eurasia	25.23	-21.19	1.04	Minster and Jordan [1978]
Africa-Eurasia	30.9	-20.4	1.7	Patriat et al [1982]
Pacific-Eurasia	60.64	78.92	-9.77	Minster and Jordan [1978]
Pacific-North America	48.77	-73.91	-8.52	Minster and Jordan [1978]
Sea of Okhotsk	52.24	140.09	-8.31	Savostin et al. [1983]
Amur-Eurasia	56.35	118.1	1.0	Zonenshain and Savostin [1979]
Sea of Okhotsk	50.57	143.93	-6.94	Savostin et al. [1983]
Sea of Okhotsk North America	44.09	144.85	-4.78	Savostin et al. [1983]
Mongol-Eurasia	57.72	98.46	0.83	Zonenshain and Savostin [1979]
Mongol-Amur	50.14	140.08	-0.62	From closure
East Sayan-Mongol	51.73	101.49		Zonenshain and Savostin [1979]
Mongol-Junggar	51.8	51.8	0.75	Zonenshain and Savostin [1979]

TABLE 11. (continued)

Plates and Microplates	Pole position N	Pole position E	Rate (10^{-7} degree/yr)	Source
1	2	3	4	5
Junggar-Eurasia	56.93	-2.77	0.135	Zonenshain and Savostin [1979]
Tarim-Tibet	14.26	99.2	2.0	Zonenshain and Savostin [1979]
Tibet-Pamir	20.23	61.60	-3.6	Zonenshain and Savostin [1979]
Afghan-Eurasia	48.38	61.22	-0.45	Zonenshain and Savostin [1979]
Pamir-Eurasia	19.71	38.46	3.1	Zonenshain and Savostin [1979]
Tarim-Pamir	21.7	31.55	-2.1	From closure
Tibet-Eurasia	8.4	120.4	-1.36	From closure
Tarim-Eurasia	15.0	52.0	1.04	From closure
Tarim-Junggar	9.2	55.6	-0.98	From closure
Afghan-India	28.60	72.40	-11.8	Sborshchikov et al. [1981]
Arabia-India	25.74	-67.6	3.0	Sborshchikov et al. [1981]
Iran-Eurasia	8.0	49.0	3.2	From closure
South Caspian-Eurasia	11.3	26.47	0.6	Vardapetian et al. [1979]
Lesser Caucasus Eurasia	33.3	-30.4	2.03	Vardapetian, [1979]
Black Sea-Eurasia	39.8	78.1	-2.9	Vardapetian [1979]
Black Sea-Lesser Caucasus	59.5	11.7	-1.9	Vardapetian [1979]
Arabia-Lesser Caucasus	20.2	25.2	1.92	From closure
Anatolian-Black Sea	18.8	35.0	2.25	McKenzie [1974]
Arabia-Anatolian	54.9	25.5	3.53	From closure
Anatolian-Eurasia	35.0	142.8	-1.96	From closure

*The first-named plate is rotated with respect to the second-named plate.
+ = Counterclockwise, - = clockwise.

some other minor plates and microplates. They calculated the approximate parameters of relative movement of these blocks. According to their calculations, the mosaic of microplates pushed by India moves northeastward and acts as a wedge, splitting the Amurian plate away from the Eurasian plate. As a result, the Baikal rift is opening near the tip of the wedge.

The Mediterranean seismic belt extends in a latitudinal direction from the Pamirs westward to Gibraltar. It separates the

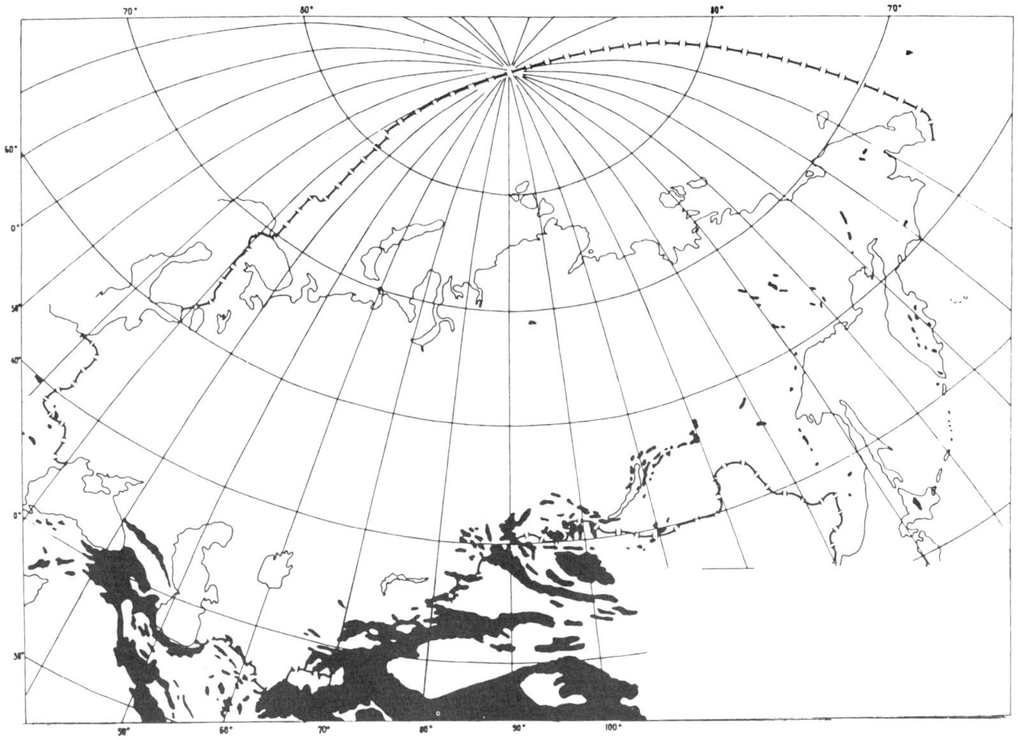

Fig. 186. Simplified topography of the USSR territory and adjacent regions. Mountains over 2000 m above sea level are shown in black.

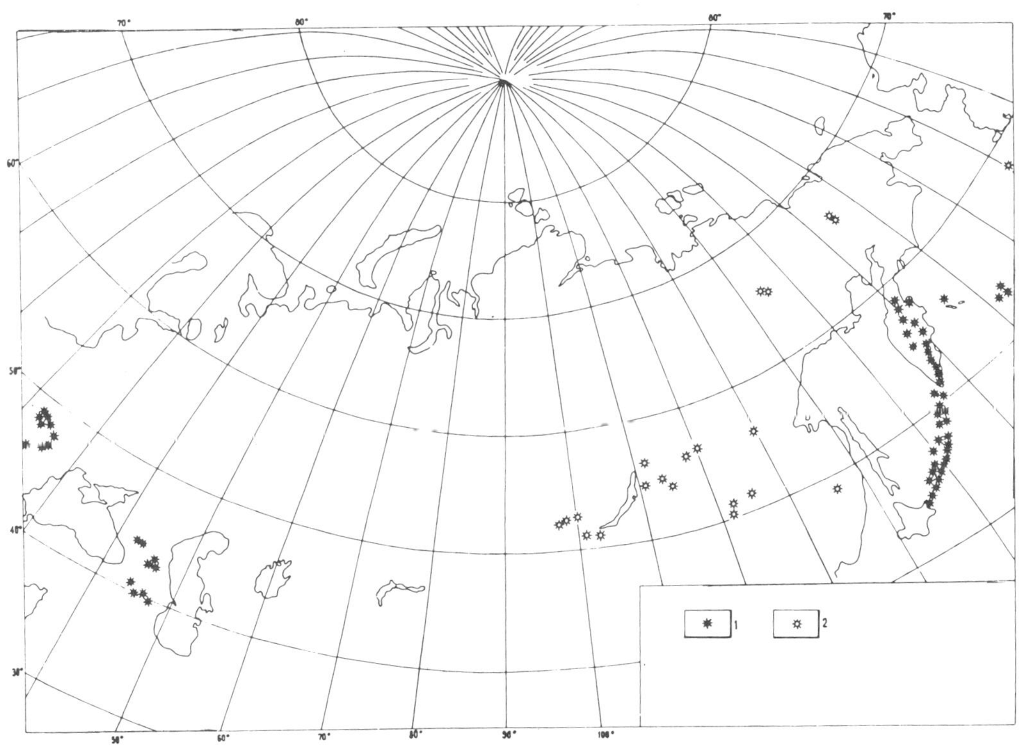

Fig. 187. Present and recent volcanism within the USSR territory and adjacent regions. 1, Calc-alkaline subduction-related volcanics; 2, rift-related and intraplate basalt.

Eurasian plate from the Arabian and African plates, and is caused by convergence and collision of these plates. Like the Central Asian belt, the Mediterranean belt is also characterized by diffused seismicity. Nevertheless, Sborshchikov et al. [1981] distinguished linear zones of high earthquake concentration within central Asia, Iran, and Afghanistan. Vardapetyan [979] made a similar analysis of the Caucasus. These zones of concentrated seismicity were perceived to outline a number of small plates--Black Sea, Turkish, Lesser Caucasus, South Caspian, Central Iran, Lut, etc. The resulting scheme of plate interaction coincides with the scenario proposed much earlier by McKenzie [1974]. The Arabian plate has indented Eurasia, and the microplates and crustal blocks diverge west and east from the Arabian promontory. Within the Soviet Union, the most significant events of recent plate interaction are: (i) subduction of the Black Sea plate under the Great Caucasus (the young calc-alkaline volcanism of the Elburz and Kazbek Mountains are the consequence of this subduction); (ii) compression in the longitudinal zone along the western coast of the South Caspian basin as the Black Sea and South Caspian plates converge; and (iii) dextral strike-slip faulting (with appreciable overthrust component) along the Kopeh-Dagh.

The parameters of relative motions within the Soviet Union are given in Table 11. If we compare the distribution of the main mountain systems more than 2000 m high (Figure 186) with Figure 184, it is evident that high mountains coincide completely with the present plate boundaries. Practically the same coincidence is seen by comparing the distribution of present-day or recent volcanoes (Figure 187) with plate boundaries. Hence, it is clear that neotectonics, mountain building, and volcanism are mainly defined by plate movements and interactions.

Chapter XIV

SUMMARY OF THE TECTONIC DEVELOPMENT OF THE USSR TERRITORY

PRECAMBRIAN HISTORY

While examining ancient platforms, we came to the conclusion that the shields originated by convergence, collision, and amalgamation of a relatively small number of minor continental blocks. We believe that by 1700 Ma two main continental landmasses--East European and Siberian--already existed, although their positions on the Earth's sphere remains unknown. Indirect evidence for the Late Precambrian oceans that separated the old continents can be seen in fully developed passive continental-margin complexes on both the East European platform (e.g., the Riphean rocks of the western Ural area) and the Siberian platform (e.g., the stratigraphic section of the eastern Siberian margin). Their formation was preceded by fracturing of the basement, which resulted in a series of aulacogens.

Available data are not sufficient to restore the tracks of East Europe and Siberia throughout the Late Precambrian, although paleomagnetic data [Khramov, 1983] show significant displacements of both continents. Still there is little doubt that plate interactions were the same as in the Phanerozoic, for accretional foldbelts formed on the continental margins. The most impressive are fold zones on the northeastern margin of the East European platform, i.e., in Timan, in the basement of the Pechora basin, and in the Polar Urals. These foldbelts mark collision of the Barentsia continent with East Europe, and they indicate significant accretion to the East European continent.

These events are in accord with the idea that a Late Precambrian Pangea-like supercontinent formed before the Vendian. It can be assumed that the Vendian-Early Paleozoic history began with break-up of this supercontinent, an event distinctly reflected in the extensive development of Vendian and Early Cambrian ophiolites. Evidence of the Vendian breakup also remains on the continents; on the East European continent, the evidence is Vendian flood basalts, alkaline plutons on the Kola Peninsula and in Scandinavia, and extensional structures of aulacogens. In Siberia the breakup is indirectly indicated by the paleo-passive margins on the south side of the platform and by a nearly ubiquitous Pre-Vendian stratigraphic hiatus.

PHANEROZOIC RECONSTRUCTIONS

The East European and Siberian continents (Figure 1) were the main rigid elements of USSR structure during Phanerozoic time. If we can determine the relative position of these blocks for each moment of time, we can obtain the framework for palinspastic reconstructions.

Motions of the East European and Siberian continents were considered in previous chapters and are shown in Figures 12 and 16. To calculate the motions three sources of data were used [Zonenshain et al., 1985]: (i) kinematic data from faults, folds, and tectonics, (ii) paleomagnetic data, and (iii) geologically deduced trajectories of continental motion over hot spots. Obviously, sea-floor magnetic anomalies could not be used for reconstructing these relative motions during Paleozoic and Early Mesozoic times, as no anomalies of such ancient age have been preserved.

On the reconstructions accompanying each of the regions described above, the boundaries of lithospheric plates are restored by noting the distribution of key rock assemblages indicative of plate boundaries. Convergent boundaries, marked by calc-alkaline magmatic belts and ophiolite sutures, can be restored most reliably. Spreading axes can be delineated only arbitrarily, based on the pattern of plate movements. Only in rare cases, for example in the Urals, do paleomagnetic data allow us to estimate the position of a spreading axis relative to an adjacent continent.

It is extremely difficult to determine the position of various small blocks (microcontinents, island-arc fragments, submarine plateaus, and other exotic terranes). Paleomagnetic data should be used for this purpose, but unfortunately they are scarce for the USSR territory. Priority is therefore given to paleoclimatic data. We have also tried to take into account lithological data indicating conditions of sedimentation and sources of clastic material. Because the evidence is scarce, the positions of many small blocks and suspect terranes are often given arbitrarily on the reconstructions. These regional reconstructions are used to compile the overall palinspastic maps (Figures 188-205).

Early-Middle Cambrian, 540 Ma (Figure 188)

According to our global reconstructions, East Europe and Siberia in the Early Cambrian were in the southern hemisphere, the Siberian continent in the tropical zone and the East European in the temperate climatic zone. Both continents were oriented very differently than at present and were compar-

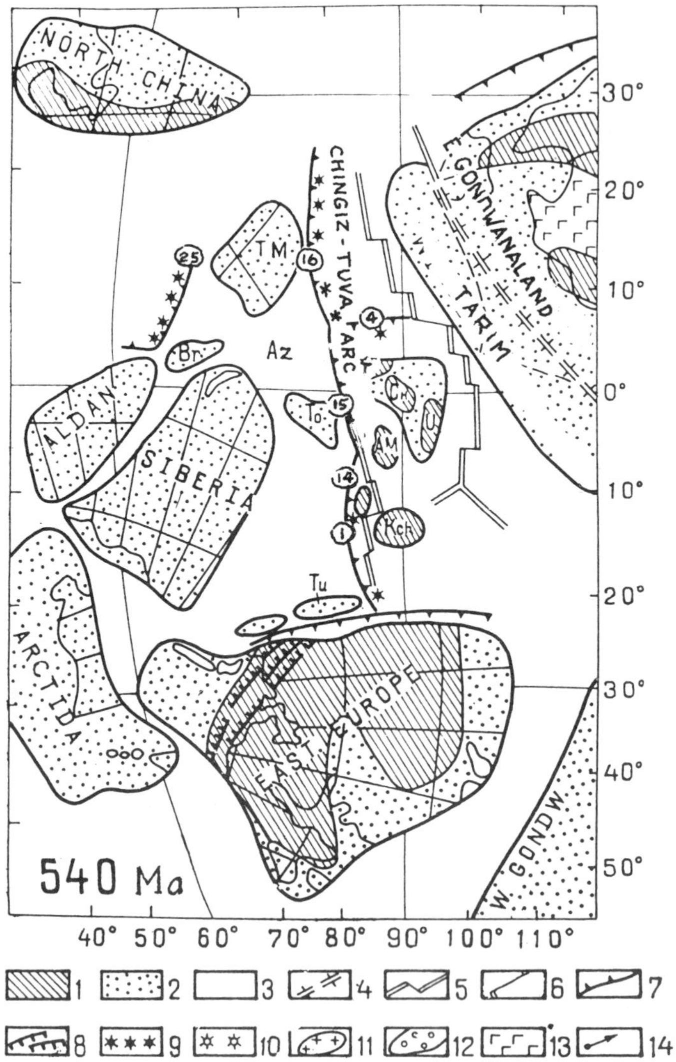

Fig. 188. Palinspastic reconstruction of the USSR territory for 540 Ma. The base map is the absolute reconstruction by Zonenshain et al. [1985]. Lambert projection centered at 0° N, 90° E.

1, Dry land; 2, shallow sea; 3, oceanic floor; 4, continental rift; 5, spreading axis; 6, transform fault; 7, subduction zone; 8, folding and metamorphism (collision zone); 9, calc-alkaline volcanics; 10, intraplate volcanics; 11, granitic batholith and gneissic-granite dome; 12, molasse basin; 13, flood basalt; 14, paleomagnetic vector; 15, strike-slip motion; 16, direction to geographical North Pole. Paleolatides are shown.

Precambrian massifs and exotic terranes: A - Alazei, Af - Afghan, Al - Altay, AM - Atasu-Mointy, B - Baisun, Br - Barguzin, Ch - Chu, Chr - Chersky, CK - Chukotka, CM - Central Mongolian, CP - Central Pamir, ES - East Sakhalin, Ju - Junggar, KhB - Khingan-Bureya, Kch - Kokchetav, Ku - Kurgovat, LK - Lesser Kuril, Ma - Maker, Mi - Moesia, Mu - Mugodzhar, Okh - Okhotsk, Ol - Olyutorsky, Om - Omolon, SO - Sea of Okhotsk, SP - South Pamirs, T - Tien Shan, TC - Transcaucasian, TM - Tuva-Mongolian, Tmr - Taimyr, To - Tom, U - Ulutau.

Oceanic basins: AK - Akdam, Az - Paleo-Asiatic, C - Canadian, Ca - Peri-Caspian, EA - Eurasian, GC - Great Caucasus, M - Makarova, MO - Mongol-Okhotsk, PT - Paleo-Tethys, SA - South Anyui, TB - Turkestan.

Volcanic arcs and belts: 1, Chingiz; 2, Stepnyak; 3, Bet-Pak-Dala; 4, North Tien Shan; 5, South Tien Shan; 6, Mariev; 7, Marginal belt of Kazakhstan; 8, North Balkhash; 9, Balkhash; 10, Gissar; 11, Beltau-Kurama; 12, Valerianovsky; 13, Ili, 14, Salair; 15, Minusinsk, 16, Tuva; 17, Rudny-Altay; 18, Zharma; 19, Saur; 20, South Mongolian; 21, Magnitogorsk; 22, Chara; 23, Tom-Kolyvan; 24, West Sayan; 25, Great Caucasus; 26, Inner Mongolian; 27, Eurasian; 28, Uyandina-Yasachnaya; 29, Sikhote-Alin; 30, Lesser Caucasus; 31, Anui; 32, Okhotsk-Chukotka; 33, Sikhote-Alin; 34, Irunei, 35, Bowers; 36, Kamchatka; 37, Alburz; 38, Carpathian; 39, Great Caucasus; 40, Hindukush; 41, Koni-Murgal; 42, Oloi; 43, Olyutorksy; 44, Anadyr-Bristrol; 45, Adjaro-Trialetian; 36, Iran; 47, Aleutian.

atively close to Australia, which was in East Gondwanaland. The wide separation of East Europe and Siberia suggests that the Late Precambrian Pangea-like supercontinent was breaking up in the Cambrian. With respect to the absolute coordinate system, East Europe and Siberia were situated in the center of the eastern hemisphere. Among other large continental blocks relevant to the USSR territory, Arctida and the North Chinese continents should be mentioned. On the reconstruction, they are plotted arbitrarily, relatively close to zones along which they later collided with main continents. The Aldan shield is shown as an independent block still not joined with Siberia. Barentsia is shown attached to East Europe, and the collisional suture is shown as the Timan foldbelt.

In the early Middle Cambrian, Siberia was nearly 3000 km from East Europe on the one side and 3000 km from East Gondwanaland on the other side. The vast area around Siberia was occupied by the Paleo-Asian Ocean, the remnants of whose floor are preserved in numerous ophiolite complexes within the Central Asian foldbelt.

The most complicated task is "to place" the numerous microcontinental blocks within the Paleo-Asian Ocean and to estimate the positions of island arcs and spreading centers. Using geological data (see corresponding chapters above), it is possible to distinguish the following Early Paleozoic microcontinents: Tuva-Mongolian, Tom, Barguzin, Atasu-Mointy, and Kokchetav. All these blocks presumably formed when the Late Precambrian Pangea-like supercontinent broke up. Therefore, in the reconstruction, a spreading axis that separated the microcontinents from Gondwanaland is plotted within the Paleo-Asian Ocean. We assume that East Gondwanaland originally included small blocks that subsequently split from it and, crossing the Paleo-Asian Ocean, joined Siberia and East Europe during the Late Paleozoic and Mesozoic. These blocks are the Tarim, Karakum, Tadzhik, Pamirs, Afghan, Iran, and other microcontinents.

The Tuva-Mongolian, Tom, and Barguzin microcontinents are shown on the reconstruction near Siberia, and the Central Kazakhstan microcontinent nearly in the center of the Paleo-Asiatic Ocean. They are postulated to be separated by a single intraoceanic subduction zone. Early-Middle Cambrian island-arc complexes are widespread in many parts of the Central Asian foldbelt (in Chingiz, Salair, Kuznetsky Alatau, West Sayan, Tuva, and North Mongolia). One might draw an independent arc in each region, but more likely they were connected in a single Chingiz-Tuva island arc (Figure 188). If so, the arc was not less than 4500 km long and extended along the 90° E meridian. In the Cambrian the subduction zone most likely dipped toward East Gondwanaland, i.e., eastward in Cambrian coordinates. Evidently, back-arc oceanic basins opened behind the Chingiz-Tuva arc, e.g., the Akdam basin.

Thus the pattern of plate interaction in the early Middle Cambrian displayed two long north-south plate boundaries: a spreading zone in the eastern part of the Paleo-Asian Ocean, and a subduction zone in the western part. The main trend of plate interaction is clear. Microcontinental massifs were torn off East Gondwanaland and moved westward toward Siberia, encountering the central subduction zone. Simultaneously, Siberia and the nearby continental blocks approached the same subduction zone. As a result, several continental margins gathered together near the subduction zone. This type of circumstance, as will be seen, prevailed through the whole Paleozoic.

Late Cambrian-Early Ordovician, 490 Ma (Figure 189)

In the second half of the Cambrian, two events are most remarkable: first, the widening of the Paleo-Asiatic ocean, and second, the collision of the Chingiz-Tuva arc with two microcontinents. In the first event, passive margins formed, e.g., on the periphery of the Kazakhstan microcontinent, and evidently the Tarim-Karakum microcontinent split from Gondwanaland. In the second event, collision of the microcontinents with the Chingiz-Tuva arc was accompanied by deformation (Salairian orogeny) and emplacement of granitic batholiths, mainly in Tuva (and in nearby North Mongolia) and in the Kuznetsky Alatau-Minusinsk region ; thus, probably only two microcontinents collided with the island arc, Tuva-Mongolian and Tom, which had formerly been west of the subduction zone. During the collision, ophiolite nappes were obducted over the basement of the microcontinents, and chaotic complexes were formed. The collisions caused reorganization of the subduction zone, changing the dip of the downgoing slab, and in Eastern Tuva the subduction zone shifted to the opposite side of the Tuva-Mongolian microcontinent.

The East European continent, moving northeastward, converged and collided with the southern flank of the Chingiz-Tuva island arc system. As a result, the former East European passive margin was deformed in the Cambrian, and the Trans-Uralian microcontinent (which contains Archaeocyatid limestones exotic for East Europe) was accreted.

The Altay submarine fan of clastic deposits began to accumulate in the Late Cambrian and Early Ordovician. Since its original position is unknown, on the reconstruction it is arbitrarily shown near the North Chinese continent.

Thus, continents and microcontinents drifted significantly throughout the Cambrian. Nevertheless, two main plate boundaries, the spreading center in the east part of the Paleo-Asicatic ocean and the subduction zone in the west, remained at approximately the same distance from each other, and in the same area of the Earth's sphere, as in previous times.

Middle Ordovician, 460 Ma (Figure 190)

In the Ordovician, East Gondwanaland moved eastward, and accordingly the Paleozoic ocean widened. In turn, Siberia moved northward, possibly as much as 1000 km. The general distribution of continents as well as the principal patterns of plate interactions resembled those of the Cambrian. Judging from global reconstructions, there should have been a long spreading ridge in the east part of the Paleo-Asian Ocean. The northern margin of East Europe was broken up in the Ordovician, and an intracontinental rift zone nearly 3000 km originated, presumbably connected with the world rift system. Later on, the Uralian paleo-ocean was opened on the site of this rift zone.

The wide distribution of Ordovician island-arc complexes, mostly in central Kazakhstan, indicates active subduction in the western part of the Paleo-Asian Ocean. We postulate the existence of four island arcs, Chingiz, Stepnyak-Bet-Pak-Dala, West Sayan, and North Tien Shan. The Chingiz arc is the only remnant of the former Chingiz-Tuva system. Its polarity must have reversed, i.e., the dip of the descending lithospheric slab changed to westward (in terms of paleo-coordinates) toward Siberia, instead of eastward as before. The Chingiz and Stepnyak-Betlak-Dala arcs extended to the south and perhaps

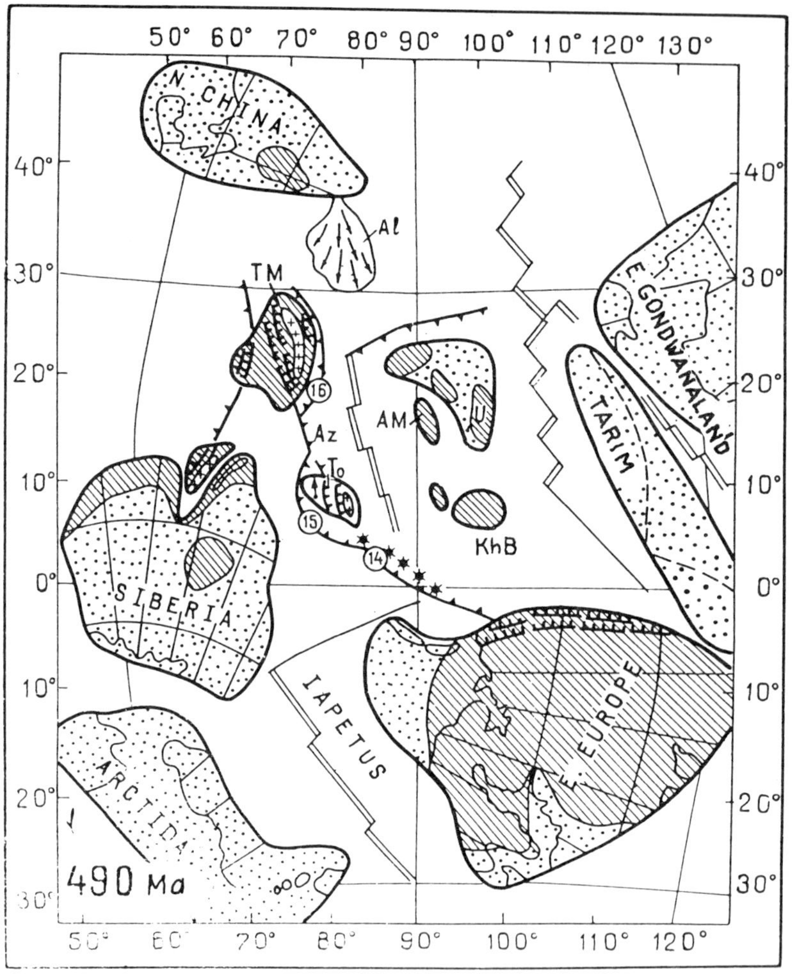

Fig. 189. Palinspastic reconstruction of the USSR territory for 490 Ma. See Figure 188 for explanation.

joined an island arc in the Iapetus Ocean, reconstituting the same 90° E island-arc system as in the Cambrian.

A new feature of this time is that a sublatitudinal subduction zone formed, largely coinciding with the West Sayan arc. The North Tien Shan arc was probably related to this sublatitudinal subduction zone. The polarity of these arcs is not reliably known, but the Benioff zones supposedly dipped toward Siberia.

Thus, in the Ordovician, in contrast to the Cambrian, along with plate boundaries trending northerly (in terms of paleocoordinates), there appeared latitudinal active zones: the Uralian rift zone and the West Sayan subduction zone.

Early Silurian, 430 Ma (Figure 191)

In the Silurian the eastern and western parts of the territory under consideration developed in a noticeably different way. In the east, as East Gondwanaland moved progressively eastward, the distance between it and East Europe and Siberia constantly increased, eventually reaching 5000 km in the Silurian. Most likely three spreading centers existed: (i) a central system responsible for formation of the Paleo-Asian Ocean crust, (ii) a southern system along which the Paleo-Ural Ocean opened, and (iii) an eastern system situated between the Tarim-Karakum microcontinent and East Gondwanaland. The latter spreading system produced the Paleo-Tethys (or Turkestan) ocean crust. In the west, on the other hand, plates converged along the subduction zone which lay between Arctida and East Europe, and between Siberia and Kazakhastan. Here, East Europe converged with Arctida and North America, thus closing the Iapetus Ocean. The Kazakhstan microcontinent was amalgamated in the center of the Paleo-Asian Ocean as numerous small blocks collided with the Stepnyak-Bet-Pak-Dala arc. Throughout the Silurian, the Kazakhstan microcontinent converged with the Chingiz island arc and collided with it prior to the Devonian.

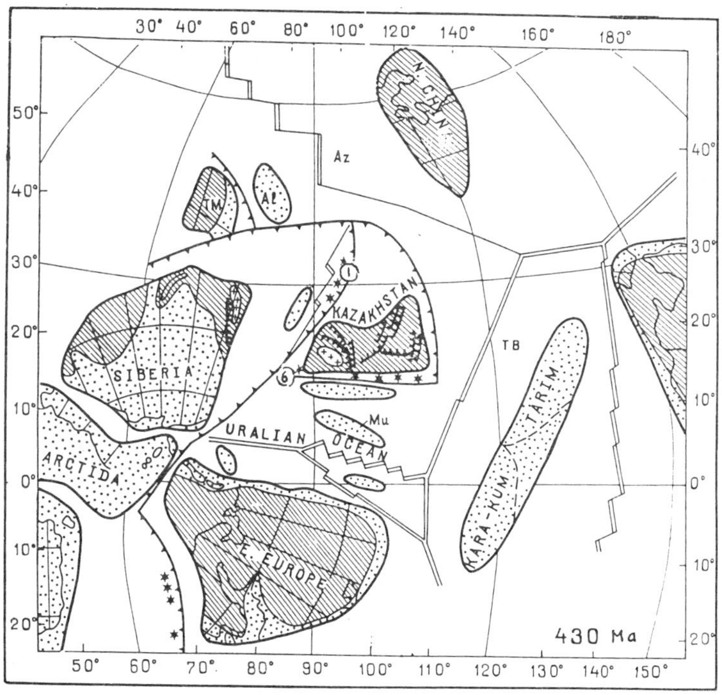

Fig. 190. Palinspastic reconstruction of the USSR territory for 460 Ma. See Figure 188 for explanation.

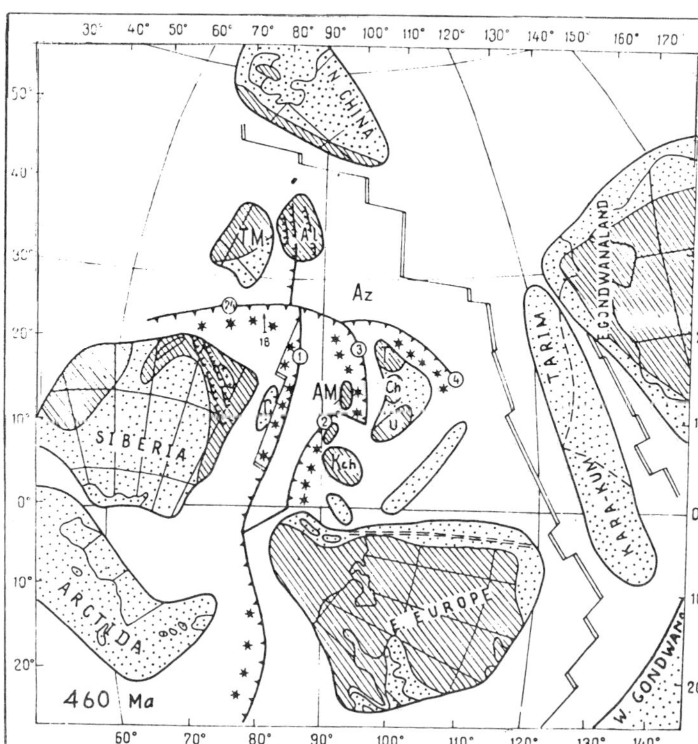

Fig. 191. Palinspastic reconstruction of the USSR territory for 430 MA. See Figure 188 for explanation.

Early Devonian, 400 Ma (Figure 192)

The data for positions of continents are much more reliable for this time than for previous ones. The oceanic space between East Europe and Siberia, on the one hand, and East Gondwanaland on the other, reached the maximum width in the Early Devonian, no less than 9000 km. The regions that would eventually be within or near the USSR included four large continents: Euramerica, Siberia, Kazakhstan, and North China. Euramerica (or Laurasia) was formed prior to the Devonian when North America and northwestern Europe collided. The collision front is well marked by the Caledonian foldbelt of Scandinavia, Scotland, Svalbard (Spitsbergen), and East Greenland. Apparently Arctida also joined East Europe by the Early Devonian. We assume that the Caledonian belt stretching from Svalbard to Severnaya Zemlya, reflects this collision. The Siberian continent was enlarged as the Tuva-Mongolian and Barguzin microcontinents were attached to it. This produced another orogenic belt that is commonly called "Caledonian" by Soviet geologists because of its appropriate age, but it is nearly 3000 km south of the true Caledonian foldbelt. It lies near the USSR-Mongolian border and extends more or less east-northeast. This orogenic belt can be traced from the Altai Mountains near the eastern tip of the Kazakh SSR through the Western Sayan Mountains into the southwestern part of the East Sayan Mountains and farther on to the Baikal highland. The Kazakhstan microcontinent was also enlarged when it collided in the Devonian with the Chingiz island arc, which then ceased to be active.

In the Devonian all the continents moved noticeably northward particularly the Siberian continent, which shifted almost to its present position in moderate and high latitudes of the northern hemisphere but was still in an inverted position relative to present coordinates.

Three oceanic basins were situated between the continents: the Paleo-Asian, Paleo-Uralian, and Paleo-Tethys (Turkestan). The Paleo-Asian Ocean, between the Siberian and North Chinese continents, was 3000 km wide. The east-west Uralian Paleo-Ocean, between the East European and Siberian continents, was 5000 km long and 2000-2500 km wide. Its north-

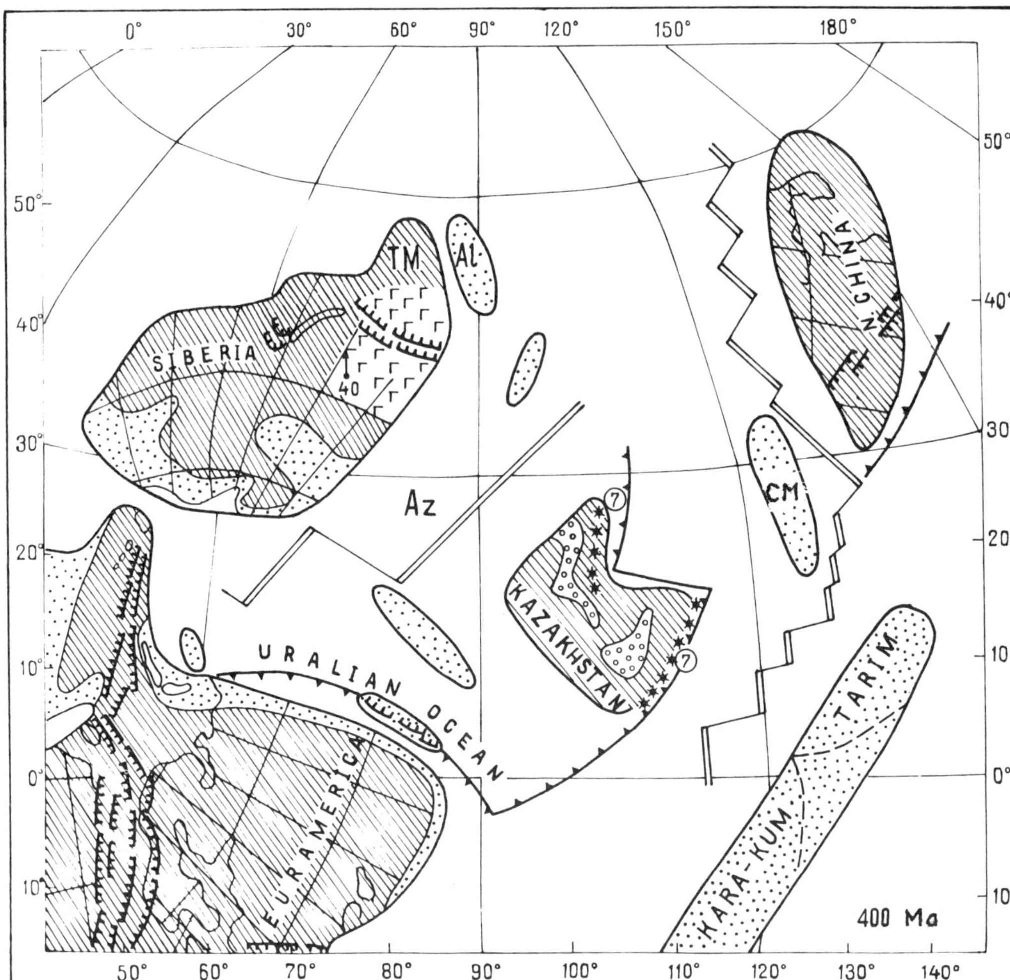

Fig. 192. Palinspastic reconstruction of the USSR territory for 400 Ma. See Figure 188 for explanation.

ern, Siberian margin was passive, judging from purely carbonate Devonian sequences of the Turukhansk-Igarka region, whereas island arcs, which had begun to develop in the Silurian, existed along its southern, East European margin. The Turkestan basin of the Paleo-Tethys ocean, as reconstructed from ophiolites of the South Tien Shan, separated the Kazakhstan and Tarim-Karakum microcontinents. In this region a subduction zone with associated terrestrial volcanoes formed along the Kazakhstanian continental margin. East of the Tarim-Karakum microcontinent, the main part of Paleo-Tethys Ocean was opening.

It should be emphasized that intraplate magmatism was wide-spread in the Early Devonian in southern Siberia, where large basaltic and alkaline volcanic fields formed in the Minusinsk, Tuva, Rybinsk and other depressions of the Altai-Sayan region.

Thus, continental divergence reached its maximum in the Devonian, and accordingly, the Paleozoic oceanic basins were as wide as they ever became within the USSR territory. Despite this vast continent dispersal, subduction zones still existed as before, concentrated in a belt that lay across the 90° E meridian (modern coordinates) near the present equator.

Middle Devonian, 370 Ma (Figure 193)

Three simultaneous events distinguish the Middle Devonian. First, all formerly existing paleo-ocean basins in the central part of the region (Paleo-Asian, Uralian and Turkestan) ceased to widen and started to close. Second, a significant phase of extension can be reconstructed within the Siberian and East European continents. Third, the Paleo-Tethys Ocean widened significantly as the North Chinese continent moved eastward. All the other continents surround a point at 30° N and 90° E (present coordinates), and the Kazakhstan continent virtually touched this point. In this whole area, plates converged, oceans closed, and numerous subduction zones appeared with accompanying magmatic arcs, e.g., the Magnitogorsk arc in the Uralian Paleo-ocean, the Rudny Altayi, Zharma, Saur, and South Mongolian arcs in the Paleo-Asian Ocean, the South Tien Shan and North Balkhash arcs in the Turkestan basin. The arcs and subduction zones can be reliably reconstructed from belts of calc-alkaline igneous rocks. Many microcontinents and exotic terranes approached and eventually collided with these subduction zones, the largest being the Salair, Tarim-Karakum, Khingan-Bureya, and

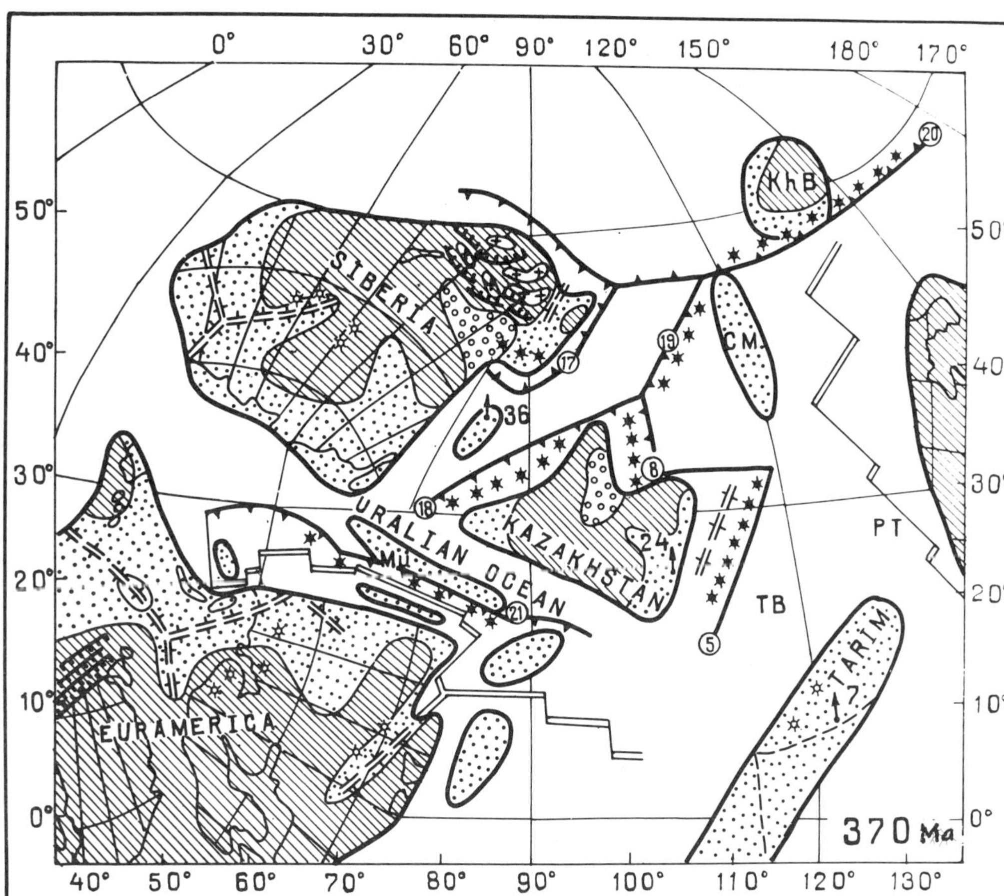

Fig. 193. Palinspastic reconstruction of the USSR territory for 370 Ma. See Figure 188 for explanation.

Mugodzhar microcontinents. Positions of the first two are derived from paleomagnetic data. Devonian deformation and granite batholith emplacement in the East Altai-Sayan area were related to the general compressional environment.

Events that were almost the opposite of the foregoing took place in the west and southwest, where continental rifting and intraplate magmatism occurred on both major continents, East European and Siberian. Two main rift systems or aulacogens (the Dneprovsk-Donetsky and Pechora-Barents Sea systems) appeared within the East European continent, and the Vilyuy rift formed on the Siberian continent. Presumably the rift zones of East Europe and Siberia were parts of the world rift system, which continued on the one side into Panthalassa, on the other side into the Paleo-Tethys Ocean. Thus, the rift system was opposite the area of plate convergence that was centered near 30° N and 90° E.

Early Carboniferous, 340 Ma (Figure 194)

The area of continental convergence centered at 30° N, 90° E that appeared in the Late Devonian is very evident in the Early Carboniferous reconstruction. From 370 to 340 Ma, Siberia and East Europe converged by 2000 km, and as a result the Uralian Paleo-ocean was almost completely closed. The Paleo-Tethys and the Turkestan oceanic basins also contracted considerably. It should be noted that after the Early Devonian the continents converged and the oceanic widths decreased mainly because Euramerica moved 2000 km northward (with respect to the hotspot frame of reference). At the same time, Siberia actually remained in the same place, although it rotated clockwise.

Subduction zones were the main tectonic elements at this time. They were concentrated within a single broad belt stretching from the equator to 60° N, mainly between meridians 70° E and 120° E. The arcs related to the subduction zones were mostly of the active continental-margin type, the largest being the Valerianovsky, South Tien Shan, and Balkhash arcs extending along the Siberian margin and Kazakhstan microcontinent. The large continents situated west of this belt of subduction zones converged with it. On the other side, i.e., the southeastern Paleo-Tethys side, the Tarim-Karakum microcontinent and Tadzhik microcontinent, situated farther east, were also converging with it, approaching the subduction zones one after another. Both microcontinents were continental blocks originally split from Gondwanaland. It should be noted that continental fragments split from Gondwanaland traveled across an ocean, and eventually were successively accreted to the Eurasian margin from time to time during almost the entire subsequent history.

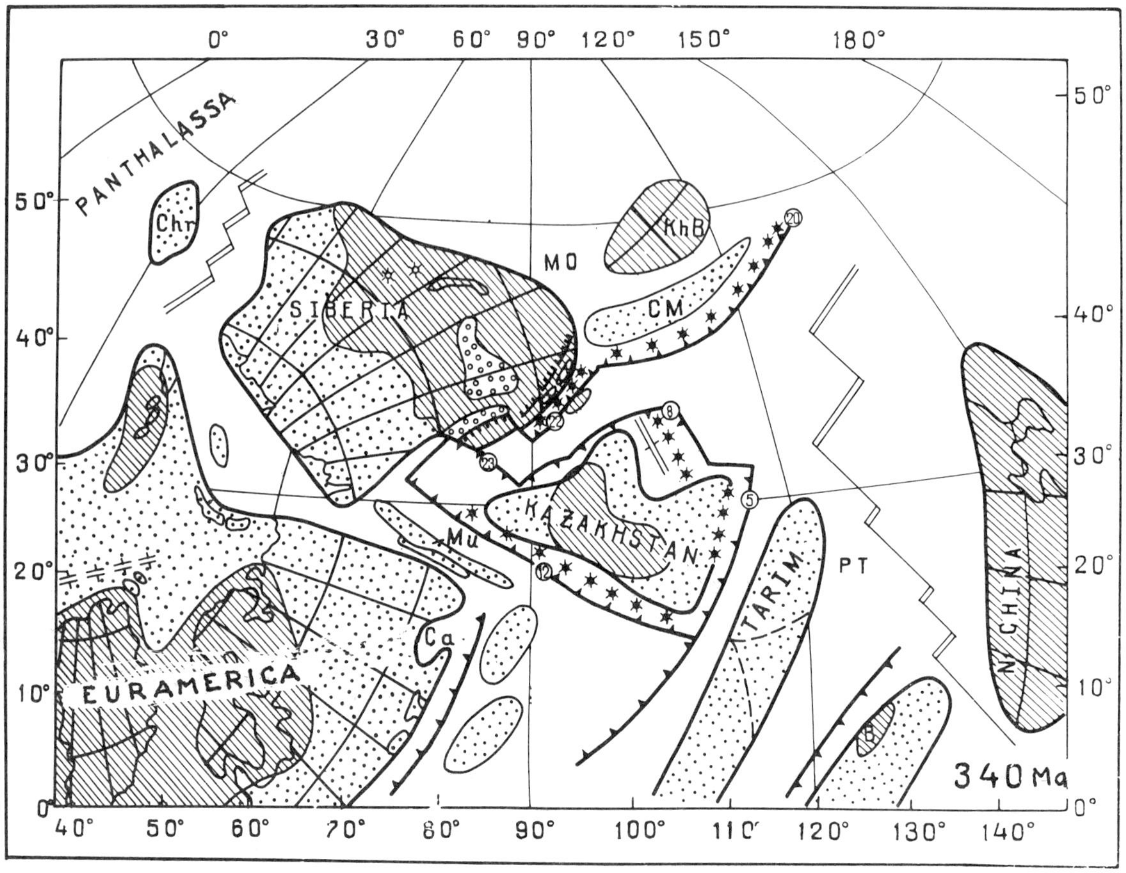

Fig. 194. Palinspastic reconstruction of the USSR territory for 340 Ma. See Figure 188 for explanation.

Middle-Late Carboniferous, 310 Ma (Figure 195)

Between 340 and 310 Ma (Figures 194 and 195) Siberia converged with East Europe, and both of them converged with the Kazakhstan continent. Only remnant oceanic basins were between them until the Middle Carboniferous; one remnant was in the site of the present West Siberian lowland, and another in the Irtysh-Zaisan zone.

The belt of subduction zones centered at 30° N, 90° E still existed throughout the Carboniferous, separating the East European and Siberian continents west of it from the Paleo-Tethys Ocean, which enclosed the Tarim-Karakum and Tadzhik microcontinents. The Kazakhstan continent remained in the center, rimmed by subduction zones on all sides.

The plate convergence and the resulting collision of continents and microcontinents with island arcs caused Carboniferous orogenic events which are sometimes said to be "Early Hercynian", in allusion to the timing. Two orogenic belts of this age are distinguished. One of them, in the basement of the Skythian platform, is truly Hercynian, being a continuation of the classic European Hercynian belt. The other is not really Hercynian sensu stricto, as it is not demonstrably connected with the belt of that name in western Europe. It is the Altai-South Mongolian belt, formed by the collision of several continental blocks along the South Mongolian subduction zone. Combined, they produced the Amuria microcontinent, a large accreted mass extending from the Gorny Altai through Mongolia and Trans-Baikal to the Khingan-Bureya and probably to the Khankai massif. Amuria extended to the east (paleo-coordinates) from the eastern (present southwestern) corner of the Siberian continent, separating a large oceanic embayment of Panthalassa,--the Mongol-Okhotsk Ocean basin, from the Paleo-Tethys.

Early Permian, 280 Ma (Figure 196)

In the Late Carboniferous-Early Permian, culminating continental collisions formed a single Laurasia. Within the USSR, the mainland trended more or less NE-SW above the former center of plate convergence at 30° N, 90° E.

By the Late Permian, the Tarim-Karakum and Tadzhik microcontinents had collided with the Kazakhstan margin, forming the thrust-fold structures of the Tien Shan orogenic belt. Continuous convergence of the Paleo-Tethys plate with Laurasia moved the accretionary mosaic of continental blocks together with the Kazakhstan continent into the wedge between Siberia and East Europe, producing orogeny and mountain building in the Urals, the Irtysh-Zaisan zone, and the basement of the West Siberian lowland. As a result of Paleo-Tethys/Laurasia convergence, an extensive (about 6000 km

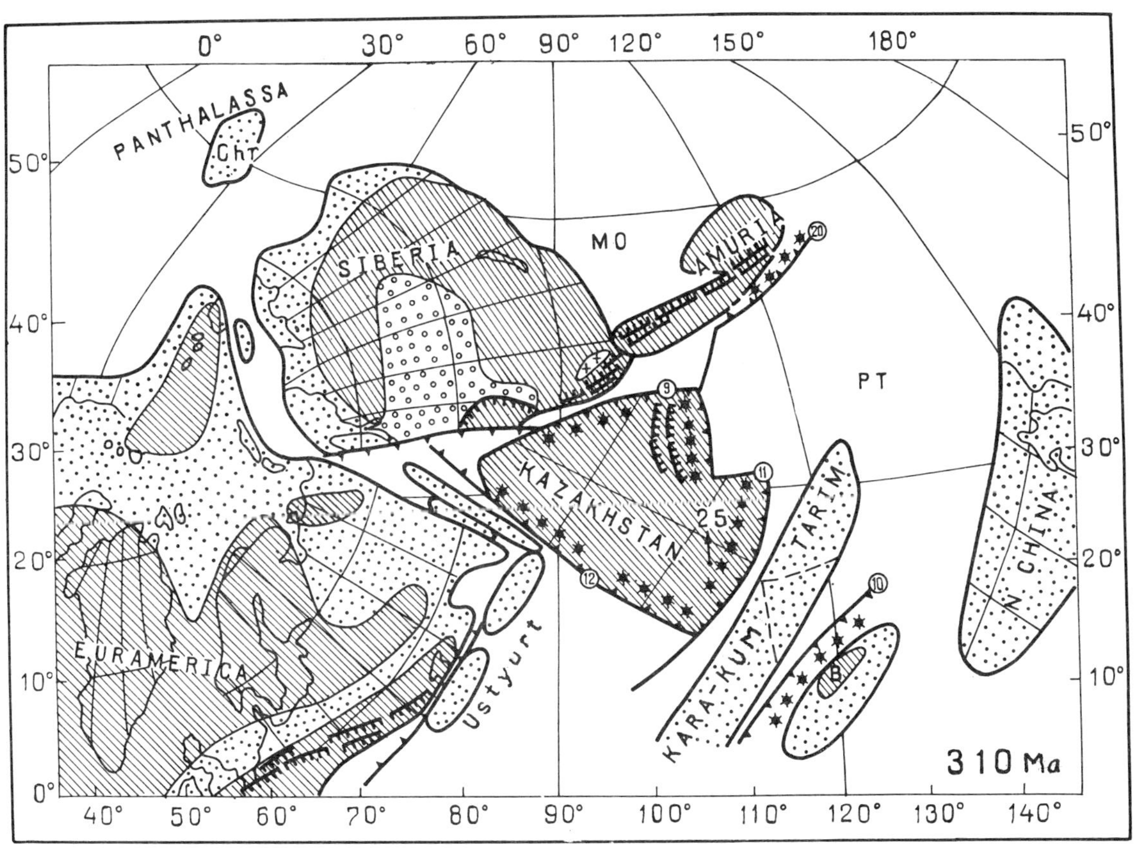

Fig. 195. Palinspastic reeconstruction of the USSR territory for 310 Ma. See Figure 188 for explanation.

long) Eurasian volcanic belt developed along the continental margin, related to the subduction zone (Figure 196).

Because boundaries of the converging continents had angular irregularities, the continents were not in complete contact; in places oceanic crust remained in gaps between them. Examples of ocean remnants are the South Anyui oceanic embayment in NW Laurasia between Siberia and Arctida, the Mongol-Okhotsk embayment between Siberia and Amuria, the Junggar-Balkhash basin between the Amuria, Kazakhstan, and Tarim-Karakum continents, and the Peri-Caspian basin between East Europe and the Usturt massif. The history of each of these basins is different; they closed at different times, and only in the Peri-Caspian basin has Devonian oceanic crust remained up to the present time.

Late Permian, 250 Ma (Figure 197)

Late Permian paleogeography was relatively simple. Laurasia, now the northern part of Pangea, was still oriented more or less NE-SW. The Paleo-Tethys ocean was situated between Laurasia and Gondwanaland. Its floor along with the North Chinese continent moved towards the Eurasian subduction zone, and thrust-fold belts continued to develop inside Laurasia, allowing the continental blocks to pack together so tightly that the oceanic gaps were closed. Thus, by the Late Permian, the Junggar-Balkhash basin was totally closed as the Tarim block collided with the Eurasian margin. At nearly the same time, the Amurian promontory started moving toward Eurasia, and the head of the Mongolian-Okhotsk oceanic embayment was closed. As compression of the supercontinent continued, the continental crust was warped, the earlier-formed thrust-fold structures were redeformed, a series of parallel dextral strike-slip faults (Talasso-Fergana, etc.) originated, and conjugate horizontal flexures were formed in Central Kazakhstan and the Tien Shan Range.

The configuration of major plate boundaries was also relatively simple. This part of Laurasia was almost surrounded on all sides by subduction zones. Several continental masses still lay within the Paleo-Tethys Ocean between Eurasia and Gondwanaland, including Moesia, Dzirula, Iran, Central Pamir, and Kurgovat. Some of them, such as the Moesia and Dzirula massifs, obviously were detatched from the Eurasia margin, but the rest came from Gondwanaland as indicated by details of their rock assemblages. Some were presumably detached from Gondwanaland comparatively early--in the Devonian or Carboniferous--but the largest, the Iran microcontinent, evidently was detached from Gondwanaland only in the Permian. Part of the newly formed Mesozoic Tethys Ocean opened be-

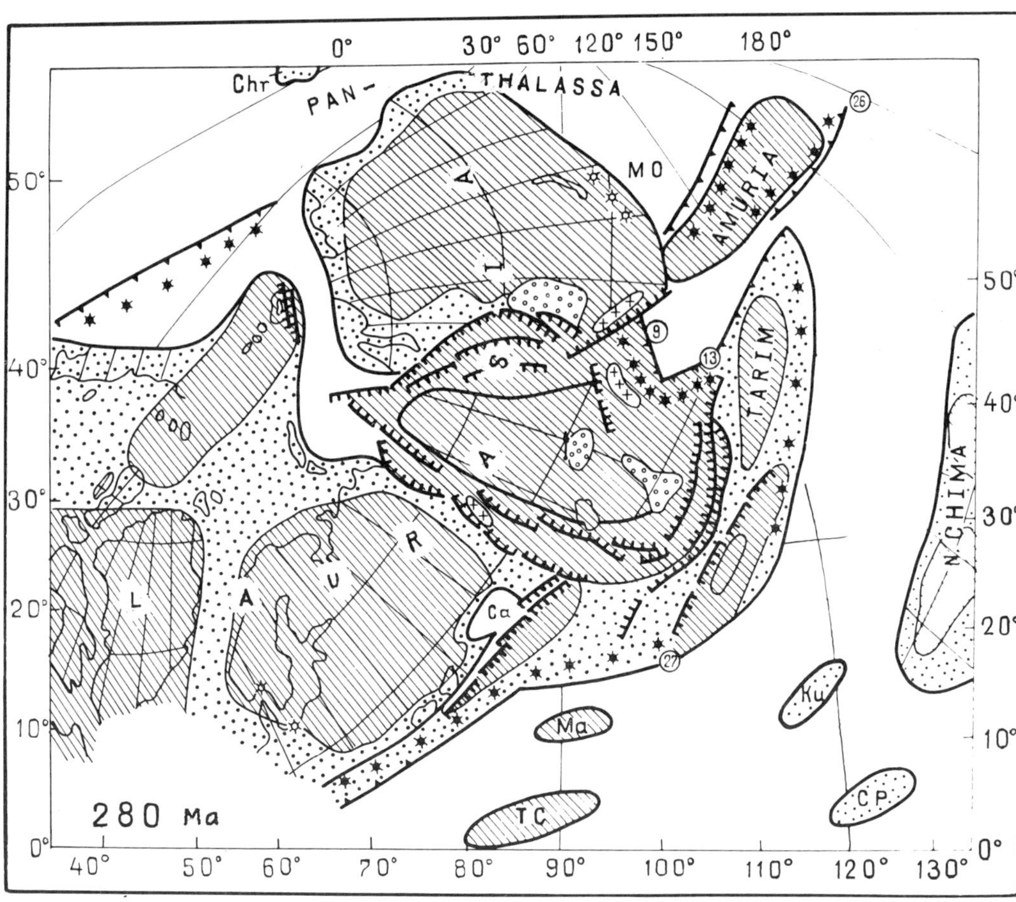

Fig. 196. Palinspastic reconstruction of the USSR territory for 280 Ma. See Figure 188 for explanation.

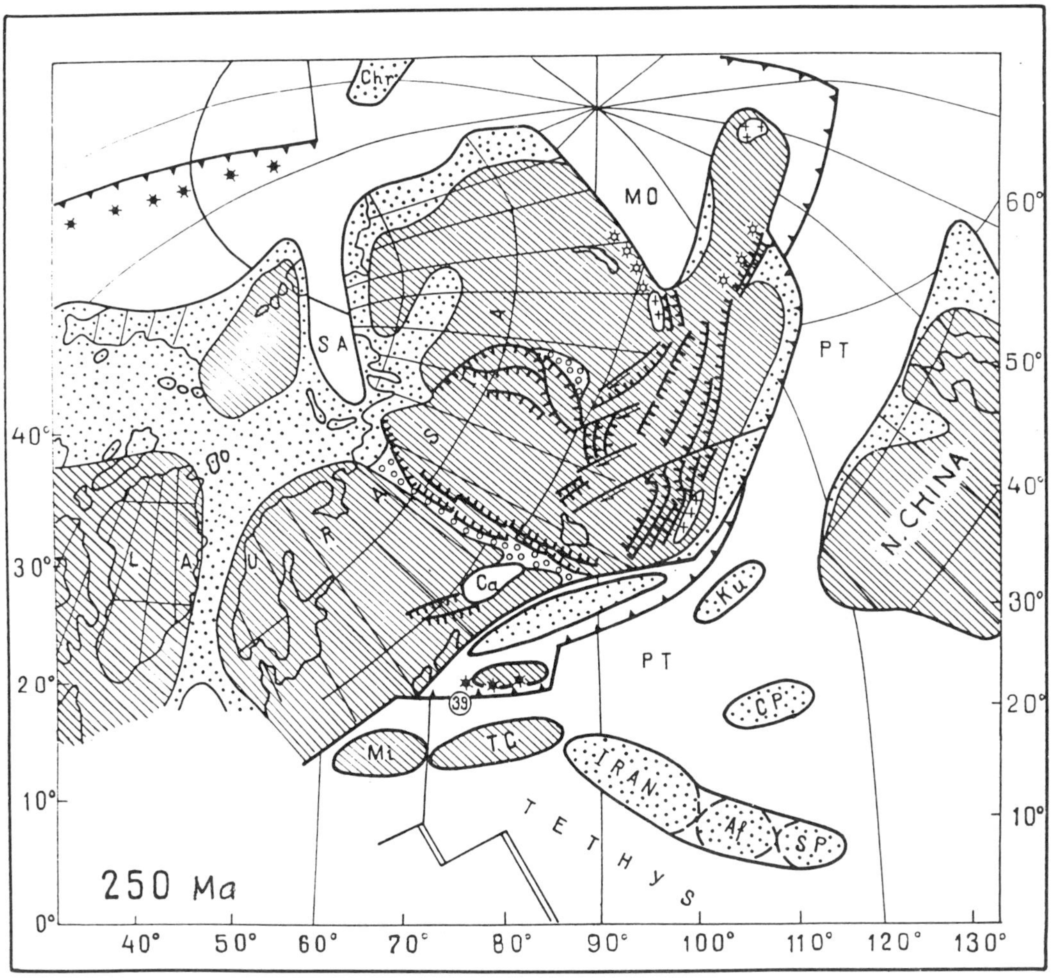

Fig. 197. Palinspastic reconstruction of the USSR territory for 250 Ma. See Figure 188 for explanation.

tween the Iran microcontinent and Gondwanaland. As the Tethys Ocean widened, the Paleo-Tethys oceanic crust was progressively consumed under the Eurasian margin.

Late Triassic, 220 Ma (Figure 198)

In the Late Triassic, the plates were considerably reorganized. First, North China and Eurasia collided and the largest part of Paleo-Tethys Ocean closed. Second, as a result of collision, the former NE-SW zones of convergence were replaced by a single sublatitudinal subduction zone. Third, compression within Eurasia ceased, and extension replaced mountain building, commonly nearly at the same places.

The part of Laurasia that was within the site of the USSR was bounded on the south by the Tethys Ocean, and on the north by the Panthalassa or the Paleo-Pacific Ocean. Two oceanic embayments, the Mongol-Okhotsk and South Anyui embayments, still invaded Laurasia from the Panthalassa Ocean. The latter contained a series of exotic terranes such as Chersky, Omolon, Okhotsk, Sergeev, and Alazei, all moving together with the oceanic plate toward Laurasia.

In the Tethys Ocean, because of convergence of the Iran and other microcontinents with Laurasia, the width of the western branch of the Paleo-Tethys Ocean progressively decreased and the Mesozoic Tethys simultaneously widened. The Eurasian active margin and its volcanic arcs can be reconstructed with considerable accuracy. In the Triassic, the Kurgovat microcontinent encountered the subduction zone in the North Pamirs, while the Iran microcontinent continued to approach this zone.

Extension within Laurasia characterized the Triassic period. This is clearly seen in West Siberia where the continent rifted and the Ob Paleo-Ocean partially opened. Furthermore, Triassic rifting affected the Barents Sea basin, and intraplate magmatism was extensive, especially where flood basalt covered vast areas of Siberia, Chukotka, and the Pechora-Barents Sea basin.

Early Jurassic, 190 Ma (Figure 199)

The unified Laurasia continent was still bounded on south and north by subduction zones in the Late Triassic and Early

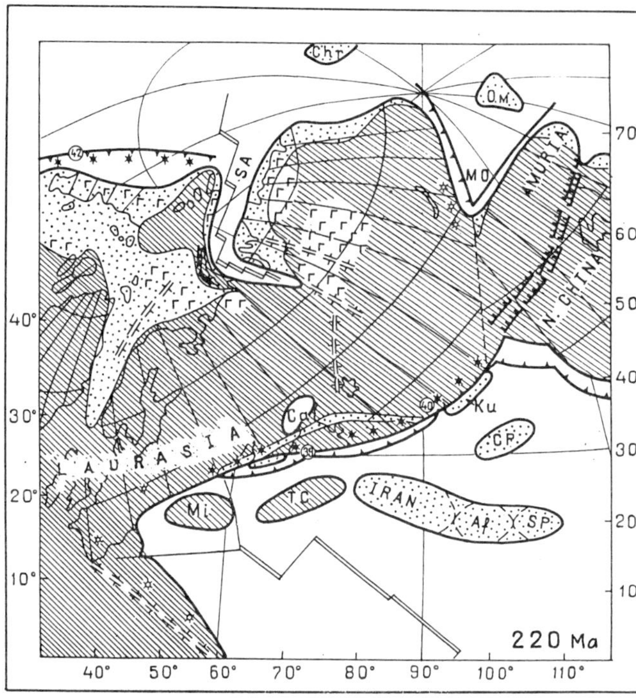

Fig. 198. Palinspastic reconstruction of the USSR territory for 220 Ma. See Figure 188 for explanation.

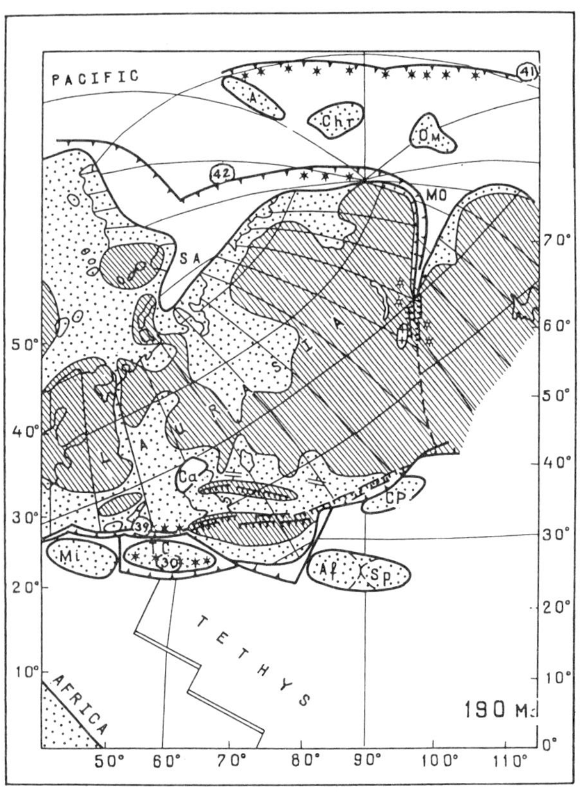

Fig. 199. Palinspastic reconstruction of the USSR territory for 190 Ma. See Figure 188 for explanation.

Jurassic. Within the continent, former rift zones began to subside, for example in West Siberia.

The Iran and Central Pamir microcontinents coming from the Tethys Ocean, arrived at the southern subduction zones. The Iran microcontinent approached the Eurasian margin obliquely; only its western part collided with Eurasia, while its eastern part, including the Afghan and South Pamir blocks, remained within the Tethys Ocean. Collision of the Iran and Central Pamir microcontinents with Eurasia caused the Cimmerian orogeny. The subduction zone was locked and jumped into a new position south of the accreted microcontinents. The Mongol-Okhotsk basin closed tightly, in the Late Triassic and Jurassic, and this was followed by folding and intrusion of granitic batholiths in Mongolia and the Trans-Baikal area.

Two intraoceanic subduction zones can be reconstructed along the Pacific margin. One of them, the Oloi zone which was closer to Siberia than the other, extended into the Uda arc of the Mongol-Okhotsk belt. The second, the Koni-Murgal zone, was evidently connected with the Sikhote-Alin arc. Together, they produced a single long belt of volcanic arcs at a distance of 2000-3000 km from the continent. The overall structural pattern resembles that of the present Philippine Sea. The following continental blocks were situated within the large marginal basin between the intra-oceanic and continental mrgin arcs: the Omolon-Okhotsk, Chersky, and Alazei blocks. In the open ocean beyond the marginal sea, other exotic terranes were situated at low latitudes in the northern hemisphere, the Sergeev terrane being the most prominent.

Late Jurassic, 160 Ma (Figure 200)

Laurasia rotated clockwise throughout the Jurassic, around a

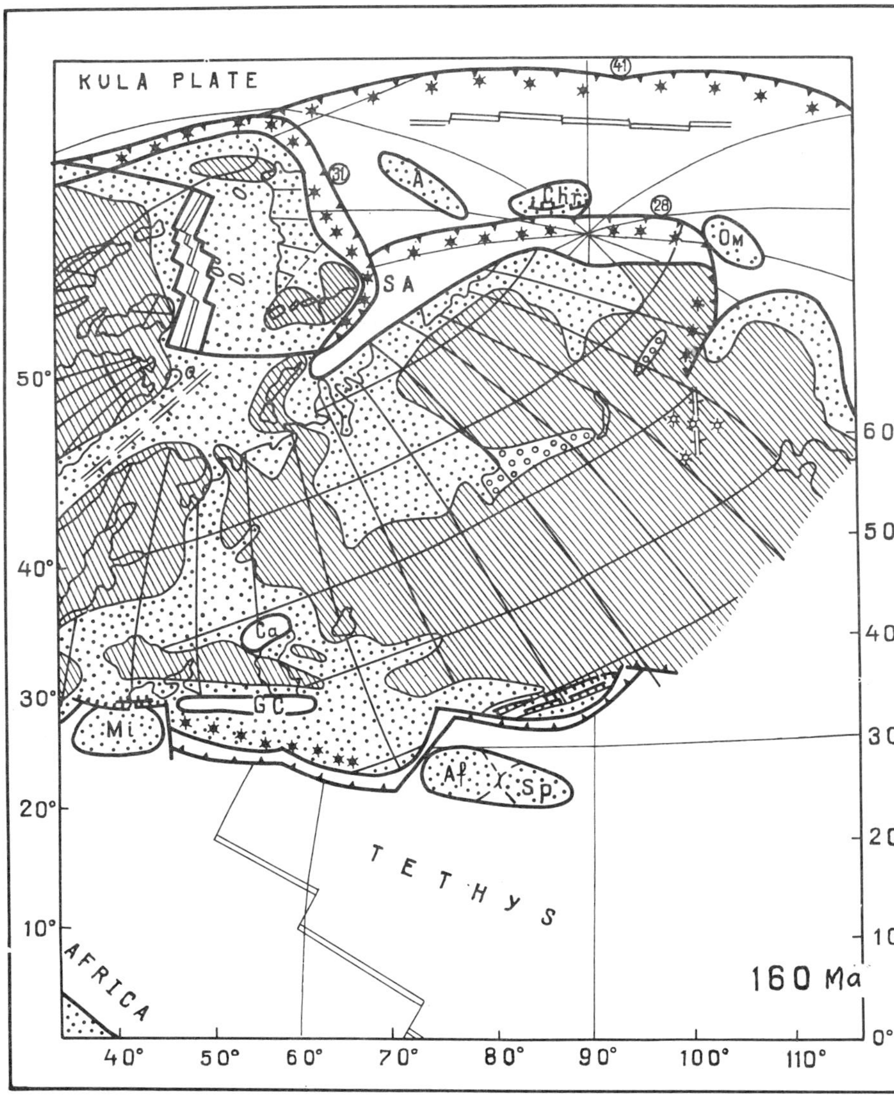

Fig. 200. Palinspastic reconstruction of the USSR territory for 160 Ma. See Figure 188 for explanation.

pole within Siberia, and both the Tethyan and the Pacific margins of the continent were bordered by island arcs, under which the floors of corresponding oceans were being consumed.

No continental blocks derived from Gondwanaland remained unaccreted except the Afghan-South Pamir block, which now approached the subduction zone near the Eurasian margin. The Moesia microcontinent collided with the Eurasian margin in the early Late Jurassic, creating the Crimea-Dobrogea orogenic zone. In the Jurassic, Pangea began to break up, and continents began to separate from Laurasia. The African-Arabian margin of Gondwanaland progressively converged with Eurasia, gradually closing the Tethys Ocean. Accordingly, the Lesser Caucasian arc and its continuation in Iran and Turkey was constantly active near the Eurasian margin. The marginal Great Caucasian basin, presumably underlain by oceanic crust, opened behind this arc in the middle of the Jurassic.

The two systems of volcanic arcs still existed along the Pacific margin of Laurasia. The Chersky and Alazei terranes reached the Uyandina-Yasachnaya arc, and as a result the subduction zone jumped oceanward to the other side of those terranes. The latter were deformed and were involved with the basement of the island arc. The Mongol-Okhotsk oceanic basin was almost entirely closed by the Late Jurassic as Amuria and Siberia converged toward one another.

The Canadian (Amerasian) oceanic basin of the future Arctic Ocean began to open as the Chukotka-Alaska (including Taimyr) block separated from North America as is shown in the reconstruction (Figure 200), closing the South Anyyui oceanic basin. As a consequence, a suduction zone appeared in front of the moving Chukotka-Alaska block, dipping under Chukotka and forming the Anyui volcanic arc.

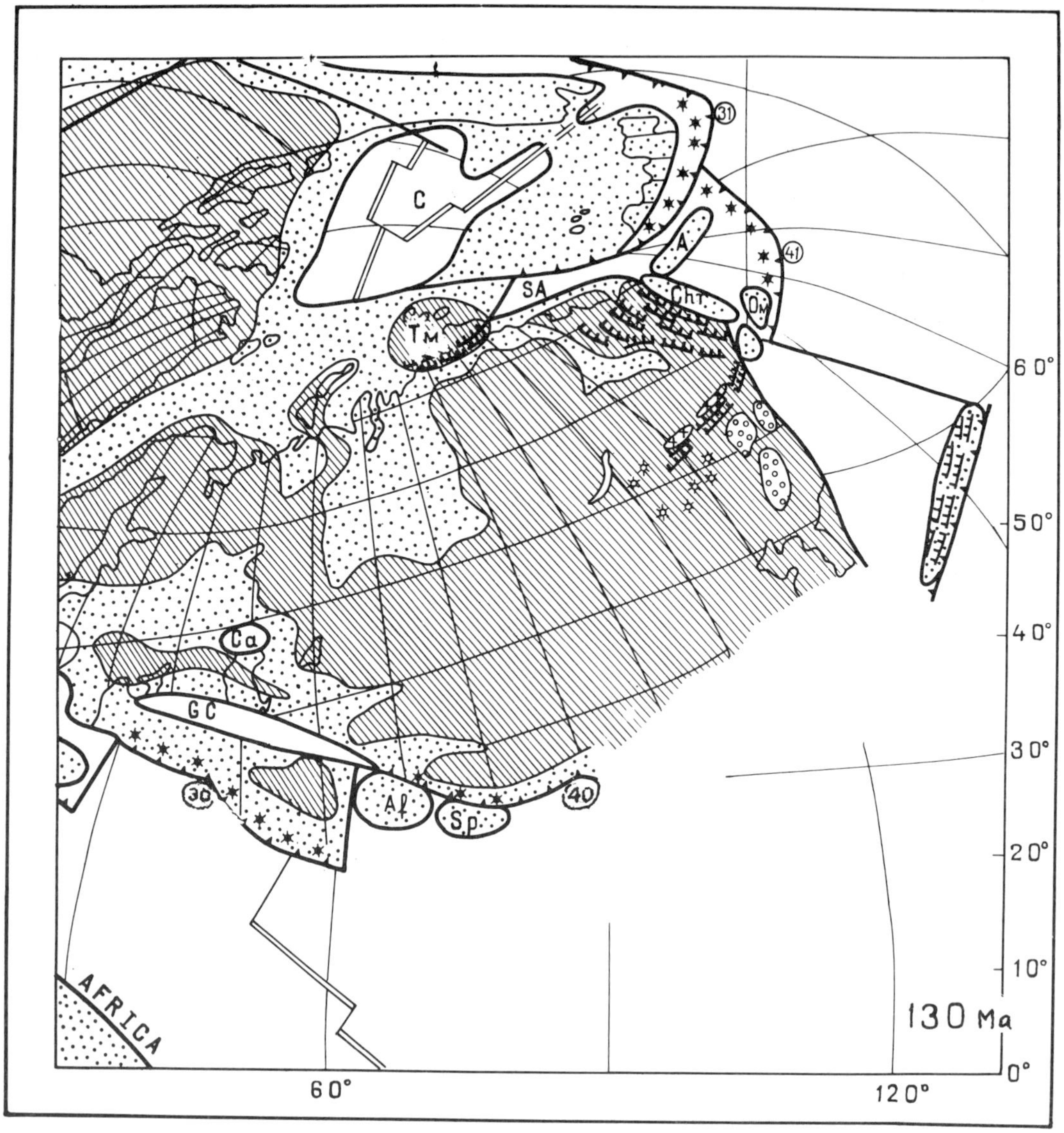

Fig. 201. Palinspastic reconstruction of the USSR territory for 130 Ma. See Figure 188 for explanation.

Early Cretaceous, 130 Ma (Figure 201)

The time span from 160 to 130 Ma was the period of climactic break-up of Pangea. All present continents, except North America and Eurasia, were either already diverging or started to separate from the supercontinent. Moreover, this period was characterized in the Pacific Basin by rapid plate motions, which increased up to 20-22 cm/yr.

By the Early Cretaceous, the USSR territory occupied approximately the same position on the Earth as nowadays. The paleomagnetic pole was situated north of Chukotka, so the present Arctic area of Siberia was already in the Arctic region. The greater part of the USSR territory was included in a single large Eurasian plate, and a considerable part was dry land. Tectonic activity, as at present, was concentrated on the continental margin.

The steady-state convergence of Africa and Arabia with Laurasia continued in the Tethys belt, and volcanic arcs, in particular the Lesser Caucasian and Hindu-Kush arcs, stretched along the Laurasian margin. The Afghanistan and South Pamir blocks (remnants of the former Iran microcontinent) approached the subduction zone in the Early Cretaceous.

In the Arctic sector, as the Canadian basin opened, the Chukotka-Alaska block converged with the Siberian margin of Laurasia, closing the South Anyui oceanic bay. The Taimyr block was the first to collide with Siberia, causing the Taimyr foldzone.

The most dramatic, although not completely understood, events occurred in the Pacific sector. By the Late Jurassic-

Early Cretaceous a number of exotic terranes primarily situated in the equatorial or tropical zones of the Pacific, approached and encountered the Sikhote-Alin and Koni-Murgal arcs, building up accretionary complexes with intricate nappe-fold structures and with ophiolite slices. The Philippine Islands seem to be the best modern analog for such diversified accretionary products. In the beginning of the Cretaceous, the intraoceanic arc system, along which the accretionary complexes were forming, was apparently broken and separated into the Sikhote-Alin and Koryak segments. The Sikhote-Alin terrane remained fixed, and the space between it and the Eurasian continent, being a marginal basin, was filled with the Cretaceous flysch charcteristic of today's East Sikhote-Alin area. The Koryak segment was attached to the Izanagi plate, and together they moved at a rate of 20 cm/yr toward the Siberian continental margin. The microcontinental Chersky block arrived at the subduction zone of the Uyandina-Yasachnaya arc. Subduction ceased; the arc, together with the Chersky block broke up, was displaced, and collided with Siberia; it was thrust over the Verkhoyansk fan, deforming rocks of the Verkhoyansk foldzone for the first time.

Mid-Cretaceous, 110 Ma (Figure 202)

As before, the spreading axis in the Tethys belt separated the passive African-Arabian and active Eurasian margins, as shown by the mid-Cretaceous isotopic ages for the Oman and Troodos ophiolites. The boundaries of the Eurasian, African-Arabian and Tethyan plates all met here, in the present Middle East region. The Tethyan plate and eventually the locus of spreading were consumed in the subduction zone near the Eurasia margin. A new spreading axis originated immediately between Africa and Arabia. A continuous subduction zone extended nearly 6000 km along the Eurasian margin. In the mid-Cretaceous, the South Pamirs and Afghan blocks collided definitively with the Eurasian margin, and the Late Cimmerian orogenic zone, e.g. , the Rushan-Pshart zone of the Pamirs with its nappes, ophiolite slices, and granitic intrusions, was formed.

In the Arctic sector, opening of the Canadian oceanic basin ceased. The Chukotka-Alaska block, already displaced far to the south, closed the South Anyui oceanic bay completely and collided with Siberia.

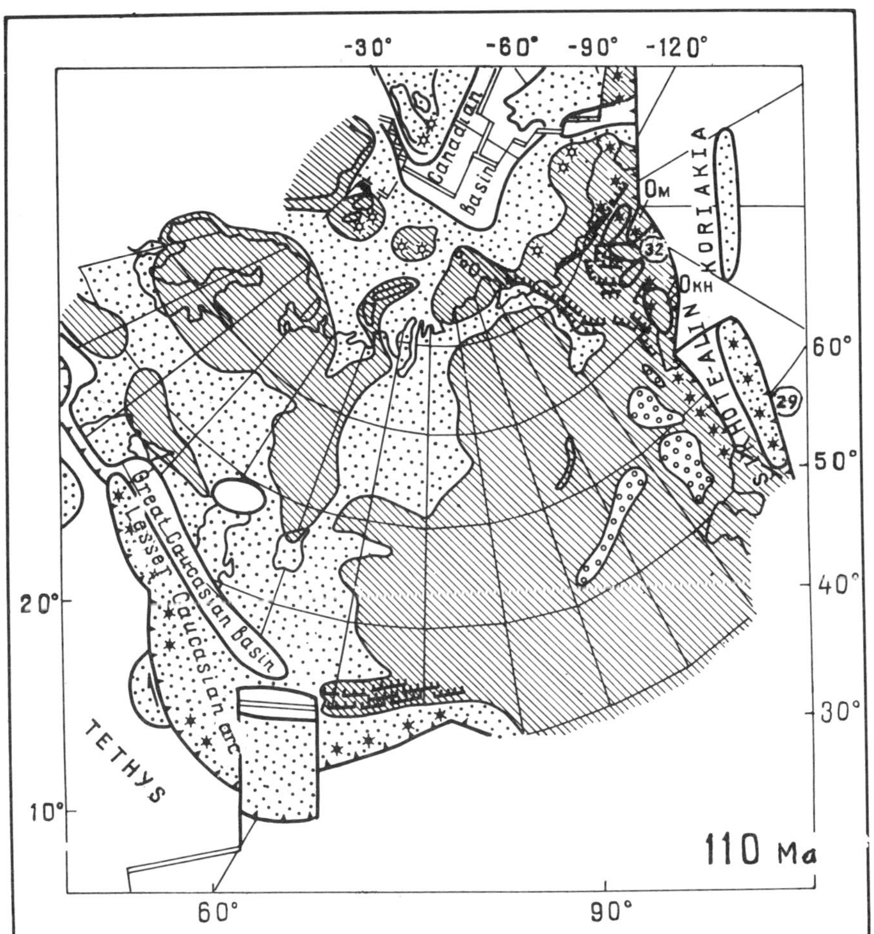

Fig. 202. Palinspastic reconstruction of the USSR territory for 110 Ma. See Figure 188 for explanation.

Approximately by the Aptian (120 Ma), all the continental and other exotic terranes in the present NE USSR had gathered together. An intricately patterned mosaic of orogenic belts arose here. Because of prolonged convergence and collision of Laurasian, Chukotka-Alaskan and Pacific plates, the area was redeformed and broken into a number of blocks, which rotated, deformed, and finally were closely packed. As a consequence, the northeast Eurasian margin was considerably enlarged by accretion. Because the Izanagi plate moved persistently towards Eurasia, a new subduction zone with the Okhotsk-Chukotka volcanic belt above it originated along a newly formed margin. To the south, in the present Sikhote-Alin area, data indicate that two parallel active volcanic arcs existed, and hence presumably two subduction zones . The marginal basin situated between them was being filled with flysch throughout the Early and mid-Cretaceous and was shortened as its floor was gradually subducted beneath the Eurasian margin. The accretional Koryak block was at that time in the Pacific Ocean, already relatively close to the Okhotsk-Chukotka subduction zone. It was pieced together from numerous small terranes in front of the former Koni-Murgal volcanic arc.

Late Cretaceous, 80 Ma (Figure 203)

In the Late Cretaceous, Laurasia was at last broken up, separating into the Eurasian and North American plates. The pole of North America/Eurasia rotation was near Greenland, so within the northeast Soviet Arctic sector, Eurasia converged with North America. The reconstruction in Figure 203 shows a plausible position of the subduction zone that separated the two plates, consuming the oceanic floor of the Canadian basin.

In the Tethyan belt the pattern of plate interaction within the USSR territory changed little as compared to early mid-Cretaceous time. It is important to mention that the Daralagez block collided with the Lesser Caucasian arc, forming the Sevano-Akera ophiolite-bearing suture zone. After collision, the subduction zone jumped inmmediately to the southern margin of the Daralagez block, and volcanism in the Lesser Caucasian arc proceeded almost without interruption.

In the Pacific sector, the Koryak accretionary block approached the subduction zone of the Okhotsk-Chukotka belt and blocked it. As a consequence, volcanism ceased in that

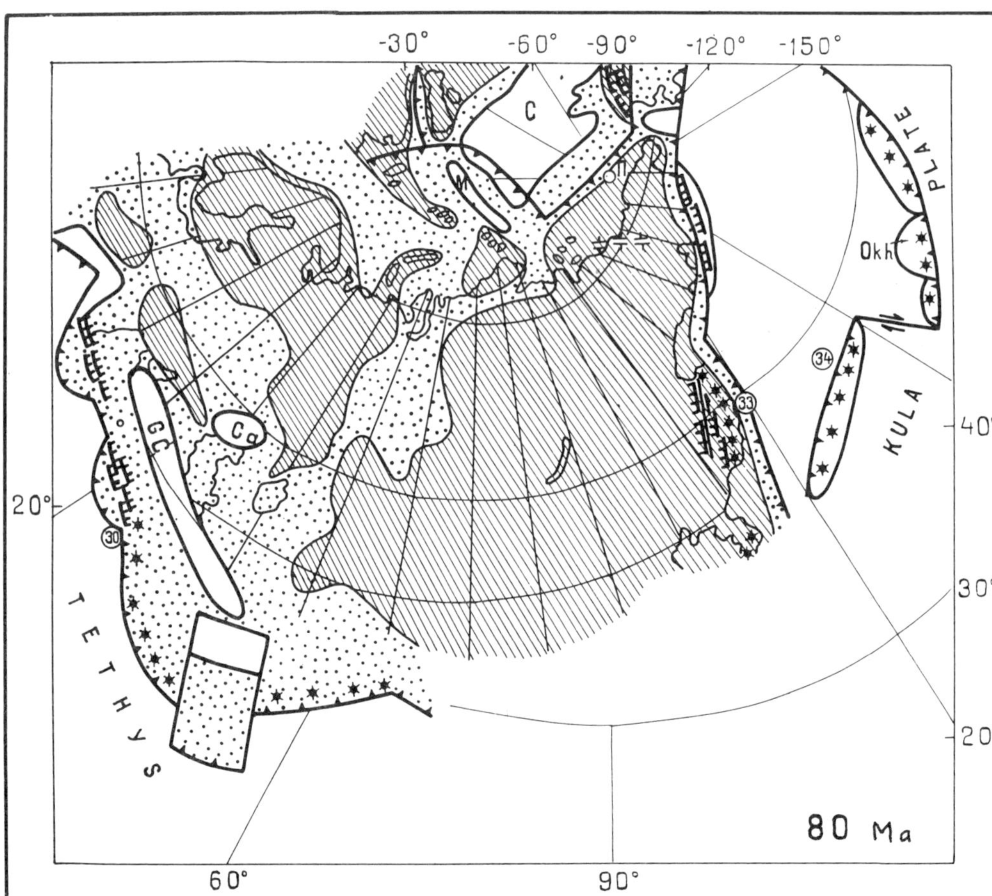

Fig. 203. Palinspastic reconstruction of the USSR territory for 80 Ma. See Figure 188 for explanation.

belt and the block itself was deformed. In the Late Cretaceous, the former Sikhote-Alin arc collided with the Eurasian margin, producing the pre-Senonian orogeny in the Sikhote-Alin area and enlarging the continental margin by accretion. Farther south also, plates continued to move from the Pacific basin toward Eurasia, forming a new subduction zone surmounted by the Sikhote-Alin continental margin volcanic belt. The present Sakhalin, Kamchatka, and Olyutorsky zones contain evidence indicating that another intraoceanic subduction zone formed in the Pacific Ocean at a distance up to 1000-1500 km from the continental margin. It was probably situated almost in the same area where the preceding Koni-Murgal-Sikhote-Alin island-arc system had existed. Various exotic terranes, including volcanic plateaus, continued to arrive from the Pacific Ocean.

The Cretaceous-Paleogene Boundary, 65 Ma (Figure 204)

The Cenozoic history of the USSR territory was characterized by nearly ubiquitous plate convergence along the Eurasian boundaries. Oceanic plates together with continental blocks moved from all sides toward Eurasia: (i) from the Atlantic Ocean on the southwest side (Africa-Arabia); (ii) from the Indian Ocean on the south side; and (iii) from the Pacific Ocean on the east side. The global spreading system penetrated into the USSR territory only in the Arctic, where at that time rifting with abundant basaltic volcanic extrusions began in the site of the future Eurasian basin.

In the Tethys belt, the width of the ocean decreased markedly at this time; the ocean was only 1000 km across in the Caucasus region. Tethys almost became a mere oceanic branch of the Indian Ocean as its floor was subducted beneath the Adjaro-Trialetan arc along the southwestern margin of Eurasia. Two important Late Cretaceous events within the Tethyan belt were; first, the obduction of ophiolites; great slices of oceanic crust such as the Semail nappe in Oman were thrust onto the continental Arabian margins; and second, the deep Black Sea basins opened behind the Adzharo-Trialetan arc as a manifestation of back-arc spreading.

In the Arctic sector, Eurasia and North America continued to converge in the Late Cretaceous and Early Paleogene. The pole of plate rotation remained near Greenland. This convergence caused compression that deformed the rocks of Herald Island (east of Wrangel Island) and the Brooks Range (Alaska) zone. The Eurasian basin began to open only in the Late Paleocene, when the North America/Eurasia pole of rotation shifted to the northeast region of the USSR.

In the Pacific sector, the western Pacific Ocean plates converged with the eastern Eurasian margin. In the Late Cretaceous, the Kula plate was subducted under Eurasia at a high rate (15-16 cm/yr). The East Sakhalin arc, which apparently

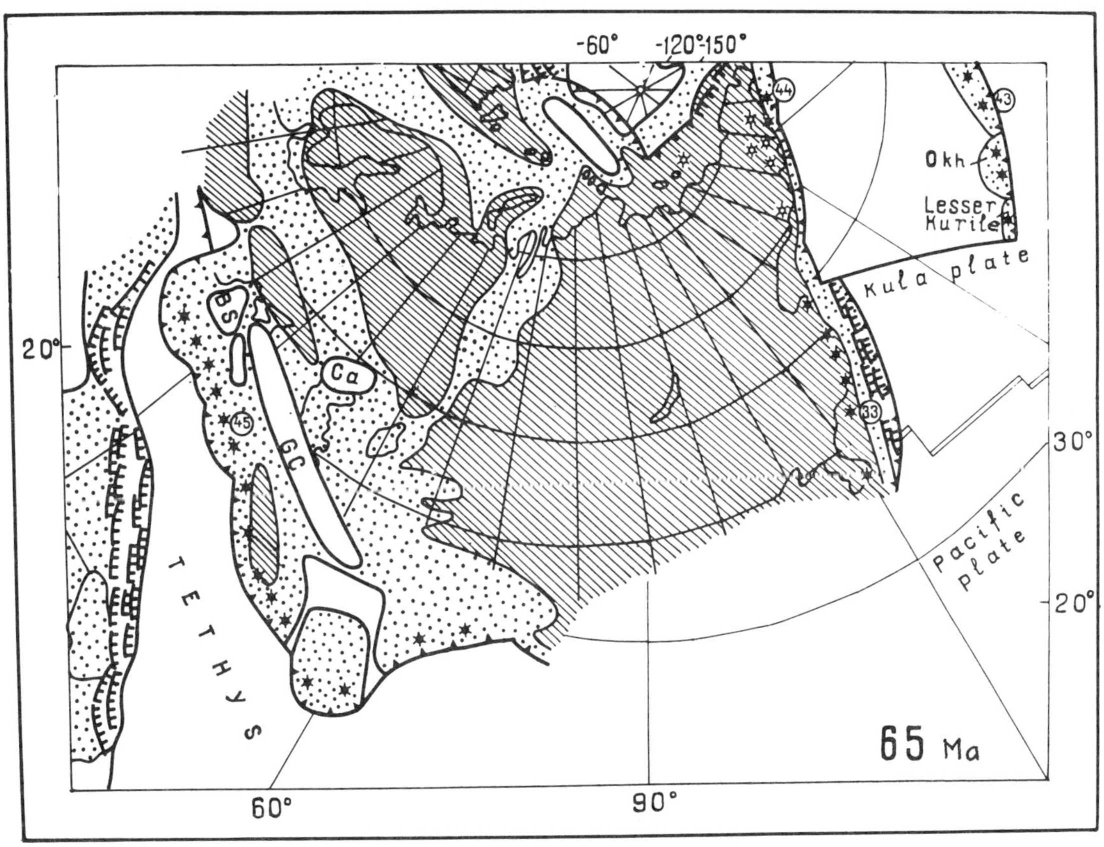

Fig. 204. Palinspastic reconstruction of the USSR territory for 65 Ma. See Figure 188 for explanation.

lay above an east-dippling subduction slab, rapidly converged with the continental margin in the Sikhote-Alin volcanic belt and collided with the margin at a time near the Cretaceous-Paleogene temporal boundary. This event induced folding in East Sakhalin, and the subduction zone jumped to the eastern side of the accreted block, which became an outer non-volcanic arc of the Sikhote-Alin belt. The Anadyr-Bristol volcanic belt originated in the Early Paleogene with a corresponding subduction zone in the Chukotka-West Alaska region. The Olyutorsky and Anadyr-Bristol arcs were connected by a NW-striking transform fault parallel to the direction that the Kula plate was moving relative to Eurasia in this locality. A portion of the Mesozoic crust of the Kula plate, which at present constitutes the floor of the Aleutian basin of the Bering Sea, moved along this fault. Throughout the Paleogene, the junction of the Pacific/Kula spreading axis with the Eurasian margin was moving northeast. At about 65 Ma it was near Hokkaido Island, while at 55 Ma it was close to the Olyutorsky arc. When the spreading center was entirely consumed and the Pacific plate began to contact Eurasia, the rate of plate convergence sharply decreased to half the previous rate. Volcanic activity in the Sikhote-Alin belt decreased in the Paleocene, possibly for this reason.

Early Oligocene, 35 Ma (Figure 205)

Although plates continued to converge along the southern and eastern boundaries of Eurasia, plate interactions largely changed in the Oligocene, mainly because convergence rates sharply decreased. They decreased from 2.5 to 1 cm/yr between Arabia and Eurasia in the Tethyan belt, and from 12 to 4 cm/yr between India and Eurasia, mainly because of the continental collisions in that region. In the Pacific Ocean, the plate convergence rates decreased from 15 to 7 cm/yr because the Kula plate had been almost completely consumed in subduction zones, and the more slowly moving Pacific plate was now interacting with Eurasia.

In the Oligocene, the Tethys Ocean was almost completely closed. After this, the oceanic crust of the Black Sea Para-Tethys back-arc basin began to be consumed. This was accompanied by shortening in the area of the Great Caucasus Mountains for the first time and the mountains started to rise.

The India/Eurasia collision was undoubtedly the most important event of this epoch. As a consequence of the progressive collision, Eurasia was broken across to its distant margin near the Sea of Okhotsk, and the mountainous belt of Inner Asia, consisting of numerous microplates and crustal blocks,

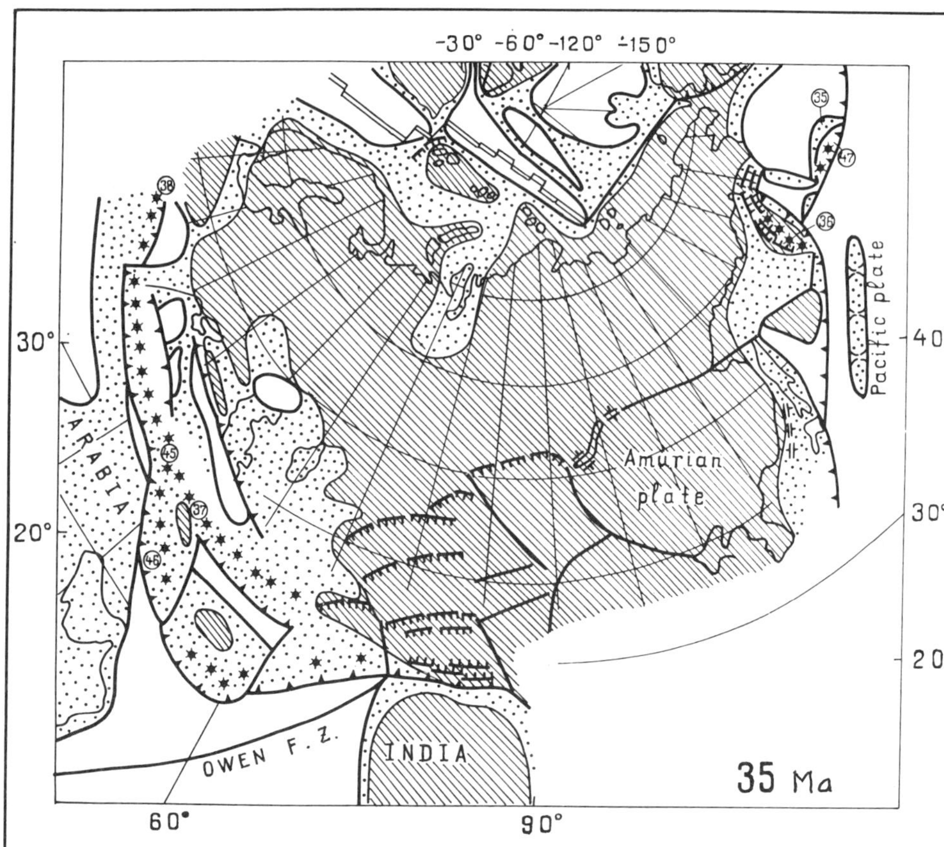

Fig. 205. Palinspastic reconstruction of the USSR territory for 35 Ma. See Figure 188 for explanation.

began to grow and was continuously deformed.

The Eurasian oceanic basin opened along the Gakkel Ridge (Arctic Mid-Ocean Ridge) in the Arctic segment, around a pole that was near the Novosibirsk (New Siberian) Islands from 35 to 20 Ma instead of in the NE USSR. Hence North America again converged with Eurasia, presumably forming the Alaska orocline in the Bering Strait region, in which the Mesozoic structures of the Brooks Range were bent like a knee.

At about 40 Ma, the Pacific realm was considerably reorganized when the Pacific plate motion changed. The former Olyutorsky arc ceased its activity, and its separate parts, the Olyutorsky and West Kamchatka zone, collided with the Eurasian margin and attached to it, as marked by the belt of Eocene orogeny in northwest Kamchatka. The Kurile-Kamchatka and Aleutian island arcs were formed along the Eurasian and North American margins at that time.

CONCLUSIONS

The USSR territory is a composite continental mass produced by numerous diverse blocks. These were formerly far from each other in different parts of the Earth, and they traveled many thousand kilometers before they eventually approached one another and amalgamated into a single large continent (Figure 206). The two main continents in this amalgamation, the East European and Siberian, that form the framework of Eurasia, were in the southern hemisphere in the Early Paleozoic, and were separated from each other and from other continents by broad oceans. The closure of oceans and numerous continental collisions are the most remarkable events in the subsequent history; they account for the predominantly accretionary tectonics of the USSR territory and for the wide distribution of foldbelts. All these events occurred as the continents moved generally to the north. Processes contrary to continental accretion and related rather to extension may also be reconstructed in the geological history of the USSR. They were responsible for the opening of paleo-oceans, e.g., Paleo-Asian, Uralian, Paleo-Tethys, Tethys, etc. The crust of these oceans was later largely consumed in subduction zones. Extensional processes are at present evident in the formation of the young oceanic basins of Arctica and in the development of various marginal and intracontinental sedimentary basins.

Data on the ancient East European and Siberian platforms indicate that their basements also formed by the convergence and collision of separate blocks. Suture zones are the evidence of collisions. They are characterized by a thrust-nappe structure and by the presence of slabs of oceanic and island-arc assemblages. The main epochs of collision in these platforms terminated about 1700 Ma, by which time the cratons were fully formed. In the subsequent Riphean and Phanerozoic history, the breaking up of the ancient basement was the main event on the platforms, as exemplified by aulacogens, failed arms of three-armed rifts, two branches of which later became active spreading axes and produced oceanic crust. It may be concluded that some continental blocks broke from the continents and drifted away; thus, the Early Riphean continents were larger than the present outlines of the platforms. The main epochs of rifting on both of the main platforms were Middle Riphean, Vendian, Devonian, and Triassic. Intra- or marginal-continental sedimentary basins formed above aulacogens and rift zones.

The plate tectonic approach may be applied with confidence to events after 800-1000 Ma, when the first true ophiolite associations appeared. Both collisional and accretional foldbelts are present in the USSR. The classical example of collisional belts is the Urals; that of accretionary belts, the Koryak-Kamchatka zone. The foldbelts are of different ages and formed at various times from the Late Precambrian to the Late Cenozoic. All formed along convergent boundaries. Some developed above subduction zones with island arcs or continental-margin volcanic belts; some in zones where exotic terranes accreted to form a mosaic of chaotic complexes; and others formed where continents collided, ophiolites were obducted, and granite-gneissic domes grew. However, each belt developed completely individually, and no single canonical scheme of foldbelt evolution can be proposed. This conclusion is of great importance as it contradicts all ad hoc schemes of universal foldbelt evolution and forces us to analyze the interrelations of the plates for each belt. The characters of each belt depend on many factors; for example, which plates (oceanic or continental) converged and collided; how long collision proceeded; what was the prehistory of the belt; whether oceanic crust was already present or originated during the development of the belt; what geological complexes were involved in collision, etc. Thus, the path to the understanding of foldbelt evolution is not straight nor paved. The hard task of analyzing the evolution of foldbelts requires strenuous efforts to put together diverse data from different sources, and requires a broad regional and sometimes global approach as well as close attention to local details. It is also necessary to emphasize that many foldbelts formed over a long time and underwent multi-phase deformation, so that later deformation distorted and obscured early structures as is exemplified by the Late Paleozoic horizontal flexures of Central Kazakhstan and the Tien Shan, or the Early Cretaceous Kolymian loop of the northeast USSR.

The succession of Phanerozoic events which formed the continental basement of the USSR can be outlined as follows:

Cambrian. - collision of Barentsia and East Europe and formation of the Timan foldbelt; growth of accretionary continental blocks within the Paleo-Asian Ocean.

Ordovician - opening of the Uralian Paleo-Ocean; collision of island arcs of the Paleo-Asian Ocean with minor continental blocks and accretion of the Kazakhstan microcontinent; beginning of the opening of the Paleo-Tethys.

Devonian - break-up of the East European and Siberian continents; formation of the Peri-Caspian oceanic basin; formation of accretional complexes in the Urals; beginning of closure of the Uralian and Paleo-Tethys Oceans.

Carboniferous - closure of the Uralian Paleo-Ocean; beginning of collision of the Kazakhstan microcontinent with Europe and the Tarim-Karakum microcontinent.

Permian - closure of the eastern arm of the Paleo-Tethys Ocean; collision of Siberia, Kazakhstan, and East Europe; formation of the Uralian and Tien Shan foldbelts; amalgamation of North Eurasia as part of Laurasia; shaping of the Mongol-Okhotsk and South Anyui oceanic bays.

Triassic-Early Jurassic - closure of the western arm of the Paleo-Tethys Ocean; closure of the Mongol-Okhotsk basin; collision of the Amurian microcontiennt with Siberia; rifting

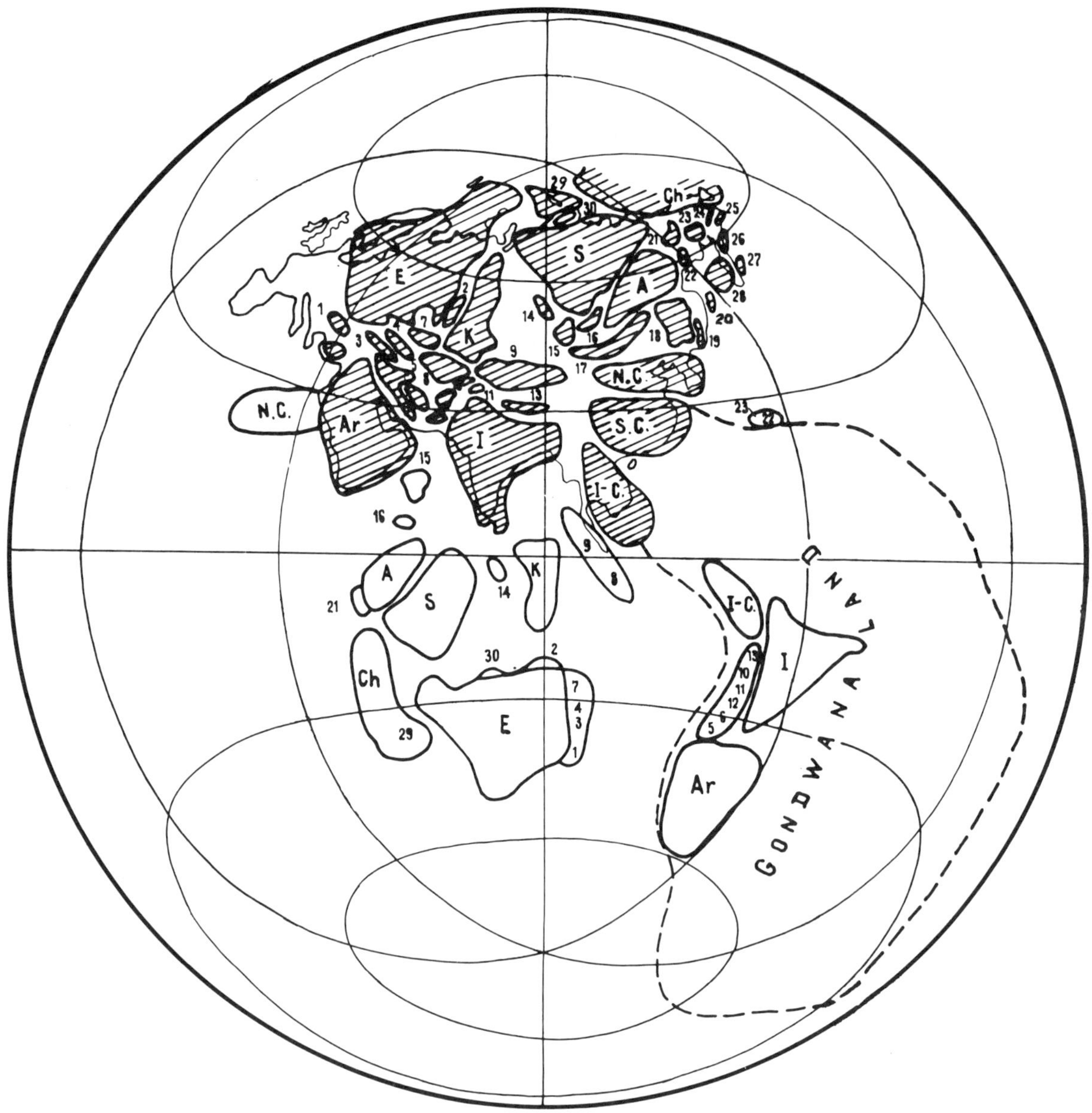

Fig. 206. Continents, microcontinents, and suspect terranes now constituting Asia and East Europe, and their initial positions in the Early Paleozoic. Lambert projection centered at 0° N, 90° E. Blocks in present positions are hatched, their Early Paleozoic positions are shown by heavy lines. The Early Paleozoic Gondwanaland is shown by dashed lines.
A - Aldan shield, Ar - Arabia, Ch - Chukotka, E - East Europe, I - India, I-C - Indo-China, K - Kazakhstan N.C. - North China, S - Siberia, S.C. - South China. Other blocks and terranes: 1, Moesia; 2, Mugodzhar; 3, Dzirula; 4, Maker; 5, Iran; 6, Lut; 7, Ustyurt; 8, Karakum; 9, Tarim; 10, Kurgovat; 11, South Pamirs; 12, Afghan; 13, Tibet; 14, Tom; 15, Tuva-Mongolian; 16, Barguzin; 17, Argun; 18, Amuria; 19, Sergeev; 20, East Sakhalin; 21, Chersky; 22, Okhotsk; 23, Omolon; 24, Talovo-Mainsky; 25, Koryakia; 26, Olyutorsky; 27, East Kamchatka; 28, Sea of Okhotsk; 29, Kara; 30, Shrenk.

in North Eurasia with formation of the West Siberian sedimentary basin.

Jurassic-Early Cretaceous - opening of the Canadian basin of the Arctic Ocean and simultaneous closure of the South Anyui oceanic bay; collision of Taimyr and Chukotka with Siberia; formation of foldbelts of the northeast USSR.

Late Cretaceous-Eocene - closure of the Tethys Ocean; start of collision of Arabia with Eurasia; formation of accretionary complexes in the Koryak-Kamchatka belt; opening of the Eurasian basin of the Arctic.

Late Cenozoic - Progressive India/Eurasia collision; formation of the belt of rejuvenated mountains in Central Asia; subduction of the Pacific plate, with formation of the system of island arcs and marginal seas of the eastern USSR.

Two very long principal stages with essentially different geodynamic characteristics are recognized in the Phanerozoic plate motions and tectonics. The first stage, which lasted through the whole Paleozoic was characterized by a more or less N-S zone of plate convergence approximately along the 90° E meridian, and by a parallel zone of plate divergence. Continental blocks were progressively brought into this zone, which was at first situated to the east of the two main continents, and were amalgamated with the latter, eventually producing the Eurasian continent. Some blocks that arrived in the convergence zone had rifted from the margin of Gondwanaland. This stage culminated in the incorporation of Eurasia as part of the Pangea supercontinent.

The second principal stage, which lasted through the Mesozoic and Cenozoic, witnessed a cardinal reorganization of plate movements. Two subduction zones along the Eurasian margins became most important: (i) a sublatitudinal zone south of Eurasia in the northern tropical belt, where subduction consumed the Tethys Ocean floor, and (ii) a meridianal subduction zone east of Eurasia, in which the Pacific Ocean floor was subducted and into which exotic terranes almost continually arrived from the ocean side. We need hardly any other reason for the plate motions to reorganize their pattern than rearrangement of the convective flow system in the Earth's mantle.

The long-range result of the first stage was amalgamation of a large continent from numerous blocks of different sizes. In the second stage, minor terranes were additionally accreted to this Paleozoic continent. Finally; some of the greater Gondwana continental fragments, e.g., Africa, Arabia, India, and Australia, have been approaching Eurasia. Thus, we anticipate a future supercontinent, one with a completely different internal arrangment of continents than that of the Late Paleozoic Pangea.

The authors know very well that the analysis presented here concerns mainly the motions of large plates. Each time an individual region was considered, there were specific data lacking, such as geochronological, paleomagnetic, paleoclimatic, biogeographic, geochemical, isotopic, and other information. Therefore, some of the regional reconstructions presented in this study are ambiguous, and in some cases alternative solutions may be proposed. We hope, however, that this material may serve as a basis for interpreting data for individual regions and for the country on the whole, using the plate-tectonic approach. The authors believe that the foregoing analysis will help field geologists to apply this approach to different regions of the Soviet Union. We are sure that we are only at the beginning of a way for geologists to interpret the territory of the USSR.

REFERENCES

Adamia, Sh. A., I. P. Gamkrelidge, G. I. Zakariadze, and M. B. Lordkipanidze, The Adzharo-Trialetan trough and formation of the Black Sea basin (in Russian), *Geotektonika, 1*, 78-94, 1974.

Adamia, Sh. A., B. Z. Asanidze, and D. M. Pechersky, Geodynamics of the Caucasus, in *The Problems of Geodynamics of the Caucasus* (in Russian), edited by M. V. Muratov, pp. 13-21, Nauka, Moscow, 1982.

Aizenshtadt, G. E., and G. I. Slepakova, Tectonic position and structure of the South-Emba uplift (in Russian), *Geotektonika, 3*, 32-41, 1979.

Akhber, D. Ya, and I. V. Mushkin, The Kyzyl-Kum - Nurata deep seated fault (Tien Shan) (in Russian), *Geotektonika, 1*, 95-102, 1976.

Akhmetjev, M. M., and G. P. Filimonova, On ages and periodicity of volcanism in eastern Sikhote-Alin (in Russian), *Izv. Acad. Sci. USSR, Geol. Ser.*, no. 1, 37-51, 1986.

Alexeev, E. S., The Kuyul serpentinite melange and structure of the Talovo-Mainsky zone (Koryak highland) (in Russian), *Geotektonika, 1*, 105-120, 1981.

Alexeev, E. S., Ophiolite complexes of the southern Koryak highland (in Russian), *Geotektonika, 4*, 87-98, 1982.

Alikhanov, E. N., *Geochemistry of the Caspian Sea* (in Russian), 190 pp., Elm Publ.House, Baku, 1978.

Al'mukhamedov, A. I., Geochemistry of tholeiitic volcanism of spreading zones and platforms (in Russian), Doct. thesis, 49 pp., Inst. Geochemistry, Irkutsk, USSR, 1985.

Antonjuk, R . M., Oceanic crust of eugeosynclinal area of eastern Central Kazakhstan, in *Tectonics of the Ural-Mongolian Foldbelt* (in Russian), edited by M. V. Muratov, pp. 67-74, Nauka, Moscow, 1974.

Antonjuk, R. M., G. F. Lyapichev, and N. G. Markova, Destructive zones of Central Kazakhstan Paleozoids, in *Tectonics of the USSR Territory* (in Russian), edited by M. V. Muratov, pp., 111-127, Nauka, Moscow, 1979.

Aplonov, S. V., *Geodynamics of the Early Mesozoic Ob Paleo-Ocean* (in Russian), Inst. Oceanol., USSR Acad. Sci., Moscow, 98 pp., 1987.

Aplonov, S., An aborted Triassic ocean in West Siberia, *Tectonics, 7*, 1103-1122, 1988.

Arkhangelsky, A. D., *Geological Structure and Geological History of the USSR* (in Russian), Pt. 1, 412 pp., Pt. 2, 371 pp., State Publ. House, Geol. Ser., Moscow, 1947.

Arkhipov, G. I., and A. E. Panskikh, Basaltoid magmatism and iron accumulation in the Dzhagdinsky eugeosyncline, in *Magmatism and Tectonics of the Far East* (in Russian), edited by L. M. Parfenov and V. A. Popeko, pp. 56-67, Inst. Tectonics and Geophysics, Khabarovsk, 1975.

Artem'ev, M. E., *Isostasy of the USSR Territory* (in Russian), 285 pp., Nauka, Moscow, 1975.

Asanidze, B. Z., D. M. Pechersky, and Sh. A. Adamia, Results of paleomagnetic studies of Paleozoic rocks in the Caucasus (in Russian), *Izv. Acad. Sci. USSR, Phys. Earth*, 34-48, 1979.

Aubouin, J., X. Le Pichon, and A. S. Monin (Eds.), *Evolution of the Tethys*, Tectonophysics, 123, 315 pp., 1986.

Avdeev, A. V., Ophiolite zones and geological history of the Kazakhstan territory: a mobilistic approach (in Russian), *Sovietskaya Geol.*, 63-72, 1984.

Balla, Z., The Carpathian loop and the Pannonian Basin: A kinematic analysis, *Geophys. Trans., 30*, 313-353, 1984.

Balla, Z., Paleotectonic reconstruction of the Central Alpine-Mediterranean belt for the Neogene, *Tectonophysics, 127*, 213-244, 1986.

Beloussov, A. F., Yu. N. Kochkin, and Z. G. Polyakova, *Volcanic Complexes of the Riphean and Paleozoic Age, Gorny Altay, Gorny Shoria, and Salair Ridge* (in Russian), 282 pp., Nauka, Moscow, 1969.

Belov, A. A., *Tectonic Development of the Alpine Fold Area in the Paleozoic* (in Russian), 212 pp., Nauka, Moscow, 1981.

Belov, A. A., The Paleozoic of the Caucasus and problem of Paleo-Tethys (in Russian), *Geotektonika*, 81-90, 1986.

Belyaev, S. Yu., Tectonics of the southern Chara nappe thrust zone (East Kazakhstan) (in Russian), *Geol. Geofiz.*, 56-64, 1982.

Belyaev, S. Yu., and B. M. Chikov, Olistostrome occurrences in the middle Onon River course (Trans-Baikal region) (in Russian), *Geol. Geofiz.*, 36-42, 1980.

Belyakov, L. N., and B. I. Dembovsky, Some tectonic features of the North Urals and Pai-Khoi (in Russian), *Geotektonika*, 2, 51-57, 1984.

Belyi, V. F., *Stratigraphy and Structures of the Okhotsk-Chukotka Volcanic Belt* (in Russian), 171 pp., Nauka, Moscow, 1977.

Berzin, N. A., Melange-olistostrome association of the Khemchik-Sistikhemsky zone of Tuva, in *Main Tectonic Complexes of Siberia* (in Russian), edited by K. V. Bogolepov, pp. 104-129, Inst. Geol. Geophys., Novosibirsk, 1979.

Bespalov, V. F., System of tectonic nappes of Kazakhstan (in Russian), *Geotektonika*. *2*, 78-94, 1980.

Bezzubstsev, V. V., The Late Precambrian structural-lithological complexes in Taimyr and some features of their formation, in *Tectonics of the Baikalian (Riphean) Megacomplex of Siberia* (in Russian), edited by K. V. Bogolepov, pp. 48-55, Inst. Geol. Geophys., Novosibirsk, 1981.

Bibikova, E. V., *Uranium-Lead Geochronology of Early Stages of the Evolution of the Ancient Shield* (in Russian), Nauka, Moscow, 1989.

Bibikova, E. V., V. A. Makarov, T. V. Gracheva, and K. V. Seslavinsky, The age of ancient rocks of the Omolon massif (in Russian), *Dokl. AN SSSR*, *241*, 434-436, 1978.

Bogatikov, O. A., *Anorthosites* (in Russian), 232 pp., Nauka, Moscow, 1979.

Bogatikov, O.A., L. P. Zonenshain, and V. I. Kovalenko, Magmatism and geodynamics, *Proc. 27 I.G.C.*, *9*, Nauka, Moscow, pp. 3-13, 1980.

Bogdanov, A. A., *Tectonics of Platform and Folded Areas* (in Russian), 340 pp, Nauka, Moscow, 1976.

Bogdanov, N. A., V. D. Chekhovich, A. N., Sukhov, and V. S. Vishnevskaya, Tectonics of the Olyutorsk zone, in *Essays on Tectonics of the Koryak Highland* (in Russian), edited by Yu. M. Pushcharovsky and S. M. Tilman, pp. 189-217, Nauka, Moscow, 1982.

Bogolepov, K. V., and O. A. Votakh, *Precambrian of the Continents. Ancient Platforms of Eurasia* (in Russian), 312 pp., Nauka, Novosibirsk, 1977.

Bomarskaya, A. M., *Geology and Hydrogeology of the Western Black Sea* (in Russian), 211 pp., Sofia, 1979.

Borukaev, Ch. B., *Precambrian Structure and Plate Tectonics* (in Russian), 200 pp., Nauka, Novosibikrsk, 1985.

Brezhnev, V. D., and V. N. Shvanov, Early troughs and rock formations in the western part of South Tien Shan (in Russian), *Geotektonika*, *4*, 63-74, 1980.

Bukharov, A. A., *Zones of Protoactivation in Ancient Platforms* (in Russian), 202 pp., Nauka, Novosibirsk, 1987.

Burtman, V. S., *Geology and Mechanics of Nappes* (in Russian), 97 pp., Nedra, Moscow, 1973.

Burtman, V. S., *Structural Evolution of the Paleozoic Fold Zones* (in Russian), 163 pp., Nauka, Moscow, 1976.

Burtman, V. S., The problem of formation of the Pamirs-Pendzhab syntaxis (in Russian), *Geotektonika*, *5*, 56-63, 1982.

Burtman, V. S., Kinematics of the Carpathian structural loop (in Russian), *Geotektonika*, *3*, 17-31, 1984.

Burtman, V. S., Origin of the structural arcs of the Carpathian-Balkan region, *Tectonophysics*, *127*, 245-260, 1986.

Burtman, V. S., and G. Z. Gurarii, On the nature of the Pamirs and Tien Shan arcs (from paleomagnetic data) (in Russian), *Geotektonika*, *2*, 62-67, 1973.

Butov, Yu. P., Discovery of Paleozoic fauna in the Oka Suite (Eastern Sayan) (in Russian), *Dokl. AN SSSR*, *252*, 167-170, 1980.

Bychkov, Yu. M., and G.K., Melnikova, First discovery of Late Triassic corals in the Koryak highland (in Russian), *Dokl. AN SSSR*, *280*, 159-161, 1985.

Byzova, S. L., and M. A. Beer, Main tectonic features of the Flysch Carpathians in the territory of the USSR (in Russian), *Geotektonika*, *6*, 81-96, 1974.

Chekhov, A. D., Tectonics of the Pavlov-Pekulnei zone, in *Essays on Tectonics of the Koryak Highland* (in Russian), edited by Yu. M. Pushcharovsky and S. M. Tilman, pp. 70-106, Nauka, Moscow, 1982.

Chekhovich, V. D., and V. A. Sukhov, Essay on the geological structure of the Goven Peninsula and Kronotsky Islands (in Russian), *Geotectonika*, (in press) 1990.

Chernov, V. I., *Volcanic Assemblages and Porphyric Intrusions of Rudny Altay*, 320 pp., Nauka, Moscow, 1974.

Churkin, M., Jr., Western boundary of the North American continental plate in Asia, *Geol. Soc. Amer. Bull.*, *83*, 1027-1036, 1972.

Churkin, M., Jr., Comment and reply on "The Siberian connection of the North American and Siberian cratons", *Geology*, *10*, 467-469, 1979.

Cooper, A. K., D. W. Scholl, and M. S. Marlow, Plate tectonic model for the evolution of the Eastern Bering Sea Basin, *Geol. Soc. Am. Bull.*, *87*, 1119-1126, 1976.

Dagis, A. A., V. I. Aneikin, and T. V. Klets, The first conodont discoveries in the Late Triassic cherty sequences in Sikhote-Alin, in *Stratigraphy, Fauna and Flora of Siberia* (in Russian), edited by V. V. Menner, pp. 22-26, Nauka, Moscow, 1984.

Dercourt, J., L. P. Zonenshain, X. Le Pichon, et al., Geological evolution of the Tethys belt from the Atlantic to the Pamirs since the Lias, *Tectonophysics*, *123*, 241-315, 1986.

Dergunov, A. B., and N. N. Kheraskov, Constructive and destructive processes of crustal formation in the Altai-Sayan area and West Mongolia (in Russian), *Geotektonika*, *6*, 97-111, 1982.

Didenko, A. N., V. G. Korinevsky, S. A. Kurenkov, A. S. Perfiliev, and D. M. Pechersky, Sheeted dike complex in Southern Mugodzhary, in *History of the Uralian Paleo-Ocean* (in Russian) edited by L. P. Zonenshain, pp. 80-100, Inst. Oceanol. USSR Acad. Sci., Moscow, 1984.

Didenko, A. N., and D. M. Pechersky, Paleomagnetism of the Shuldak sheeted dike complex (South Mugodzhary) (in Russian), *Izv. Acad. Sci. USSR Phys. Earth*, *9*, 75-80, 1986.

Didenko, A. N., and Pechersky, D. M., Paleomagnetism of Middle Paleozoic ophiolitic rocks from Alai Ridge (in Russian), *Geotectonika*, *4*, 56-68, 1988.

Dobretsov, N. L., On nappe 'tectonics' of East Sayan (in Russian), *Geotektonika*, *1*, 39-50, 1985.

Dobretsov, N. L., Yu. E. Moldavantsev, and A. P. Kazak, *Petrology and Metamorphism of Ancient Ophiolites with Emphasis on the Polar Urals and West Sayan* (in Russian), 221 pp., Nauka, Novosibirsk, 1977.

Dobretsov, N. L., V. G. Kartavchenko, and N. F. Gabov, Metamorphosed Riphean olistostromes in the North Baikal region (in Russian), *Geol. Geofiz.*, *5*, 22-33, 1983.

Dotduyev, S. I., The nappe structure of the Great Caucasus (in Russian), *Geotektonika*, *5*, 94-105, 1986.

Drugova, G. M., S. A. Bushman,. and A. V. Sochava, Trough complexes of the western part of the Vitim-Aldan shield, in *Precambrian Trough Structures of the Baikal-Amur Region and Their Metallogeny* (in Russian), edited by V. A. Kuznetsov, pp. 102-111, Nauka, Novosibirsk, 1985.

Druzhinin, V. S., V. S. Sosedkov, and A. B. Karmanov, Structure of the upper crust in the vicinity of the Tyumen superdeep drillhole (in Russian), *Sov. Geol.*, *4*, 70-79,

1988.
Engebretson, D. C., A. Cox, and R. G. Gordon, Relative motions between oceanic and continental plates in the Pacific Basin, *Geol. Soc. Am. Spec. Paper 206*, 59 pp, 1985.
Ermolov, P. V., N. V. Polyansky, and N. L. Dobretsov, Ophiolites of the Chara zone, in *Ophiolites* (in Russian), edited by E. K. Patalakha, pp. 103-172, Nauka, Alma-Ata, 1981.
Ermolov, P. V., A. G. Vladimirov, and A. E. Izokh, *Orogenic Magmatism of Ophiolite Belts* (in Russian), 207 pp., Nauka, Novosibirsk, 1983.
Fedchin, F. G., V. P. Simonenko, and S. N. Zalevsky, *Geochemistry of the East Sikhote Alin and Okhotsk Volcanic Belts* (in Russian), 187 pp., Nauka, Moscow, 1981.
Fedorchuk, A. V., N. I. Gul'ko, and A. T. Savichev, Geochemistry of basalts from volcano-clastic associations of the Olyutorsky Range (Koryak highland) (in Russian), *Izv. Acad. Sci. USSR*, Ser. Geol. 8, 57-67, 1988.
Fedorovsky, V. S., *The Lower Proterozoic of the Baikal Mountain Area* (in Russian), 200 pp., Nauka, Moscow, 1985.
Fershtater, G. B., and N. S. Borodina, *Petrology of Magmatic Granitoids* (in Russian), 288 pp., Nauka, Moscow, 1975.
Filatova, N. I., Cretaceous-Paleogene volcanism in the transitional zone between the Verkhoyansk-Chukotka and Koryak-Kamchatka area (in Russian), *Geotektonika, 5*, 98-115, 1979.
Filatova, N. I., and V. I. Bush, The Hercynian geosynclinal structures of the Junggar Alatau (in Russian), *Geotektonika, 3*, 49-60, 1965.
Fomenko, K. E., Crustal structure of the Peri-Caspian depression according to geological and geophysical data (in Russian), *Bull. Moscow Soc. Naturalists, 47*, 103-111, 1972.
Fomichev, V. D., and L. E. Alexeeva, *Geological Essay on Salair* (in Russian), 218 pp., Gosgeoltekhizdat, Moscow, 1961.
Frolova, T. I., and I. A. Burikova, *Geosynclinal Volcanism, Eastern Slope of the South Urals* (in Russian), 264 pp., Moscow State Univ., Moscow, 1977.
Frolova, T. I., I. A. Burikova, and A. V. Gushchin, *Origin of Island Arc Volcanic Series* (in Russian), 275 pp., Nedra, Moscow, 1985.
Fujita, K., Pre-Cenozoic tectonic evolution of northern Siberia, *J. Geol., 86*, 159-171, 1978.
Gaal, G., and R. Gorbatchev, An outline of the Precambrian evolution of the Baltic Shield, *Precambrian Res., 35*, 15-52, 1987.
Gafarov, R. A., A. M. Leites, and V. S. Fedorovsky, Tectonic zoning of the Siberian platform basement and stages of continental crust development (in Russian), *Geotektonika, 1*, 43-58, 1978.
Gaiduk, V. V., *The Viluy Middle Paleozoic Rift System* (in Russian), Cand. thesis, 16 pp., Inst. Earth's Crust, Irkutsk, USSR, 1985.
Gamkrelidze, I. P., Mobilism and problems of tectonics of the Caucasus, in *Geodynamic Problems of the Caucasus* (in Russian), edited by M. V. Muratov, pp. 408, Nauka, Moscow, 1982.
Gamkrelidze, I. P., Geodynamic evolution of the Caucasus and adjacent areas in Alpine time, *Tectonophysics, 127*, 261-278, 1986.
Garkalenko, I. A., V. A. Bondarenko, A. V. Zhuravlev, and S. A. Ushakov, New data on crustal structure of the Kuril-Kamchatka trench (in Russian), *Dokl. AN USSR, 234*, 135-137, 1977.
Ged'ko, M. I., The Late Jurassic Uyandino-Jasachnaya island arc (NE USSR) (in Russian), *Geotektonika, 3*, 88-100, 1988.
Gegelyants, A. A., E. N. Galperin, I. P. Kosmynskaya, and B. M. Krafshina, The crustal structure of the central Caspian Sea from deep seismic soundings data (in Russian), *Dokl. AN SSSR, 123*, 520-522, 1958.
Geodekyan, A. A., G. B. Udintsev, and B. V. Baranov, Bedrock in the central part of the Sea of Okhotsk (in Russian), *Sovjetskaya Geol., 6*, 12-31, 1976.
Glukhovsky, M. Z., V. I. Moralev, and M. I. Kuzmin, Tectonics and petrogenesis of the Katarchean complex (Aldan shield) related to the proto-ophiolite problem (in Russian), *Geotektonika, 6*, 103-117, 1977.
Gordienko, I. V., Paleozoic magmatism and geodynamics of the northern part of Central Asian belt (in Russian), *Geol. Geofiz.*, 19-31, 1984.
Gorokhov, I. M., M. M. Rubinstein, E. P. Kutyavin, and E. S. Varshavskaya, Application of the Rb-Sr isotopic methods for age determinations of some Pre-Mesozoic rocks in Georgia, in *Geochronology of the East European Platform and its Junction with the Caucasus - Carpathian System* (in Russian), edited by I. M. Gorokhov, pp. 35-55, Nauka, Moscow, 1978.
Grachevsky, M. M., On possible Permian reefs in the marginal part of the North Caspian region (in Russian), *Novosti Neftjanoi i Gazovoi Tekhniki, Geol. Ser., 11*, 16-18, 1961.
Grantz, A., S. Eittreim, and C. T. Whitney, Geology and physiography of the continental margin north of Alaska and implication for the origin of the Canada basin, in *The Ocean Basins and Margins, vol. 5, The Arctic Ocean*, edited by E. Nairn, Plenum Press, New York, 1981.
Hamilton, W., The Uralides and motion of the Eurasian and Siberian platforms, *Geol. Soc. Am. Bull., 81*, 2553-1476, 1970.
Harbert, W., L. S. Frei, A. Cox, and D. C. Engebretson, Relative motions between Eurasia and North America in the Bering Sea region, *Tectonophysics, 134*, 239-261, 1987.
Herron, E., J. E. Dewey, and W. C. Pitman, Plate tectonic model for the Arctic, *Geology, 2*, 377-380, 1974.
Hiraoka, K., T. Nakajima, and H. Sakai, Accretion tectonics inferred from paleomagnetic measurements of Paleozoic and Mesozoic rocks in central Japan, in *Accretion Tectonics in Circum-Pacific Regions*, edited by H. Hashimoto and S. Uyeda, pp. 179-194, Terra Pub., 1983.
Irving, E., Drift of the major continental blocks since the Devonian, *Nature, 270*, 304-309, 1977.
Irving, E., Fragmentation and assembling of the continents, mid-Carboniferous to present, *Geophys. Surv., 5*, 299-333, 1983.
Ivanov, S. N., A. A. Efimov, and L. M. Minkin, The nature of the Uralian geosyncline (in Russian), *Dokl. AN SSSR, 206*, 1177-1180, 1972.
Ivanov, S. N., V. G. Korinevsky, and G. P. Belyanina, The relics of the oceanic rift valley in the Urals (in Russian), *Dokl. AN SSSR*, 211, 939-942, 1973.

Ivanov, S. N., A. S. Perfiliev, and V. M. Necheukhin, The Paleozoic history of the Urals, in *Magmatism, Metamorphism and Ore Formation in the Geological History of the Urals* (in Russian), edited by S. N. Ivanov and D. S. Shteinberg, pp. 13-38, Sverdlovsk, 1974.

Jackson, G. R., and G. L. Johnson, Structure and evolution of the Amerasian Basin, *Proc. 27th I.G.C., Coll. 04, 4*, 119-127, Moscow, 1984.

Kalyaev, G. I., Tectonics of the Ukranian shield and its position in the East European platform, in *Tectonics of the Basement of Ancient Platforms* (in Russian), edited by M. V. Muratov, pp. 50-60, Moscow, 1973.

Kalyaev, G. I., E.G. Glevassky, and G. Kh. Dimitrova, *Paleotectonics and Crustal Structure of the Iron Bearing Province of the Ukraine* (in Russian), 239 pp., Naukova, Dumka, Kiev, 1984.

Kamaletdinov, M. A., *Nappe Structures of the Urals* (in Russian), 230 pp., Nauka, Moscow, 1974.

Karasik, A. M., Main features of the history and structure of the Arctic basin from aeromagnetic data, in *Marine Geology, Sedimentation, Sedimentary Petrology, and Geology of the Ocean* (in Russian), edited by I. S. Gramberg, pp. 178-193, Nedra, Leningrad, 1980.

Karasik, A. M., and N. A. Sochevanova, *The Magnetic Lineation Map of the World Ocean* (in Russian), IZMIRAN, Leningrad, 1981.

Karyaev, V. A., On the history of the Central Kazakhstan Paleozoids (in Russian), *Geotektonika*, *4*, 46-60, 1984.

Kasymov, M. A., R. R. Khaibullin, and V. N. Matvienko, On the problem of the Mugodzhar Precambrian, in *Precambrian-Early Paleozoic History of the Urals* (abstract) (in Russian), pp. 28-29, Sverdlovsk, 1980.

Katyukha, Yu. P., and A. M. Rogachev, On the age of the Mangatgol, Dabanmalga and Oka Suites of East Sayan (in Russian), *Geol. Geofiz.*, *5*, 68-78, 1983.

Kazmin, V. G., Collisions and rifting in the history of the Tethys (in Russian), *Geotektonika, 5*, 14-23, 1989.

Kazmin, V. G., I. M. Sbortshikov, L. E. Ricou, L. P. Zonenshain, J. Boulin, and A. L. Knipper, Volcanic belts as markers of the Mesozoic-Cenozoic active margin of Eurasia, *Tectonophysics*, *123*, 83-122, 1986.

Keilman, G. A., *Migmatitic Complexes of Mobile Belts* (in Russian), 200 pp., Nedra, Moscow, 1974.

Keller, B. M., M. A. Semikhatov, and N. M. Chumakov, Typical stratigraphic sections of the upper eratema of the Proterozoic (in Russian), *Proc. 27th I.G.C.*, *1*, 56-75, Nauka, Moscow, 1984.

Kent, D. V., X. Zeng, W. J. Zhang, and N. D. Opdyke, Widespread Late Mesozoic to recent remagnetization of Paleozoic and Lower Triassic sedimentary rocks from South China, *Tectonophysics*, *139*, 133-144, 1987.

Khain, V. E., Main stages of the tectono-magmatic development of the Caucasus: Geodynamic interpretation (in Russian), *Geotektonika, 1*, 13-27, 1975.

Khain, V. E., *Regional Geotectonics, Outer-Alpine Europe and West Asia* (in Russian), 359 pp., Nedra, Moscow, 1977.

Khain, V. E., *Regional Geotectonics, Outer-Alpine Asia and Australia* (in Russian), 356 pp., Nedra, Moscow, 1979a.

Khain, V. E., The Late Triassic North Caucasian-Turkmen-North Afghan volcanic belt and opening of the North Tethys zone (in Russian), *Dokl. AN SSSR, 249*, 1190-1192, 1979b.

Khain, V. E., *Regional Geotectonics. Alpine Mediterranean Belt* (in Russian), 344 pp., Nedra, Moscow, 1984.

Khain, V. E., *Regional Geotectonics, Oceans, Synthesis* (in Russian), 292 pp., Nedra, Moscow, 1985.

Khain, V. E., *Geology of the USSR*, 385 pp., Schneizerbart-Borenträger, West Berlin, Stuttgart, 1987.

Khalilov, E. N., Sh. F. Mekhtiev, and V. E. Khain, Some geophysical data confirming collisional origin of the Great Caucasus (in Russian), *Geotektonika*, *2*, 54-60, 1987.

Khanchuk, A. I., Panchenko, I. V., and Kemkin, I. V., *Geodynamic Evolution of Sikhote-Alin and Sakhalin in the Paleozoic and Mesozoic*, 56 pp., Far East Branch of the USSR Acad. Sci., Vladivostok, 1988.

Kheraskov, N. N., *Rock Assemblages and Early Stages of Geosynclinal Development of West Sayan* (in Russian), 119 pp., Nauka, Moscow, 1979.

Kheraskov, N. P., and A. S. Perfiliev, Main features of geosynclinal structures in the Urals (in Russian), *Proc. Geol. Inst. USSR Acad. Sci.*, 92, 25-30, 1963.

Khomentovsky, V. V., V. Yu. Shenfil, and A. S. Gibshert, *Geological Structure of the Mana Trough and its Position in the Sayan-Altay Caledonides* (in Russian), 224 pp., Nauka, Novosibirsk, 1978.

Khoreva, B. Ya., and B. A. Bljuman, *The Polymetamorphic Complexes of the Precambrian Basement of Paleozoic and Mesozoic Fold Areas of Middle Asia* (in Russian) 275 pp., Nedra, Leningrad, 1974.

Khramov, A. N., *Paleomagnetology* (in Russian), 312 pp., Nedra, Leningrad, 1982.

Khramov, A. N., Global reconstructions of the position of ancient cratons in the Late Precambrian, in *Paleomagnetism of the Upper Precambrian* (in Russian) edited by A. N. Khramov, pp. 127-137, VNIGRI, Leningrad, 1983.

Khramov, A. N., (ed.), Paleomagnetic directions and paleomagnetic poles (USSR), (in Russian) *Materials Intern. Info. Ctr.*, 38 pp. Moscow, 1986.

Khristov, E. V., and A. V. Mikolaichuk, On the pre-geosynclinal basement of the Fergana-Kokshaal Hercynides (in Russian), *Geotektonika*, *3*, 76-86, 1983.

Khvorova, I. V., T. A. Voznesenskaya, B. P. Zolotarev et al., *Rock Assemblages of the Sakmara Allochthon (South Urals)* (in Russian), 230 pp., Nauka, Moscow, 1978.

Kirilova, G L., and M. T. Turbin, *Rock Assemblages and Tectonics of the Dzhagda Zone, Mongol-Okhotsk Foldbelt* (in Russian), 115 pp., Nauka, Moscow, 1979.

Kiryukhin, L. M., *Tectonic Map of the Peri-Caspian Depression* (in Russian), 43 pp., USSR Min. Geol., Moscow, 1982.

Klitin, K.,A., T. G. Pavlova, and E. S. Postelnikov, *Baikalids of NE Siberia* (in Russian), 114 pp., Nauka, Moscow, 1970.

Knipper, A. L., Oceanic crust in the structure of Alpine foldbelts (South Europe, West Asia and Cuba) (in Russian), *Proc. Geol. Inst. USSR Acad. Sci.*, 267, 208 p., Nauka, Moscow, 1975.

Kolesov, E. V., Paleomagnetic characteristics of Middle Paleozoic rocks of the Omolon massif, in *Rock Magnetism and Paleomagnetic Stratigraphy of Eastern and Northern Asia* (in Russian), edited by T. I. Linkova, pp. 68-74, Magadan, 1981.

Kononov, M. B., Absolute movement of the Pacific plate during the last 120 Ma (in Russian), *Okeanologia*, *293*, 484-492, 1984.

Kopp, M. L., and I. G. Shcherba, History of the Late Alpine development of the East Caucasus (in Russian), *Geotektonika, 6*, 94-108, 1985.

Korago, A. E., and A. P. Chukhonin, Granitoid formations of Novya Zemlya (in Russian), *Izv. ANSSSR, 10*, 28-35, 1988.

Korinevsky, V. G., Volcanogenic rocks of the Lower Silurian in the Sakmara zone of Mugodzhary, in *Petrological Problems of the Uralian Volcanics* (in Russian), edited by D. S. Shteinberg, pp. 58-71, Sverdlovsk, 1975.

Korinevsky, V. G., Geological essay about South Mugodzhary, in *The History of the Uralian Paleo-Ocean* (in Russian), edited by L. P. Zonenshain, pp. 57-79, Inst. Oceanol. USSR Acad. Sci., Moscow, 1984.

Koronovsky, N. V., *Geology of the USSR* (in Russian), 334 pp., Moscow State Univ., Moscow, 1984.

Koroteev, V. A., T. V. Dianova, and L. Ya. Kabanova, *The Middle Paleozoic Volcanism of the East Urals* (in Russian), 130 pp., Nauka, Moscow, 1979.

Koshkin, V. Ya., Tectonic position of the Hercynian Balkhash-Iliiski volcanic belt in *Tectonics of the Ural-Mongolian Foldbelt* (in Russian), edited by M. V. Muratov, pp. 112-133, Nauka, Moscow, 1974.

Kovalev, A. A., and Yu. V. Karyakin, The Zaisan fold system: A new view on the development (in Russian), *Proc. Geol. Inst. Kazakh Acad. Sci., 35*, 75-135, Nauka, Alma-Ata, 1975.

Kozlovsky, E. A. (ed.), *The Soviet Arctic Seas. Geological Structure of the USSR and Patterns of Mineral Deposit Distribution* (in Russian), 280 pp., Nedra, Leningrad, 1984.

Kratz, K. O., O. A. Levchenko, and I. K. Shuleshko, The age boundaries of the Yatulian complex in Karelia (in Russian), *Dokl. AN SSSR, 231*, 1191-1194, 1976.

Kotlyar, G. V., and L. I. Popeko, Carboniferous deposits of the Trans-Baikal region, in *The Paleozoic of the Far East* (in Russian), edited by L. I. Krasny and L. I. Popeko, pp. 140-156, Inst. Tectonics and Geophys., Khabarovsk, 1974.

Kropotkin, P. N., Eurasia as a composite continent, *Trans. Am. Geophys. Union, 53*, 180, 1967.

Krylov, I. N., O. A. Levchenko, and S. B. Lobach-Zhuchenko, Heterogenous structure and development of the Archean lithosphere in the Karelian granite-greenstone area (in Russian), *Proc. 27 I.G.C., 5*, pp. 100-105, Nauka, Moscow, 1984.

Kurbanov, M. K., and V. A. Rzhanitsin, Crustal structure of Western Turkmenistan from geological and geophysical data (in Russian), *Geotektonika, 5*, 64-73, 1982.

Kurchavov, A. M., *Zonation of Orogenic Volcanics in Kazakhstan* (in Russian), 205 pp., Nauka, Moscow, 1984.

Kuzebny, V. S., *Magmatic Rocks of the SW Altay and Their Metallogeny* (in Russian), 342 pp., Nauka, Alma-Ata, 1975.

Kuzmin, M. I., *Geochemistry of Magmatic Rocks of the Phanerozoic Mobile Belts* (in Russian), 199 pp., Nauka, Novosibirsk, 1985.

Kuzmin, M. I., and I. B. Filippova, The history of the Mongol-Okhotsk belt in the Middle - Late Paleozoic and Mesozoic, in *Lithospheric Plate Structure* (in Russian), edited by L. P. Zonenshain, pp. 189-226, Inst. Oceanol. USSR Acad. Sci., Moscow, 1979.

Kuzmin, M. I., and A. I. Al'mukhamedov, Chemical and rare-element composition of basaltoids along the Shuldak River (South Mugodzhary), in *The History of the Uralian Paleo-Ocean* (in Russian), edited by L. P. Zonenshain, pp. 126-139, Inst. Oceanol. USSR Acad. Sci., Moscow, 1984.

Kuzmin, V. I., and A. P. Chekhonin, On the Precambrian age of gneisses in the Kamchatka massif (in Russian), *Dokl. AN SSSR, 251*, 932-935, 1980.

Lavrovich, N. N., On the history of formation of the Precambrian Udokan series of Siberia (in Russian), *Izv. Acad. Sci. USSR, Geol. Ser., 11*, 147-153, 1970.

Lazarev, Yu. I., Tectonic development of early Karelides of Karelia (in Russian), *Geotektonika, 5*, 58-64, 1973.

Legler, V. A., and I. A. Przhiyalgovskaya, Interaction of the Indian and Asiatic plates and tectonics of the Tadjik depression, in *Lithospheric Plate Structure. Plate Interaction and Formation of Crustal Structures* (in Russian), edited by L. P. Zonenshain and O. G. Sorokhtin, pp. 125-188, Inst. Oceanol. USSR Acad. Sci., Moscow, 1979.

Lennykh, V. I., *Eclogite-Glaucophane Schist Belt of the South Urals* (in Russian), 185 pp., Nauka, Moscow, 1977.

Lennykh, V. I., Pre-Uralides in the zone of the East European platform and Urals junction, in *Metamorphism and Tectonics of Western Uralian Zones* (in Russian), edited by K. O. Krats, pp. 21-42, Sverdlovsk, 1984.

Leont'ev,. A. N., B. A. Litvinovsky, S. P. Gavrilova, and A. A. Zakharov, *The Paleozoic Granitoid Magmatism of Central Asiatic Foldbelt* (in Russian), 316 pp., Nauka, Novosibirsk, 1981.

Lepezin, G. G., *Metamorphic Complexes of the Altai Sayan Fold Area* (in Russian), 231 pp., Nauka, Novosibirsk, 1978.

Levashov, K. K., Magmatism of South Verkhoyanie, in *Tectonics, Magmatism and Pattern of Location of Ore Deposits* (in Russian), edited by V. I. Smirnov, pp. 157-169, Nauka, Moscow, 1964.

Librovich, L. S., Geological structure of the Kizim-Urtazym region in the South Urals (in Russian), *Proc. TSNIGRI, 81*, 208 p., 1936.

Litvinovsky, B. A., and A. N. Zanvilevich, *The Paleozoic Granitoid Magmatism of the West Trans-Baikal Region* (in Russian), 141 pp., Nauka, Novosibirsk, 1976.

Lomize, M. G., Alpine geosyncline of the Caucasus in a global context (in Russian), *Geotektonika, 2*, 14-23, 1987.

Lordkipanidze, M. B., *The Alpidic Volcanism and Geodynamics of the Central Segment of the Mediterranean Foldbelt* (in Russian), 162 pp., Metsnierba, Tbilisi, 1980.

Lozhkina, N. V., Magnetic properties of Middle-Upper Triassic rocks of the Northern Okhotsk massif, in *Rock Magnetism and Paleomagnetic Stratigraphy of Eastern and Northern Asia* (in Russian), edited by T. I. Lin'kova, pp. 59-67 Magadan, 1981a.

Lozhkina, N. V., Remanent magnetization of Upper Triassic rocks of the Omolon massif, in *Rock Magnetism and Paleomagnetic Stratigraphy of Eastern and Northern Asia* (in Russian), edited by T. I. Lin'kova, pp. 97-106, Magadan, 1981b.

Mamontov, Yu. A., and V. A. Popeko, The Paleozoic basaltic magmatism of the eastern Mongol-Okhotsk belt, in *The Problems of Magmatism and Tectonics of the Far East* (in Russian), edited by L. M. Parfenov and V. A. Popeko, pp. 68-82, Inst. Tectonics Geophys., Khabarovsk, 1975.

Markevich, P. V., *Flysch Complexes of the Northwestern Pacific Foldbelt* (in Russian), 143 pp., Nauka, Moscow, 1978.

Markova, N. G. (ed.) *Tectonics of Kazakhstan. Explanatory Notes for the Tectonic Map of East Kazakhstan, Scale 1:2,500,000* (in Russian), 137 pp., Nauka, Moscow, 1982.

Markovsky, V. A., and A. A. Makarjev, Ordovician deposits of the Severnaya Zemlya archipelago, in *Geology of the Severnaya Zemlya Archipelago* (in Russian), edited by V. Ya. Kabanikov and N. P. Lazarenko, pp. 22-39, Sevmorgeologia, Leningrad, 1982.

Marlow, M. S., and D. W. Scholl, Structure and evolution of Bering Sea shelf south of St. Lawrence Island, *Am. Assoc. Petrol. Bull., 60,* 161-183, 1976.

Martynova, V. P., Formation of continental crust in the Peri-Ladoga region (in Russian), *Geotektonika, 4,* 18-28, 1980.

Maslov, V. A., *The Devonian of the Eastern Slope of the South Urals* (in Russian), 224 pp., Nauka, Moscow, 1980.

Matveenkov, V. V., and V. V. Khain, The Karaksak paleo-volcanic structure (South Mugodzhary): An example of studies of the paleo-oceanic floor, in *History of the Uralian Paleo-Ocean* (in Russian), edited by L. P. Zonenshain, pp. 112-125, Inst. Oceanol. USSR Acad. Sci., Moscow, 1984.

Mazarovich, A. O., *Tectonic Development of South Primorje in the Paleozoic and Early Mesozoic* (in Russian), 102 pp., Nauka, Moscow, 1985.

McElhinny, M. W.., B. J. J. Embleton, X. H. Ma, and Z. K. Zhang, Fragmentation of Asia in the Permian, *Nature, 293,* 212-216, 1981.

McKenzie, D., Active tectonics of the Mediterranian region, *Geophys. J. Roy Soc. London, 30,* 104-185, 1974.

Melankholina, E. N., Gabbroids and parallel dikes in the structure of Shikotan Island (Lesser Kurile Islands) (in Russian), *Geotektonika, 3,* 128-136, 1978.

Melnikov, N. G., and V. V. Golozubov, Olistostrome sequences and syn-sedimentary nappes in Sikhote-Alin (in Russian), *Geotektonika, 1,* 95-106, 1980.

Melnikov, N. G., and L. A. Izosov, Structural-lithologcial zones of Primorje (in Russian), *Tikhookeanskaya Geol.,* 53-611, 1984.

Milanovsky, E. E., *Geology of the USSR* (in Russian), I, 415 pp., Moscow State Univ. Publ. House, Moscow, 1987.

Milovsky, A. V., R. V. Retling, and A. T. Zverev, *The Precambrian and Lower Paleozoic of West Kazakhstan* (in Russian), 268 pp., Moscow State Univ. Publ. House, Moscow, 1977.

Minster, J. B., and T. H. Jordan, Present-day plate motions, *J. Geophys. Res., 83,* 5331-5354, 1978.

Mirchink, M. F., The nature of the marginal scarp of the Peri-Caspian depression (in Russian), *Geol. Nefti i Gasa, 1,* 21-30, 1973.

Misnik, Yu. F., and V. V. Shevchuk, *The Shilka Zone of the Mongol-Okhotsk Deep Fault* (in Russian), 160 pp., Lvov State Univ., Lvov, 1975.

Misnik, Yu. F., and V. V. Shevchuk, The East Trans-Baikal ancient block and its role in formation of the regional structure (in Russian), *Geotektonika, 5,* 25-37, 1980.

Mitrofanov, G. L., The structural evolution and stages of continental crust formation in the northwestern Transbaikal region, in *Tectonics and Metallogeny of Eastern Siberia* (in Russian), edited by V. A. Naumov, pp. 38-58, Irkutsk State Univ., Irkutsk, 1978.

Mitrofanov, G. L., and N. I. Mitrofanova, The Riphean volcano-plutonic association of the South Muya ridge and its metallogeny (in Russian), *Geol. Geofiz., 1,* 52-60, 1980.

Mokshantsev, K. B., D. K. Gornstein, and G. S. Gusev, *The Tectonics of Yakutia* (in Russian), 198 pp., Nauka, Novosibirsk, 1975.

Molnar, P., and P. Tapponier, Cenozoic tectonics of Asia: Effects of a continental collision, *Science, 189,* 419-426, 1975.

Moralev, V. M., Tectonics and petrogenesis of Early Precambrian complexes of the Aldan Shield, in *Precambrian Plate Tectonics,* edited by A. K. Kröner, pp. 237-260, Elsevier, Amsterdam, 1981.

Moralev, V. M., and G. F. Grigorash, Geodynamic setting and metallogeny of the Late Cretaceous volcanism in the Penzhina sector of the Okhotsk-Chukotka belt, in *Deep Structure, Magmatism and Metallogeny of Pacific Volcanic Belts* (in Russian), edited by G. M. Fremd, pp. 387-388, Vladivostok, 1976.

Morkovsky, V. A., and M. A. Smirnova, Silurian deposits of the Severnaya Zemlya archipelago, in *Geology of Severnaya Zemlya Archipelago* (in Russian), edited by V. Ya. Kabanjkov and N. P. Lazarenko, pp. 39-61, Sevmorgeologia, Leningrad, 1982.

Moskovchenko, N. I., The sequence of development of structures and the evolution of polymetamorphic complexes of the basement of Phanerozoic fold areas, with emphasis on the south-western Pamirs (in Russian), *Geotektonika, 3,* 84-99, 1978.

Mossakovsky, A. A., *Orogenic Structures and Volcanism of the Eurasian Paleozoids* (in Russian), 318 pp., Nauka, Moscow, 1975.

Mossakovsky, A. A., and A. B. Dergunov, Caledonides of Kazakhstan and Central Asia (tectonic structure, history and paleotectonic setting) (in Russian), *Geotektonika, 2,* 16-33, 1983.

Nalivkin, D. V., *Geological structure of the USSR* (in Russian), 384 pp., Gosgeoltekhizdat, Moscow, 1958.

Natal'in, B. A., *The Early Mesozoic Eugeosynclinal System of the Northern Pacific Mobile Belt* (in Russian), 195 pp., Nauka, Novosibirsk, 1984.

Natapov, L. M., L. P. Zonenshain, and V. S. Shulgina, Geological development of the Kolyma-Indigirka region and the problem of the Kolymian massif (in Russian), *Geotektonika, 4,* 18-31, 1977.

Natapov, L. M., and A. P. Stavsky, Geodynamic model of the north-east USSR and its application to metallogenic analysis (in Russian), *Sovetskaya Geol., 3,* 70-78, 1985.

Nekrasov, G. E., *Tectonics and Magmatism of Taigonos and NW Kamchatka* (in Russian), 159 pp., Nauka, Moscow, 1976.

Neprochnov, Yu. P., Deep crustal structure under the Black Sea from seismic data (in Russian) *Bull. Moscow Naturalist Soc., 4,* 15-29, 1966.

Neprochnov, Yu. P., V. V. Sedov, L. P. Merklin, and B. V. Baranov, Tectonic structure of the Shirshov Ridge (Bering Sea) (in Russian), *Geotektonika, 3,* 21-37, 1985.

Nesterov, I. I., N. K. Kulakhmetov, and V. S. Bochkarev, Geology and oil- and gas-productivity of the Yamal-Gydan syneclise in West Siberia (in Russian), *Proc. 27th I.G.C., Arctic Geol., 4.* 80-87, Nauka, Moscow, 1984a.

Nesterov, I. I., L. L. Podsosova, M. Ya. Rudkevich, and A. A. Trofimuk, Geological structure of the West-Siberian platform (in Russian), *Proc. 27th I.G.C., Geol. of the USSR, 1,* 44-50, Nauka, Moscow, 1984b.

Nevolin, N. V., Deep structure of the Peri-Caspian depression (in Russian), *Geotektonika, 3,* 49-60, 1978.

Novikova, A. S., *Zones of Metabasites in the Basement of the East European Platform* (in Russian), 153 pp., Nauka, Moscow, 1975.

Olivet, J. L., J. Bonnin, P. Beazart, and J. M . Auzende, *Cinematique de l'Atlantique nord et central*, 108 pp., Raport Sci., CNEXO, Paris, 1984.

Palandzjan, S. A., Ophiolite belts in the Koryak upland, north-east Asia, *Tectonophysics*, *127*, 341-360, 1986.

Parfenov, L. M., *Continental Margins and Island Arcs of the Mesozoides of NE Asia* (in Russian), 192 pp., Nauka, Novosibirsk, 1984.

Parfenov, L., Two stages of Mesozoic folding in the North Verkhoyansk Mountains (in Russian), *Geol. Geoph.*, *4*, 3-10, 1988.

Pashkov, B. P., and V. A. Shvolman, Rifted margins of the Tethys in the Pamirs (in Russian), *Geotektonika*, *6*, 42-70, 1979.

Patriat, P., J. Segoufin, R. Schlich et al., Les mouvements relatifs de l'Inde, de l'Afrique et de l'Eurasia, *Bull. Soc. Geol. France*, *24*, 363-373, 1982.

Pavlov, N. V., and I. N. Grigorieva-Chuprynina, *The Pattern of Formation of Chrome Deposits* (in Russian), 199 pp., Nauka, Moscow, 1973.

Peive, A. A., *Composition and structural position of ophiolites in the Koryak Ridge* (in Russian), 100 pp., Proc. Geol. Inst. USSR Acad. Sci., Nauka, Moscow, 1984.

Peive, A. V., (ed.), *Guidebook for the International Symposium Excursions* (in Russian), 49 pp., Nauka, Moscow, 1973.

Peive, A. V. (ed.), *Tectonics of North Eurasia* (in Russian), 220 pp., Nauka, Moscow, 1980.

Peive, A. V., S. N. Ivanov, and V. M. Necheukhin, *Tectonics of the Urals. Explanatory Note for the Tectonic Map of the Urals* (in Russian), 119 pp., Nauka, Moscow, 1977.

Peive, A. V., V. E. Khain, and M. V. Muratov (eds.), *Tectonics of Europe and adjacent areas. Ancient platforms, Baikalides and Caledonides* (in Russian), 422 pp., Nauka, Moscow, 1978.

Perfiliev, A. S., *Formation of the Earth's crust in the Uralian eugeosyncline* (in Russian), 188 pp., Nauka, Moscow, 1979.

Pevnev, A. K., T. V. Guseva, N. N. Odinov, and G. V. Saprykin, Crustal deformation pattern within the Pamirs-Tien Shan junction (in Russian), in *Active Crustal Tectonics*, edited by N. A. Florensov, pp. 86-92, Nauka, Novosibirsk, 1978.

Piskunov, B. N., *Volcanism of the Great Kurile Arc and Petrology of High-Aluminous Series* (in Russian), 178 pp., Nauka, Novosibirsk, 1975.

Pistrak, R. M., and N. T. Pashkova, Structural position of volcanic localities in the Pripiat-Donetsk trough (in Russian), *Geotektonika*, *4*, 33-38, 1974.

Pitman, W. G., III, and M. Talwani, Sea floor spreading in the North Atlantic, *Geol. Soc. Am. Bull.*, *83*, 619-646, 1972.

Pogrebitsky, Yu. E., *Paleotectonic Analysis of the Taimyr Foldsystem* (in Russian), 248 pp., Nedra, Leningrad, 1971.

Polyansky, N. V., N. L. Dobretsov, P. V. Ermolov, and V. S. Kuzebny, Structure and history of the Chara ophiolite belt (in Russian), *Geol. Geofiz.*, *4*, 66-79, 1979.

Popeko, V. A., V. E. Pilatsky, and V. A. Kaidalov, The Upper Triassic and Lower Jurassic basalts of Sikhote-Alin (in Russian), *Tikhookeanskaya Geol.*, *6*, 21-31, 1983.

Popeko, V. A., and L. V. Zozulenko, Pre-Mesozoic basaltoids of the western Mongol-Okhotsk belt, in *Problems of Tectonics and Magmatism: Far East* (in Russian), edited by Yu. A. Kosygin, pp. 43-55, Far East Branch USSR Acad. Sci., Vladivostok, 1975.

Popkov, V. I., Tectonics of the Pre-Jurassic sedimentary complex of the western Turan plate (in Russian), *Geotektonika*, *4*, 106-116, 1986.

Popolitov, E. I., and O. N. Volynets, *Geochemical and Petrogenetic Features of the Quaternary Volcanism in the Kurile Kamchatka Island Arc* (in Russian), 181 pp., Nauka, Novisibirk, 1981.

Porshnyakov, G. S., *The Hercynides of the Alay Ridge and Adjacent Regions of South Tien-Shan* (in Russian), 216 p., Leningrad State Univ. Pub. House, Leningrad, 1973.

Portnyagin, E. A., N. A. Gnutenko, and I. A. Kovalchuk, Carboniferous volcanism and some tectonic problems of Gissar (South Tien Shan) (in Russian), *Bull. Moscow Naturalist Soc., Geol. Ser.*, *48*, 32-93, 1973.

Prozorovsky, V. A., The role of Cimmerian movements in the formation of structures in western Middle Asia (in Russian), *Geotektonika*, *6*, 86-93, 1985.

Puchkov, V. N., *Structural Relations of the Polar Urals and Russian Platform* (in Russian), 203 pp., Nauka, Leningrad, 1975.

Puchkov, V. N., *Passive Margin Bathyal Complexes in Geosyncline Areas* (in Russian), 260 pp., Nauka, Moscow, 1979.

Puchkov, V. N., I. F. Tavrin, and L. A. Karsten, Structure of the major Uralian deep-seated fault and the problem of junction of the Urals and East European platform, in *Abstr. Conf.: Tectonics, Magmatism, Metamorphism and Metallogeny of the Urals*, pp. 29-31, Sverdlovsk, 1985.

Pushcharovsky, Yu. M. (ed.), *Essays on Koryak Highland Tectonics* (in Russian), 220 pp., Nauka, Moscow, 1982.

Pushcharovsky, Yu. M., V. P. Zinkevich, and O. A. Mazarovich, Thrust-nappe structures in the northwest Pacific margin (in Russian), *Geotektonika*, *6*, 30-45, 1983.

Raznitsin, Yu. N., Comparative tectonics of ultramafic belts of the Schmidt Peninsula (Sakhalin), Papua (New Guinea), and Sabakh (Kalimantan) (in Russian), *Geotektonika*, *9*, 68-84, 1975.

Raznitsin, Yu. N., Serpentinite melange and olistostrome of the south-eastern part of East Sakhalin Mountains (in Russian), *Geotektonika*, *1*, 96-108, 1978.

Raznitsin, Yu. N., S. A. Khubunaja, and N. V. Tsukanov, Tectonics of the eastern Kononotsky Peninsula (Kamchatka) and provenance of basalts (in Russian), *Geotektonika*, *1*, 88-101, 1985.

Rodnikov, A. V., (ed.), *Tectonics of the Northwest Pacific Ocean* (in Russian), 118 pp., Nauka, Moscow, 1983.

Ross, D. A., E. Uchupi, and C. O. Bowin, Shallow structure of the Black Sea, in *Black Sea - Geology, Chemistry and Biology*, edited by E. T. Degens and D. A. Ross, pp. 11-34, Am. Assoc. Petrol. Geol., Tulsa, Okla., 1974.

Rotarash, I. A., and E. A. Gredjushko, History of formation and structure of serpentinite melange of the Zaisan fold area (in Russian), *Geotektonika*, *4*, 73-79, 1974.

Rotarash, I. A., N. A. Klepikov, and E. A. Gredjushko, The Lower Paleozoic olistostromes of the Chingiz-Tarbagatay anticlinorium (in Russian), *Geotektonika*, *4*, 52-67, 1980.

Rotarash, I. A., S. G. Samygin, and E. A. Gredjushko, The Devonian active continental margin in southwest Altai (in

Russian), *Geotektonika*, *1*, 33-59, 1982.
Rotarash, I. A., and L. M. Trubnikov, The Rudny Altai volcanic belt (in Russian), *Geol. Geofiz.*, *12*, 29-36, 1983.
Rozhdestvensky, V. S., Evolution of the Sakhalin fold system, *Tectonophysics*, *127*, 331-340, 1986.
Rozhdestvensky, V. S., and A. N. Rechkin, Evolution of the ophiolite magmatism of Sakhalin (in Russian), *Tikhookeanskaya Geol.*, *2*, 40-44, 1982.
Rutshtein, I. G., Structure of the Early Mesozoic troughs and structural conditions for formation of the Triassic and Jurassic granitic intrusions in the central part of the East Trans-Baikal region (in Russian), Cand. thesis, 23 pp., Leningrad State Univ., Leningrad, 1970.
Ruzhentsev, S. V., *Tectonic Development of the Eastern Pamirs and Role of Horizontal Movements in the Formation of Its Alpine Structure* (in Russian), 156 pp., Nauka, Moscow, 1968.
Ruzhentsev, S. V., *Marginal Ophiolite Allochthons* (in Russian), 171 pp., Nauka, Moscow, 1976.
Ruzhentsev, S. V., I. I. Pospelov, and A. N. Sukhov, Tectonics of the Kalaikhumb-Sauksai zone of the North Pamirs (in Russian), *Geotektonika*, *4*, 68-80, 1977.
Ruzhentsev, S. V., S. G. Bjalobzhesky, and V. N. Grigorjev, Koryak Ridge tectonics, in *Essays on the Koryak Highland Tectonics* (in Russian), edited by Yu. M. Pushcharovsky and S. M. Tilman, pp. 136-189, Nauka, Moscow, 1982.
Sabdjushev, Sh. Sh., and K. R. Usmanov, Tectonic nappes, melange, and ancient oceanic crust in Tamdytau (West Uzbekistan) (in Russian), *Geotektonika*, *5*, 27-36, 1971.
Samygin, S. G., *The Chingiz Strike-Slip Fault and its Role in the Central Kazakhstan Structure* (in Russian), 204 pp., Nauka, Moscow, 1974.
Samygin, S. G., The Late Cambrian and Early Paleozoic history of the continent - ocean boundary in the Urals, in *Abst. Conf.: Precambrian - Early Paleozoic History of the Urals* (in Russian), pp. 57-7, Sverdlovbsk, 1980.
Sasajima, S., and M. Torii, Paleomagnetism and moving continents with special reference to the opening of the Japan Sea, *J. Geography*, *92*, 38-47, 1983.
Satiani, M. A., *The Ophiolite Troughs of the Meso-Tethys* (in Russian), 195 pp., Armenian Publ. House, Erevan, 1984.
Saveljev, A. A., and G. N. Saveljeva, The ophiolites of the Voikaro-Synjnsky massif, the Polar Urals (in Russian), *Geotektonika*, *6*, 41-61, 1977.
Savostin , L. A., L. P. Zonenshain, and B. V. Baranov, Geology and plate tectonics of the Sea of Okhotsk, in *Geodynamics of the Western Pacific - Indonesian Region*, edited by T. Hilde and S. Uyeda, pp. 189-222, Am. Geophys. Union Geodynamics Ser. 11, 1983.
Savostin, L. A., and A. M. Kheifets, Paleomagnetism of the Maastrichtian-Paleocene island arc rocks of the Olyutorsky zone (South Koryakya), in *Paleomagnetism and Accretionary Tectonics* (in Russian), edited by A. N. Khramov, pp. 127-140, VNIGRI, USSR Min. Geol., 1988.
Sborshchikov, I. M., L. A. Savostin, and L. P. Zonenshain, Recent plate tectonics from Turkey to Tibet, *Tectonophysics*, *78*, 209-230, 1981.
Segedin, R. M., (ed.), *Explanatory Notes for the Geological Map of the Kazakhstan SSR, Scale 1:500,000, Turgai-Mugodjary Series* (in Russian), 228 pp., Min. Geol., Alma-Ata, 1981.
Semenov, D. F., *Magmatic Assemblages of Pacific Fold Areas with Emphasis on Sakhalin* (in Russian), 168 pp., Nauka, Moscow, 1982.
Sengor, A. M. C., The dual nature of the Alpine-Himalayan system: progress, problems and prospects, *Tectonophysics*, *127*, 177-196, 1986.
Seslavinsky, K. B., Structure and evolution of the South Anyui suture trough (in Russian), *Geotektonika*, *5*, 73-85, 1970.
Shantser, A. E., M. N. Shapiro, and A. V. Koloskov, Structural evolution of the Leskovsky uplift and surrounding areas in the Cenozoic (in Russian), *Tikhookeanskaya Geol.*, *4*, 66-74, 1985.
Shapiro, M. N., *Tectonic Development of Eastern Kamchatka* (in Russian), 123 pp., Nauka, Moscow, 1976.
Shapiro, M. N., Yu. N. Raznitsin, A. E. Shentser, and A. V. Lander, Structure of the northeast margin of the Median Ridge metamorphic massif of Kamchatka, in *Geological Essays on the East USSR* (in Russian), edited by Yu. M. Pushcharovsky, pp. 5-21, Nauka, Moscow, 1986.
Shatsky, N. S., Main features of the structure and development of the East European platform (in Russian) *Izv. Acad. Sci. USSR, Geol. Ser. 1*, 3-21, 1956.
Shcherbak, P. P., and E. V. Bibikova, Stratigraphy and geochronology of the Early Precambrian, USSR (in Russian), in *Proc. 27th I.G.C., Precambrian Geol.*, *5*, pp. 3-14, Nauka, Moscow, 1984.
Shikhalibeili, E. Sh., R. N. Abdullaev, and A. A. Ali-Zade, Geologic results of the Saatla super drillhole (in Russian), *Sov. Geol.*, *11*, 61-66, 1988.
Shipilov, E. V., and B. V. Senin, Deep crustal structure of the Barents sea floor (in Russian), *Geotektonika*, *6*, 96-110, 1988.
Shlygin, E. D., *Explanatory Notes for Geological map of the Kazakhstan SSR, Scale 1:500,000, Central Kazakhstan Series* (in Russian), 324 pp., Min. Geol., Alma-Ata, 1981.
Shmidt, O. A., *Tectonics of the Komandor Islands and Structure of the Aleutian Arc* (in Russian), 285 pp., Nauka, Moscow, 1978.
Shtekh, G. I., *Deep Structure and Tectonic History of the Viljuy Depression* (in Russian), 124 pp., Nauka, Moscow, 1965.
Shuldiner, V. I., M. I. Stetsjuk, and S. M. Sinitsa, The Upper Precambrian and Cambrian of the Borshchovochny Ridge, in *Proterozoic Complexes of the East Trans-Baikal* (in Russian), edited by E. A. Radkevich, pp. 52-75, Vladivostok, 1977.
Shults, S. S., *Geological Structure of the Urals and Tien Shan Junction* (in Russian), 205 pp., Nedra, Moscow, 1972.
Shults, S. S., *Geology of the Junction Between the Urals and Tien Shan* (in Russian), 207 pp.,. Nedra, Moscow, 1974.
Shvolman, V. A., *Tectonic Evolution of the Pamirs in the Cretaceous and Paleogene* (in Russian), 156 pp., Nauka, Moscow, 1977.
Shvolman, V. A., The Mesozoic ophiolite complex in the Pamirs (in Russian), *Geotektonika*, *5*, 72-81, 1980.
Sidorenko, A. B. (ed.), *Geology of the USSR, Vol. XIX (Khabarovsk and Amur Regions)* (in Russian), 736 pp., Nauka, Moscow, 1966a.
Sidorenko, A. V. (ed.), *Geology of the USSR, Vol. XXIX (Tuva Region)* (in Russian), 458 pp., Nedra, Moscow, 1966b.
Sidorenko, A. V., (ed.), *Geology of the USSR, Vol. XXVI (The Middle Urals)* (in Russian), 360 pp., Nedra, Moscow, 1970a.
Sidorenko, A. V. (ed.), *Geology of the USSR, Vol. XXX (NE USSR)* (In Russian), 548 pp., Nedra, Moscow, 1970b.
Sidorenko, A. V. (ed.), *Geology of the USSR, Vol. XXXIII*

(Sakhalin) (in Russian), 594 pp., Nedra, Moscow, 1970c.

Sidorenko, A. V. (ed.), *Geology of the USSR, Vol. XXXV (Kirghiz SSR)* (in Russian), 226 pp., Nedra, Moscow, 1972.

Sinitsa, S. M., *Gneissic Domes of the Nerchinsk Ridge in the East Trans-Baikal Region* (in Russian), 128 pp., Nauka, Novosibirsk, 1975.

Sinitsa, S. M., Deformation and metamorphism in the ancient basement of the East Trans-Baikal region related to Mesozoic diastrophism, in *Proterozoic Complexes of East Trans-Baikal* (in Russian), edited by E. A. Radkevich, pp. 117-133, Far East Geol. Inst., Vladivostok, 1977.

Sivoronov, A. A., A. B. Bobrov, and B. I. Maljuk, Morphological types of Early Precambrian greenstone structures of the middle Dnieper region (Ukranian shield) (in Russian), *Geotektonika, 5*, 22-37, 1984.

Sizykh, A. I., *Petrology of North Baikal Metamorphic Belts* (in Russian), 120 pp., Nauka, Novosibirsk, 1985.

Sljusar, B. S., Compressional structures in North Dobrogea (in Russian), *Geotektonika, 4*, 90-105, 1984.

Sokolov, S. D., V. N. Grigoriev, and V. N. Zinkevich, Accretion and tectonic layering in the structure of the Koryak highlands and Kamchatka (in Russian), *Proc. XX All Union Tectonic, Conf.*, pp. 61-63, Geol. Inst., Moscow, 1987.

Sokolov, V. A., *Gas productivity of the Peri-Caspian depression* (in Russian), 55 pp., Spec. Rev., VIEMS, Moscow 1970.

Sokolov, V. B., The upper crustal structure of the western edge of the Uralian geosyncline from seismic studies, in *Abstr. Conf., Tectonics, Magmatism, Metamorphism and Metallogeny of the Urals-East European Platform Junction* (in Russian), pp. 41-43, Sverdlovsk, Miass, 1985.

Somin, M. L., *Pre-Jurassic Basement of the Glavny Ridge and Southern Slope of the Great Caucasus* (in Russian), 246 pp., Nauka, Moscow, 1971.

Spiridonov, E. M., Geosynclinal basitic complexes of NE Central Kazakhstan and their metallogeny (in Russian), *Proc. Moscow State Univ.*, 19, 102-122, Moscow, 1980.

Stavsky, A. P., V. D. Chekhovich, V., Kononov, and L. P. Zonenshain, Plate tectonics and palinspastic reconstructions of the Anadyr-Koryakya region, northeast USSR, *Tectonics, 9*, 81-101, 1990.

Streltsov, M. I., *Dislocations of the Southern Part of the Kurile Island Arc* (in Russian), 131 pp., Nauka, Moscow, 1976.

Stupka, O. S., The main stages of origin of the continental crust, and structure of the southern margin of the East European platform in Pre-Riphean time (in Russian), *Geotektonika, 4*, 3-17, 1980.

Stupka, O. S., On the tectonic regime in the Archean with emphasis on the Ukranian shield (in Russian), *Geotektonika, 1*, 46-59, 1984.

Sukhanov, M. K., Comparative study of autochthonous anorthositic massifs, *Izv. Akad. Nauk SSSR, 7*, 3-18, 1988.

Surkov, V. S., A. A. Trofimuk, and O. G. Shero, The Triassic rift system of the West Siberian platform and its influence on structure and oil productivity of the Meso-Cenozoic cover (in Russian), *Geol. Geofiz., 8*, 3-15, 1982.

Tarakanov, R. Z., Structure of the focal zone in the Kurile-Kamchatka island arc, in *Earth's Crust of Island Arcs and Far East Seas* (in Russian), edited by N. A. Belyaevsky, A. G. Gainanov and A. G. Rodnikov, pp. 215-234, Nauka, Moscow, 1972.

Terekhova, G. P., and O. G. Epshtein, New data on stratigraphy and lithology of chert-volcanic Paleozoic in the eastern Koryak highland (in Russian), *Izv. Acad. Sci. USSR, Geol. Ser., 3*, 132-135, 1979.

Tikhomirov, V. G., *Paleozoic Magmatism and Tectonics of Central Kazakhstan* (in Russian), 146 pp., Nedra, Moscow, 1975.

Tilman, S. M., *Comparative Tectonics of Mesozoides of the North Circum-Pacific* (in Russian), 326 pp., Nauka, Novosibirsk, 1973.

Tilman, S. M., Accretional tectonics and metallogeny of the Koryak highland (NE USSR) (in Russian), *Dokl. AN SSSR, 292*, 1220-1222, 1987.

Tsvetkov, A. A., Magmatic evolution of the Aleutian arc and paragenesis of island arc igneous rocks (in Russian), *Izv. Acad. Sci. USSR, Geol. Ser., 4*, 3-19, 1983.

Tugolesov, D. A., A. A. Gorshkov, L. B. Meisher, V. V. Solovjov, and E. M. Khakhalev, *Tectonics of the Meso-Cenozoic of the Black Sea Depression* (in Russian), 214 pp., Nedra, Moscow, 1985.

Ushakov, S. A., Yu. I. Galushkin, and O. P. Ivanov, Nature of folding in the sedimentary pile of the Black Sea floor near the transition between the Crimea and the Caucasus (in Russian), *Dokl. AN SSSR, 233*, 932-934, 1977.

Utkin, V. P., *Strike-Slip Faults and Methodology of Their Study* (in Russian), 144 pp., Nauka, Moscow, 1980.

Vail, P. R., R. M. Mitchum, Jr., and S. Thompson, III, Seismic stratigraphy and global changes of sea level, Pt. 4; Global cycles of relative changes of sea level, in *Seismic Stratigraphy Applications to Hydrocarbron Exploration*, edited by C. E. Payton, pp. 83-97, Mem. Am. Assoc. Petrol. Geol., Tulsa, 1977.

Vakar, V. A., P. S. Voronov, and B. Kh. Egiazarov, The Taimyr-Severozemelsky fold area, in *Geological Structure of the USSR* (in Russian), edited by A. P. Markovsky, 3, Tectonics, pp. 350-365, Gosgeoltekhizdat, Moscow, 1958.

Vardapetyan, A. N., The Late Cenozoic plate tectonics of the Black Sea-Caspian region (in Russian), *Okeanologia, 19*, 1066-1075, 1979.

Varlakov, V. S., *Petrography, Petrochemistry and Geochemistry of Ultramafic Rocks of the Orenburg Urals* (in Russian), 239 pp., Nauka, Moscow, 1978.

Veevers, J. J. (ed.), *Phanerozoic Earth History of Australia*, 418 pp., Clarendon Press, Oxford, 1986.

Vinogradov, V. I., V. S. Grigoriev, and A. M. Leites, Age of metamorphism of Median Ridge rocks in Kamchatka (in Russian), *Izv. Acad. Sci. USSR, Geol. Ser., 9*, 30-38, 1988.

Vishnevskaya, V. S., and V. V. Bernard, The age and conditions of formation of the Cenozoic cherty rocks of Kamchatka, in *Geological Essays on the East USSR* (in Russian), edited by Yu. M. Pushcharovsky, pp. 35-41, Nauka, Moscow, 1986.

Vogt, P. R., P. T. Taylor, J. Kovach and B. C. Johnson, Detailed aeromagnetic investigation of the Arctic Basin, I, *J. Geophys. Res., 84*, 1071-1089, 1979.

Vogt, P. R., L. K. Kovach, R. K. Perry, and P.T. Taylor, The Amerasian Basin: The Arctic: Magnetic lineation and its interpretation, in *Proc. 27th I.G.C., Coll. 04, 4*, pp. 128-135, Nauka, Moscow, 1984.

Voitovich, V. S., On the deep seated overthrust in the basement of the East European platform (in Russian), *Geotektonika*, *6*, 123-33, 1980.

Volk, V. E., G. I. Gaponenko, E. N. Zatsepin, Yu. G. Kiselev, N. N. Lastochkina, A. M. Malyavkin, V. N. Shamarayev, V. G. Shchelovanov, and A. G. Krayev, Crustal structure of the Arctic according to geophysical data (in Russian), in *Proc. 27th I.G.C. 4, Geology of the Arctic*, pp. 26-36, Nauka, Moscow, 1984.

Westphal, M., M. L. Bazhenov, J. P. Lauer, D. M. Pechersky, and J. C. Sibuet, Paleomagnetic implications for the evolution of the Tethys belt from the Atlantic Ocean to the Pamirs since the Triassic, *Tectonophysics*, *123*, 37-82, 1986.

Yanshin, A. L., V. P. Avrov, and Yu. A. Volosh, Structure and paleotectonic setting of Paleozoic rocks south of the Peri-Caspian depression (in Russian), *Geotektonika*, 45-57, 1980a.

Yanshin, A. L., Sh. A. Bakenjants, A. N. Pilichenko, and A. E. Shlezinger, New data on the time of formation of the deep Black Sea depression (in Russian), *Dokl. AN SSSR*, *25*, 223-227, 1980b.

Yudin, V. V., *The Variscides of the North Urals* (in Russian), 174 pp., Nauka, Leningrad, 1983.

Yuzvitsky, A. Z., New data on the Tomsky nappe (in Russian), *Geotektonika*, *4*, 67-71, 1976.

Zabiyaka, A. I., I. D. Zabiyaka, and V. A. Vernikovsky, *Geological Structure and Tectonic Development of the Northeastern Taimyr* (in Russian), 144 pp., Nauka, Novosibirsk, 1986.

Zakharov, S. A., On the nature of the Pamir - Pendzhab syntaxis, in *Tectonics. Geology of Alpidaes of the Tethian provenance* (in Russian), edited by M. V. Muratov, pp. 198-206, Nauka, Moscow, 1980.

Zhivkovich, A. E., and P. A. Chekhovich, *The Paleozoic Rock Formations and Tectonics of the Ufimsky Amphitheater* (in Russian), 183 pp., Nauka, Moscow, 1985.

Zhulanova, I. L., The role of Pre-Riphean gneissic domes in the tectonic development of the NE USSR Mesozoides (in Russian), *Geotectonika*, *1*, 78-92, 1987.

Zonenshain, L. P., *The Geosynclinal Theory and its Application to the Central Asiatic Foldbelt* (in Russian), 239 pp., Nedra, Moscow, 1972.

Zonenshain, L. P., The evolution of the Central Asian geosynclines through sea floor spreading, *Tectonophysics*, *19*, 201-235, 1973.

Zonenshain, L. P., Tectonics of intracontinental foldbelts, *Proc. 27th I.G.C., Tectonics* (in Russian), pp. 48-59, Nauka, Moscow, 1984.

Zonenshain, L. P., M. I. Kuzmin, and V. M. Moralev, *Global Tectonics, Magmatism and Metallogeny* (in Russian), 231 pp., Nedra, Moscow, 1976.

Zonenshain, L. P., L. N. Natapov, A. A. Savostin, and L. P. Stavsky, Recent plate tectonics of northeastern Asia related to opening of the North Atlantic and Arctic basins, *Okeanologia*, *5*, 38-47, 1978.

Zonenshain, L. P., and L. A. Savostin, *Introduction to Geodynamics* (in Russian), 311 pp., Nedra, Moscow, 1979.

Zonenshain, L. P., and L. A. Savostin, Geodynamics of the Baikal Rift Zone and plate tectonics of Asia, *Tectonophysics*, *76*, 133-171, 1981.

Zonenshain, L. P., L. A. Savostin, and B. V. Baranov, Present plate tectonics within the USSR, *Bull. 32, Comm. Geol. Map World*, June, 1983, pp. 129-132, 1983.

Zonenshain, L. P., V. G. Korinevsky, V. G. Kazmin, V. V. Matveenkov, and V. V. Khain, Plate tectonic model of the South Urals development, *Tectonophysics*, *109*, 95-135, 1984.

Zonenshain, L. P., M. I. Kuzmin, and M. V. Kononov, Absolute reconstructions of the Paleozoic oceans, *Earth and Planet. Sci. Lett.*, *74*, 103-116, 1985.

Zonenshain, L. P., M. V. Kononov, and L. A. Savostin, Pacific and Kula/Eurasia relative motions during the last 130 Ma and their bearing on orogenesis in North-East Asia, in *Circum-Pacific Orogenic Belts and Evolution of the Pacific Ocean*, edited by J. Monger and J. Francheteau, Am. Geophys. Union, Geodynamics Ser., 18, pp. 29-58, 1987.

Zonenshain, L. P., and L. M. Natapov, Tectonic history of the Arctic: From Ordovician to Cretaceous, in *The Arctic Sea*, edited by Y. Herman, pp. 829-862, Van Nostrand Reinhold, N.Y., 1990.

Zubtsov, E. I., N. S. Porshnyakov, and A. V. Yagovsky, New scheme of the Pre-Mesozoic tectonics of Tien Shan (in Russian) *Dokl. AN SSSR*, *217*, 1153-1156, 1974.